# Applications of Random Matrices in Physics

# NATO Science Series

*A Series presenting the results of scientific meetings supported under the NATO Science Programme.*

The Series is published by IOS Press, Amsterdam, and Springer in conjunction with the NATO Public Diplomacy Division.

*Sub-Series*

| | |
|---|---|
| **I. Life and Behavioural Sciences** | IOS Press |
| **II. Mathematics, Physics and Chemistry** | Springer |
| **III. Computer and Systems Science** | IOS Press |
| **IV. Earth and Environmental Sciences** | Springer |

The NATO Science Series continues the series of books published formerly as the NATO ASI Series.

The NATO Science Programme offers support for collaboration in civil science between scientists of countries of the Euro-Atlantic Partnership Council. The types of scientific meeting generally supported are “Advanced Study Institutes” and “Advanced Research Workshops”, and the NATO Science Series collects together the results of these meetings. The meetings are co-organized by scientists from NATO countries and scientists from NATO’s Partner countries — countries of the CIS and Central and Eastern Europe.

**Advanced Study Institutes** are high-level tutorial courses offering in-depth study of latest advances in a field.
**Advanced Research Workshops** are expert meetings aimed at critical assessment of a field, and identification of directions for future action.

As a consequence of the restructuring of the NATO Science Programme in 1999, the NATO Science Series was re-organized to the four sub-series noted above. Please consult the following web sites for information on previous volumes published in the Series.

http://www.nato.int/science
http://www.springer.com
http://www.iospress.nl

**Series II: Mathematics, Physics and Chemistry – Vol. 221**

# Applications of Random Matrices in Physics

edited by

Édouard Brézin
Laboratoire de Physique Théorique,
Ecole Normale Supérieure, Paris, France

Vladimir Kazakov
Laboratoire de Physique Théorique
de l'Ecole Normale Supérieure,
Université Paris-VI, Paris, France

Didina Serban
Service de Physique Théorique, CEA Saclay,
Gif-sur-Yvette Cedex, France

Paul Wiegmann
James Frank Institute,
University of Chicago, Chicago, IL, U.S.A.

and

Anton Zabrodin
Institute of Biochemical Physics, Moscow, Russia
and ITEP, Moscow, Russia

Springer

Proceedings of the NATO Advanced Study Institute on
Applications of Random Matrices in Physics
Les Houches, France
6–25 June 2004

A C.I.P. Catalogue record for this book is available from the Library of Congress.

ISBN-10 1-4020-4530-1 (PB)
ISBN-13 978-1-4020-4530-1 (PB)
ISBN-10 1-4020-4529-8 (HB)
ISBN-13 978-1-4020-4529-5 (HB)
ISBN-10 1-4020-4531-X (e-book)
ISBN-13 978-1-4020-4531-8 (e-book)

---

Published by Springer,
P.O. Box 17, 3300 AA Dordrecht, The Netherlands.

*www.springer.com*

*Printed on acid-free paper*

---



Printed in the Netherlands.

# Contents


Preface ix

Random Matrices and Number Theory 1
*J. P. Keating*
1 Introduction 1
2 $\zeta(\frac{1}{2}+it)$ and $\log\zeta(\frac{1}{2}+it)$ 9
3 Characteristic polynomials of random unitary matrices 12
4 Other compact groups 17
5 Families of $L$-functions and symmetry 19
6 Asymptotic expansions 25

References 30

2D Quantum Gravity, Matrix Models and Graph Combinatorics 33
*P. Di Francesco*
1 Introduction 33
2 Matrix models for 2D quantum gravity 35
3 The one-matrix model I: large $N$ limit and the enumeration of planar graphs 45
4 The trees behind the graphs 54
5 The one-matrix model II: topological expansions and quantum gravity 58
6 The combinatorics beyond matrix models: geodesic distance in planar graphs 69
7 Planar graphs as spatial branching processes 76
8 Conclusion 85

References 86

Eigenvalue Dynamics, Follytons and Large $N$ Limits of Matrices 89
*Joakim Arnlind, Jens Hoppe*

References 93

Random Matrices and Supersymmetry in Disordered Systems 95
*K.B. Efetov*
1 Supersymmetry method 104
2 Wave functions fluctuations in a finite volume. Multifractality 118
3 Recent and possible future developments 126
4 Summary 134

Acknowledgements 134

References 134

Hydrodynamics of Correlated Systems 139
*Alexander G. Abanov*
1 Introduction 139
2 Instanton or rare fluctuation method 142
3 Hydrodynamic approach 143
4 Linearized hydrodynamics or bosonization 145
5 EFP through an asymptotics of the solution 147
6 Free fermions 148
7 Calogero-Sutherland model 150
8 Free fermions on the lattice 152
9 Conclusion 156

Acknowledgements 157
Appendix: Hydrodynamic approach to non-Galilean invariant systems 157
Appendix: Exact results for EFP in some integrable models 158

References 160

QCD, Chiral Random Matrix Theory and Integrability 163
*J.J.M. Verbaarschot*
1 Summary 163
2 Introduction 163
3 QCD 166
4 The Dirac spectrum in QCD 174
5 Low energy limit of QCD 176
6 Chiral RMT and the QCD Dirac spectrum 182
7 Integrability and the QCD partition function 188
8 QCD at finite baryon density 200
9 Full QCD at nonzero chemical potential 211
10 Conclusions 212

Acknowledgements 213

References 214

Euclidean Random Matrices: Solved and Open Problems 219
*Giorgio Parisi*
1 Introduction 219
2 Basic definitions 222
3 Physical motivations 224
4 Field theory 226
5 The simplest case 230
6 Phonons 240

References 257

Matrix Models and Growth Processes 261
*A. Zabrodin*
1 Introduction 261
2 Some ensembles of random matrices with complex eigenvalues 264

3 Exact results at finite $N$ 274
4 Large $N$ limit 282
5 The matrix model as a growth problem 298

References 316

Matrix Models and Topological Strings 319
*Marcos Mariño*
1 Introduction 319
2 Matrix models 323
3 Type B topological strings and matrix models 345
4 Type A topological strings, Chern-Simons theory and matrix models 366

References 374

Matrix Models of Moduli Space 379
*Sunil Mukhi*
1 Introduction 379
2 Moduli space of Riemann surfaces and its topology 380
3 Quadratic differentials and fatgraphs 383
4 The Penner model 388
5 Penner model and matrix gamma function 389
6 The Kontsevich Model 390
7 Applications to string theory 394
8 Conclusions 398

References 400

Matrix Models and 2D String Theory 403
*Emil J. Martinec*
1 Introduction 403
2 An overview of string theory 406
3 Strings in D-dimensional spacetime 408
4 Discretized surfaces and 2D string theory 413
5 An overview of observables 421
6 Sample calculation: the disk one-point function 425
7 Worldsheet description of matrix eigenvalues 434
8 Further results 441
9 Open problems 446

References 452

Matrix Models as Conformal Field Theories 459
*Ivan K. Kostov*
1 Introduction and historical notes 459
2 Hermitian matrix integral: saddle points and hyperelliptic curves 461
3 The hermitian matrix model as a chiral CFT 470
4 Quasiclassical expansion: CFT on a hyperelliptic Riemann surface 477
5 Generalization to chains of random matrices 483

References 486

Large $N$ Asymptotics of Orthogonal Polynomials from Integrability to Algebraic Geometry 489
*B. Eynard*
1 Introduction 489
2 Definitions 489
3 Orthogonal polynomials 490
4 Differential equations and integrability 491
5 Riemann-Hilbert problems and isomonodromies 492
6 WKB–like asymptotics and spectral curve 493
7 Orthogonal polynomials as matrix integrals 494
8 Computation of derivatives of $F^{(0)}$ 495
9 Saddle point method 496
10 Solution of the saddlepoint equation 497
11 Asymptotics of orthogonal polynomials 507
12 Conclusion 511

References 511

# Preface

Random matrices are widely and successfully used in physics for almost 60-70 years, beginning with the works of Wigner and Dyson. Initially proposed to describe statistics of excited levels in complex nuclei, the Random Matrix Theory has grown far beyond nuclear physics, and also far beyond just level statistics. It is constantly developing into new areas of physics and mathematics, and now constitutes a part of the general culture and curriculum of a theoretical physicist.

Mathematical methods inspired by random matrix theory have become powerful and sophisticated, and enjoy rapidly growing list of applications in seemingly disconnected disciplines of physics and mathematics.

A few recent, randomly ordered, examples of emergence of the Random Matrix Theory are:

- universal correlations in the mesoscopic systems,
- disordered and quantum chaotic systems;
- asymptotic combinatorics;
- statistical mechanics on random planar graphs;
- problems of non-equilibrium dynamics and hydrodynamics, growth models;
- dynamical phase transition in glasses;
- low energy limits of QCD;
- advances in two dimensional quantum gravity and non-critical string theory, are in great part due to applications of the Random Matrix Theory;
- superstring theory and non-abelian supersymmetric gauge theories;
- zeros and value distributions of Riemann zeta-function, applications in modular forms and elliptic curves;
- quantum and classical integrable systems and soliton theory.

In these fields the Random Matrix Theory sheds a new light on classical problems.

On the surface, these subjects seem to have little in common. In depth the subjects are related by an intrinsic logic and unifying methods of theoretical physics. One important unifying ground, and also a mathematical basis for the Random Matrix Theory, is the concept of integrability. This is despite the fact that the theory was invented to describe randomness.

The main goal of the school was to accentuate fascinating links between different problems of physics and mathematics, where the methods of the Random Matrix Theory have been successfully used.

We hope that the current volume serves this goal. Comprehensive lectures and lecture notes of seminars presented by the leading researchers bring a reader to frontiers of a broad range of subjects, applications, and methods of the Random Matrix Universe.

We are gratefully indebted to Eldad Bettelheim for his help in preparing the volume.

EDITORS

# RANDOM MATRICES AND NUMBER THEORY

J. P. Keating
*School of Mathematics,*
*University of Bristol,*
*Bristol, BS8 1TW*
*UK*

## 1. Introduction

My purpose in these lecture notes is to review and explain some recent results concerning connections between random matrix theory and number theory. Specifically, I will focus on how random matrix theory has been used to shed new light on some classical problems relating to the value distributions of the Riemann zeta-function and other $L$-functions, and on applications to modular forms and elliptic curves.

This may all seem rather far from Physics, but, as I hope to make clear, the questions I shall be reviewing are rather natural from the random-matrix point of view, and attempts to answer them have stimulated significant developments within that subject. Moreover, analogies between properties of the Riemann zeta function, random matrix theory, and the semiclassical theory of quantum chaotic systems have been the subject of considerable interest over the past 20 years. Indeed, the Riemann zeta function might be viewed as one of the best testing grounds for those theories.

In this introductory chapter I shall attempt to paint the number-theoretical background needed to follow these notes, give some history, and set some context from the point of view of Physics. The calculations described in the later chapters are, as far as possible, self-contained.

### 1.1 Number-theoretical background

The Riemann zeta function is defined by

$$\zeta(s) = \sum_{n=1}^{\infty} \frac{1}{n^s} = \prod_p \left(1 - \frac{1}{p^s}\right)^{-1} \tag{1}$$

for Re$s > 1$, where $p$ labels the primes, and then by analytic continuation to the rest of the complex plane. It has a single simple pole at $s = 1$, zeros at $s = -2, -4, -6$, etc., and infinitely many zeros, called the *non-trivial zeros*,



*E. Brezin et al. (eds.), Applications of Random Matrices in Physics,* 1–32.

in the *critical strip* $0 < \text{Re}s < 1$. It satisfies the *functional equation*

$$\pi^{-s/2}\Gamma\left(\frac{s}{2}\right)\zeta(s) = \pi^{-(1-s)/2}\Gamma\left(\frac{1-s}{2}\right)\zeta(1-s). \tag{2}$$

The *Riemann Hypothesis* states that all of the non-trivial zeros lie on the *critical line* $\text{Re}s = 1/2$ (i.e. on the symmetry line of the functional equation); that is, $\zeta(1/2 + it) = 0$ has non-trivial solutions only when $t = t_n \in \mathbb{R}$ [33]. This is known to be true for at least 40% of the non-trivial zeros [6], for the first 100 billion of them [36], and for batches lying much higher [29].

In these notes I will, for ease of presentation, assume the Riemann Hypothesis to be true. This is not strictly necessary – it simply makes some of the formulae more transparent.

The mean density of the non-trivial zeros increases logarithmically with height $t$ up the critical line. Specifically, the *unfolded* zeros

$$w_n = t_n \frac{1}{2\pi} \log \frac{|t_n|}{2\pi} \tag{3}$$

satisfy

$$\lim_{W\to\infty} \frac{1}{W} \# \{w_n \in [0, W]\} = 1; \tag{4}$$

that is, the mean of $w_{n+1} - w_n$ is 1.

The zeta function is central to the theory of the distribution of the prime numbers. This fact follows directly from the representation of the zeta function as a product over the primes, known as the *Euler product*. Essentially the nontrivial zeros and the primes may be thought of as Fourier-conjugate sets of numbers. For example, the number of primes less than $X$ can be expressed as a harmonic sum over the zeros, and the number, $N(T)$, of non-trivial zeros with heights $0 < t_n \leq T$ can be expressed as a harmonic sum over the primes. Such connections are examples of what are generally called *explicit formulae*. Ignoring niceties associated with convergence, the second takes the form

$$N(T) = \overline{N}(T) - \frac{1}{\pi} \sum_p \sum_{r=1}^{\infty} \frac{1}{rp^{r/2}} \sin(rT \log p), \tag{5}$$

where

$$\overline{N}(T) = \frac{T}{2\pi} \log \frac{T}{2\pi} - \frac{T}{2\pi} + \frac{7}{8} + O\left(\frac{1}{T}\right) \tag{6}$$

as $T \to \infty$. This follows from integrating the logarithmic derivative of $\zeta(s)$ around a rectangle, positioned symmetrically with respect to the critical line and passing through the points $s = 1/2$ and $s = 1/2 + iT$, using the functional equation. (Formulae like this can be made to converge by integrating both sides against a smooth function with sufficiently fast decay as $|T| \to \infty$.)

It will be a crucial point for us that the Riemann zeta-function is but one example of a much wider class of functions known as $L$-functions. These $L$-functions all have an Euler product representation; they all satisfy a functional equation like the one satisfied by the Riemann zeta-function; and in each case their non-trivial zeros are subject to a generalized Riemann hypothesis (i.e. they are all conjectured to lie on the symmetry axis of the corresponding functional equation).

To give an example, let

$$\chi_d(p) = \left(\frac{d}{p}\right) = \begin{cases} +1 & \text{if } p \nmid d \text{ and } x^2 \equiv d \pmod p \text{ solvable} \\ 0 & \text{if } p|d \\ -1 & \text{if } p \nmid d \text{ and } x^2 \equiv d \pmod p \text{ not solvable} \end{cases} \quad (7)$$

denote the Legendre symbol. Then define

$$\begin{aligned} L_D(s,\chi_d) &= \prod_p \left(1 - \frac{\chi_d(p)}{p^s}\right)^{-1} \\ &= \sum_{n=1}^{\infty} \frac{\chi_d(n)}{n^s}, \end{aligned} \quad (8)$$

where the product is over the prime numbers. These functions form a family of $L$-functions parameterized by the integer index $d$. The Riemann zeta-function is itself a member of this family.

There are many other ways to construct families of $L$-functions. It will be particularly important to us that elliptic curves also provide a route to doing this. I will give an explicit example in the last chapter of these notes.

## 1.2 History

The connection between random matrix theory and number theory was first made in 1973 in the work of Montgomery [28], who conjectured that

$$\lim_{W\to\infty} \frac{1}{W} \#\{w_n, w_m \in [0,W] : \alpha \le w_n - w_m < \beta\} = \int_\alpha^\beta \left(\delta(x) + 1 - \frac{\sin^2(\pi x)}{\pi^2 x^2}\right) dx. \quad (9)$$

This conjecture was motivated by a theorem Montgomery proved in the same paper that may be restated as follows:

$$\lim_{N\to\infty} \frac{1}{N} \sum_{n,m\le N} f(w_n - w_m) = \int_{-\infty}^{\infty} f(x) \left(\delta(x) + 1 - \frac{\sin^2(\pi x)}{\pi^2 x^2}\right) dx \quad (10)$$

for all test functions $f(x)$ whose Fourier transforms

$$\widehat{f}(\tau) = \int_{-\infty}^{\infty} f(x) \exp(2\pi i x \tau) dx \tag{11}$$

have support in the range (-1, 1) and are such that the sum and integral in (10) converge. The generalized form of the Montgomery conjecture is that (10) holds for all test functions such that the sum and integral converge, without any restriction on the support of $\widehat{f}(\tau)$. The form of the conjecture (9) then corresponds to the particular case in which $f(x)$ is taken to be the indicator function on the interval $[\alpha, \beta)$ (and so does not fall within the class of test functions covered by the theorem).

The link with random matrix theory follows from the observation that the pair correlation of the nontrivial zeros conjectured by Montgomery coincides precisely with that which holds for the eigenvalues of random matrices taken from either the Circular Unitary Ensemble (CUE) or the Gaussian Unitary Ensemble (GUE) of random matrices [27] (i.e. random unitary or hermitian matrices) in the limit of large matrix size. For example, let $A$ be an $N \times N$ unitary matrix, so that $A(A^T)^* = AA^\dagger = I$. The eigenvalues of $A$ lie on the unit circle; that is, they may be expressed in the form $e^{i\theta_n}$, $\theta_n \in \mathbb{R}$. Scaling the eigenphases $\theta_n$ so that they have unit mean spacing,

$$\phi_n = \theta_n \frac{N}{2\pi}, \tag{12}$$

the two-point correlation function for a given matrix $A$ may be defined as

$$R_2(A;x) = \frac{1}{N} \sum_{n=1}^{N} \sum_{m=1}^{N} \sum_{k=-\infty}^{\infty} \delta(x + kN - \phi_n + \phi_m), \tag{13}$$

so that

$$\frac{1}{N} \sum_{n,m} f(\phi_n - \phi_m) = \int_0^N R_2(A;x) f(x) dx. \tag{14}$$

$R_2(A;x)$ is clearly periodic in $x$, so can be expressed as a Fourier series:

$$R_2(A;x) = \frac{1}{N^2} \sum_{k=-\infty}^{\infty} |\mathrm{Tr} A^k|^2 e^{2\pi i k x/N}. \tag{15}$$

The CUE corresponds taking matrices from $U(N)$ with a probability measure given by the normalized Haar measure on the group (i.e. the unique measure that is invariant under all unitary transformations). It follows from (15) that the CUE average of $R_2(A;x)$ may be evaluated by computing the corresponding average of the Fourier coefficients $|\mathrm{Tr} A^k|^2$. This was done by Dyson

[14]:

$$\int_{U(N)} |\mathrm{Tr} A^k|^2 d\mu_{Haar}(A) = \begin{cases} N^2 & k = 0 \\ |k| & |k| \leq N \\ N & |k| > N. \end{cases} \tag{16}$$

There are several methods for proving this. One reasonably elementary proof involves using Heine's identity

$$\int_{U(N)} f_c(\theta_1, \ldots, \theta_N) d\mu_{Haar}(A)$$
$$= \frac{1}{(2\pi)^N} \int_0^{2\pi} \cdots \int_0^{2\pi} f_c(\theta_1, \ldots, \theta_N) \det(e^{i\theta_n(n-m)}) d\theta_1 \cdots d\theta_N \tag{17}$$

for *class functions* $f_c(A) = f_c(\theta_1, \theta_2, \ldots, \theta_N)$ (i.e. functions $f_c$ that are symmetric in all of their variables) to give

$$\int_{U(N)} |\mathrm{Tr} A^k|^2 d\mu_{Haar}(A) = \frac{1}{(2\pi)^N} \int_0^{2\pi} \cdots \int_0^{2\pi} \sum_j \sum_l e^{ik(\theta_j - \theta_l)}$$
$$\times \begin{vmatrix} 1 & e^{-i\theta_1} & \cdots & e^{-i(N-1)\theta_1} \\ e^{i\theta_2} & 1 & \cdots & e^{-i(N-2)\theta_2} \\ \vdots & \vdots & \ddots & \vdots \\ e^{i(N-1)\theta_N} & e^{i(N-2)\theta_N} & \cdots & 1 \end{vmatrix} d\theta_1 \cdots d\theta_N. \tag{18}$$

The net contribution from the diagonal ($j = l$) terms in the double sum is $N$, because the measure is normalized and there are $N$ diagonal terms. Using the fact that

$$\frac{1}{2\pi} \int_0^{2\pi} e^{in\theta} d\theta = \begin{cases} 1 & n = 0 \\ 0 & n \neq 0 \end{cases}, \tag{19}$$

if $k \geq N$ then the integral of the off-diagonal terms is zero, because, for example, when the determinant is expanded out and multiplied by the prefactor there is no possibility of $\theta_1$ cancelling in the exponent. If $k = N - s$, $s = 1, \ldots, N-1$, then the off-diagonal terms contribute $-s$; for example, when $s = 1$ only one non-zero term survives when the determinant is expanded out, multiplied by the prefactor, and integrated term-by-term – this is the term coming from multiplying the bottom-left entry by the top-right entry and all of the diagonal entries on the other rows. Thus the combined diagonal and off-diagonal terms add up to give the expression in (16), bearing in mind that when $k = 0$ the total is just $N^2$, the number of terms in the sum over $j$ and $l$.

Heine's identity itself may be proved using the Weyl Integration Formula [35]

$$\int_{U(N)} f_c(A) d\mu_{Haar}(A) = \frac{1}{(2\pi)^N\, N!} \int_0^{2\pi} \cdots \int_0^{2\pi} f_c(\theta_1, \ldots, \theta_N) \times \prod_{1\le j<k\le N} |e^{i\theta_j} - e^{i\theta_k}|^2 d\theta_1 \cdots d\theta_N \quad (20)$$

for class functions $f_c(A)$, the Vandermonde identity

$$\prod_{1\le j<k\le N} |e^{i\theta_j} - e^{i\theta_k}|^2 = \det\left[MM^\dagger\right] \quad (21)$$

where

$$M = \begin{pmatrix} 1 & 1 & \cdots \\ e^{i\theta_1} & e^{i\theta_2} & \cdots \\ \vdots & \ddots & \cdots \\ e^{i(N-1)\theta_1} & e^{i(N-1)\theta_2} & \cdots \end{pmatrix}, \quad (22)$$

the fact that

$$\det\left[MM^\dagger\right] = \det\left[\sum_{\ell=1}^{N} e^{i\theta_\ell(n-m)}\right], \quad (23)$$

and then by performing elementary manipulations of the rows in this determinant.

The Weyl Integration formula will play a central role in these notes. One way to understand it is to observe that, by definition, $d\mu_{Haar}(A)$ is invariant under $A \to \tilde{U}A\tilde{U}^\dagger$ where $\tilde{U}$ is any $N \times N$ unitary matrix, and that $A$ can always be diagonalized by a unitary transformation; that is, it can be written as

$$A = U \begin{pmatrix} e^{i\theta_1} & \cdots & 0 \\ \vdots & \ddots & \vdots \\ 0 & \cdots & e^{i\theta_N} \end{pmatrix} U^\dagger, \quad (24)$$

where $U$ is an $N \times N$ unitary matrix. Therefore the integral over $A$ can be written as an integral over the matrix elements of $U$ and the eigenphases $\theta_n$. Because the measure is invariant under unitary transformations, the integral over the matrix elements of $U$ can be evaluated straightforwardly, leaving the integral over the eigenphases (20).

Henceforth, to simplify the notation, I shall drop the subscript on the measure $d\mu(A)$ – in all integrals over compact groups the measure may be taken to be the Haar measure on the group.

It follows from Dyson's theorem (16) that

$$\int_{U(N)} R_2(A;x)d\mu(A) = \frac{1}{N^2}\sum_{k=-\infty}^{\infty} e^{2\pi ikx/N}\begin{cases} N^2 & k=0 \\ |k| & |k|<N \\ N & |k|\geq N \end{cases} \tag{25}$$

$$= \sum_{j=-\infty}^{\infty} \delta(x-jN) + 1 - \frac{\sin^2(\pi x)}{N^2\sin^2(\frac{\pi x}{N})}. \tag{26}$$

Hence, for test functions $f$ such that $f(x) \to 0$ as $|x| \to \infty$,

$$\lim_{N\to\infty}\int_{U(N)}\int_{-\infty}^{\infty} f(x)R_2(A;x)dxd\mu(A)$$

$$= \int_{-\infty}^{\infty} f(x)\left(\delta(x) + 1 - \frac{\sin^2(\pi x)}{\pi^2 x^2}\right)dx. \tag{27}$$

For example,

$$\lim_{N\to\infty}\int_{U(N)} \frac{1}{N}\#\{\phi_n,\phi_m : \alpha \leq \phi_n - \phi_m \leq \beta\}d\mu(A)$$

$$= \int_{\alpha}^{\beta}\left(\delta(x) + 1 - \frac{\sin^2(\pi x)}{\pi^2 x^2}\right)dx. \tag{28}$$

The key point is now that the right-hand sides of (27) and (28) coincide precisely with those of (10) and (9) respectively. That is, the pair correlation of the Riemann zeros, in the limit as the height up the critical line tends to infinity, is, conjecturally, the same as that of the eigenphases of random unitary (or hermitian) matrices in the limit as the matrix size tends to infinity.

It is important to note that the proof of Montgomery's theorem does not involve any of the steps in the derivation of the CUE pair correlation function. It is instead based entirely on the connection between the Riemann zeros and the primes. In outline, the proof involves computing the pair correlation function of the derivative of $N(T)$. Using the explicit formula (5), this pair correlation function can be expressed as a sum over pairs of primes, $p$ and $q$. The diagonal terms, for which $p = q$, obviously involve only single primes. Their sum can then be evaluated using the Prime Number Theorem, which governs the asymptotic density of primes. (Roughly speaking, the Prime Number Theorem guarantees that prime sums $\sum_p F(p)$ may, for appropriate functions $F$, be approximated by $\int F(x)/\log x dx$.) The off-diagonal terms ($p \neq q$) cannot be summed rigorously. However, it can be shown that these terms do not contribute to the limiting form of the pair correlation function for test functions $f(x)$ in (10) whose Fourier transforms have support in (-1, 1). This follows from the fact that the separation between the primes is bounded from below (by one!), which in turn means that the off-diagonal terms oscillate sufficiently quickly that they are killed for test functions satisfying the support condition by the averaging inherent in the definition of the correlation function.

In order to prove Montgomery's conjecture for all test function $f(x)$ it would be necessary to evaluate the off-diagonal terms in the sum over prime pairs, and this would require significantly more information about the pair correlation of the primes than is currently available rigorously. Nevertheless, there are conjectures about correlations between the primes due to Hardy and Littlewood [17] which can be used to provide a heuristic verification [22].

Perhaps the most compelling evidence in support of Montgomery's conjecture comes, however, from Odlyzko's numerical computations of large numbers of zeros very high up on the critical line [29]. The pair correlation of these zeros is in striking agreement with (9).

Montgomery's conjecture and theorem generalize immediately to higher order correlations between the Riemann zeros. The most general theorem, which holds for all $n$-point correlations and for test functions whose Fourier transforms are supported on restricted sets, is due to Rudnick and Sarnak [31]. Again, the conjectures are supported by Odlyzko's numerical computations [29] and by heuristic calculations for all $n$-point correlations based on the Hardy-Littlewood conjectures and which make no assumptions on the test functions [2, 3].

The results and conjectures described above extend straightforwardly to other $L$-functions – the zeros of each individual $L$-function are, assuming the generalized Riemann Hypothesis, believed to be correlated along the critical line in the same way as the eigenvalues of random unitary matrices in the limit of large matrix size [31]. They extend in a much more interesting way, however, when one considers families. It was suggested by Katz and Sarnak [20, 21] that statistical properties of the zeros of $L$-functions computed by averaging over a family, rather than along the critical line, should coincide with those of the eigenvalues of matrices from one of the classical compact groups (e.g. the unitary, orthogonal or symplectic groups); which group depends on the particular symmetries of the family in question. For example, the family defined in (7) is believed to have symplectic symmetry. I will give an example later in these notes which has orthogonal symmetry. In both these examples, the zeros of the $L$-functions come in pairs, symmetrically distributed around the centre of the critical strip (where the critical line intersects the real axis), just as the eigenvalues of orthogonal and symplectic matrices come in complex conjugate pairs. In the case of the $L$-functions, this pairing is a consequence of the functional equation. The differences between the various groups show up, for example, when one looks at the zero/eigenvalue distribution close to the respective symmetry points [30].

I mention in passing that one can define analogues of the $L$-functions over finite fields. These are polynomials. In this case Katz and Sarnak were able to prove the connection between the distribution of the zeros and that of the eigenvalues of random matrices associated with the classical compact groups, in the limit as the field size tends to infinity.

## 1.3 Physics

Much of the material in these lectures may seem rather far removed from Physics, but in fact there are a number of remarkable similarities and analogies that hint at a deep connection. Many of these similarities have been reviewed elsewhere [22, 1], and so I shall not discuss them in detail here. However, I shall make a few brief comments that I hope may help orient some readers in the following sections.

Underlying the connection between the theory of the zeta function and Physics is a suggestion, due originally to Hilbert and Polya, that one strategy to prove the Riemann Hypothesis would be to identify the zeros $t_n$ with the eigenvalues of a self-adjoint operator. The Riemann Hypothesis would then follow immediately from the fact that these eigenvalues are real. One might thus speculate that the numbers $t_n$ are the energy levels of some quantum mechanical system.

In quantum mechanics there is a semiclassical formula due to Gutzwiller [15] that relates the counting function of the energy levels to a sum over the periodic orbits of the corresponding classical system. In the case of strongly chaotic systems that do not possess time-reversal symmetry, this formula is very closely analogous to (5), the primes being associated with periodic orbits.

The fact that the analogy is with chaotic systems that are not time-reversal invariant is consistent with the conjecture that the energy level statistics of such systems should, generically, in the semiclassical limit, coincide with those of the eigenvalues of random matrices from one of the ensembles that are invariant under unitary transformations, such as the CUE or the GUE, in the limit of large matrix size [4].

The appearance of random matrices associated with the orthogonal and symplectic groups in the statistical description of the zeros statistics within families of $L$-functions is analogous to the appearance of these groups in Zirnbauer's extension of Dyson's three-fold way to include systems of disordered fermions (see, for example, [37]).

# 2. $\zeta(\frac{1}{2}+it)$ and $\log\zeta(\frac{1}{2}+it)$

The background reviewed in the introduction relates to connections between the statistical distributions of the zeros of the Riemann zeta function and other $L$-functions and those of the eigenvalues of random matrices associated with the classical compact groups, on the scale of the mean zero/eigenvalue spacing. My goal in the remainder of these notes is to focus on more recent developments that concern the value distribution of the functions $\zeta(\frac{1}{2}+it)$ and $\log\zeta(\frac{1}{2}+it)$ as $t$ varies. I will then go on to describe the value distribution of $L$-functions within families, and applications of these results to some other important questions in number theory.

The basic ideas I shall be reviewing were introduced in [24], [25], and [7]. The theory was substantially developed in [8, 9]. The applications I shall describe later were initiated in [12] and [13]. Details of all of the calculations I shall outline can be found in these references.

I shall start by reviewing what is known about the value distribution of $\log\zeta(1/2+it)$. The most important general result, due originally to Selberg, is that this function satisfies a central limit theorem [33]: for any rectangle $B$ in the complex plane,

$$\lim_{T\to\infty}\frac{1}{T}\text{meas.}\{T\le t\le 2T : \frac{\log\zeta(\frac{1}{2}+it)}{\sqrt{\frac{1}{2}\log\log\frac{t}{2\pi}}}\in B\}$$
$$=\int\int_B e^{-\frac{1}{2}(x^2+y^2)}dxdy. \quad (29)$$

Odlyzko has investigated the value distribution of $\log\zeta(1/2+it)$ numerically for values of $t$ around the height of the $10^{20}$th zero. Surprisingly, he found a distribution that differs markedly from the limiting Gaussian. His data are plotted in Figures 1 and 2. The CUE curves will be discussed later.

In order to quantify the discrepancy illustrated in Figures 1 and 2, I list in Table 1 the moments of $\mathrm{Re}\log\zeta(1/2+it)$, normalized so that the second moment is equal to one, calculated numerically by Odlyzko in [29]. The data in the second and third columns relate to two different ranges near the height of the $10^{20}$th zero. The difference between them is therefore a measure of the fluctuations associated with computing over a finite range near this height. The data labelled $U(42)$ will be explained later.

Next let us turn to the value distribution of $\zeta(1/2+it)$ itself. Its moments satisfy the long-standing and important conjecture that

$$\lim_{T\to\infty}\frac{1}{(\log\frac{T}{2\pi})^{\lambda^2}}\frac{1}{T}\int_0^T|\zeta(\tfrac{1}{2}+it|^{2\lambda}dt$$
$$=f_\zeta(\lambda)\prod_p\left[(1-\tfrac{1}{p})^{\lambda^2}\sum_{m=0}^{\infty}\left(\frac{\Gamma(\lambda+m)}{m!\Gamma(\lambda)}\right)^2p^{-m}\right] \quad (30)$$

This can be viewed in the following way. It asserts that the moments grow like $(\log\frac{T}{2\pi})^{\lambda^2}$ as $T\to\infty$. Treating the primes as being statistically independent of each other would give the right hand side with $f_\zeta(\lambda)=1$. $f_\zeta(\lambda)$ thus

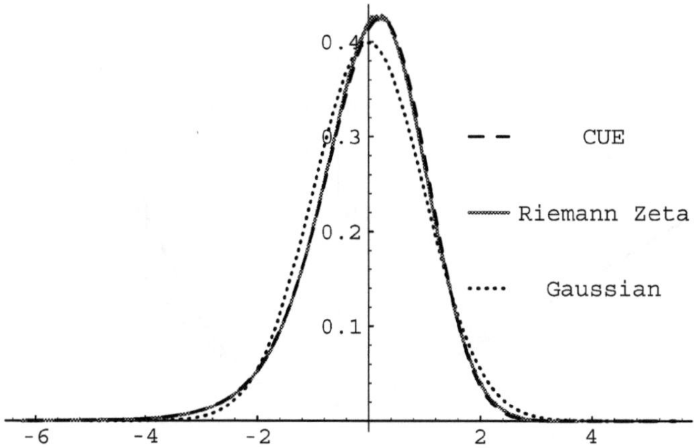


*Figure 1.* Odlyzko's data for the value distribution of $\text{Re}\log\zeta(1/2+it)$ near the $10^{20}$th zero (taken from [29]), the value distribution of $\text{Re}\log Z$ with respect to matrices taken from $U(42)$, and the standard Gaussian, all scaled to have unit variance. (Taken from [24].)

*Table 1.* Moments of $\text{Re}\log\zeta(1/2+it)$, calculated by Odlyzko over two ranges (labelled a and b) near the $10^{20}$th zero ($t \simeq 1.520 \times 10^{19}$) (taken from [29]), compared with the moments of $\text{Re}\log Z$ for $U(42)$ and the Gaussian (normal) moments, all scaled to have unit variance.

| *Moment* | $\zeta$ a) | $\zeta$ b) | $U(42)$ | *Normal* |
|---|---|---|---|---|
| 1 | 0.0 | 0.0 | 0.0 | 0 |
| 2 | 1.0 | 1.0 | 1.0 | 1 |
| 3 | -0.53625 | -0.55069 | -0.56544 | 0 |
| 4 | 3.9233 | 3.9647 | 3.89354 | 3 |
| 5 | -7.6238 | -7.8839 | -7.76965 | 0 |
| 6 | 38.434 | 39.393 | 38.0233 | 15 |
| 7 | -144.78 | -148.77 | -145.043 | 0 |
| 8 | 758.57 | 765.54 | 758.036 | 105 |
| 9 | -4002.5 | -3934.7 | -4086.92 | 0 |
| 10 | 24060.5 | 22722.9 | 25347.77 | 945 |

quantifies deviations from this simple-minded ansatz. Assuming that the moments do indeed grow like $(\log \frac{T}{2\pi})^{\lambda^2}$, the problem is then to determine $f_\zeta(\lambda)$.

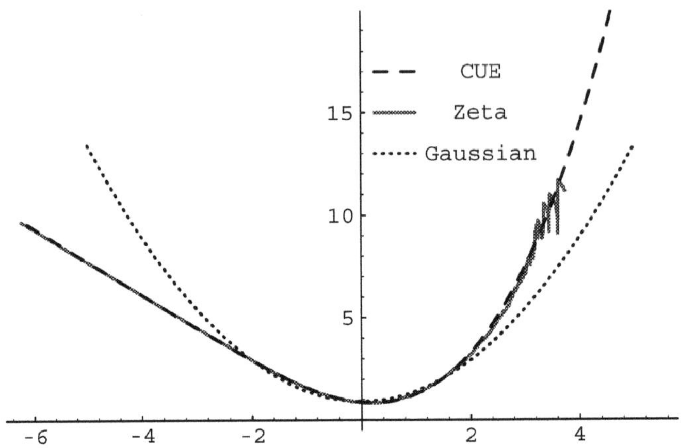


*Figure 2.* The logarithm of the inverse of the value distribution plotted in Figure 1. (Taken from [24].)

The conjecture is known to be correct in only two non-trivial cases, when $\lambda = 1$ and $\lambda = 2$. It was shown by Hardy and Littlewood in 1918 that $f_\zeta(1) = 1$ [16] and by Ingham in 1926 that $f_\zeta(2) = \frac{1}{12}$ [18]. On number-theoretical grounds, Conrey and Ghosh have conjectured that $f_\zeta(3) = \frac{42}{9!}$ [10] and Conrey and Gonek that $f_\zeta(4) = \frac{24024}{16!}$ [11].

We shall now look to random matrix theory to see what light, if any, it can shed on these issues.

## 3. Characteristic polynomials of random unitary matrices

Our goal is to understand the value distribution of $\zeta(1/2 + it)$. Recalling that the zeros of this function are believed to be correlated like the eigenvalues of random unitary matrices, we take as our model the functions whose zeros are these eigenvalues, namely the characteristic polynomials of the matrices in question.

Let us define the characteristic polynomial of a matrix $A$ by

$$\begin{aligned} Z(A,\theta) &= \det(I - Ae^{-i\theta}) \\ &= \prod_n (1 - e^{i(\theta_n - \theta)}). \end{aligned} \tag{31}$$

Consider first the function

$$P_N(s,t) = \int_{U(N)} |Z(A,\theta)|^t e^{is\mathrm{Im}\log Z(A,\theta)} d\mu(A). \tag{32}$$

This is the moment generating function of $\log Z$: the joint moments of $\mathrm{Re}\log Z$ and $\mathrm{Im}\log Z$ are obtained from derivatives of $P$ at $s = 0$ and $t = 0$, and

$$\int_{U(N)} \delta(x - \mathrm{Re}\log Z)\delta(y - \mathrm{Im}\log Z) d\mu(A) \tag{33}$$

$$= \frac{1}{4\pi^2} \int_{-\infty}^{\infty} \int_{-\infty}^{\infty} e^{-itx - isy} P(s, it) ds dt. \tag{34}$$

Written in terms of the eigenvalues,

$$P_N(s,t) = \int_{U(N)} \prod_{n=1}^{N} |1 - e^{i(\theta_n - \theta)}|^t e^{-is\sum_{m=1}^{\infty} \frac{\sin[(\theta_n - \theta)m]}{m}} d\mu(A). \tag{35}$$

Since the integrand is a class function, we can use Weyl's integration formula (20) to write

$$\begin{aligned} P_N(s,t) = \frac{1}{(2\pi)^N N!} \int_0^{2\pi} \cdots \int_0^{2\pi} \prod_{n=1}^{N} |1 - e^{i(\theta_n - \theta)}|^t \\ \times e^{-is\sum_{m=1}^{\infty} \frac{\sin[(\theta_n - \theta)m]}{m}} \prod_{1 \le j < k \le N} |e^{i\theta_j} - e^{i\theta_k}|^2 d\theta_1 \cdots d\theta_N. \end{aligned} \tag{36}$$

This integral can then be evaluated using a form of Selberg's integral described in [27], giving [24]

$$P_N(s,t) = \prod_{j=1}^{N} \frac{\Gamma(j)\Gamma(t+j)}{\Gamma(j + \frac{t}{2} + \frac{s}{2})\Gamma(j + \frac{t}{2} - \frac{s}{2})}. \tag{37}$$

Note that the result is independent of $\theta$. This is because the average over $U(N)$ includes rotations of the spectrum and is itself therefore rotationally invariant.

## 3.1 Value distribution of log Z

Consider first the Taylor expansion

$$P_N(s,t) = e^{\alpha_{00}+\alpha_{10}t+\alpha_{01}s+\alpha_{20}t^2/2+\alpha_{11}ts+\alpha_{02}s^2/2+\cdots}. \tag{38}$$

The $\alpha_{m0}$ are the cumulants of $\operatorname{Re}\log Z$ and the $\alpha_{0n}$ are $i^n$ times the cumulants of $\operatorname{Im}\log Z$. Expanding (37) gives:

$$\alpha_{10} = \alpha_{01} = \alpha_{11} = 0; \tag{39}$$

$$\alpha_{20} = -\alpha_{02} = \tfrac{1}{2}\log N + \tfrac{1}{2}(\gamma+1) + O(\tfrac{1}{N^2}); \tag{40}$$

$$\alpha_{mn} = O(1) \text{ for } m+n \geq 3; \tag{41}$$

and more specifically,

$$\alpha_{m0} = (-1)^m(1 - \tfrac{1}{2^{m-1}})\Gamma(m)\zeta(m-1) + O(\tfrac{1}{N^{m-2}}), \text{ for } m \geq 3. \tag{42}$$

This leads to the following theorem [24]: for any rectangle $B$ in the complex plane

$$\lim_{N\to\infty} \text{meas.}\left\{A \in U(N) : \frac{\log Z(A,\theta)}{\sqrt{\frac{1}{2}\log N}} \in B\right\} = \frac{1}{2\pi}\int\int_B e^{-\frac{1}{2}(x^2+y^2)}dxdy. \tag{43}$$

Comparing this result to (29), one sees that $\log\zeta(1/2+it)$ and $\log Z$ both satisfy a central limit theorem when, respectively, $t\to\infty$ and $N\to\infty$. Note that the scalings in (29) and (43), corresponding to the asymptotic variances, are the same if we make the identification

$$N = \log\tfrac{t}{2\pi}. \tag{44}$$

This is the same as identifying the mean eigenvalue density with the mean zero density; c.f. the unfolding factors in (12) and (3).

The identification (44) provides a connection between matrix sizes and heights up the critical line. The central limit theorems imply that when both of these quantities tend to infinity $\log\zeta(1/2+it)$ and $\log Z$ have the same limit distribution. This supports the choice of $Z$ as a model for the value distribution of $\zeta(1/2+it)$ when $t\to\infty$. It is natural then to ask if it also constitutes a useful model when $t$ is large but finite; that is, whether it can explain the deviations from the limiting Gaussian seen in Odlyzko's data.

The value of $t$ corresponding to the height of the $10^{20}$th zero should be associated, via (44), to a matrix size of about $N = 42$. The moments and value distribution of $\log Z$ for any size of matrix can be obtained directly from the formula for the moment generating function (37). The value distribution when $N = 42$ is the CUE curve plotted in Figures 1 and 2. Values of the moments are listed in Table 1. The obvious agreement between the results for random $42 \times 42$ unitary matrices and Odlyzko's data provides significant further support for the model. It suggests that random matrix theory models not just the limit distribution of $\log \zeta(1/2 + it)$, but the rate of approach to the limit as $t \to \infty$.

## 3.2 Moments of $|Z|$

We now turn to the more important problem of the moments of $|\zeta(1/2+it)|$. It is natural to expect these moments to be related to those of the modulus of the characteristic polynomial $Z$, which are defined as

$$\begin{aligned} \int_{U(N)} |Z(A,\theta)|^{2\lambda} d\mu(A) &= P(0, 2\lambda) \\ &= \prod_{j=1}^{N} \frac{\Gamma(j)\Gamma(j+2\lambda)}{(\Gamma(j+\lambda))^2} \qquad (45) \\ &= e^{\sum_{m=0}^{\infty} \alpha_{m0} (2\lambda)^n / n!}. \qquad (46) \end{aligned}$$

Therefore

$$\lim_{N\to\infty} \frac{1}{N^{\lambda^2}} \int_{U(N)} |Z(A,\theta)|^{2\lambda} d\mu(A) = e^{\lambda^2(\gamma+1) + \sum_{m=3}^{\infty} (-2\lambda)^m \frac{2^{m-1}-1}{2^{m-1}} \frac{\zeta(m-1)}{m}}, \qquad (47)$$

for $|\lambda| < \frac{1}{2}$. Note that since we are identifying $Z$ with $\zeta(1/2 + it)$ and $N$ with $\log \frac{t}{2\pi}$, the expression on the left-hand side of (47) corresponds precisely to that in (30).

We now recall some properties of the Barnes' $G$-function. This is an entire function of order 2 defined by

$$G(1+z) = (2\pi)^{z/2} e^{-[(1+\gamma)z^2+z]/2} \prod_{n=1}^{\infty} \left[ (1+z/n)^n e^{-z+z^2/(2n)} \right]. \qquad (48)$$

It satisfies

$$G(1) = 1, \qquad (49)$$

$$G(z+1) = \Gamma(z) G(z) \qquad (50)$$

and

$$\log G(1+z) = (\log 2\pi - 1)\frac{z}{2} - (1+\gamma)\frac{z^2}{2} + \sum_{n=3}^{\infty} (-1)^{n-1} \zeta(n-1) \frac{z^n}{n}. \qquad (51)$$

Thus we have that

$$\lim_{N\to\infty} \frac{1}{N^{\lambda^2}} \int_{U(N)} |Z(A,\theta)|^{2\lambda} d\mu(A) = f_U(\lambda), \tag{52}$$

with

$$f_U(\lambda) = \frac{G^2(1+\lambda)}{G(1+2\lambda)}. \tag{53}$$

Using (50) we further have that for positive integers $k$

$$f_U(k) = \prod_{j=0}^{k-1} \frac{j!}{(j+k)!}. \tag{54}$$

In particular, $f_U(1) = 1$, $f_U(2) = \frac{1}{12}$, $f_U(3) = \frac{42}{9!}$ and $f_U(4) = \frac{24024}{16!}$, which match the values of $f_\zeta$ listed after (30). This then motivates the conjecture [24] that

$$f_\zeta(\lambda) = f_U(\lambda) \tag{55}$$

for all $\lambda$ such that $\mathrm{Re}\lambda > -\frac{1}{2}$.

## 3.3 Value distribution of $|Z|$

Let us now define the value distribution of $|Z(A,0)|$ by

$$\int_{U(N)} \delta(|Z(A,0)| - w) d\mu(A) = \rho_U(w,N). \tag{56}$$

Obviously,

$$\int_{U(N)} |Z|^t d\mu(A) = \int_0^\infty \rho_U(w,N) w^t dw; \tag{57}$$

that is, the moments of $|Z|$ are given by the Mellin transform of the value distribution we seek to evaluate. Therefore, using (45),

$$\rho_U(w,N) = \frac{1}{2\pi i} \int_{c-i\infty}^{c+i\infty} \prod_{j=1}^{N} \frac{\Gamma(j)\Gamma(j+t)}{(\Gamma(j+\frac{t}{2}))^2} \frac{1}{w^{t+1}} dt, \tag{58}$$

where $c > 0$. As $N \to \infty$ this can be approximated using the method of stationary phase. In the limit as $w \to 0$ an expansion in increasing powers of $w$ can be formed by considering the residues from the poles at the negative integers. The rightmost pole is at $t = -1$, and so $\rho_U(w,N) \to$ constant as $w \to 0$.

Values of $\rho_U(w,N)$ when $N = 12$ are plotted and compared with the value distribution of $|\zeta(\frac{1}{2}+it)|$ when $t = 10^6$ in Figure 3.

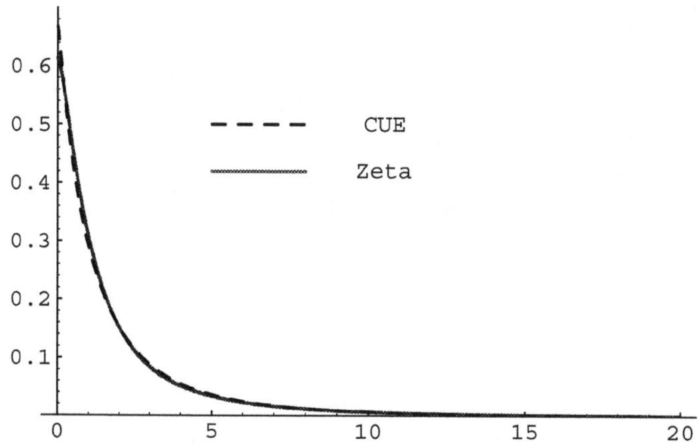


*Figure 3.* The CUE value distribution of $|Z|$ corresponding to $N = 12$ (that is, $U(12)$) (dashed), with numerical data for the value distribution of $|\zeta(1/2 + it)|$ (solid) near $t = 10^6$.

## 4. Other compact groups

We have seen so far that the characteristic polynomials of random unitary matrices may be used to model the moments and value distribution of the Riemann zeta function on its critical line. As discussed in the introduction, Katz and Sarnak [20, 21] have shown that the distribution of the zeros within families of $L$-functions is related to averages over the various classical compact groups, the particular group in question being determined by symmetries of the family. This suggests that the moments and value distribution of $L$-functions within a family may be understood by extending the calculations for the unitary group described above to the other classical compact groups.

Consider a matrix $A \in USp(2N)$ or $A \in O(N)$. In both cases there is a symmetry in the spectrum not present for general unitary matrices: the complex eigenvalues come in complex conjugate pairs, $e^{\pm i\theta_n}$. For example, in the case of $USp(2N)$ and $O(2N)$ the characteristic polynomial is

$$Z(A, \theta) = \prod_{n=1}^{N} (1 - e^{i(\theta_n - \theta)})(1 - e^{i(-\theta_n - \theta)}). \tag{59}$$

Our goal now is to determine the moments and value distribution of the characteristic polynomials with respect to averages over these groups. Like for $U(N)$, averages are understood to be computed with respect to the relevant Haar measure. Unlike for $U(N)$, in both cases the symmetry in the spectrum means that the results depend on $\theta$. We will focus on the symmetry point $\theta = 0$, as this is where the differences are greatest.

## 4.1 Moments

To calculate the moments of the characteristic polynomials with respect to averages over $O(N)$ or $USp(2N)$ we need the two key ingredients used in the calculation for $U(N)$: the Weyl integration formula for these groups [35] and appropriate forms of the Selberg integral (c.f. [27], chapter 17). Following the steps detailed above (for further details, see [25]) we then find for the symplectic group that

$$\begin{aligned}\int_{USp(2N)} Z(A,0)^s d\mu(A) &= 2^{2Ns} \prod_{j=1}^{N} \frac{\Gamma(1+N+j)\Gamma(\frac{1}{2}+s+j)}{\Gamma(\frac{1}{2}+j)\Gamma(1+s+N+j)} \\ &\equiv M_{Sp}(s;N). \end{aligned} \tag{60}$$

It follows that $\log Z$ again satisfies a central limit theorem and that

$$\lim_{N\to\infty} \frac{1}{N^{s(s+1)/2}} \int_{USp(2N)} Z(A,0)^s d\mu(A) \tag{61}$$

$$= 2^{s^2/2} \frac{G(1+s)\sqrt{\Gamma(1+s)}}{\sqrt{G(1+2s)\Gamma(1+2s)}} \equiv f_{Sp}(s). \tag{62}$$

For positive integers $n$

$$f_{Sp}(n) = \frac{1}{\prod_{j=1}^{n}(2j-1)!!} = \frac{1}{(2n-1)(2n-3)^2(2n-5)^3\cdots}. \tag{63}$$

(This last result was obtained independently in [5].)

The distribution of values of $Z(A,0)$ for $A \in USp(2N)$ is given by

$$\int_{USp(2N)} \delta(Z(A,0)-w)d\mu(A) = \frac{1}{2\pi i}\int_{c-i\infty}^{c+i\infty} M_{Sp}(s)\frac{ds}{w^{s+1}}. \tag{64}$$

As $w \to 0$ it vanishes like $w^{1/2}$, because, from (60), the rightmost pole of $M_{Sp}(s;N)$ is at $s = -3/2$.

As a second example we take the orthogonal group $SO(2N)$. In this case

$$\begin{aligned}\int_{SO(2N)} Z(A,0)^s d\mu(A) &= 2^{2Ns} \prod_{j=1}^{N} \frac{\Gamma(N+j-1)\Gamma(s+j-1/2)}{\Gamma(j-1/2)\Gamma(s+j+N-1)} \\ &\equiv M_O(s;N), \end{aligned} \tag{65}$$

$\log Z$ again satisfies a central limit theorem, and

$$\lim_{N\to\infty} \frac{1}{N^{s(s-1)/2}} \int_{SO(2N)} Z(A,0)^s d\mu(A) \tag{66}$$

$$= 2^{s^2/2} \frac{G(1+s)\sqrt{\Gamma(1+2s)}}{\sqrt{G(1+2s)\Gamma(1+s)}} \equiv f_O(s). \tag{67}$$

For positive integers $n$ we have

$$f_O(n) = 2^n f_{Sp}(n-1). \tag{68}$$

(This last result was also obtained independently in [5].)

(I note in passing the following rather interesting relationship between the leading order moment coefficients for the three compact groups discussed:

$$f_O(s) f_{Sp}(s) = 2^{s^2} f_U(s).) \tag{69}$$

The value distribution of the characteristic polynomials is again given by

$$\int_{SO(2N)} \delta(Z(A,0) - w) d\mu(A) = \frac{1}{2\pi i} \int_{c-i\infty}^{c+i\infty} M_O(s;N) \frac{1}{w^{s+1}} ds. \tag{70}$$

In this case it diverges like $w^{-1/2}$ as $w \to 0$.

## 5. Families of *L*-functions and symmetry

I now describe how the results listed above may be applied to $L$-functions within families. The main ideas will be illustrated by focusing on two representative examples.

### 5.1 Example 1: Dirichlet *L*-functions

Let

$$\chi_d(p) = \left(\frac{d}{p}\right) = \begin{cases} +1 & \text{if } p \nmid d \text{ and } x^2 \equiv d \pmod p \text{ solvable} \\ 0 & \text{if } p|d \\ -1 & \text{if } p \nmid d \text{ and } x^2 \equiv d \pmod p \text{ not solvable} \end{cases} \tag{71}$$

denote the Legendre symbol. That is, $\chi_d(p)$ is a real quadratic Dirichlet character. Then

$$\begin{aligned} L_D(s, \chi_d) &= \prod_p \left(1 - \frac{\chi_d(p)}{p^s}\right)^{-1} \\ &= \sum_{n=1}^{\infty} \frac{\chi_d(n)}{n^s}, \end{aligned} \tag{72}$$

where the product is over the prime numbers. These functions form a family of $L$-functions parameterized by the integer index $d$. (This is the family mentioned in the Introduction.)

## 5.2 Example 2: $L$-functions associated with elliptic curves

Consider the function

$$\begin{aligned} f(z) &= e^{2\pi iz}\prod_{n=1}^{\infty}(1-e^{2\pi inz})^2(1-e^{22\pi inz})^2 \\ &= \sum_{n=1}^{\infty} a_n e^{2\pi inz}, \end{aligned} \tag{73}$$

where the integers $a_n$ are the Fourier coefficients of $f$. This function may be shown to satisfy

$$f\left(\frac{az+b}{cz+d}\right) = (cz+d)^2 f(z) \tag{74}$$

for every $\begin{pmatrix} a & b \\ c & d \end{pmatrix} \in SL_2(\mathbb{Z})$ with $11|c$. That is, $f(z)$ is a cusp form of weight 2 for $\Gamma_0(11)$. It is important to note that the weight is an integer.

Now consider the elliptic curve

$$E_{11}: \ y^2 = 4x^3 - 4x^2 - 40x - 79. \tag{75}$$

Let

$$N_p = \#\{(x,y) \in \mathbb{F}_p^2 : y^2 = 4x^3 - 4x^2 - 40x - 79\}. \tag{76}$$

Then

$$a_p = p - N_p; \tag{77}$$

that is, the Fourier coefficients of $f$ determine the number of solutions of $E_{11}$. One can construct a zeta function

$$\zeta_{E_{11}}(s) = \sum_{n=1}^{\infty} \frac{a_n}{n^s} \tag{78}$$

and then a family of $L$-functions by twisting with the Dirichlet characters defined in (71):

$$L_{E_{11},d}(s) = \sum_{n=1}^{\infty} \frac{a_n \chi_d(n)}{n^s}. \tag{79}$$

This family is again parameterized by the integer index $d$. The $L$-functions satisfy the following functional equation:

$$\begin{aligned} \Phi_{E_{11},d}(s) &\equiv \left(\frac{2\pi}{\sqrt{11}|d|}\right)^{-s} \Gamma(s) L_{E_{11},d}(s) \\ &= \chi_d(-11)\Phi_{E_{11},d}(2-s). \end{aligned} \tag{80}$$

We will here focus on those $L$-functions associated with characters satisfying

$$\chi_d(-11) = +1, \tag{81}$$

i.e. those that do not vanish trivially at $s = 1$.

This example is fully representative in that in every aspect it generalizes to all other elliptic curves: it follows from Wiles' work relating to Fermat's Last Theorem that every elliptic curve is associated with an integer-weight modular form whose Fourier coefficients determine the number of rational solutions, as in (77), and may be used to form $L$-functions, as in (79).

In both of the above examples the $L$-functions satisfy a Riemann Hypothesis. In the first example, this places their complex zeros on the (*critical*) line $\mathrm{Re}s = 1/2$; in the second, it places them on the line $\mathrm{Re}s = 1$ (this is merely a matter of conventional normalization rather than a significant difference). In each case the zeros high up on the critical line are believed to be distributed like the eigenvalues of random unitary matrices [31], and so the results obtained for the Riemann zeta function extend, conjecturally, to every individual (principal) $L$-function.

Rather than fixing the $L$-function and averaging along the critical line, we can instead fix a height on the critical line and average through the family; that is, average with respect to $d$. In this way one can therefore examine the distribution of the zeros nearest the critical point, $s = 1/2$ or $s = 1$, within these families.

It was conjectured by Katz and Sarnak [20, 21] that the zero statistics around the critical point are related to the eigenvalue statistics of one of the compact groups described above near to a spectral symmetry point (if one exists). The particular group in question is determined by symmetries of the family. There is now extensive numerical and theoretical evidence in support of this [30].

The first example of a family of $L$-functions given above (the Dirichlet $L$-functions) is conjectured to have symplectic symmetry and so the zeros behave like the eigenvalues of matrices from $USp(2N)$. The family of elliptic curve $L$-functions in the second example is conjectured to have orthogonal symmetry. Their zeros behave like the eigenvalues of $SO(2N)$ matrices.

Following the Katz-Sarnak philosophy, it is natural to believe that random matrix theory can predict the moments of $L$-functions in families like those described here; that is, it is natural to conjecture that the moments

$$\frac{1}{X^*} \sum_{0<d<X}^{*} (L_D(\tfrac{1}{2}, \chi_d))^s$$

(where the sum is over fundamental discriminants $d$, and $X^*$ is the number of terms in the sum) are modelled by

$$\int_{USp(2N)} (Z(A,0))^s dA,$$

whereas the moments

$$\frac{1}{X^*} \sum_{\substack{0<d<X \\ \chi_{-d}(-11)=+1}}^{*} (L_{E_{11},d}(1))^s$$

are modelled by

$$\int_{SO(2N)} (Z(A,0))^s dA.$$

For example, the factors corresponding to $f_\zeta$ in the moments of the $L$-functions are conjectured to be given by $f_{Sp}(s)$ and $f_O(s)$. This agrees with all previous results and conjectures for the integer moments (see, for example, [7, 25]). (These factors must be multiplied by arithmetical contributions to give the moments.) Furthermore, the value distributions of the $L$-functions with respect to varying $d$ are expected to be related to the value distributions of the associated characteristic polynomials. Numerical evidence in support of this is illustrated in Figure 4 for the family of $L$-functions associated with $E_{11}$, described in the second example.

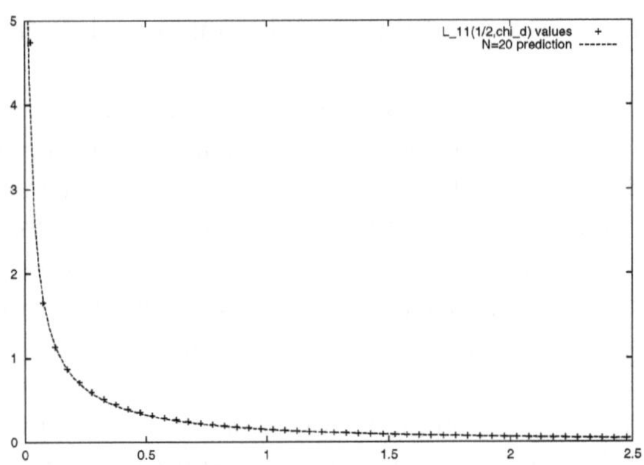


*Figure 4.* The value distribution of $L_{E_{11},d}(1)$, for prime $|d|$, $-788299808 < d < 0$, even functional equation, compared to equation (70), with $N = 20$. Note the square-root divergence as $w \to 0$. The $L$-function values have been normalized so that they have the same means as the random matrix value distributions. (From [12].)

The key question is obviously: what use can be made of the random matrix model for the value distribution of $L$-functions? I will now outline some applications that are currently being explored for the $L$-functions associated with elliptic curves. These exploit certain explicit formulae for the values at the central point $s = 1$. The approach is general, but for simplicity I shall describe it in the specific context of the family defined in example 2.

The formula for $L_{E_{11},d}(1)$ that we shall exploit is an example of a general class of formulae developed by Shimura [32], Waldspurger [34] and Kohnen-Zagier [26]. For $d < 0$ and $\chi_d(-11) = +1$ it asserts that

$$L_{E_{11},d}(1) = \frac{\kappa c_{|d|}^2}{\sqrt{|d|}}, \tag{82}$$

where

$$\kappa = \text{constant } (= 2.91763\ldots)$$

and

$$g(z) = \sum_{n=1}^{\infty} c_n e^{2\pi i n z} \tag{83}$$

satisfies

$$g\left(\frac{az+b}{cz+d}\right) = \epsilon(a,b,c,d)(cz+d)^{3/2} g(z) \tag{84}$$

for every $\begin{pmatrix} a & b \\ c & d \end{pmatrix} \in SL_2(\mathbb{Z})$ such that $44|c$; that is the numbers $c_n$ are the Fourier coefficients of a three-halves-weight form for $\Gamma_0(44)$. Note that the $L$-functions were originally defined using the Fourier coefficients $a_n$ of an integer-weight form (weight-two in our example), but that at the central point their values are related to the Fourier coefficient of a half-integer weight form. One important point to notice is that

$$c_n \in \mathbb{Z}.$$

I will now describe two conjectural implications that follow from combining this formula with the random-matrix model.

## 5.3 Generalization of the Sato-Tate law to half-integer weight modular forms

The Sato-Tate law describes the value distribution of the Fourier coefficients $a_p$ defined in (72). According to the theorems of Hasse and Deligne these satisfy $|a_p| \leq 2\sqrt{p}$ and so may be written

$$\frac{a_p}{\sqrt{p}} = 2\cos\theta_p, \quad 0 \leq \theta \leq \pi.$$

The question then is: how are the angles $\theta_p$ distributed as the prime $p$ varies? This is the subject of the celebrated Sato-Tate law:

$$\lim_{x\to\infty} \frac{1}{\pi(x)} \#\{p < x : \alpha < \theta_p \leq \beta\} = \frac{2}{\pi} \int_\alpha^\beta \sin^2\theta d\theta, \tag{85}$$

Given that the Fourier coefficients of integer-weight forms satisfy a simple limit distribution law, it is natural to ask whether the Fourier coefficients of half-integer-weight forms do as well.

Combining (82) with the random matrix model for the value distribution of $L_{E_{11},d}(1)$ provides a conjectural answer to this question [13]. For example, it follows from the central limit theorem for the logarithm of the characteristic polynomial that one should expect that [13]

$$\lim_{D\to\infty}\frac{1}{D^*}\#\{2<d\le D:\chi_{-d}(-11)=1,$$
$$\frac{2\log|c_{|d|}|-\frac{1}{2}\log d+\frac{1}{2}\log\log d}{\sqrt{\log\log d}}\in(\alpha,\beta)\}=\frac{1}{\sqrt{2\pi}}\int_\alpha^\beta e^{-x^2/2}dx,$$

where

$$D^*=\#\{2<d\le D:\chi_{-d}(-11)=1\}. \tag{86}$$

Furthermore, the value distribution of $c_{|d|}$ should be related to that described in (70).

## 5.4 Frequency of vanishing of $L$-functions

I now turn to the question of the frequency of vanishing of $L$-functions at the central point. In the light of the Birch & Swinnerton-Dyer conjecture, which relates the order of vanishing at this point to the number of rational points on the corresponding elliptic curve, this is an issue of considerable importance.

The formula (82) for $L_{E_{11},d}(1)$ implies a discretization (or quantization) of its values. So if $L_{E_{11},d}(1)<\frac{\kappa}{\sqrt{|d|}}$ then in fact $L_{E_{11},d}(1)=0$. Pushing the random matrix model to the very limits of the range where it can be justified (and hopefully not beyond), the probability that $L_{E_{11},d}(1)<\frac{\kappa}{\sqrt{|d|}}$ may be estimated by integrating the probability density (70) from 0 to $\frac{\kappa}{\sqrt{|d|}}$. Using the fact that the probability density has a square-root singularity at the origin then motivates the following two conjectures due to Conrey, Keating, Rubinstein & Snaith [12]:

$$\#\{p\le D:\chi_{-p}(-11)=1,L_{E_{11},-p}(1)=0\}\asymp\frac{D^{3/4}}{(\log D)^{5/8}}; \tag{87}$$

and if

$$R_p(D)=\frac{\#\{d<D:\chi_{-d}(-11)=1,\chi_{-d}(p)=1,L_{E_{11},d}(1)=0\}}{\#\{d<D:\chi_{-d}(-11)=1,\chi_{-d}(p)=-1,L_{E_{11},d}(1)=0\}}, \tag{88}$$

then

$$R_p=\lim_{D\to\infty}R_p(D)=\sqrt{\frac{p+1-a_p}{p+1+a_p}}. \tag{89}$$

Preliminary data relating to the first conjecture are plotted in Figure 5. These would appear to support the dependence on $D^{3/4}$, but do not cover a large enough range to determine the power of $\log D$.

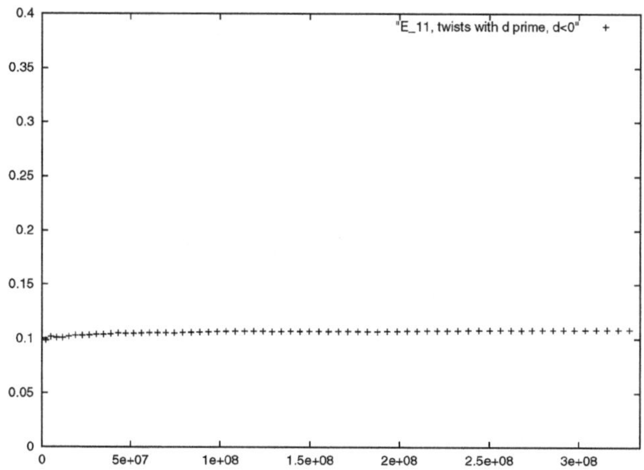


*Figure 5.* The l.h.s. of (87) divided by $D^{3/4}(\log D)^{-5/8}$. The calculations include only twists with $d < 0$, $d$ prime, and cases with even functional equation. While the picture is reasonably flat, $\log(D)$ is almost constant for most of the interval in question. The flatness observed therefore reflects the main dependence on $D^{3/4}$. (From [12].)

Data in support of the second conjecture are listed in Table 2 and are plotted in Figure 6. In this case the agreement with the conjecture is striking.

## 6. Asymptotic expansions

The limit (30) may be thought of as representing the leading-order asymptotics of the moments of the zeta function, in that it implies that

$$\frac{1}{T}\int_0^T |\zeta(1/2+it)|^{2\lambda}dt \sim f(\lambda)a(\lambda)\log^{\lambda^2}\left(\frac{T}{2\pi}\right) \tag{90}$$

as $T \to \infty$. Very little is known about lower order terms (in powers of $\log T$) in the asymptotic expansion of these moments. Does random matrix theory suggest what form these should take?

When $\lambda$ is an integer, it does. Note first that it follows from (45) that

$$\int_{U(N)} |Z(A,\theta)|^{2k}dA = \prod_{j=0}^{k-1}\frac{j!}{(j+k)!}\prod_{i=1}^{k}(N+i+j) = Q_k(N) \tag{91}$$

where $Q_k(N)$ is a polynomial in $N$ of degree $k^2$. This is consistent with the natural guess that the $2k$th moment of the zeta function should be a polynomial

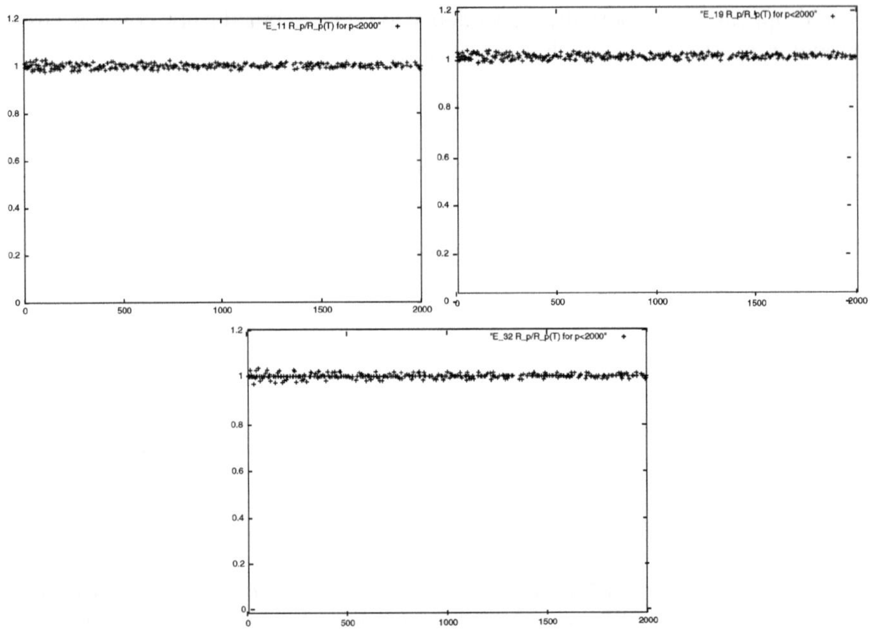


*Figure 6.* Pictures depicting $R_p/R_p(D)$, for $p < 2000$, $D$ as in Table 2. (From [12].)

of degree $k^2$ in $\log(T/2\pi)$ (modulo terms that vanish faster than any inverse power of $\log T/2\pi$ as $T \to \infty$).

Unfortunately it is not easy to see directly how to combine the coefficients in (91) with arithmetical information to guess the form of the coefficients of the lower-order terms in the moments of the zeta function. The expression in (91) can, however, be re-expressed in the form [8]

$$\int_{U(N)} |Z(A,\theta)|^{2k} dA = \frac{(-1)^k}{k!^2(2\pi i)^{2k}} \oint \cdots \oint e^{\frac{N}{2}\sum_{j=1}^{k} z_j - z_{j+k}} \times \frac{G(z_1,\ldots,z_{2k})\Delta^2(z_1,\ldots,z_{2k})}{\prod_{i=1}^{2k} z_i^{2k}} dz_1 \cdots dz_{2k}, \quad (92)$$

where the contours are small circles around the origin,

$$\Delta(z_1,\ldots,z_m) = \prod_{1\le i<j\le m} (z_j - z_i) \quad (93)$$

*Table 2.* A table in support of the conjecture (89), comparing $R_p$ v.s. $R_p(D)$ for three elliptic curves $E_{11}$, $E_{19}$, $E_{32}$ ($D$ equal to 333605031, 263273979, 930584451 respectively). More of this data, for $p < 2000$, is depicted in the Figure 6. The 0 entries for $p = 11$ and $p = 19$ are explained by the fact that we are restricting ourselves to twists with even functional equation. Hence for $E_{11}$ and $E_{19}$, we are only looking at twists with $\chi_d(11) = \chi_d(19) = -1$. (From [12].)

| $p$ | conjectured $R_p$ for $E_{11}$ | data for $E_{11}$ | conjectured $R_p$ for $E_{19}$ | data for $E_{19}$ | conjectured $R_p$ for $E_{32}$ | data for $E_{32}$ |
|---|---|---|---|---|---|---|
| 3 | 1.2909944 | 1.2774873 | 1.7320508 | 1.7018241 | 1 | 0.99925886 |
| 5 | 0.84515425 | 0.84938811 | 0.57735027 | 0.57825622 | 1.4142136 | 1.4113424 |
| 7 | 1.2909944 | 1.288618 | 1.1338934 | 1.134852 | 1 | 1.0003445 |
| 11 | | 0 | 0.77459667 | 0.76491219 | 1 | 1.0001457 |
| 13 | 0.74535599 | 0.73266305 | 1.3416408 | 1.3632977 | 0.63245553 | 0.61626177 |
| 17 | 1.118034 | 1.1282072 | 1.183216 | 1.196637 | 0.89442719 | 0.88962298 |
| 19 | 1 | 1.000864 | | 0 | 1 | 1.0006726 |
| 23 | 1.0425721 | 1.0470095 | 1 | 0.99857962 | 1 | 1.0000812 |
| 29 | 1 | 0.99769402 | 0.81649658 | 0.80174375 | 1.4142136 | 1.4615854 |
| 31 | 0.80064077 | 0.78332934 | 1.1338934 | 1.143379 | 1 | 1.0008405 |
| 37 | 0.92393644 | 0.91867671 | 0.9486833 | 0.94311279 | 1.0540926 | 1.0603105 |
| 41 | 1.2126781 | 1.2400086 | 1.1547005 | 1.1683113 | 0.78446454 | 0.76494748 |
| 43 | 1.1470787 | 1.1642671 | 1.0229915 | 1.0229106 | 1 | 1.0006774 |
| 47 | 0.84515425 | 0.82819492 | 1.0645813 | 1.0708874 | 1 | 0.99951502 |
| 53 | 1.118034 | 1.1332312 | 0.79772404 | 0.77715638 | 0.76696499 | 0.74137107 |
| 59 | 0.91986621 | 0.91329134 | 1.1055416 | 1.1196252 | 1 | 0.99969828 |
| 61 | 0.82199494 | 0.79865031 | 1.0162612 | 1.0199932 | 1.1766968 | 1.1996892 |
| 67 | 1.1088319 | 1.1216776 | 1.0606602 | 1.0705574 | 1 | 1.0002831 |
| 71 | 1.0425721 | 1.0497774 | 0.91986621 | 0.90939741 | 1 | 0.99992715 |
| 73 | 0.94733093 | 0.94345043 | 1.099525 | 1.1110782 | 1.0846523 | 1.0950853 |
| 79 | 1.1338934 | 1.1562237 | 0.90453403 | 0.8922209 | 1 | 0.99882039 |
| 83 | 1.0741723 | 1.0854551 | 0.8660254 | 0.84732408 | 1 | 0.99979996 |
| 89 | 0.84515425 | 0.82410673 | 0.87447463 | 0.85750248 | 0.89442719 | 0.88154899 |
| 97 | 1.0741723 | 1.0877289 | 0.92144268 | 0.90867892 | 0.8304548 | 0.80811684 |
| 101 | 0.98058068 | 0.97846254 | 0.94280904 | 0.93032086 | 1.0198039 | 1.0229108 |
| 103 | 1.1677484 | 1.1976448 | 0.87333376 | 0.855721 | 1 | 1.0004009 |
| 107 | 0.84515425 | 0.82186438 | 1.183216 | 1.2153554 | 1 | 1.0009282 |
| 109 | 0.91287093 | 0.89933354 | 1.1577675 | 1.1844329 | 0.94686415 | 0.94015124 |
| 113 | 0.92393644 | 0.9146531 | 0.9486833 | 0.93966595 | 1.1313708 | 1.1534106 |
| 127 | 0.93933644 | 0.93052596 | 0.98449518 | 0.98005032 | 1 | 0.99904006 |
| 131 | 1.1470787 | 1.171545 | 1.1208971 | 1.1413931 | 1 | 0.99916309 |
| 137 | 1.052079 | 1.0603352 | 1.0219806 | 1.0285831 | 1.1744404 | 1.2066518 |
| 139 | 0.93094934 | 0.91532106 | 1.0975994 | 1.1176423 | 1 | 1.0000469 |
| 149 | 1.069045 | 1.0833831 | 0.86855395 | 0.84844439 | 0.91064169 | 0.89706709 |

and

$$G(z_1,\ldots,z_{2k}) = \prod_{\substack{1\le \ell \le k\\ k+1\le q\le 2k}} (1-e^{z_q-z_\ell})^{-1} \tag{94}$$

(c.f. the similar expressions found for the Gaussian ensembles by Brezin and Hikami [5].) Note that $G$ has simple poles when $z_i = z_j$, $i \neq j$. An evaluation of the contour integral in terms of residues confirms the identity by giving (91). This formula has a natural generalization to the zeta function [9]:

$$\frac{1}{T}\int_0^T |\zeta(1/2+it)|^{2k}dt = \frac{1}{T}\int_0^T W_k(\log \tfrac{t}{2\pi})(1+O(t^{-\frac{1}{2}+\epsilon}))dt, \tag{95}$$

where

$$W_k(x) = \frac{(-1)^k}{k!^2(2\pi i)^{2k}}\oint\cdots\oint e^{\frac{x}{2}\sum_{j=1}^k z_j - z_{j+k}} \times \frac{\tilde G(z_1,\ldots,z_{2k})\Delta^2(z_1,\ldots,z_{2k})}{\prod_{i=1}^{2k} z_i^{2k}}\,dz_1\ldots dz_{2k}, \tag{96}$$

the path of integration being the same as in (92), and

$$\tilde G(z_1,\ldots,z_{2k}) = A_k(z_1,\ldots,z_{2k})\prod_{i=1}^{k}\prod_{j=1}^{k}\zeta(1+z_i-z_{j+k}), \tag{97}$$

with

$$A_k(z) = \prod_p\prod_{i=1}^k\prod_{j=1}^k\left(1-\frac{1}{p^{1+z_i-z_{j+k}}}\right) \times\int_0^1\prod_{j=1}^k\left(1-\frac{e(\theta)}{p^{1/2+z_j}}\right)^{-1}\left(1-\frac{e(-\theta)}{p^{1/2-z_{j+k}}}\right)^{-1}d\theta \tag{98}$$

and $e(\theta) = \exp(2\pi i\theta)$. Note that $\tilde G$ has the same pole structure as $G$. An evaluation of this integral in terms of residues shows that $W_k$ is a polynomial of degree $k^2$ and allows the coefficients to be computed. For example,

$$W_2(x) = 0.0506605918\,x^4 + 0.6988698848\,x^3 + 2.4259621988\,x^2 + 3.2279079649\,x + 1.3124243859 \tag{99}$$

and

$$W_3(x) = 0.0000057085\,x^9 + 0.0004050213\,x^8 + 0.0110724552\,x^7 + 0.1484007308\,x^6 + 1.0459251779\,x^5 + 3.9843850948\,x^4 + 8.6073191457\,x^3 + 10.2743308307\,x^2 + 6.5939130206\,x + 0.9165155076 \tag{100}$$

(we quote here numerical approximations for the coefficients, rather than the analytical expressions, which are rather cumbersome).

These polynomials describe the moments of the zeta function to a very high degree of accuracy [9]. For example, when $k = 3$ and $T = 2350000$, the left hand side of (95) evaluates to 1411700.43 and the right hand side to 1411675.64. Note that the coefficient of the leading order term is small. This explains the difficulties, described at length by Odlyzko [29], associated with numerical tests of (30).

Alternatively, one can also compare

$$\int_0^\infty |\zeta(1/2 + it)|^{2k} \exp(-t/T) dt \tag{101}$$

with

$$\int_0^\infty W_k(\log(t/2\pi)) \exp(-t/T) dt. \tag{102}$$

This is done in Table 3.

*Table 3.* Smoothed moments, (101) and (102), when $T = 10000$.

| k | (101) | (102) | difference |
|---|---|---|---|
| 1 | 79499.9312635 | 79496.7897047 | 3.14156 |
| 2 | 55088332.55512 | 55088336.43654 | -3.8814 |
| 3 | 708967359.4 | 708965694.5 | 1664.9 |
| 4 | 143638308513.0 | 143628911646.6 | 9396866.4 |

Similar asymptotic expansions have been derived for the moments of families of $L$-functions, using expressions analogous to (92) [9].

## 6.1 Extension to other compact Lie groups

It is interesting that the ideas reviewed above concerning connections between the value distribution of $L$-functions and averages over the classical compact groups extend to other Lie groups, such as the exceptional Lie groups [23]. For example, consider $G_2$. This is a 14-dimensional group of rank 2 (it is the automorphism group of the octonions), with an embedding into $SO(7)$. In the 7-dimensional representation, the characteristic polynomial associated with the corresponding unitary matrix $U$ factorizes as

$$Z(U, \theta) = \det(I - Ue^{-i\theta}) = (1 - e^{-i\theta})\tilde{Z}(U, \theta). \tag{103}$$

The moments of $\tilde{Z}(U,\theta)$ with respect to an average over the group can be calculated as for the classical compact groups using the corresponding Weyl integration formula and one of MacDonald's constant term identities (which plays the role of the Selberg integral). The result is that [23]

$$\int_{G_2} |\tilde{Z}(U,0)|^s d\mu(U) = \frac{\Gamma(3s+7)\Gamma(2s+3)}{\Gamma(2s+6)\Gamma(s+4)\Gamma(s+3)\Gamma(s+2)}. \tag{104}$$

We note in this context that Katz [19] has found a one-parameter family of $L$-functions over a finite field whose value distribution in the limit as the size of the field grows is related to $G_2$. Thus the random matrix moments (104) determine the value distribution of these $L$-functions.

The random matrix calculations extend straightforwardly to all of the exceptional Lie groups. It would be very interesting indeed to know whether the others also describe families of $L$-functions over finite fields.

## References

[1] M.V. Berry and J.P. Keating, The Riemann zeros and Eigenvalue Asymptotics, *SIAM Rev.* **41**, 236–266, 1999.

[2] E.B. Bogomolny and J.P. Keating, Random matrix theory and the Riemann zeros I: three- and four-point correlations, *Nonlinearity* **8**, 1115–1131, 1995.

[3] E.B. Bogomolny and J.P. Keating, Random matrix theory and the Riemann zeros II: $n$-point correlations, *Nonlinearity* **9**, 911–935, 1996.

[4] O. Bohigas, M.J. Giannoni and C. Schmit, Characterization of chaotic quantum spectra and universality of level fluctuation laws, *Phys. Rev. Lett.* **52**, 1-4, 1984.

[5] E. Brezin and S. Hikami, Characteristic Polynomials of Random Matrices *Commun. Math. Phys.* **214**, 111–135, 2000.

[6] J.B. Conrey, More than 2/5 of the zeros of the Riemann zeta function are on the critical line, *J. Reine. Ang. Math.* **399**, 1–26 (1989).

[7] J.B. Conrey and D.W. Farmer, Mean values of $L$-functions and symmetry, *Int. Math. Res. Notices* **17**, 883–908, 2000.

[8] J.B. Conrey, D.W. Farmer, J.P. Keating, M.O. Rubinstein, and N.C. Snaith, Autocorrelation of random matrix polynomials, *Commun. Math. Phys.* **237**: 365–395, 2003.

[9] J.B. Conrey, D.W. Farmer, J.P. Keating, M.O. Rubinstein, and N.C. Snaith, Integral moments of $L$-functions, *Proc. Lond. Math. Soc*, to appear.

[10] J.B. Conrey and A. Ghosh, On mean values of the zeta-function, iii, *Proceedings of the Amalfi Conference on Analytic Number Theory, Università di Salerno*, 1992.

[11] J.B. Conrey and S.M. Gonek, High moments of the Riemann zeta-function, *Duke Math. J.* **107**, 577–604, 2001.

[12] J.B. Conrey, J.P. Keating, M.O. Rubinstein, and N.C. Snaith, On the frequency of vanishing of quadratic twists of modular $L$-functions, In *Number Theory for the Millennium I: Proceedings of the Millennial Conference on Number Theory*; editor, M.A. Bennett et al., pages 301–315. A K Peters, Ltd, Natick, 2002, arXiv:math.nt/0012043.

[13] J.B. Conrey, J.P. Keating, M.O. Rubinstein, and N.C. Snaith, Random matrix theory and the Fourier coefficients of half-integral weight forms, *preprint*, 2004.

[14] F.J. Dyson, Statistical theory of the energy levels of complex systems, i, ii and iii, *J. Math. Phys.* **3**, 140–175, 1962.

[15] M.C. Gutzwiller, Periodic orbits and classical quantization conditions *J. Math. Phys.* **12**, 343–358, 1971.

[16] G.H. Hardy and J.E. Littlewood, Contributions to the theory of the Riemann zeta-function and the theory of the distribution of primes, *Acta Mathematica* **41**, 119–196, 1918.

[17] G.H. Hardy and J.E. Littlewood, Some problems in "Partitio Numerorum" III: on the expression of a number as a sum of primes, *Acta Mathematica* **44**, 1–70, 1923.

[18] A.E. Ingham, Mean-value theorems in the theory of the Riemann zeta-function, *Proceedings of the London Mathematical Society (2)* **27**, 273–300, 1926.

[19] N. Katz, *Exponential sums and differential equations (Annals of Mathematics Studies 124)*, Princeton, NJ: Princeton University Press, 1990.

[20] N.M. Katz and P. Sarnak, *Random Matrices, Frobenius Eigenvalues and Monodromy*, American Mathematical Society Colloquium Publications, 45. American Mathematical Society, Providence, Rhode Island, 1999.

[21] N.M. Katz and P. Sarnak, Zeros of zeta functions and symmetry, *Bull. Amer. Math. Soc.* **36**, 1–26, 1999.

[22] J.P. Keating, The Riemann zeta function and quantum chaology, in *Quantum Chaos*; editors, G. Casati, I Guarneri, and U. Smilansky, pages 145–85. North-Holland, Amsterdam (1993).

[23] J.P. Keating, N. Linden, and Z. Rudnick, Random matrix theory, the exceptional Lie groups, and $L$-functions, *J. Phys. A-Math. Gen.* **36**, 2933–44, 2003.

[24] J.P. Keating and N.C. Snaith, Random matrix theory and $\zeta(1/2 + it)$, *Commun. Math. Phys.* **214**, 57–89, 2000.

[25] J.P. Keating and N.C. Snaith, Random matrix theory and $L$-functions at $s = 1/2$, *Commun. Math. Phys* **214**, 91–110, 2000.

[26] W. Kohnen and D. Zagier, Values of $L$-series of modular forms at the center of the critical strip, *Invent. Math.* **64**, 175–198, 1981.

[27] M.L. Mehta, *Random Matrices*, Academic Press, London, second edition, 1991.

[28] H.L Montgomery, The pair correlation of zeros of the zeta function, *Proc. Symp. Pure Math.* **24**, 181–193, 1973.

[29] A.M. Odlyzko, The $10^{20}$th zero of the Riemann zeta function and 70 million of its neighbors, *Preprint*, 1989.

[30] M.O. Rubinstein, *Evidence for a spectral interpretation of the zeros of L-functions*, Ph.D. thesis, Princeton University, 1998.

[31] Z. Rudnick and P. Sarnak, Zeros of principal $L$-functions and random matrix theory, *Duke Math. J.* **81**, 269–322, 1996.

[32] G. Shimura, On modular forms of half integral weight, *Ann. Math.* **97**(2), 440–481, 1973.

[33] E.C. Titchmarsh, *The Theory of the Riemann Zeta Function*, Oxford University Press, 1986.

[34] J.-L. Waldspurger, Sur les coefficients de Fourier des formes modulaires de poids demi-entier, *J. Math. Pures Appl.* **60**(9), 375–484, 1981.

[35] H. Weyl, *Classical Groups*, Princeton University Press, 1946.

[36] See http://www.zetagrid.net/zeta/rh.html

[37] P. Heinzner, A. Huckleberry and M.R. Zirnbauer, Symmetry Classes of Disordered Fermions *Commun. Math. Phys.* **257**, 725–771, 2005.

# 2D QUANTUM GRAVITY, MATRIX MODELS AND GRAPH COMBINATORICS

P. Di Francesco
philippe@spht.saclay.cea.fr
*Service de Physique Théorique, CEA/DSM/SPhT*
*Unité de recherche associée au CNRS*
*CEA/Saclay*
*91191 Gif sur Yvette Cedex, France*

## 1. Introduction

### 1.1 Matrix models *per se*

The purpose of these lectures is to present basic matrix models as practical combinatorial tools, that turn out to be "exactly solvable". In short, a matrix model is simply a statistical ensemble of matrices with some specific measure, here given as an invariant weight, to be integrated over the relevant matrix ensemble. So solving a matrix model really amounts to computing integrals over matrix ensembles.

The lectures will be divided into two steps: first we show how to interpret such matrix integrals in terms of discrete two-dimensional quantum gravity, namely in terms of graphs with prescribed topology and valences, carrying also configurations of statistical "matter" models; in a second step, we show how to compute these integrals explicitly. The main difficulty here is that the immense power of matrix integrals allows to get right and simple answers, but gives no really good reason for such simplicity, except for technical miracles that are sometimes called "integrability". To compensate for this lack of understanding, we will always try to develop parallelly to the matrix model techniques and calculations some purely combinatorial reading of the various results.

The simplest combinatorial objects in many respects are trees, and we will see, at least in the planar case, how graphs representing discrete surfaces of genus zero are reducible to decorated trees. This eventually explains the simplicity of the corresponding matrix model results. By pushing these ideas a little further, we will be able to investigate refined properties of discrete surfaces (graphs), involving their intrinsic geometry. For instance we will compute correlation functions for surfaces with marked points at a prescribed geodesic distance from one-another.



*E. Brezin et al. (eds.), Applications of Random Matrices in Physics,* 33–88.

Having collected many exact solutions for models of discrete geometry, it is natural to go to the continuum limit, which displays a rich singularity structure: indeed singularities may arise from the graphs themselves, say when parameters coupled to valences reach some critical values, and the contribution from large graphs start dominating the statistical sum. They may also arise from criticity of the matter statistical models defined on these already critical graphs, in which case collective behaviors start dominating configurations. The matrix models allow for taking both limits simultaneously (the so-called *double-scaling limit*) while keeping track of all genera. The continuum model is expected to be described by conformally invariant matter field theories [1] coupled to 2D quantum gravity, i.e. defined on random surfaces [2]. Similarly we will write continuum correlation functions of the geodesic distance on the corresponding random surfaces.

## 1.2 A brief history

Planar graphs first arose in combinatorics, in the groundbreaking works of Tutte [3] in the 60's, who was able to compute generating functions for many classes of such objects, usually called maps by combinatorists. Higher genus was not considered then, and came up only later in physics works. The intrusion of matrix models in this subject occurred with the fundamental observation, due to t'Hooft [4] in the 70's, that planar graphs appearing in QCD with a large number of colors could be viewed as Feynman diagrams for matrix models, and that moreover the size of the matrices could serve as an expansion parameter to keep track of the topology of these diagrams. This caused the interest for matrix model to immediately rise, and led to the basic work of Brezin, Itzykson, Parisi and Zuber [5], who used various techniques to compute these matrix integrals, and among other things made the contact with Tutte's enumeration results. The matrix model techniques were then perfected by a number of people, whose list would be too long and probably not exhaustive. Then came the invention of continuum and discrete quantum gravity [6], as the coupling of matter theories to fluctuations of the underlying space, both in field-theoretical and matrix languages. This second life of matrix models came to a climax in 1990 with three quasi-simultaneous papers [7] making drastic progress in two-dimensional quantum gravity, as a toy model for low-dimensional non-critical strings, via the double-scaling limit of matrix models. This started a new matrix crazyness, and certainly helped develop matrix model theory a great deal (see [8] for a review and references). Remarkably, new areas of mathematics got infected by the matrix virus, thanks to Witten and Kontsevich [9], who formulated a mathematically rigorous approach to the moduli space of punctured Riemann surfaces using matrix models, and set the ground for a little revolution in enumerative geometry.

On the combinatorics front, it was only recently understood how to continue Tutte's work for higher genus graphs or more complicated planar cases [10], but a good relation to matrix model results is still to be found. For planar graphs however, the simplicity of the matrix model solutions has finally been explained combinatorially by Schaeffer [11], who found various bijections between planar graphs and trees, allowing for a simple enumeration, and a precise contact with the matrix model solutions [12]. A remarkable by-product of this approach is that one may keep track on the trees of some features of the planar graphs, such as geodesic distances between vertices or faces [13] [14], a task beyond the reach of matrix models so far.

## 2. Matrix models for 2D quantum gravity

### 2.1 Discrete 2D quantum gravity

The purpose of quantum gravity is to incorporate in a field-theoretical setting the interactions between matter fields and the fluctuations of the underlying space. In Euclidian 2D quantum gravity, the latter are represented by dynamical surfaces $\Sigma$ endowed with a Riemannian metric $g$ and scalar curvature $R$, and for which the Einstein action of General Relativity reads

$$\begin{aligned} S_E &= \Lambda \int_\Sigma \sqrt{g} d^2\xi + \mathcal{N} \int_\Sigma \sqrt{g} R d^2\xi \\ &= \Lambda A(\Sigma) + \mathcal{N}\chi(\Sigma) \end{aligned} \tag{1}$$

made of a cosmological term, in which the coslological constant $\Lambda$ is coupled to the area of the surface $A(\Sigma)$ and of the Newton term, in which the Newton constant $\mathcal{N}$ is coupled to the Euler characteristic $\chi(\Sigma)$ of the surface. The dynamical surfaces are then discretized in the form of graphs with prescribed topology.

We will now explain how matrix integrals can be used to generate such graphs, while precisely keeping track of their area and their Euler characteristic. For pedagogical purposes, we start with some simple remark on ordinary Gaussian integration, before going into the diagrammatics of Gaussian matrix integrals.

### 2.2 Gaussian integralís diagrammatics

Consider the following Gaussian average

$$\langle x^{2n} \rangle = \frac{1}{\sqrt{2\pi}} \int_{-\infty}^{\infty} e^{-\frac{x^2}{2}} x^{2n} dx = (2n-1)!! = \frac{(2n)!}{2^n n!} \tag{2}$$

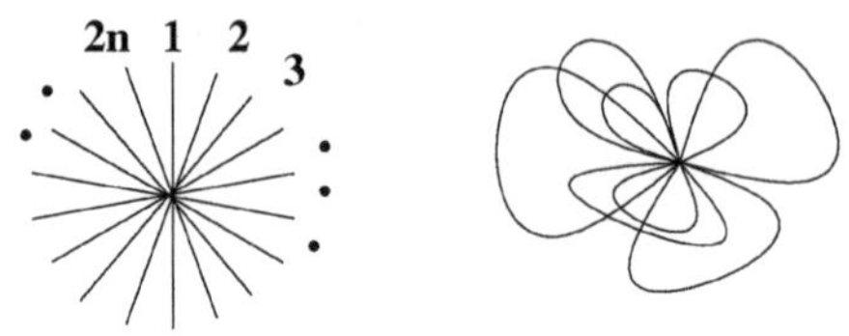


*Figure 1.* A star-diagram with one vertex and $2n$ out-coming half-edges stands for the integrand $x^{2n}$. In the second diagram, we have represented one non-zero contribution to $\langle x^{2n}\rangle$ obtained by taking derivatives of $\Sigma(s)$ by pairs represented as the corresponding pairings of half-edges into edges.

Among the many ways to compute this integral, let us pick the so-called source integral method, namely define the source integral

$$\Sigma(s) = \langle e^{xs}\rangle = \frac{1}{\sqrt{2\pi}}\int_{-\infty}^{\infty} e^{-\frac{x^2}{2}+sx}dx = e^{\frac{s^2}{2}} \tag{3}$$

Then the average (2) is obtained by taking $2n$ derivatives of $\Sigma(s) = e^{\frac{s^2}{2}}$ w.r.t. $s$ and by setting $s = 0$ in the end. It is then immediate to see that these derivatives must be taken by *pairs*, in which one derivative acts on the exponential and the other one on the prefactor $s$. Parallelly, we note that $(2n-1)!! = (2n-1)(2n-3)...3.1$ is the total number of distinct combinations of $2n$ objects into $n$ pairs. We may therefore formulate pictorially the computation of (2) as follows. We first draw a star-graph (see Fig.2.2), with one central vertex and $2n$ outcoming half-edges labelled 1 to $2n$ clockwise, one for each $x$ in the integrand (this amounts to labelling the $x$'s in $x^{2n}$ from 1 to $2n$). Now the pairs of derivatives taken on the source integral are in one-to-one correspondence with pairs of half-edges in the pictorial representation. Moreover, to get a non-zero contribution to $\langle x^{2n}\rangle$, we must saturate the set of $2n$ legs by taking $n$ pairs of them. Let us represent each such saturation by drawing the corresponding edges as in Fig.2.2. We get exactly $(2n-1)!!$ distinct labeled closed star-graphs with one vertex. This is summarized in the one-dimensional version of Wick's theorem:

$$\langle x^{2n}\rangle = \sum_{\text{pairings}} \prod \langle x^2\rangle \tag{4}$$

where the sum extends over all pairings saturating the $2n$ half-edges, and the weight is simply the product over all the edges thus formed of the corresponding averages $\langle x^2\rangle = (d^2/ds^2)\Sigma(s)|_{s=0} = 1$. Each saturation forms a Feynman diagram of the Gaussian average. The edge pairings are propagators (with value 1 here). This may appear like a complicated way of writing a rather trivial result, but it suits our purposes for generalization to matrix models and graphs.

## 2.3 Gaussian matrix integral and more diagrammatics

Let us now repeat the calculations of the previous section with the following Gaussian Hermitian matrix average of an arbitrary function $f$

$$\langle f(M)\rangle = \frac{1}{Z_0(N)}\int dM e^{-NTr\frac{M^2}{2}} f(M) \tag{5}$$

where the integral extends over Hermitian $N \times N$ matrices, with the standard Haar measure $dM = \prod_i dM_{ii}\prod_{i<j} dRe(M_{ij})dIm(M_{ij})$, and the normalization factor $Z_0(N)$ is fixed by requiring that $\langle 1\rangle = 1$ for $f = 1$. Typically, we may take for $f$ a monomial of the form $f(M) = \prod_{(i,j)\in I} M_{ij}$, $I$ a finite set of pairs of indices. Note the presence of the normalization factor $N$ (=the size of the matrices) in the exponential. Note that the case of the previous section is simply the particular case of integration over $1 \times 1$ Hermitian matrices (i.e. real numbers) here.

Like before, for a given Hermitian $N \times N$ matrix $S$, let us introduce the source integral

$$\Sigma(S) = \langle e^{Tr(SM)}\rangle = e^{\frac{Tr(S^2)}{2N}} \tag{6}$$

easily obtained by completing the square $M^2 - N(SM + MS) = (M - NS)^2 - N^2S^2$ and performing the change of variable $M' = M - NS$. We can use (6) to compute any average of the form

$$\langle M_{ij}M_{kl}...\rangle = \frac{\partial}{\partial S_{ji}}\frac{\partial}{\partial S_{lk}}...\ \Sigma(S)\big|_{S=0} \tag{7}$$

Note the interchange of the indices due to the trace $Tr(MS) = \sum M_{ij}S_{ji}$. As before, derivatives w.r.t. elements of $S$ must go by pairs, one of which acts on the exponential and the other one on the $S$ element thus created. In particular, a fact also obvious from the parity of the Gaussian, (7) vanishes unless there are an even number of matrix elements of $M$ in the average. In the simplest case of two matrix elements, we have

$$\langle M_{ij}M_{kl}\rangle = \frac{\partial}{\partial S_{lk}}\frac{1}{N}S_{ij}e^{\frac{Tr(S^2)}{2N}}\bigg|_{S=0} = \frac{1}{N}\delta_{il}\delta_{jk} \tag{8}$$

Hence the pairs of derivatives must be taken with respect to $S_{ij}$ and $S_{ji}$ for some pair $i, j$ of indices to yield a non-zero result. This leads naturally to the Matrix Wick's theorem:

$$\langle \prod_{(i,j)\in I} M_{ij}\rangle = \sum_{\text{pairings P}}\ \prod_{(ij),(kl)\in P} \langle M_{ij}M_{kl}\rangle \tag{9}$$

where the sum extends over all pairings saturating the (pairs of) indices of $M$ by pairs.

We see that in general, due to the restrictions (8) many terms in (9) will vanish. Let us now give a pictorial interpretation for the non-vanishing contributions to (9). We represent a matrix element $M_{ij}$ as a half-edge (with a marked end) made of a double-line, each of which is oriented in an opposite direction. We decide that the line pointing from the mark carries the index $i$, while the other one, pointing to the mark, carries the index $j$. This reads

$M_{ij} \quad \leftrightarrow$ 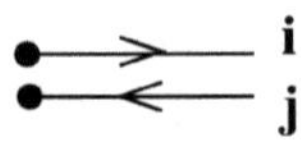 (10)



The two-element result (8) becomes simply the construction of an edge (with both ends marked) out of two half-edges $M_{ij}$ and $M_{kl}$, but is non-zero only if the indices $i$ and $j$ are conserved along the oriented lines. This gives pictorially

$\langle M_{ij} M_{ji} \rangle \quad \leftrightarrow$ 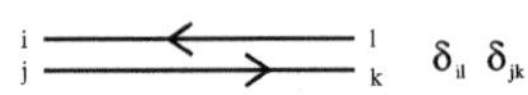 (11)



Similarly, an expression of the form $\mathrm{Tr}(M^n)$ will be represented as a star-diagram with one vertex connected to $n$ double half-edges in such a way as to respect the identification of the various running indices, namely

$$\mathrm{Tr}(M^n) = \sum_{i_1, i_2, \ldots, i_n} M_{i_1 i_2} M_{i_2 i_3} \ldots M_{i_n i_1} \leftrightarrow$$ 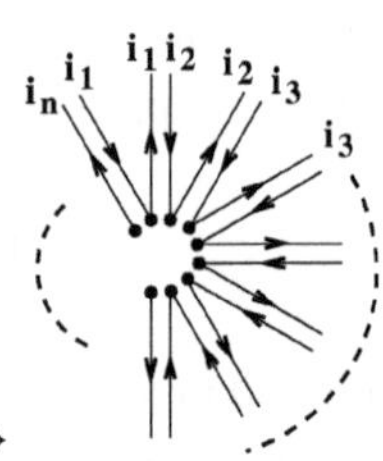 (12)



As a first application of this diagrammatic interpretation of the Wick theorem (9), let us compute the large $N$ asymptotics of $\langle \mathrm{Tr}(M^n) \rangle$. To compute $\langle \mathrm{Tr}(M^n) \rangle$, we must first draw a star-diagram as in (12), then apply (9) to express the result as a sum over the saturations of the star with edges connecting its outcoming half-edges by pairs. To get a non-zero result, we must clearly have $n$ even, say $n = 2p$. Again, there are $(2p-1)!!$ such pairings, and indeed we recover the case of previous section by taking $N = 1$. But if instead we take $N$ to be large, we see that only a fraction of these $(2p-1)!!$ pairings will contribute at leading order. Indeed, assume first we restrict the set of pairings to *planar* ones (see Fig.(2.3) (a)), namely such that the saturated star diagrams have a petal structure in which the petals are either juxtaposed or included into one-another (with no edges-crossings). We may compute the genus of the petal diagrams by noting that they form a tessellation of the sphere (=plane plus point at infinity). This tessellation has $V = 1$ vertex (the star), $E = p$

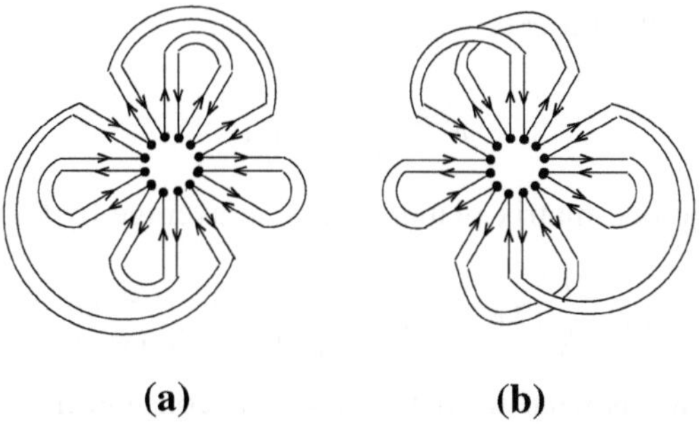

*Figure 2.* An example of planar (petal) diagram (a) and a non-planar one (b). Both diagrams have $n = 2p = 12$ half-edges, connected with $p = 6$ edges. The diagram (a) has $p + 1 = 7$ faces bordered by oriented loops, whereas (b) only has 3 of them. The Euler characteristic reads $2 - 2h = F - E + 1$ ($V = 1$ in both cases), and gives the genus $h = 0$ for (a), and $h = 2$ for (b).

edges, and $F$ faces, including the "external" face containing the point at infinity. The planarity of the diagram simply expresses that its genus $h$ vanishes, namely

$$2 - 2h = 2 = F - E + V = F + 1 - p \;\Rightarrow\; F = p + 1 \tag{13}$$

Such diagrams receive a total contribution $1/N^p$ from the propagators (weight $1/N$ per connecting edge), but we still have to sum over the remaining matrix indices $j_1, j_2, ..., j_{p+1}$ running over the $p + 1$ oriented loops we have created, which form the boundaries of the $F = p + 1$ faces. This gives a weight $N$ per face of the diagram, hence a total contribution of $N^{p+1}$. So all the petal diagrams contribute the same total factor $N^{p+1}/N^p = N$ to $\langle \mathrm{Tr}(M^n) \rangle$. Now any non-petal (i.e. non-planar, see Fig.(2.3) (b)) diagram must have at least *two less* oriented loops. Indeed, its Euler characteristic is negative or zero, hence it has $F \leq E - V = p - 1$ and it contributes at most for $N^{F-p} \leq 1/N$. So, to leading order in $N$, only the genus zero (petal) diagrams contribute. We simply have to count them. This is a standard problem in combinatorics: one may for instance derive a recursion relation for the number $c_p$ of petal diagrams with $2p$ half-edges, by fixing the left end of an edge (say at position 1), and summing over the positions of its right end (at positions $2j$, $j = 1, 2, ..., p$), and noting that the petal thus formed may contain $c_{j-1}$ distinct petal diagrams and be next to $c_{p-j}$ distinct ones. This gives the recursion relation

$$c_p = \sum_{j=1}^{p} c_{j-1} c_{p-j} \qquad c_0 = 1 \tag{14}$$

solved by the Catalan numbers

$$c_p = \frac{(2p)!}{(p+1)!p!} \tag{15}$$

Finally, we get the one-matrix planar Gaussian average by taking the large $N$ limit:

$$\lim_{N\to\infty} \frac{1}{N}\langle \mathrm{Tr}(M^n)\rangle = \begin{cases} c_p & \text{if } n = 2p \\ 0 & \text{otherwise} \end{cases} \tag{16}$$

This exercise shows us what we have gained by considering $N \times N$ matrices rather than numbers: we have now a way of discriminating between the various genera of the graphs contributing to Gaussian averages. This fact will be fully exploited in the next example.

## 2.4 Model building I: using one-matrix integrals

Let us apply the matrix Wick theorem (9) to the following generating function $f(M) = \exp(N\sum_{i\geq 1} g_i \mathrm{Tr}(M^i)/i)$, to be understood as a formal power series of the $g_i$, $i = 1, 2, 3, 4, \ldots$

$$\begin{aligned} Z_N(g_1, g_2, \ldots) &= \langle e^{N\sum_{i\geq 1} g_i \mathrm{Tr}(\frac{M^i}{i})}\rangle \\ &= \sum_{n_1,n_2,\ldots\geq 0}\prod_{i\geq 1}\frac{(Ng_i)^{n_i}}{i^{n_i}n_i!}\langle\prod_{i\geq 1}\mathrm{Tr}(M^i)^{n_i}\rangle \\ &= \sum_{n_1,n_2,\ldots\geq 0}\prod_{i\geq 1}\frac{(Ng_i)^{n_i}}{i^{n_i}n_i!}\sum_{\substack{\text{all labelled fatgraphs } \Gamma \\ \text{with } n_i\ i\text{-valent vertices}}} N^{-E(\Gamma)}N^{F(\Gamma)} \end{aligned} \tag{17}$$

by direct application (9).

In (17), we have first represented pictorially the integrand $\prod_i(\mathrm{Tr}(M^i))^{n_i}$ as a succession of $n_i$ $i$-valent star diagrams like that of (12), $i = 1, 2, \ldots$. Then we have summed over all possible saturations of all the marked half-edges of all these stars, thus forming (non-necessarily connected) ribbon or fatgraphs $\Gamma$ with some labelling of their half-edges (see Fig.(2.4) for an example of connected fatgraph). In (17), we have denoted by $E(\Gamma)$ the total number of edges of $\Gamma$, connecting half-edges by pairs, i.e. the number of propagators needed (yielding a factor $1/N$ each, from (8)). The number $F(\Gamma)$ is the total number of faces of $\Gamma$. The faces of $\Gamma$ are indeed well-defined because $\Gamma$ is a fatgraph, i.e. with edges made of doubly oriented parallel lines carrying the corresponding matrix indices $i = 1, 2, \ldots N$: the oriented loops we have created by the pairing process are interpreted as face boundaries, in one-to-one correspondence with faces of $\Gamma$. But the traces of the various powers of $M$ still have to be taken, which means all the indices running from 1 to $N$ have to be summed

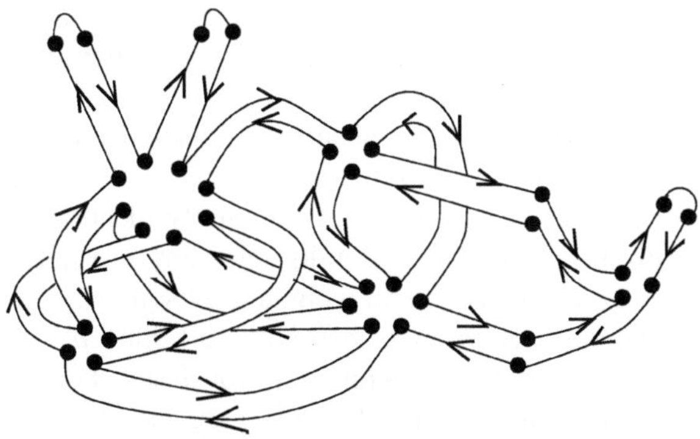

*Figure 3.* A typical connected fatgraph $\Gamma$, corresponding to the average $\langle \mathrm{Tr}(M)^3 \mathrm{Tr}(M^2)^2 \mathrm{Tr}(M^3) \mathrm{Tr}(M^4)^2 \mathrm{Tr}(M^6) \mathrm{Tr}(M^8) \rangle$. The graph was obtained by saturating the ten star-diagrams corresponding to the ten trace terms, namely with $n_1 = 3$ univalent vertices, $n_2 = 2$ bi-valent ones, $n_3 = 1$ tri-valent one, $n_4 = 2$ four-valent ones, $n_6 = 1$ six-valent one and $n_8 = 1$ eight-valent one, hence a total of $V = 10$ vertices. This graph corresponds to some particular Wick pairing for which we have drawn the $E = 16$ connecting edges, giving rise to $F = 2$ oriented loops bordering the faces of $\Gamma$.

over all these loops. This results in the factor $N$ per face of $\Gamma$ in (17). Finally, the sum extends over all (possibly disconnected) fatgraphs $\Gamma$ with labelled half-edges. Each such labelled graph corresponds to exactly one Wick pairing of (9). Summing over all the possible labellings of a given un-labelled fatgraph $\Gamma$ results in some partial cancellation of the symmetry prefactors $\prod_i 1/(i^{n_i} n_i!)$, which actually leaves us with the inverse of the order of the symmetry group of the un-labelled fatgraph $\Gamma$, denoted by $1/|Aut(\Gamma)|$. This gives the final form

$$Z_N(g_1, g_2, ...) = \sum_{\substack{\text{fatgraphs} \\ \Gamma}} \frac{N^{V(\Gamma)-E(\Gamma)+F(\Gamma)}}{|Aut(\Gamma)|} \prod_{i \geq 1} g_i^{n_i(\Gamma)} \tag{18}$$

where $n_i(\Gamma)$ denotes the total number of $i$-valent vertices of $\Gamma$ and $V(\Gamma) = \sum_i n_i(\Gamma)$ is the total number of vertices of $\Gamma$. To restrict the sum in (18) to only connected graphs, we simply have to formally expand the logarithm of $Z_N$, resulting in the final identity

$$F_N(g_1, g_2, ...) = \mathrm{Log}\, Z_N(g_1, g_2, ...) = \sum_{\substack{\text{connected} \\ \text{fatgraphs } \Gamma}} \frac{N^{2-2h(\Gamma)}}{|Aut(\Gamma)|} \prod_i g_i^{n_i(\Gamma)} \tag{19}$$

where we have identified the Euler characteristic $\chi(\Gamma) = F - E + V = 2 - 2h(\Gamma)$, where $h(\Gamma)$ is the genus of $\Gamma$ (number of handles). Eqn.(19) gives a clear geometrical meaning to the Gaussian average of our choice of $f(M)$: it amounts to computing the generating function for fatgraphs of given genus and given vertex valencies. Such a fatgraph $\Gamma$ is in turn dual to a tessellation $\Gamma^*$ of a Riemann surface of same genus, by means of $n_i$ $i$-valent polygonal tiles, $i = 1, 2, \ldots$.

The result (19) is therefore a statistical sum over discretized random surfaces (the tessellations), that can be interpreted in physical terms as the free energy of a model of discrete 2D quantum gravity. It simply identifies the Gaussian matrix integral with integrand $f(M)$ as a discrete sum over configurations of tessellated surfaces of arbitrary genera, weighted by some exponential factor. More precisely, imagine only $g_3 = g \neq 0$ while all other $g_i$'s vanish. Then (19) becomes a sum over fatgraphs with cubic (or 3-valent) vertices, dual to triangulations $T$ of Riemann surfaces of arbitrary genera. Assuming these triangles have all unit area, then $n_3(\Gamma) = A(T)$ is simply the total area of the triangulation $T$. Hence (19) becomes

$$F_N(g) = \sum_{\text{connected triangulations } T} \frac{g^{A(T)} N^{2-2h(T)}}{|Aut(T)|} \tag{20}$$

and the summand $g^A N^{2-2h} = e^{-S_E}$ is nothing but the exponential of the discrete version of Einstein's action for General Relativity in 2 dimensions (1), in which we have identified the two invariants of $\Sigma$: its area $A(\Sigma)$ and its Euler characteristic $\chi(\Sigma) = 2 - 2h(\Sigma)$. The contact with (20) is made by setting $g = e^{-\Lambda}$ and $N = e^{-\mathcal{N}}$.

If we now include all $g_i$'s in (19) we simply get a more elaborate discretized model, in which we can keep track of the valencies of vertices of $\Gamma$ (or tiles of the dual $\Gamma^*$). These in turn may be understood as discrete models of matter coupled to 2D quantum gravity. This is best seen in the case of the Hard-Dimer model on random 4-valent graphs [15]. The configurations of the model are made of arbitrary 4-valent fatgraphs of arbitrary genus (the underlying discrete fluctuating space) and of choices of edges occupied by dimers, with the hard-core condition that no two adjacent edges may be simultaneously occupied (see Fig.(3) for an illustration in the case of a planar graph). These matter configurations are given an occupation energy weight $z$ per dimer, while the space part receives the standard weight $g$ per 4-valent vertex, and the overall weight $N^{2-2h}$ for each graph of genus $h$. We then note that any occupied dimer may be shrunk to naught, thus creating a 6-valent vertex by the fusion of its two 4-valent adjacent vertices. Comparing the configurations of the Hard-Dimer model on 4-valent graphs and those of graphs with only 4- and 6-valent vertices, we see that there is a one-to-three correspondence between those, as there are exactly three ways of decomposing a 6-valent vertex into two adjacent 4-valent ones connected by a dimer (see the bottom line of Fig.(3)). The Hard-Dimer model is therefore generated by an integral of the form (18), with only $g_4$ and $g_6$ non-zero, and more precisely $g_4 = g$ and $g_6 = 3g^2 z$ (=three

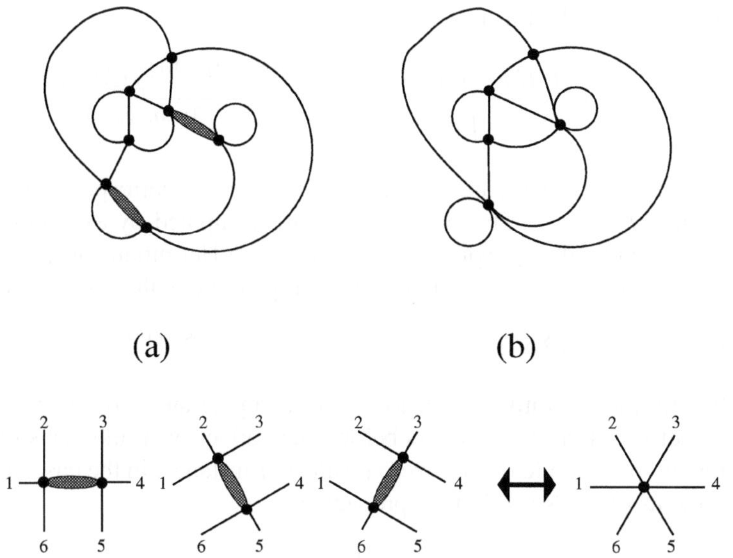


*Figure 4.* A 4-valent planar graph with hard dimers, represented by thickened edges. The corresponding graph obtained by shrinking the dimers (b) has both 4-valent and 6-valent vertices. The correspondence is three-to-one per dimer, as shown.

decompositions into two 4-valent vertices and one dimer). This is the simplest instance of matter coupled to 2D quantum gravity we could think of, and it indeed corresponds to graphs with specific valence weights.

Going back to the purely mathematical interpretation of (19), we start to feel how simple matrix integrals can be used as tools for generating all sorts of graphs whose duals tessellate surfaces of arbitrary given topology. The size $N$ of the matrix relates to the genus, whereas the details of the integrand relate to the structure of vertices. An important remark is also that the large $N$ limit of (19) extracts the genus zero contribution, namely that of planar graphs. So as a by-product, it will be possible to extract results on planar graphs from asymptotics of matrix integrals for large size $N$.

## 2.5 Model building II: using multi-matrix integrals

The results of previous section can be easily generalized to multiple Gaussian integrals over several Hermitian matrices. More precisely, let $M_1$, $M_2$, ... $M_p$ denote $p$ Hermitian matrices of same size $N \times N$, and $Q_{a,b}$, $a, b = 1, 2, ..., p$ the elements of a positive definite form $Q$. We consider the multiple

Gaussian integrals of the form

$$\langle f(M_1,...,M_p)\rangle = \frac{\int dM_1...dM_p e^{-\frac{N}{2}\sum_{a,b=1}^{p}\mathrm{Tr}(M_a Q_{ab} M_b)} f(M_1,...,M_p)}{\int dM_1...dM_p e^{-\frac{N}{2}\sum_{a,b=1}^{p}\mathrm{Tr}(M_a Q_{ab} M_b)}} \tag{21}$$

The one-Hermitian matrix case of the previous section corresponds simply to $p = 1$ and $Q_{1,1} = 1$. The averages (21) are computed by extending the source integral method of previous section: for some Hermitian source matrices $S_1, ..., S_p$ of size $N \times N$, we define and compute the multi-source integral

$$\Sigma(S_1,...,S_p) = \langle e^{\sum_{a=1}^{p}\mathrm{Tr}(S_a M_a)}\rangle = e^{\frac{1}{2N}\sum_{a,b=1}^{p}\mathrm{Tr}(S_a (Q^{-1})_{a,b} S_b)} \tag{22}$$

and apply multiple derivatives w.r.t. to $S_a$'s to compute any expression of the form (21), before taking $S_a \to 0$. As before, derivatives w.r.t. elements of the $S$'s must go by pairs to yield a non-zero result. For instance, in the case of two matrix elements of $M_a$'s we find the propagators

$$\langle (M_a)_{ij}(M_b)_{kl}\rangle = \frac{1}{N}\delta_{il}\delta_{jk}(Q^{-1})_{a,b} \tag{23}$$

In general we will apply the multi-matrix Wick theorem

$$\langle \prod_{(a,i,j)\in J} (M_a)_{ij}\rangle = \sum_{\substack{\text{pairings}\\ P}} \prod_{\substack{\text{pairs}\\ (aij),(bkl)\in P}} \langle (M_a)_{ij}(M_b)_{kl}\rangle \tag{24}$$

expressing the multi-matrix Gaussian average of any product of matrix elements of the $M$'s as a sum over all pairings saturating the matrix half-edges, weighted by the corresponding value of the propagator (8)mu. Note that half-edges must still be connected according to the rule (8)ag, but that in addition, depending on the form of $Q$, some matrices may not be allowed to connect to one another (e.g. if $(Q^{-1})_{ab} = 0$ for some $a$ and $b$, then $\langle M_a M_b\rangle = 0$, and in such a case, there cannot be any edge connecting a matrix with index $a$ to one with index $b$).

This gives us much freedom in cooking up multi-matrix models to evaluate generating functions of graphs with specific decorations such as colorings, spin models, etc... This is expected to describe the coupling of matter systems (e.g. a spin model usually defined on a regular lattice) to 2D quantum gravity (by letting the lattice fluctuate into tessellations of arbitrary genera). Famous examples are the O(n) model [16], the q-states Potts model [17], both including the Ising model as particular cases. Other models of interest require to use different types of matrices, to best represent their degrees of freedom. This is the case for the 6 vertex model expressed in terms of complex matrices, and for the so-called IRF (interaction round a face) models, expressed in terms of complex rectangular arrays [18] [19].

# 3. The one-matrix model I: large $N$ limit and the enumeration of planar graphs

In this section, we will mainly cover the one-matrix integrals defined in Sect.2.4. Multi-matrix techniques are very similar, and we will present them in a concluding section. More precisely, we will study the one-matrix integral

$$Z_N(V) = \frac{\int dM e^{-N \mathrm{Tr}\, V(M)}}{\int dM e^{-N \mathrm{Tr}\, V_0(M)}} \tag{25}$$

with an arbitrary polynomial potential, say

$$V(x) = \frac{x^2}{2} - \sum_{i=1}^{d} \frac{g_i}{i} x^i, \quad \text{and} \quad V_0(x) = \frac{x^2}{2} \tag{26}$$

This contains as a limiting case the partition function (17) of Sect.2.4. Note also that we are not worrying at this point about convergence issues for these integrals, as they must be understood as formal tools allowing for computing well-defined coefficients in formal series expansions in the $g$'s.

## 3.1 Eigenvalue reduction

The step zero in computing the integral (**25**) is the reduction to a $N$-dimensional integral, namely over the real eigenvalues $m_1, ..., m_N$ of the Hermitian matrix $M$. This is done by performing the change of variables $M \to (m, U)$, where $m = \mathrm{diag}(m_1, ..., m_N)$, and $U$ is a unitary diagonalization matrix such that $M = UmU^\dagger$, hence $U \in U(N)/U(1)^N$ as $U$ may be multiplied by an arbitrary matrix of phases. The Jacobian of the transformation is readily found to be the squared Vandermonde determinant

$$J = \Delta(m)^2 = \prod_{1 \le i < j \le N} (m_i - m_j)^2 \tag{27}$$

A simple derivation consists in expressing the differential $dM$ in terms of $dU$ and $dm$ in the vicinity of $U = I$, namely $dM = dUm + dm + mdU^\dagger$, but noting that $UU^\dagger = I$, we get $dU^\dagger = -dU$, and finally $dM = dm + [dU, m]$, or $dM_{ij} = dm_i \delta_{ij} + (m_i - m_j) dU_{ij}$, from which we directly read the Jacobian (27). Performing the change of variables in both the numerator and denominator of (25) we obtain

$$Z_N(V) = \frac{\int_{\mathbb{R}^N} dm_1 ... dm_N \Delta(m)^2 e^{-N \sum_{i=1}^N V(m_i)}}{\int_{\mathbb{R}^N} dm_1 ... dm_N \Delta(m)^2 e^{-N \sum_{i=1}^N \frac{m_i^2}{2}}} \tag{28}$$

## 3.2 Large size: the saddle-point technique

Starting from the $N$-dimensional integral (28), we rewrite

$$Z_N(V) = \frac{\int dm_1...dm_N e^{-N^2 S(m_1,...,m_N)}}{\int dm_1...dm_N e^{-N^2 S_0(m_1,...,m_N)}} \tag{29}$$

where we have introduced the actions

$$\begin{aligned} S(m_1,...,m_N) &= \frac{1}{N}\sum_{i=1}^{N} V(m_i) - \frac{1}{N^2}\sum_{1\le i\ne j\le N} \mathrm{Log}|m_i - m_j| \\ S_0(m_1,...,m_N) &= \frac{1}{N}\sum_{i=1}^{N} V_0(m_i) - \frac{1}{N^2}\sum_{1\le i\ne j\le N} \mathrm{Log}|m_i - m_j| \end{aligned} \tag{30}$$

For large $N$ the numerator and denominator of (29) are dominated by the semi-classical (or saddle-point) minima of $S$ and $S_0$ respectively. For $S$, the saddle-point equations read

$$\frac{\partial S}{\partial m_j} = 0 \Rightarrow V'(m_j) = \frac{2}{N}\sum_{\substack{1\le i\le N \\ i\ne j}} \frac{1}{m_j - m_i} \tag{31}$$

for $j = 1, 2, ..., N$. Introducing the discrete resolvent

$$\omega_N(z) = \frac{1}{N}\sum_{i=1}^{N} \frac{1}{z - m_i} \tag{32}$$

evaluated at the solution $m_1, .., m_N$ to (31), multiplying (31) by $1/(N(z - m_j))$ and summing over $j$, we easily get the equation

$$\begin{aligned} V'(z)\omega_N(z) &+ \frac{1}{N}\sum_{j=1}^{N} \frac{V'(m_j) - V'(z)}{z - m_j} \\ &= \frac{1}{N^2}\sum_{1\le i\ne j\le N} \frac{1}{m_j - m_i}\left(\frac{1}{z - m_j} - \frac{1}{z - m_i}\right) \\ &= \frac{1}{N^2}\sum_{1\le i\ne j\le N} \frac{1}{(z - m_i)(z - m_j)} \\ &= \omega_N(z)^2 + \frac{1}{N}\omega_N'(z) \end{aligned} \tag{33}$$

Assuming $\omega_N$ tends to a differentiable function $\omega(z)$ when $N \to \infty$ we may neglect the last derivative term, and we are left with the quadratic equation

$$\omega(z)^2 - V'(z)\omega(z) + P(z) = 0$$
$$P(z) = \lim_{N\to\infty} \frac{1}{N} \sum_{j=1}^{N} \frac{V'(z) - V'(m_j)}{z - m_j} \tag{34}$$

where $P(z)$ is a polynomial of degree $d-2$, $d$ the degree of $V$. The existence of the limiting resolvent $\omega(z)$ boils down to that of the limiting density of distribution of eigenvalues

$$\rho(z) = \lim_{N\to\infty} \frac{1}{N} \sum_{j=1}^{N} \delta(z - m_j) \tag{35}$$

normalized by the condition

$$\int_{\mathbb{R}} \rho(z)dz = 1 \tag{36}$$

as there are exactly $N$ eigenvalues on the real axis. This density is related to the resolvent through

$$\omega(z) = \int \frac{\rho(x)}{z-x}dx = \sum_{m=1}^{\infty} \frac{1}{z^m} \int_{\mathbb{R}} x^{m-1}\rho(x)dx \tag{37}$$

where the expansion holds in the large $z$ limit, and the integral extends over the support of $\rho$, included in the real line. Conversely, the density is obtained from the resolvent by use of the discontinuity equation across its real support

$$\rho(z) = \frac{1}{2i\pi} \lim_{\epsilon\to 0} \omega(z + i\epsilon) - \omega(z - i\epsilon) \qquad z \in \mathrm{supp}(\rho) \tag{38}$$

Solving the quadratic equation (34) as

$$\omega(z) = \frac{V'(z) - \sqrt{(V'(z))^2 - 4P(z)}}{2} \tag{39}$$

we must impose the large $z$ behavior inherited from (36)-(37), namely that $\omega(z) \sim 1/z$ for large $z$. For $d \geq 2$, the polynomial in the square root has degree $2(d-1)$: expanding the square root for large $z$ up to order $1/z$, all the terms cancel up to order 0 with $V'(z)$, and moreover the coefficient in front of $1/z$ must be 1 (this fixes the leading coefficient of $P$). The other coefficients of $P$ are fixed by the higher moments of the measure $\rho(x)dx$.

For instance, when $k = 2$ and $V = V_0$, we get $P = 1$ and

$$\omega_0(z) = \frac{1}{2}(z - \sqrt{z^2 - 4}) \tag{40}$$

It then follows from (38) that the density has the compact support $[-2, 2]$ and has the celebrated "Wigner's semi-circle law" form

$$\rho_0(z) = \frac{1}{2\pi}\sqrt{4 - z^2} \tag{41}$$

The resolvent $\omega_0$ is the generating function for the moments of the measure whose density is $\rho_0$ (via the expansion (37)), from which we immediately identify

$$\int_{\mathbb{R}} x^n \rho_0(x)dx = \begin{cases} c_p & \text{if } n = 2p \\ 0 & \text{otherwise} \end{cases} \tag{42}$$

with $c_p$ as in (15). Indeed, due to the quadratic recursion relation (14), the generating function $C(x) = \sum_{p\geq 0} x^p c_p$ satisfies $xC(x)^2 = C(x) - 1$, and therefore we have $\omega_0(z) = C(1/z^2)/z$. The coefficients (42) are nothing but the planar limit of the Gaussian Hermitian matrix averages (with potential $V_0(x) = x^2/2$), namely $\lim_{N\to\infty}\langle \frac{1}{N}\mathrm{Tr}M^n\rangle_{V_0} = \int_{\mathbb{R}} x^n \rho_0(x)dx$, hence our analytical result (42) is an alternative for that already obtained combinatorially in (16).

In the general case, the density reads

$$\rho(z) = \frac{1}{2\pi}\sqrt{4P(z) - (V'(z))^2} \tag{43}$$

and may have a disconnected support, made of a union of intervals (the so-called multicut solutions). It is however interesting to restrict oneself to the case when the support of $\rho$ is made of a single real interval $[a, b]$, as this will always be the preferred saddle-point solution for generating the correct formal series expansions of the all-genus free energy. For supports made of more than one interval, resonances may occur as eigenvalues tunnel from one interval to another, and oscillations develop in the $N$ dependence, which cause the large $N$ expansion to break down, unless some strong conditions are imposed on say complex contour integrals for the eigenvalues. The one-cut hypothesis will be justified *a posteriori* in Sect.4 below, when we revisit the problem from a purely combinatorial perspective.

In the one-cut case, the polynomial $V'(z)^2 - 4P(z)$ has single roots at say $z = a$ and $z = b$ and all other roots have even multiplicities. In other words, we may write the limiting resolvent as

$$\omega(z) = \frac{1}{2}(V'(z) - Q(z)\sqrt{(z-a)(z-b)}) \tag{44}$$

where $Q(z)$ is a polynomial of degree $k-2$, entirely fixed in terms of $V$ by the asymptotics $\omega(z) \sim 1/z$ for large $|z|$. More precisely, let us introduce $H(z) = V'(z)/\sqrt{(z-a)(z-b)}$, considered as a series expansion for large $z$, then $Q(z)$ is nothing but the part of this series that is polynomial in $z$, denoted as $H_+(z)$. Writing moreover $H(z) = H_+(z) + H_-(z)$, we finally get

$$\omega(z) = \frac{1}{2} H_-(z) \sqrt{(z-a)(z-b)} \tag{45}$$

Writing $H_-(z) = \sum_{i\geq 1} H_{-i} z^{-i}$, we get that $\omega(z) \sim 1/z$ iff $H_{-1} = 0$ and $H_{-2} = 2$. These coefficients are expressed as residue integrals at infinity, namely

$$H_{-m}(z) = \oint \frac{dz}{2i\pi} z^{m-1} \frac{V'(z)}{\sqrt{(z-a)(z-b)}} \tag{46}$$

The square root term is uniformized by the change of variables $z = w + S + R/w$, with $S = \frac{a+b}{2}$ and $R = \left(\frac{b-a}{4}\right)^2$, and

$$H_{-m}(z) = \oint \frac{dw}{2i\pi w} (w + S + R/w)^{m-1} V'(w + S + R/w) \tag{47}$$

so that finally $H_{-1} = V_0'$ and $H_{-2} = V_{-1}' + SV_0' + RV_1'$, where the shorthand notation $V_m'$ stands for the coefficient of $w^m$ in the large $w$ expansion of $V'(w + S + r/w)$. Performing the change of variables $w \to R/w$ allows to relate $V_{-m}' = R^m V_m'$. Finally, the asymptotic condition $\omega(z) = 1/z + O(1/z^2)$ at large $z$ boils down to

$$\begin{aligned} V_0' &= 0 = S - \sum_{i\geq 1} g_i \sum_{j=0}^{[(i-1)/2]} S^{i-2j-1} R^j \frac{(i-1)!}{(j!)^2 (i-2j-1)!} \\ V_{-1}' &= 1 = R - \sum_{i\geq 1} g_i \sum_{j=0}^{[i/2]} S^{i-2j} R^j \frac{(i-1)!}{j!(j-1)!(i-2j)!} \end{aligned} \tag{48}$$

These equations simplify drastically in the case of even potentials, where $g_i = 0$ for all odd $i$. The parity of $V$ indeed induces that of $\rho$, and we have $S = (a+b)/2 = 0$ as the support of the density is symmetric w.r.t. the origin. This leaves us with only one equation

$$1 = R - \sum_{i\geq 1} g_{2i} R^i \binom{2i-1}{i} \tag{49}$$

for $R = a^2/4$. In the particular case of the gaussian potential $V = V_0$, this reduces to $R = 1$ and $S = 0$, in agreement with $b = -a = 2$ (40). Expanding the solutions of (48) as formal power series of the $g_i$'s, the conditions $R = 1 +$

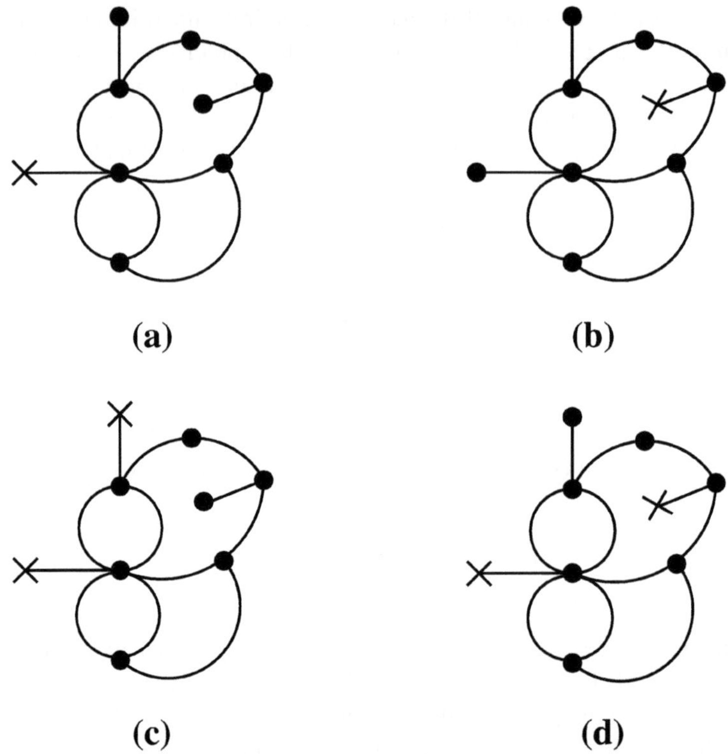


*Figure 5.* Samples of planar graphs with external legs (univalent vertices marked with a cross) and arbitrary valences, with respectively (a) one leg in the external face (b) one leg (anywhere) (c) two legs in the same (external) face (d) two-legs (one in the external face, the other anywhere).

$O(\{g_i\})$ and $S = O(\{g_i\})$ determine them uniquely. These in turn determine $a$ and $b$ and therefore $\rho$ and $\omega$ completely.

The planar free energy $f = F - F_0 = \lim_{N\to\infty} \frac{1}{N^2}\text{Log}\left(Z_N(V)/Z_N(V_0)\right)$ is finally obtained by substituting the limiting densities $\rho, \rho_0$ in the saddle point actions $S$ and $S_0$, with the result $F - F_0 = S_0 - S$. It is however much simpler to evaluate some derivatives of the free energy, by directly relating them to the planar resolvent $\omega(z)$, the subject of next section.

## 3.3 Enumeration of planar graphs with external legs

Let us first consider the generating function $\Gamma_1$ for planar graphs with weights $g_i$ per i-valent vertex, and with one external (univalent) leg, represented in the external face on the plane (see Fig.(2.3)graphs (a)):

$$\Gamma_1 = \partial f/\partial g_1 = \lim_{N\to\infty} \frac{1}{N}\langle \mathrm{Tr}(M)\rangle_V = \omega_{-2} = \frac{1}{2}(H_{-3} - SH_{-2} - 2RH_{-1}) \tag{50}$$

where use has been made of (37), and as before $\omega_{-m}$ denotes the coefficient of $z^{-m}$ in the large $z$ expansion of $\omega(z)$. From the large $z$ asymptotics of $\omega(z)$, we know that $H_{-1} = 0$ and $H_{-2} = 2$, and we must now evaluate $H_{-3} = V'_{-2} + 2SV'_{-1} + (S^2+2R)V'_0 + 2RSV'_1 + R^2V'_2 = 2(V'_{-2} + 2SV'_{-1}) + (S^2+2R)V'_0 = 2(V'_{-2} + 2S)$, leaving us with

$$\Gamma_1 = V'_{-2} + S \tag{51}$$

Analogously, we may compute the connected two-leg-in-the-same-face graph generating function $\Gamma_2 = \omega_{-3} - \Gamma_1^2$ (see Fig.(2.3)graphs (c)), in which we subtract the contributions from disconnected pairs of one-leg graphs. We get $\omega_{-3} = \partial f/\partial g_2 = R + S^2 + V'_{-3} + 2SV'_{-2}$ and finally

$$\Gamma_2 = R + V'_{-3} - (V'_{-2})^2 \tag{52}$$

Another quantity of interest is the connected two-leg graph generating function $\Gamma_{1,1} = \partial^2 f/\partial g_1^2 = \partial\omega_{-2}/\partial g_1$ (see Fig.(2.3)graphs (d)). This turns into

$$\Gamma_{1,1} = \frac{\partial\omega_{-2}}{\partial g_1} = \frac{\partial S}{\partial g_1} + \frac{\partial V'_{-2}}{\partial g_1} = \frac{\partial S}{\partial g_1}(1 + V''_{-2}) + \frac{\partial R}{\partial g_1}V''_{-1} \tag{53}$$

Let us first replace the term 1 in factor of $\partial S/\partial g_1$ by $1 = V'_{-1}$, the second equation of (48). Note that the residue of a total differential always vanishes, hence in particular

$$\big(d/dw(wV'(w + S + R/w))\big)_{-1} = 0 = V'_{-1} + V''_{-2} - RV''_0 \tag{54}$$

This allows to rewrite

$$\Gamma_{1,1} = \frac{\partial S}{\partial g_1}RV''_0 + \frac{\partial R}{\partial g_1}V''_{-1} \tag{55}$$

Finally, differentiating the equation $V'_0 = 0$ w.r.t. $g_1$ yields $0 = \partial S/\partial g_1 V''_0 + \partial R/\partial g_1 V''_1 - 1$, where the last term comes from the explicit derivation w.r.t. $g_1$ of $V'(x) = x - g_1 - g_2 x - g_3 x^2/2 - ...$ Multiplying this by $R$, and noting as before that $RV''_1 = V''_{-1}$, we get $R\partial S/\partial g_1 V''_0 + \partial R/\partial g_1 V''_{-1} = R$ and finally

$$\Gamma_{1,1} = R \tag{56}$$

This result holds for even potentials as well, upon setting all $g_{2i+1} = 0$ in the end. Eq.(56) gives a straightforward combinatorial interpretation of $R$ as the generating function for planar graphs with two external (univalent) legs, not necessarily in the same face.

To conclude the section, let us now give a combinatorial interpretation for $S$. Let us show that $S$ is the generating function for one-leg planar graphs. By this we mean that the leg need not be adjacent to the external face, as was the case for $\Gamma_1$ (see Fig.(2.3)graphs (b)). Comparing with the definition of $\Gamma_1$, we must show that $S$ is the generating function for one-leg planar graphs (with the leg in the external face), and with a marked face (chosen to be the new external face). This amounts to the identity

$$S = z\partial_z \Gamma_1|_{z=1} \tag{57}$$

where we have included a weight $z$ per *face* of the graph, to be set to 1 in the end. Due to Euler's relation $F = 2 + E - V$, where $E$ is the total number of edges, and $V$ that of vertices of the one-leg graphs at hand, and noting that $2E = 1 + \sum iV_i$ while $V = 1 + \sum V_i$, where $V_i$ is the number of internal $i$-valent vertices, so that $2E - V = \sum(i-1)V_i$, we see that $z\partial_z\Gamma_1 = (2+t\partial_t)\Gamma_1$, if we attach a weight $1/t$ per edge and $t^{i-1}$ per $i$-valent vertex (with a net resulting weight $t^{2E-V-E} = t^{E-V}$). Modifying the propagator and vertex weights of the matrix model accordingly, this simply amounts to replacing $V'(x)$ by $V'(tx) = tx - \sum g_i t^{i-1}x^{i-1}$ in all the above formulas, and setting $t = 1$ after differentiation. This yields

$$(2+\partial_t)\Gamma_1|_{t=1} = 2S + 2V'_{-2} + \frac{\partial S}{\partial t}|_{t=1}(1+V''_{-2}) + \frac{\partial R}{\partial t}|_{t=1}V''_{-1} + V''_{-3} + SV''_{-2} + RV''_{-1} \tag{58}$$

We now use the above trick (54) that the residue of a derivative vanishes, but this time with

$$\left(d/dw(w^2V'(w+S+R/w))\right)_{-1} = 0 = 2V'_{-2} + V''_{-3} - RV''_{-1} \tag{59}$$

and we use this to eliminate $V''_{-3}$ from (58), as well as (54) to rewrite the factor of $\partial S/\partial t$ as $1 + V''_{-2} = V'_{-1} + V''_{-2} = RV''_0$, with the result

$$(2+\partial_t)\Gamma_1|_{t=1} = 2S + SV''_{-2} + 2RV''_{-1} + RV''_0\frac{\partial S}{\partial t}|_{t=1} + V''_{-1}\frac{\partial R}{\partial t}|_{t=1} \tag{60}$$

Let us now differentiate w.r.t. $t$ the equation $0 = V'_0$, and then set $t = 1$ and multiply it by $R$. This gives

$$\begin{aligned} 0 &= R(V''_{-1} + SV''_0 + RV''_1 + V''_0\frac{\partial S}{\partial t}|_{t=1} + V''_1\frac{\partial R}{\partial t}|_{t=1}) \\ &= RSV''_0 + 2RV''_{-1} + RV''_0\frac{\partial S}{\partial t}|_{t=1} + V''_{-1}\frac{\partial R}{\partial t}|_{t=1} \end{aligned} \tag{61}$$

and allows to rewrite (60) as

$$(2+\partial_t)\Gamma_1|_{t=1} = 2S+SV''_{-2}-RSV''_0 = S+S(V'_{-1}+V''_{-2}-RV''_0) = S \quad (62)$$

by replacing $1 \to V'_{-1}$ and using again the equation (54). This completes the identification of $S$ as the generating function for one-leg planar graphs, with the leg not necessarily in the external face.

That the generating functions for both one- and two-leg planar graphs should satisfy a system of two algebraic equations (48), looks like magic at first sight. It is the purpose of Sect.4 below to unearth the combinatorial grounds for this apparent miracle.

### 3.4 The case of 4-valent planar graphs

Before going into this, let us conclude with the case of the quartic potential say $V(z) = \frac{z^2}{2} - g\frac{z^4}{4}$, for which we have $S = 0$ and eq.(49) reduces to

$$1 = R - 3gR^2 \quad \Rightarrow \quad R = \frac{a^2}{4} = \frac{1}{6g}(1 - \sqrt{1-12g}) \qquad (63)$$

as $R$ is the unique solution with the power series expansion $R = 1 + O(g)$. The corresponding resolvent and density of eigenvalues read respectively

$$\omega(z) = \frac{1}{2}(z - gz^3 - (1 - g\frac{a^2}{2} - gz^2)\sqrt{z^2 - a^2})$$
$$\rho(z) = \frac{1}{2\pi}(1 - g\frac{a^2}{2} - gz^2)\sqrt{a^2 - z^2} \qquad (64)$$

The two-leg-in-the-same-face graph generating function $\Gamma_2$ of eq.(52) reads here

$$\Gamma_2 = R - gR^3 = \frac{R(4-R)}{3} \qquad (65)$$

where we have used eq.(63) to eliminate $g$. But any planar 4-valent graph with two external legs in the same face is obtained by cutting an arbitrary edge in any closed planar 4-valent graph. As the two legs are distinguished, and as there are exactly twice as many edges than vertices in a closed 4-valent graph, we have $\Gamma_2 = 1 + 4g\partial f/\partial g$. The contribution 1 comes from the unique graph made of one loop, with one edge and no vertex, not counted in $f$. This gives the differential equation

$$4g\frac{df}{dg} = \frac{(R-1)(3-R)}{3} \qquad (66)$$

and eliminating $g = (R-1)/(3R^2)$ from (63), we finally get $\frac{df}{dg} = R^2(3-R)/4$. Changing variables to $R$, this turns into $\frac{df}{dR} = (2-R)(3-R)/(12R)$, easily integrated into

$$f = \frac{1}{2}\mathrm{Log}\, R + \frac{1}{24}(R-1)(R-9) \tag{67}$$

where the constant of integration is fixed by requiring that $f = 0$ when $R = 1$ (Gaussian case $V = V_0$). Substituting the expansion $R = 1 + 3g + 18g^2 + \ldots$ into (67) yields the expansion

$$f = \frac{g}{2} + \quad \frac{9}{8}g^2 \quad + \ldots \tag{68}$$

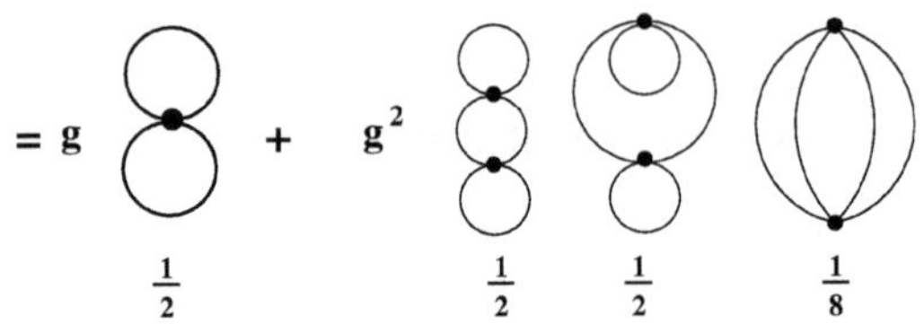


where we have represented the planar 4-valent graphs with up to 2 vertices, together with their inverse symmetry factors.

# 4. The trees behind the graphs

Using the above interpretation of $R$ as the generating function for planar graphs with two distinguished external legs not necessarily in the same face, let us now establish a general bijection between such graphs and suitably decorated trees, also called blossom-trees.

## 4.1 4-valent planar graphs and blossom trees

For reasons of simplicity, let us start with the case of 4-valent graphs. Given a two-leg such graph $G$ (see Fig.(4.1) for an illustration), we represent it in the plane by picking the external face to be adjacent to the first (in-coming) leg. We now visit all edges bordering this external face in counterclockwise direction, and cut them iff the resulting graph remains connected. We then replace the two halves of the cut edges by respectively a black and a white leaf. This "first passage" has merged a number of faces of the initial graph with the external one. We now repeat the algorithm with the new external face, and so on until all faces are merged. The resulting graph is a 4-valent tree $T$ (by construction, it has only one face and is connected). The tree is then rooted at its second (outcoming) leg, while its incoming one is replaced with a white leaf. Attaching a charge $+1$ (resp. $-1$) to white (resp. black) leaves, we obtain a tree with total charge $+1$. It is easy to convince oneself that the resulting 4-valent tree has exactly one black leaf at each vertex.

This is best proved by showing that its descendent subtrees not reduced to a black leaf all have charge $+1$. To see why, consider any edge of the blossom

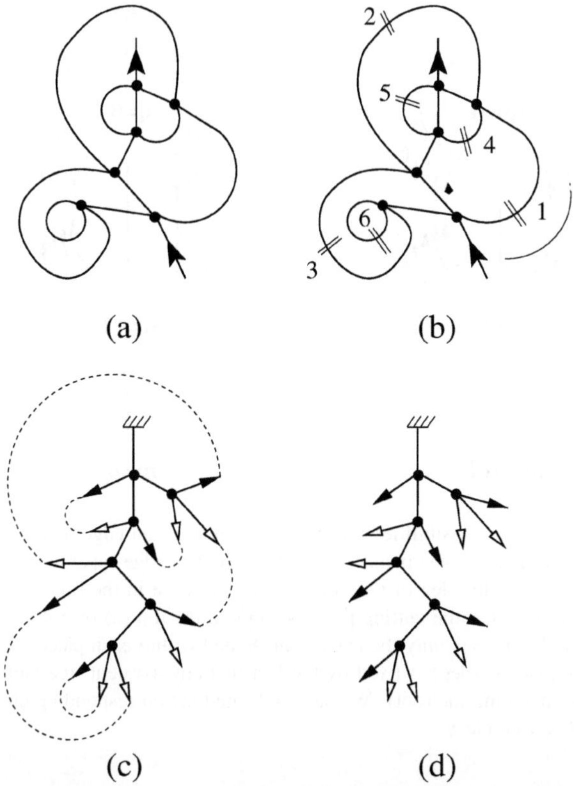


*Figure 6.* Illustration of the bijection between two-leg planar 4-valent graphs and rooted blossom trees. Starting from a two-leg graph (a), we apply the iterative cutting procedure, which here requires turning twice around the graph. In (b), the indices indicate the order in which the edges are cut during the 1st turn $(1, 2, 3)$ and 2nd turn $(4, 5, 6)$. Each cut edge is replaced by a black/white leaf pair (c), while the in-coming leg is replaced by a leaf and the out-coming one by a root, finally leading to a blossom tree (d). Conversely, the matching of black and white leaves of the blossom tree (d) rebuilds the edges of (a).

tree, not directly attached to a black leaf. It separates the tree into two (top and bottom) pieces as depicted in Fig.(6). As a result of the above iterative cutting procedure, we may keep track of the $m$ and $p$ cut edges encompassing this edge, respectively lying on its right and left, and connecting the top and bottom pieces. Assuming the first leg was in the bottom part, and as the cutting process travels in counterclockwise direction, we may only have $m = p + 1$ or $m = p$ according to whether the cutting process stopped in the top or bottom piece. But as the top and bottom pieces are trees with only 4-valent inner

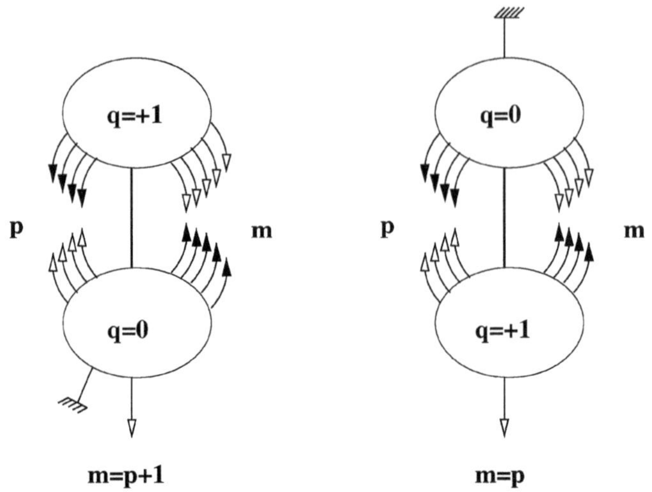


*Figure 7.* The only two possibilities for the environment of an edge in a 4-valent blossom-tree, obtained by cutting a two-leg planar 4-valent graph. The edge separates the tree into a top and a bottom piece. The first leg of the graph is chosen to be in the bottom piece. The two cases correspond to whether the cutting process stops in the top (a) or bottom (b) piece. We have represented in both cases only the leaves unmatched within each piece. In each case, the position of the root (second leg) is fixed by the fact that any 4-valent tree must have an even number of leaves (including the root). We have indicated the corresponding charges $q = 0$ or $+1$ of the top and bottom pieces.

vertices, they must have an even number of leaves, including the root, and the cut edge. Eliminating those matched by black/white pairs within each piece, we are respectively left with: in case (a), $2p + 2$ leaves on top and $2p + 3$ on the bottom, hence the root must be in the bottom; in case (b), $2p + 1$ leaves on top and $2p + 2$ on the bottom, hence the root must be on top. Adding up the charges, we see that the descendent piece (not containing the root) always has charge $q = +1$.

Let us now define *rooted blossom-trees* as rooted planar 4-valent trees with black and white leaves, a total charge $+1$, and exactly one black leaf at each vertex (or equivalently such that each subtree not reduced to a black leaf has charge $+1$). Then the rooted blossom-trees are in bijection with the two-leg 4-valent planar graphs. The inverse mapping goes as follows. Starting from a rooted blossom-tree $T$, we build a two-leg 4-valent planar graph by connecting in counterclockwise direction around the tree all pairs of black/white leaves immediately following one-another, and by repeating this until all black leaves are exhausted. This leaves us with one unmatched white leaf, which we replace by the first leg, while the root becomes the second leg. The order in which leaves are connected exactly matches the inverse of that of the above cutting procedure. This bijection now allows for a direct and simple counting of 2-leg 4-valent planar graphs, as we simply have to count rooted blossom-trees. Decomposing such trees according to the environment of the first vertex attached to their root, we get the following equation for their generating function

$$R = \quad 1 \quad + \quad 3gR^2 \tag{69}$$

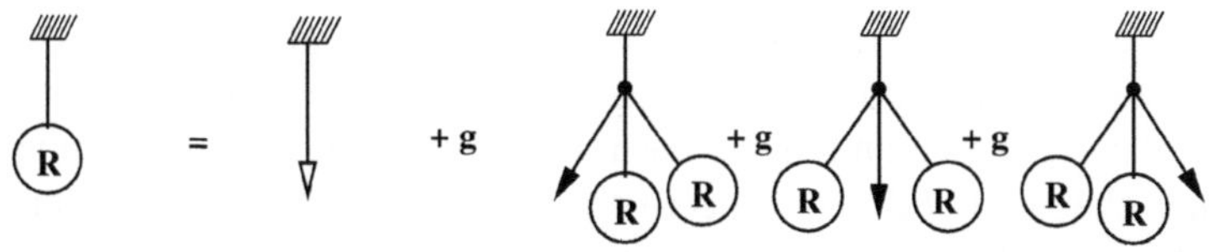


where the first term corresponds to no vertex (and a white leaf directly connected to the root), and the three others to a vertex with one black leaf and two descendent blossom-trees, each receiving a weight $g$ for the decomposed vertex. Note that eq.(69) trivially amounts to the first equation of (63). We have therefore found a purely combinatorial re-derivation of the one-cut large $N$ matrix model result for planar 4-valent graphs, which confirms its validity.

## 4.2 Generalizations

More generally, the above bijection may be adapted to two-leg planar graphs with arbitrary even vertex valences. Repeating the above cutting procedure on such a two-leg planar graph leaves us with a rooted tree with only even vertex valences, with black and white leaves, and a total charge $+1$, but with now exactly $k-1$ black leaves attached to each of its $2k$-valent vertices. This is again a consequence of the equivalent property that any subtree not reduced to a black leaf has charge $+1$, a fact proved exactly in the same manner as before (actually, Fig.(6) is still valid for the case of arbitrary even valences). This suggests a straightforward generalization of rooted blossom-trees with arbitrary even vertex valences, with black and white leaves and such that any subtree not reduced to a black leaf has charge $+1$. The latter are again in bijection with the two-leg planar graphs with even valences, and are easily enumerated by considering the environment of the vertex attached to the root, with the result for the generating function $R$, including weights $g_{2k}$ per $2k$-valent vertex:

$$R = 1 + \sum_{k\geq 1} g_{2k} \binom{2k-1}{k} R^k \tag{70}$$

where the first term corresponds as in (69) to the tree with no vertex, while the $k$-th term in the sum corresponds to the $\binom{2k-1}{k}$ ways of picking the $k-1$ black leaves among the $2k-1$ descendents of the $2k$-valent vertex attached to the root, the remaining descendents being themselves trees of charge $+1$ generated by $R$. The equation (70) is nothing but (49), written in a different fashion.

Finally, the bijection may be adapted so as to also include arbitrary (both even or odd) valences, but then requires the introduction of one-leg graphs as well. Such graphs are represented in the plane with their unique leg not necessarily adjacent to the external face, hence are not generated by $\Gamma_1 = S + V'_{-2}$, but, as we showed in the previous section, by $S$ itself. The graphs are again cut according to the above procedure, to produce rooted trees. The system of equations (48) is nothing but that obeyed by the rooted blossom-trees of two kinds corresponding to cutting one- and two-leg graphs, respectively generated by $S$ and $R$, and defined as rooted trees with black and white leaves, and total charge $0$ and $+1$ respectively, and whose descendent subtrees not reduced to a black leaf all have charge $0$ or $+1$. A simple way of recovering all combinatorial factors in the two lines of (48) is to note that in a rooted blossom tree of charge $0$ (resp. $+1$), the $i-1$ descendents subtrees of any $i$-valent vertex attached to the root may be either black leaves (charge $-1$), blossom trees of charge $0$, or blossom trees of charge $1$, the total charge being $0$ (resp. $+1$). These subtrees are generated respectively by the functions $1$, $S$ and $R$. Denoting by $j$ in both cases the total number of descendent subtrees of charge $+1$, we must have $j$ (resp. $j-1$) black leaves to ensure the correct total charge, and the remaining $i-2j-1$ (resp. $i-2j$) descendents have charge $0$. The combinatorial factors of (48) account for the possible choices of these among the $i-1$ descendents.

This combinatorial interpretation sheds light on the algebraicity of the equations obtained in the large $N$ limit for the general one-matrix model: trees are indeed archetypical objects whose generating functions obey algebraic relations, and we have shown that the planar graphs generated by the large $N$ matrix model could be represented by (blossom) trees. This correspondence will be fully exploited in Sect.6 to investigate the intrinsic geometry of planar graphs.

## 5. The one-matrix model II: topological expansions and quantum gravity

We now turn to higher genus contributions to the one-matrix model free energy. This is best done by use of the so-called orthogonal polynomial technique [20].

## 5.1 Orthogonal polynomials

The standard technique of computation of (28) uses orthogonal polynomials. The idea is to disentangle the Vandermonde determinant squared interaction between the eigenvalues. The solution is based on the following simple lemma: if $p_m(x) = x^m + \sum_{j=0}^{m-1} p_{m,j} x^j$ are monic polynomials of degree $m$, for $m = 0, 1, ..., N-1$, then

$$\Delta(m) = \det(m_i^{j-1})_{1\le i,j\le N} = \det(p_{j-1}(m_i))_{1\le i,j\le N} \tag{71}$$

easily derived by performing suitable linear combinations of columns. Let us now introduce the unique set of monic polynomials $p_m$, of degree $m = 0, 1, ..., N-1$, that are orthogonal w.r.t. the real one-dimensional measure $d\mu(x) = \exp(-NV(x))dx$, namely such that

$$(p_m, p_n) = \int_{\mathbb{R}} p_m(x) p_n(x) d\mu(x) = h_m \delta_{m,n} \tag{72}$$

These allow us to rewrite the numerator of (28), using (71), as

$$\sum_{\sigma,\tau\in S_N} \epsilon(\sigma\tau) \prod_{i=1}^{N} \int_{\mathbb{R}} d\mu(m_i) p_{\sigma(i)-1}(m_i) p_{\tau(i)-1}(m_i) e^{-NV(m_i)} = N! \prod_{j=0}^{N-1} h_j \tag{73}$$

We may apply the same recipee to compute the denominator, with the result $N! \prod_{j=0}^{N-1} h_j^{(0)}$, where the $h_j^{(0)}$ are the squared norms of the orthogonal polynomials w.r.t. the Gaussian measure $d\mu_0(x) = \exp(-Nx^2/2)dx$. Hence the $h$'s determine $Z_N(V)$ entirely through

$$Z_N(V) = \prod_{i=0}^{N-1} \frac{h_i}{h_i^{(0)}} \tag{74}$$

To further compute the $h$'s, let us introduce the two following operators $Q$ and $P$, acting on the polynomials $p_m$:

$$\begin{aligned} Q p_m(x) &= x p_m(x) \\ P p_m(x) &= \frac{d}{dx} p_m(x) \end{aligned} \tag{75}$$

with the obvious commutation relation

$$[P, Q] = 1 \tag{76}$$

Using the self-adjointness of $Q$ w.r.t.the scalar product $(f, g) = \int f(x) g(x) d\mu(x)$, it is easy to prove that

$$Qp_m(x) = xp_m(x) = p_{m+1}(x) + s_m p_m(x) + r_m p_{m-1}(x) \tag{77}$$

for some constants $r_m$ and $s_m$, and that $s_m = 0$ if the potential $V(x)$ is even. The same reasoning yields

$$r_m = \frac{h_m}{h_{m-1}}, \quad m = 1, 2, ... \tag{78}$$

and we also set $r_0 = h_0$ for convenience.

Moreover, expressing both $(Pp_m, p_m)$ and $(Pp_m, p_{m-1})$ in two ways, using integration by parts, we easily get the master equations

$$\begin{aligned} \frac{m}{N} &= \frac{(V'(Q)p_m, p_{m-1})}{(p_{m-1}, p_{m-1})} \\ 0 &= (V'(Q)p_m, p_m) \end{aligned} \tag{79}$$

which amount to a recursive system for $s_m$ and $r_m$. Note that the second line of (79) is automatically satisfied if $V$ is even: it vanishes as the integral over $\mathbb{R}$ of an odd function. Assuming for simplicity that $V$ is even, the first equation of (79) gives a non-linear recursion relation for the $r$'s, while the second is a tautology, due to the vanishing of all the $s$'s:

$$\begin{aligned} \frac{m}{N} &= \frac{(V'(Q)p_m, p_{m-1})}{(p_{m-1}, p_{m-1})} = \sum_{k\geq 1} g_{2k} \frac{(Q^{2k-1}p_m, p_{m-1})}{(p_{m-1}, p_{m-1})} \\ &= \sum_{k\geq 1} g_{2k} \sum_{\substack{\text{paths } p|p(1)=m,\ p(2k-1)=m-1 \\ p(i+1)-p(i)=\pm 1}} \prod_{i=1}^{2k-2} w(p(i), p(i+1)) \end{aligned} \tag{80}$$

where the sum extends over the paths $p$ on the non-negative integer line, with $2k-1$ steps $\pm 1$, starting at $p(1) = m$ and ending at $p(2k-1) = m-1$, and the weight reads $w(p, q) = 1$ if $q = p+1$, and $w(p, q) = r_p$ if $q = p-1$. For up to 6-valent graphs this reads

$$\begin{aligned} \frac{n}{N} &= r_n(1-g_2) - g_4 r_n(r_{n+1} + r_n + r_{n-1}) \\ &\quad - g_6(r_{n+1}r_{n+2} + r_{n+1}r_{n-1} + r_{n-1}r_{n-2} + r_n^2 + r_{n+1}^2 + r_{n-1}^2 \\ &\quad + 2r_n(r_{n+1} + r_{n-1}) \end{aligned} \tag{81}$$

In general, the degree $d$ of $V$ fixes the number $d-1$ of terms in the recursion. So, we need to feed the $d-2$ initial values of $r_0, r_1, r_2, ..., r_{d-3}$ into the recursion relation, and we obtain the exact value of $Z_N(V)$ by substituting $h_i = r_0 r_1 ... r_i$ in both the numerator and the denominator of (74). Note that for $V_0(x) = x^2/2$ the recursion (79) reduces simply to

$$\frac{m}{N} = \frac{(Qp_m^{(0)}, p_{m-1}^{(0)})}{(p_{m-1}^{(0)}, p_{m-1}^{(0)})} = r_m^{(0)} \tag{82}$$

and therefore $h_m^{(0)} = h_0^{(0)} m!/N^m = \sqrt{2\pi} m!/N^{m+1/2}$. The $p_m^{(0)}$ are simply the (suitably normalized) Hermite polynomials.

Finally, the full free energy of the model (25) reads

$$F_N(V) = \mathrm{Log}\, Z_N(V) = N\, \mathrm{Log}\, r_0 \sqrt{\frac{N}{2\pi}} + \sum_{i=1}^{N-1} (N-i) \mathrm{Log} \frac{N r_i}{i} \tag{83}$$

in terms of the $r$'s.

## 5.2 Large $N$ limit revisited

In view of the expression (83), it is straightforward to get large $N$ asymptotics for the free energy, by first noting that as $h_0 \sim \sqrt{\frac{2\pi}{N}}$, the first term in (83) doesn't contribute to the leading order $N^2$ and then by approximating the sum by an integral of the form

$$f = \lim_{N\to\infty} \frac{1}{N} \sum_{i=1}^{N-1} (1 - \frac{i}{N}) \mathrm{Log} \frac{r_i}{i/N} = \int_0^1 dz (1-z) \mathrm{Log} \frac{r(z)}{z} \tag{84}$$

where we have assumed that the sequence $r_i$ tends to a function $r_i \equiv r(i/N)$ of the variable $z = i/N$ when $N$ becomes large. This assumption, wrong in general, basically amounts to the one-cut hypothesis encountered in Sect.3.2. The limiting function $r(z)$ in (84) is then determined by the equations (79), that become polynomial in this limit. In the case $V$ even for instance, we simply get

$$z = r(z) - \sum_{k\geq 1} \binom{2k-1}{k} g_{2k} r(z)^k \tag{85}$$

The function $r(z)$ is the unique root of this polynomial equation that tends to $z$ for small $z$ (it can be expressed using the Lagrange inversion method for instance, as a formal power series of the $g$'s), and the free energy follows from (84). To relate this expression to our former results, let us again attach an extra weight $t$ per face of the graphs. As before, it amounts to replacing $V'(x) \to V'(tx) = tx - \sum_{k\geq 1} g_{2k} t^{2k-1} \binom{2k-1}{k} x^{2k-1}$, and to rescale $f \to t^2 f$. Setting $\rho(z) = t^2 r(z)$, we arrive at

$$tz = \rho(z) - \sum_{k\geq 1} \binom{2k-1}{k} g_{2k} \rho(z)^k \equiv \varphi(\rho(z)) \tag{86}$$

and $f = \int_0^1 dz(1-z)\mathrm{Log}\frac{\rho(z)}{tz}$, $\rho(z)$ being determined by $\varphi(\rho(z)) = tz$. Let us perform in this integral the change of variables $z \to \rho$, with $dz = \varphi'(\rho)/t d\rho$, and integration bounds $\rho(0) = 0$ and $\rho(1) = \mathbf{r}$, solution of $t = \varphi(\mathbf{r})$. We obtain:

$$t^2 f = \int_0^{\mathbf{r}} d\rho\varphi'(\rho)(t - \varphi(\rho)) \operatorname{Log} \frac{\rho}{\varphi(\rho)} \tag{87}$$

We now take derivatives w.r.t. $t$: as the dependence on $t$ is either via $\mathbf{r}$ or explicit in the integrand, there are only two terms involved. But the integrand vanishes at the upper bound, as $t - \varphi(\mathbf{r}) = 0$,hence only the explicit derivative contributes, and we have

$$\begin{aligned} \partial_t(t^2 f) &= \int_0^{\mathbf{r}} d\rho\varphi'(\rho) \operatorname{Log} \frac{\rho}{\varphi(\rho)} \\ \partial_t^2(t^2 f) &= \partial_t \mathbf{r} \varphi'(\mathbf{r}) \operatorname{Log} \frac{\mathbf{r}}{\varphi(\mathbf{r})} = \operatorname{Log} \frac{\mathbf{r}}{t} \end{aligned} \tag{88}$$

Note that $\mathbf{r}$ may be interpreted in the light of Sect.4 as the generating function for rooted blossom trees with a weight $t$ per white leaf (easily read off the relation $\mathbf{r} = t + \sum_{k\geq 1} g_{2k} \binom{2k-1}{k} \mathbf{r}^k$). Finally, setting $t = 1$, we may rewrite

$$\partial_t^2(t^2 f)|_{t=1} = \operatorname{Log} R = -\operatorname{Log}\big(1 - \sum_{k\geq 1} g_{2k} \binom{2k-1}{k} R^{k-1}\big) \tag{89}$$

as $\mathbf{r}$ reduces to $R$ at $t = 1$. This expresses the generating function for planar graphs with even valences and with two distinct marked faces (as each derivative amounts to a marking) as the logarithm of the generating function for blossom trees. This formula will become combinatorially clear in Sect.7.2 below.

## 5.3 Singularity structure and critical behavior

In the even potential case, according to (88), the singularities of $\mathbf{r}$ govern those of the free energy. $\mathbf{r}$ attains a first critical singularity at some $t = t_c$ where $\mathbf{r} = \mathbf{r}_c$ with $\varphi(\mathbf{r}_c) = t_c$ and $\varphi'(\mathbf{r}_c) = 0$. We may then Taylor-expand

$$t_c - t = \varphi(\mathbf{r}_c) - \varphi(\mathbf{r}) = -\frac{1}{2}(\mathbf{r} - \mathbf{r}_c)^2 \varphi''(\mathbf{r}_c) + O((\mathbf{r} - \mathbf{r}_c)^3) \tag{90}$$

As $t$ is an activity per face of the graphs, we may consider the number of faces as a measure of the area of the associated discrete surface, therefore the singularity $\mathbf{r}_{sing} \sim (t_c - t)^{1/2}$ is immediately translated via (88) into a singularity of the planar free energy $f_{sing} \sim (t_c - t)^{2-\gamma}$, with a string susceptibility exponent $\gamma = -1/2$. Alternatively, upon Laplace-transforming the result, this exponent also governs the large area behavior of $f_A \sim \text{const. } t_c^{-A}/A^{3-\gamma}$, the planar free energy for fixed area ($A$=number of faces here). This is the generic singularity expected from a model describing space without matter, such as that of the pure 4-valent graphs studied above.

We may reach more interesting multicritical points with different universality classes and exponents by fine-tuning the parameters $g_{2k}$ so as to ensure that a higher order singularity is attained at some $t = t_c$ such that $\mathbf{r} = \mathbf{r}_c$, while $\varphi'(\mathbf{r}_c) = \varphi''(\mathbf{r}_c) = ... = \varphi^{(m)}(\mathbf{r}_c) = 0$, while $\varphi^{(m+1)}(\mathbf{r}_c) \neq 0$. Taylor-expanding now yields

$$t_c - t = \varphi(\mathbf{r}_c) - \varphi(\mathbf{r}) = -\frac{\varphi^{(m+1)}(\mathbf{r}_c)}{(m+1)!}(\mathbf{r} - \mathbf{r}_c)^{m+1} + O((\mathbf{r} - \mathbf{r}_c)^{m+2}) \quad (91)$$

This translates into a singularity of the free energy with string susceptibility exponent $\gamma = -\frac{1}{m+1}$. This is characteristic of non-unitary matter conformal field theory with central charge $c(2, 2m+1)$ coupled to 2D quantum gravity [1] [2]. The first example of this is the Hard Dimer model introduced in Sect.2.4 above, for which

$$\varphi_{HD}(\mathbf{r}) = \mathbf{r} - 3g\mathbf{r}^2 - 30zg^2\mathbf{r}^3 \quad (92)$$

Writing $\varphi'_{HD}(\mathbf{r}) = \varphi''_{HD}(\mathbf{r}) = 0$ yields $z_c = -1/10$, $g\mathbf{r}_c = 1/3$, and $gt_c = 1/3$, with a critical exponent $\gamma = -1/3$, corresponding to the Lee-Yang edge singularity (conformal field theory with central charge $c(2,5) = -22/5$) coupled to 2D quantum gravity.

The inclusion of vertices of odd valences does not give any additional multicritical singularities. This is why we choose to stick here and in the following to the even case as much as possible.

## 5.4 Higher genus

To keep the full fledge of the model, we must keep track of all shifts of indices in (80). This is easily done by still introducing $r(z = m/N) \equiv r_m$, but by also keeping track of finite shifts of the index $m \to m + a$, namely, setting $\epsilon = 1/N$, via $r(z + a\epsilon) \equiv r_{m+a}$. In other words, as $N \to \infty$, we still assume that $r_m$ becomes a smooth function of $z = m/N$, but keep track of finite index shifts. Solving eq.(80) order by order in $1/N$ involves writing the "genus" expansion

$$r(z) = \sum_{k \geq 0} \epsilon^{2k} r^{(k)}(z), \quad (93)$$

implementing all finite index shifts by the corresponding $\epsilon$ shifts of the variable $z$, and solving for the $r^{(k)}$'s order by order in $\epsilon^2$. We finally have to substitute the solution back into the free energy (83), with $r_i = r(i/N)$. This latter expression must then be expanded order by order in $\epsilon$ using the Euler-MacLaurin formula. Setting $h(x) = (1-x)\text{Log}(r(x)/x)$, this gives

$$\frac{F_N(V)}{N^2} = \frac{1}{N}\sum_{i=1}^{N} h\left(\frac{i}{N}\right) = \int_0^1 h(z)dz + \frac{\epsilon}{2}(h(1)-h(0)) + \frac{\epsilon^2}{12}(h'(1)-h'(0)) + \ldots \tag{94}$$

in which we must also expand $r(x)$ according to (93). The result is the genus expansion $F_N(V) = \sum N^{2-2h} F^{(h)}(V)$, where $F^{(h)}$ is the generating function for graphs of genus $h$. For illustration, in the 4-valent case, we have

$$r(z) - gr(z)(r(z+\epsilon) + r(z) + r(z-\epsilon)) = z \tag{95}$$

Writing $r(z) = r^{(0)}(z) + \epsilon^2 r^{(1)}(z) + O(\epsilon^4)$, we find that

$$r^{(1)}(z)(1 - 6gr^{(0)}(z)) = gr^{(0)}(z)r^{(0)''}(z) \tag{96}$$

at order 2 in $\epsilon$, while $r^{(0)}(z) = (1-\sqrt{1-12gz})/(6g)$, and $(1-6gr^{(0)}(z))r^{(0)'}(z) = 1$, so that $r^{(1)}(z) = gr^{(0)}(z)r^{(0)'}(z)r^{(0)''}(z)$. At to order 2 in $\epsilon$, this gives

$$F^{(1)} = \frac{1}{12}(h_0'(1) - h_0'(0)) + g\int_0^1 dx(1-x)r^{(0)'}(x)r^{(0)''}(x) \tag{97}$$

where $h_0(x) = (1-x)\mathrm{Log}(r^{(0)}(x)/x)$, namely

$$\begin{aligned} F^{(1)} &= \frac{1}{24}\sum_{n\geq 1} \frac{g^n}{n} 3^n (4^n - \binom{2n}{n}) \\ &= \frac{g}{4} + \qquad \frac{15}{8} g^2 + \ldots \end{aligned} \tag{98}$$

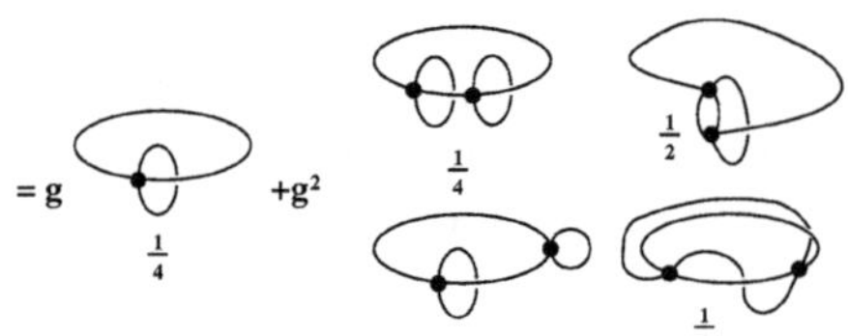


where we have displayed the genus one 4-valent graphs with up to two vertices, together with their inverse symmetry factors.

## 5.5 Double-scaling limit

The idea behind the double-scaling limit is to combine the large $N$ limit and the singularity structure of the free energy at all genera into a single scaling function. Let us first consider the 4-valent case (95). We wish to approach the critical value $g = g_c = 1/12$ displayed by the planar solution $R = (1 - \sqrt{1-12g})/(6g)$ at $t = 1$, *at the same time* as $N \to \infty$. Setting $\rho(z) = gr(z)$, we have

$$\begin{aligned} gz &= \rho(z)(1 - (\rho(z+\epsilon) + \rho(z) + \rho(z-\epsilon))) \\ g_c &= \rho_c(1 - 3\rho_c) \end{aligned} \tag{99}$$

with $g_c = 1/12$ and $\rho_c = 1/6$. Subtracting both lines of (99) and expanding up to order 2 in $\epsilon$ yields

$$g_c - gz = \rho_c(1-3\rho_c) - \rho(z)(1-3\rho(z)) - \epsilon^2\rho(z)\rho''(z) + O(\epsilon^4) \tag{100}$$

This suggests to introduce rescaled variables and functions $g_c - gz = a^2 g_c y$, while $\rho(z) = \rho_c(1 - au(y))$, and to expand up to order 2 in $a$ as well:

$$a^2 g_c y = 3\rho_c^2 a^2 u(y)^2 - \epsilon^2 \rho_c^2 a^{-3} u''(y) \tag{101}$$

where we have noted that $dz = -a^2 dy$ at $g = g_c$. The large $N$ limit of Sect.5.2 is recovered by taking $\epsilon = 0$, in which case we are left with $u(y) = \sqrt{y}$, another way of expressing the planar singularity of the free energy $u(y) = y^{-\gamma}$, with $\gamma = -1/2$. For non-zero $\epsilon$, all terms in (100)a will contribute if we take $\epsilon^2 = a^5$. We then have

$$y = u(y)^2 - \frac{1}{3}u''(y) \tag{102}$$

which is nothing but the Painleve I equation. Moreover, the singular part of the free energy reads

$$F \equiv F_{sing} = N^2 \int_0^1 dz(1-z)\mathrm{Log}(\frac{\rho(z)}{gz})|_{sing} = N^2 a^5 \int_{a^{-2}}^{x} (y-x)u(y)dy \tag{103}$$

where $g_c - g = a^2 g_c x$. Differentiating twice w.r.t. $x$ yields $u(x) = -F''(x)$. To summarize, if we take simultaneously $N \to \infty$ and $g \to g_c$, but keep the quantity

$$N^{\frac{4}{5}}\left(\frac{g_c - g}{g_c}\right) = x \tag{104}$$

fixed, then the singular parts of the free energy at all genera recombine into a single scaling function $F(x)$, whose second derivative satisfies the Painleve I differential equation. To recover the leading singularity at genus $h$, we simply have to expand the solution of (102) at large $x$ as $u(x) = \sum_{h\geq 0} u_h x^{\frac{1}{2}(1-5h)}$ and solve the resulting recursion relation for $u_h$. This is the so-called double-scaling limit of pure 2D quantum gravity.

We may repeat this exercise with the multicritical models of Sect.5.3, however algebra becomes cumbersome. Let us instead look at the scaling limits of the operators $P$ and $Q$ acting on the orthogonal polynomials (75). Let us rescale the orthogonal polynomials $p_n$ to make them orthonormal, namely set $\tilde{p}_n = p_n/\sqrt{h_n}$, so that (77) (with $s_n = 0$) becomes more symmetric

$$(Q\tilde{p})_n = \lambda \tilde{p}_n = \sqrt{r_{n+1}}\tilde{p}_{n+1} + \sqrt{r_n}\tilde{p}_{n-1} \tag{105}$$

or equivalently

$$Q_{n,m} = (\tilde{p}_m, Q\tilde{p}_n) = \sqrt{r_{n+1}}\delta_{m,n+1} + \sqrt{r_n}\delta_{m,n-1} \tag{106}$$

Let us now take the large $N$ limit. Setting $\epsilon = 1/N$ as before, we note that the shift operator $\delta_{m,n+1}$, acting on sequences $(\alpha_m)$ can be generated as $e^{\epsilon d/dz}$, acting on the continuum limit of $(\alpha_m)$, i.e. a function $\alpha(z) = \alpha_m$, for $z = m/N$. Indeed, one just has to write

$$\sum_m \delta_{m,n+1}\alpha_m = \alpha_{n+1} = e^{\frac{d}{dn}}\alpha_n \simeq e^{\epsilon\frac{d}{dz}}\alpha(z) \tag{107}$$

Setting again $r(z) = r_c(1 - au(y))$, this permits to rewrite $Q$ as

$$\begin{aligned} Q &\simeq \sqrt{r(z)}\left(e^{\epsilon\frac{d}{dz}} + e^{-\epsilon\frac{d}{dz}}\right) \\ &= \sqrt{r_c(1-au(y))}(2 + \epsilon^2\frac{d^2}{dz^2} + O(\epsilon^4)) \\ &= 2\sqrt{r_c} - \sqrt{r_c}(au - (\epsilon\frac{d}{dz})^2 + O(\epsilon^4, a^2)) \end{aligned} \tag{108}$$

In the general multicritical case, we must set

$$t_c - tz = a^{m+1}t_c y \tag{109}$$

and $y = x$ at $z = 1$, so that $dz \sim a^{m+1}dy$. The two terms in the r.h.s. of (108) are of the same order $a$ provided $(\epsilon d/dz)^2 = (\epsilon a^{-m-1}d/dy)^2$ is of order $a$, and we obtain the double-scaling condition that $\epsilon^2 a^{-2m-2} = a$, hence

$$N^2 = a^{-2m-3} \tag{110}$$

or equivalently

$$N^{\frac{2m+2}{2m+3}}\left(\frac{t_c - t}{t_c}\right) = x \tag{111}$$

remains fixed while $N \to \infty$ and $t \to t_c$. Retaining only the coefficient of $a$, we find that

$$Q \to \frac{d^2}{dy^2} - u(y) \tag{112}$$

in the double-scaling limit. This limit is a differential operator, acting on functions of the rescaled variable $y$.

Let us now turn to $P$. It will be useful to slightly change the definition of the operator $P$, in the following way

$$\begin{aligned} P_{n,m} &= \int_{-\infty}^{\infty} d\lambda \tilde{p}_m(\lambda) e^{-N\frac{V(\lambda)}{2}} \frac{d}{d\lambda} e^{-N\frac{V(\lambda)}{2}} \tilde{p}_n(\lambda) \\ &= -\frac{N}{2}(\tilde{p}_m, V'(Q)\tilde{p}_n) + (\tilde{p}_m, \tilde{p}_n') \\ &= -\frac{N}{2} V'(Q)_{n,m} + A_{n,m} \end{aligned} \tag{113}$$

where $A$ is a lower triangular matrix $A_{nm} = 0$ if $n \leq m$. Upon an integration by parts we may as well write

$$\begin{aligned} P_{n,m} &= -\frac{N}{2} V'(Q)_{n,m} + A_{n,m} \\ &= \frac{N}{2} V'(Q)_{n,m} - A^t_{n,m} \end{aligned} \tag{114}$$

where the matrix $A^t$ is upper triangular. Eq.(114) permits to compute the matrix elements of $P$ in terms of those of $Q$ only, by using the first equation when $n \leq m$ ($A_{nm} = 0$) and the second one when $n \geq m$ ($A^t_{nm} = 0$). This can be summarized by the following operator relation:

$$P = \frac{N}{2}(V'(Q)_+ - V'(Q)_-) \tag{115}$$

where the index $+$ (resp $-$) indicates that we retain only the upper (resp. lower) triangular part. In particular, as it is expressed polynomially in terms of $Q$, $P$ has a finite range, namely $P_{n,m} = 0$ if $|n - m| > B$, $B$ some uniform bound, independent of $N$ ($B$ depends only on the degree of $V$). This bound ensures that $P$ goes over in the double scaling limit to a differential operator of finite degree $p$, of the form

$$P = \frac{1}{a\sqrt{r_c}}(d^p + v_2 d^{p-2} + v_3 d^{p-3} + .... + v_p) \tag{116}$$

to ensure the correct normalization of $[P, Q] = 1$. From the precise form of $P$ (115), and as each derivative $d$ w.r.t. $y$ carries a prefactor $\epsilon a^{-m-1} = \sqrt{a}$, we must have $N \times a^{p/2} = 1/a$, which together with the double-scaling condition (110) fixes the degree

$$\deg(P) = 2m + 1 \tag{117}$$

We must finally write the canonical commutation relation (76) $[P, Q] = 1$, with the renormalized values $P = d^{2m-1} + v_2 d^{2m-2} + .... + v_{2m}$ and $Q = (d^2 - u)$. Let us introduce the square root $L$ of $Q$, namely the unique pseudo-differential operator

$$L = d + \sum_{i\geq 1} \ell_i d^{-i} \tag{118}$$

such that $L^2 = Q$. This equation is expressed as a triangular system for the $\ell$'s, provided we normal-order the result by pushing all functions to the left of powers of the differential $d$, by means of the Leibnitz formula $d^{-i}f(y) = \sum_{j\geq 0}(-1)^j\binom{i+j-1}{j}f^{(j)}(y)d^{-i-j}$. Let us now express $P$. Solving $[\tilde{P}, Q] = 0$ rather than $[P, Q] = 1$ makes no difference as far as we only write the equations for the coefficients of positive powers of $d$: solving these equations precisely allows to express $P$ as a function of $Q$. As the solution to $[\tilde{P}, Q] = 0$ for a pseudo-differential operator $\tilde{P}$ of degree $2m+1$ is nothing but $\tilde{P} = L^{2m+1}$, we simply have $P = (L^{2m+1})_+$, where the subscript $+$ indicates that we have retained only the differential polynomial part. So far, we have solved all the equations obtained by setting to 0 the coefficients of all positive powers of $d$ in $[P, Q] = 1$. We still have to write the $d^0$ coefficient. Writing $(L^{2m+1})_- = L^{2m+1} - (L^{2m+1})_+ = R_{m+1}[u]d^{-1} + O(d^{-2})$, this last equation reads simply

$$2R_{m+1}[u]' = 1 \qquad \Rightarrow \qquad 2R_{m+1}[u] = y \tag{119}$$

This is nothing but a higher order generalizations of the Painlevé I equation, related to the so-called KdV hierarchy. From their definition, the "KdV residues" $R_m[u]$ satisfy the recursion relation

$$R_{m+1}[u]' = \frac{1}{4}R_m[u]''' - \frac{1}{2}u'R_m[u] - uR_m[u]' \tag{120}$$

obtained by writing $(L^{2m+1})_- = ((L^{2m-1})_- Q)_- = (Q(L^{2m-1})_-)_-$, while the initial term reads $R_1[u] = -u/2$. Again, plugging the large $y$ expansion $u(y) = \sum_{h\geq 0} u_h y^{\frac{1}{m+1}(1-(2m+3)h)}$ into eq.(119) yields a recursion relation for the $u_h$ and gives acces to the all genus singular part of the free energy via the relation $F_{sing}(x)'' = -u(x)$.

The actual general solution of $[P, Q] = 1$ involves integration constants which we have all set to zero for convenience, hence the most general solution for a degree $2m+1$ differential operator $P$ reads

$$2\sum_{j=1}^{m+1} \mu_j R_j[u] = y \tag{121}$$

for some integration constants $\mu_j$. This equation interpolates between the various matter critical points $\mu_j = \delta_{j,k+1}$, corresponding to the various multi-critical points already identified as $c(2, 2k+1)$ CFT coupled to 2D quantum gravity. From the point of view of the $m+1$-critical model, the $\mu$'s are just dimensionful parameters coupled to the order parameters of the theory.

## 5.6 Generalization to multi-matrix models

A large class of multi-matrix models turns out to be solvable by exactly the same techniques as those developed in the previous sections for the one-matrix model. It corresponds to matrices $M_1, ..., M_p$ with a chain-like interaction, namely involving a quadratic form $Q_{a,b}$ as in (21), for which only the elements $Q_{a,a}$, $a = 1, 2, ..., p$ and say $Q_{a,a+1}$, $a = 1, 2, ..., p - 1$ are non-vanishing. In this particular case only, the unitary group integrations may be disentangled from the eigenvalue integrations for all $M$'s and we may still reduce the integral to one over eigenvalues of the different matrices. Once this step is performed, the orthogonal polynomial technique is easily adapted and a complete solution follows from considering again operators $P_a$ and $Q_a$ of differentiation w.r.t. or multiplication by an eigenvalue of the matrix $M_a$, $a = 1, 2, ..., p$. Note that the saddle-point technique with several matrices is more subtle.

One is eventually left with solving an equation of the form $[P_1, Q_1] = 1$, the scaling function $u$ such that $u'' = -F$ being identified with some coefficient of $Q_1$. The remarkable fact is that both $P_1$ and $Q_1$ remain of uniformly bounded range, the latter depending only on the degrees of the potentials for the various matrices. This implies that in a suitable double scaling limit where the size of the matrices is sent to infinity and the parameters of the potentials go to some (multi-) critical values, the operators $P_1$ and $Q_1$ still become differential operators of finite degree say $p$ and $q$, two coprime integers. The resulting differential system $[P_1, Q_1] = 1$ governs the all-genus singular part of the free energy of the general $c(p, q)$ minimal conformal field theory coupled to 2D quantum gravity. This completes the picture of critical behaviors covered by matrix models solvable by orthogonal polynomial techniques: it exhausts all minimal CFT's with $c < 1$, according to the famous ADE classification thereof [1]footThis statement is not completely correct: only the A-type CFT's are covered by the standard multi-matrix models. A proposal for D-type CFT's was given in [21], based on D-type generalizations of the KP hierarchy [22], but no direct relation to solvable matrix models was found. Not to speak about E-type solutions....

From a combinatorial point of view, the bijection presented in Sect.4.1 may be generalized to the case of two-matrix models [23], and presumably to all cases solvable by orthogonal polynomial techniques, which all lead to algebraic systems, henceforth suggest tree-like interpretations.

# 6. The combinatorics beyond matrix models: geodesic distance in planar graphs

In this section, we return to the bijections between planar graphs and trees to investigate more refined properties of the discrete random surfaces generated by matrix models, involving their intrinsic geometry. In particular, we will derive in a purely combinatorial manner sets of closed equations for generating functions of planar graphs with marked points at a given geodesic distance, a task still eluding the matrix model description.

## 6.1 Keeping track of the geodesic distance: the 4-valent case

Let us return to the bijection between two-leg 4-valent planar graphs and rooted blossom trees shown in Sect.4. Looking at Fig.(4.1), we see that the bijection allows to keep track of the geodesic distance between the two legs, namely the smallest possible number of edges of the graph crossed by a curve joining them. Indeed, this distance is nothing but the number of edges encompassing the root of the corresponding blossom-tree, when black and white leaves are re-connected. Loosely speaking, the geodesic distance between the legs corresponds in the blossom tree language to the number of black leaves "in excess", which require encompassing the root to be connected to their white alter ego in counterclockwise direction. Let us now derive simple relations for the generating function $R_n$ for two-leg diagrams with geodesic distance at most $n$ between the legs. To get the more interseting generating function $G_n$ for graphs with two legs at geodesic distance equal to $n$, we just have to write $G_n = R_n - R_{n-1}$. Alternatively, $R_n$ can be thought of as the generating function for blossom trees with at most $n$ black leaves in excess. As such, it obeys the following recursion relation:

$$R_n = \quad 1 \quad + \quad gR_{n+1}R_n \quad + \quad gR_n^2 \quad + \quad gR_nR_{n-1} \tag{122}$$

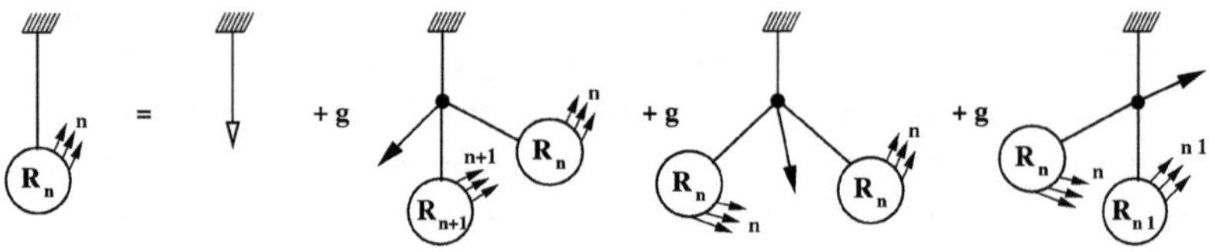


This is just a refinement of eq.(69) in which we have kept track of the maximal numbers of excess black leaves. The presence of single black leaves around the vertex connected to the root lowers by 1 the maximal number of excess leaves of any object on its left, while as each blossom tree has one white leaf in excess, it always absorbs one excess black leaf from objects on its left: these two facts are responsible for the shifts of the index $n$.

The recursion relation (122) holds for all $n \geq 0$ provided the term involving $R_{-1}$ is dropped. Let us therefore supplement the recursion relation with the initial value $R_{-1} = 0$. Moreover, the function $R_n$ should go over to the function $R$ of (69) in the limit $n \to \infty$, which amounts to suppressing the constraint on the distance between the two legs. $R_n$ is the unique solution to (122) such that $R_{-1} = 0$ and $\lim_{n\to\infty} R_n = R$. If we are only interested in the power series expansion of $R_n$ in $g$, we may solve (122) order by order in $g$, starting with $R_{-1} = 0$ at all orders and $R_n = 1 + O(g)$ for all $n \geq 0$. To any given order in $g$, the system for the series coefficients is indeed triangular, and moreover $R - R_n = O(g^{n+1})$, which guarantees the convergence condition. In the next section, we actually display the exact solution $R_n$ in a very compact form.

An important remark is in order. The relation (122) is strikingly reminiscent of that for the orthogonal polynomials (81) say with $g_2 = g_6 = 0$, $g_4 = g$, except that the l.h.s. of (81) is now replaced by 1, and $r_n = h_n/h_{n-1}$ by $R_n$. One may wonder whether eq.(122) may be derived from some matrix model solution. The answer is not known to this day, but the boundary condition that $R_{-1} = 0$ would mean in matrix model language that some norm of orthogonal polynomial must vanish, hence if there is such a matrix model formulation, it must be very singular. As to the r.h.s. of (81), its similarity with that of (122) suggests to express the rules for the possible subtrees encountered around the vertex attached to the root in *counterclockwise order* in terms of a "Q-operator" acting on a formal orthonormal basis $|n\rangle$, $\langle m|n\rangle = \delta_{m,n}$ for $m, n \geq 0$ and $|n\rangle = 0$ for $n < 0$, via

$$Q|n\rangle = |n+1\rangle + R_n|n-1\rangle \tag{123}$$

The first term is interpreted as the contribution of a single black leaf, while the second corresponds to a blossom tree with at most $n$ excess black leaves. Then the r.h.s. of (122) is nothing but $1 + g\langle n-1|Q^3|n\rangle$.

## 6.2 Exact solution

To solve (122), we use the convergence condition to write $R_n = R - \rho_n$ at large $n$, and expand (122) at first order in $\rho_n$. This gives the linear recursion relation

$$gR(\rho_{n+1}^{(1)} + \rho_{n-1}^{(1)}) - \rho_n^{(1)}(1 - 4gR) = 0 \tag{124}$$

This has the characteristic equation

$$x + \frac{1}{x} + 4 = \frac{1}{gR} \tag{125}$$

with $R$ given by (63). Picking the solution $x$ with modulus less than 1, we find that $R_n = R(1 - \lambda_1 x^n + O(x^{2n}))$ for some integration constant $\lambda_1$. We may next expand $R_n = R(1 - \sum_{j\geq 1} \lambda_j x^{jn})$, and (122) turns into a recursion relation for the coefficients $\lambda_j$:

$$\lambda_{j+1}\left(x^{j+1} + \frac{1}{x^{j+1}} - x - \frac{1}{x}\right) = \sum_{i=1}^{j} \lambda_i \lambda_{j+1-i}\left(x^i + \frac{1}{x^i}\right) \tag{126}$$

solved recursively as

$$\lambda_j = \lambda_1 \left( \frac{\lambda_1 x}{(1-x)(1-x^2)} \right)^{j-1} \frac{1-x^j}{1-x} \tag{127}$$

Picking $\lambda_1 = x(1-x)(1-x^2)\lambda$, $R_n$ is easily resummed into

$$R_n = R \frac{(1-\lambda x^{n+1})(1-\lambda x^{n+4})}{(1-\lambda x^{n+2})(1-\lambda x^{n+3})} \tag{128}$$

Further imposing the initial condition $R_{-1} = 0$ fixes $\lambda = 1$, hence finally

$$R_n = R \frac{(1-x^{n+1})(1-x^{n+4})}{(1-x^{n+2})(1-x^{n+3})} \tag{129}$$

with $|x| < 1$ solving (125).

This gives an explicit formula for the generating function of 4-valent two-leg graphs with geodesic distance at most $n$ between the legs. In particular, for $n = 0$, this gives the generating function for graphs with the two legs in the same face (also called $\Gamma_2$ in Sect.3.3), namely

$$R_0 = G_0 = \Gamma_2 = R \frac{1+x^2}{1+x+x^2} = R \frac{(1-4gR)}{(1-3gR)} = R - gR^3 \tag{130}$$

where we have used (125) and (69) to simplify the result. This is in perfect agreement with the matrix model result (65).

## 6.3 Integrability

The equation (122) is intergable in the classical sense that there exists an "integral of motion", namely a conserved quantity $f(R_n, R_{n+1})$ =const. which implies (122). More precisely, defining

$$f(x,y) = xy(1-gx-gy) - x - y \tag{131}$$

we have

$$f(R_n, R_{n+1}) - f(R_{n-1}, R_n) = (R_{n+1} - R_{n-1})\big(R_n - 1 - gR_n(R_{n+1} + R_n + R_{n-1})\big) \tag{132}$$

We deduce that if $f(R_n, R_{n+1})$is a constant independent of $n$, then $R_n$ obeys (122): $f$ is an integral of motion of the equation (122).

Using $f$, we may write in a compact way the condition $\lim_{n\to\infty} R_n = R$ for solutions of (122). Indeed, we simply have to write

$$f(R_n, R_{n+1}) = f(R, R) = R^2(1-2gR) - 2R = -(R - gR^3) \tag{133}$$

All solutions to (133) are also solutions of (122), and they moreover converge to $R$ as $n \to \infty$. As an immediate application of (133), we may recover $R_0$, by imposing that $R_{-1} = 0$:

$$R_0 = -f(R_{-1}, R_0) = R - gR^3 \tag{134}$$

in agreement with (130).

## 6.4 Fractal dimension

The advantage of having an exact formula like (129) is that we may also extract the "fixed area" coefficient $R_{n,A}$ of $g^A$ in $R_n$ via the contour integral

$$R_{n,A} = \oint \frac{dg}{2i\pi g^{A+1}} R_n \tag{135}$$

with $R_n$ given by (129). This gives access to asymptotic properties at large area $A$. In particular, the ratio

$$B_n \equiv \lim_{A\to\infty} \frac{R_{n,A}}{R_{0,A}} \tag{136}$$

may be taken as a good estimate of the average number of points at a geodesic distance less or equal to $n$ from a given point in random 4-valent graphs of infinite area. It is expected to behave like

$$B_n \sim n^{d_F} \quad \text{for large } n \tag{137}$$

where $d_F$ is the fractal dimension of the random surfaces. Performing in (135) the change of variables $v = gR$, i.e. $g = v(1-3v)$, we obtain

$$R_{n,A} = \oint \frac{dv(1-6v)}{2i\pi(v(1-3v))^{A+1}} \frac{1}{1-3v} \frac{(1-x(v)^{n+1})(1-x(v)^{n+4})}{(1-x(v)^{n+2})(1-x(v)^{n+3})} \tag{138}$$

where we have used $R(g(v)) = 1/(1-3v)$ and the expression $x = x(v) \equiv (1-4v-\sqrt{1-8v+12v^2})/(2v)$. The large $A$ behavior is obtained by a saddle-point approximation, as the integral is dominated by the vicinity of $v = v_c = 1/6$, corresponding to the critical point $g = g_c = 1/12$, where $x \to 1$. Making the change of variables $v = v_c(1 + i\frac{\xi}{\sqrt{A}})$, expanding all terms in powers of $1/\sqrt{A}$ and integrating over $\xi$, we finally get the leading behavior

$$R_{n,A} \sim \text{const.} \frac{(12)^A}{A^{\frac{5}{2}}} \frac{(n+1)(n+4)}{(n+2)(n+3)} (140 + 270n + 179n^2 + 50n^3 + 5n^4) \tag{139}$$

which finally gives the ratio

$$B_n = \frac{3}{280} \frac{(n+1)(n+4)}{(n+2)(n+3)} (140+270n+179n^2+50n^3+5n^4) \sim \frac{3}{56} n^4 \tag{140}$$

hence $d_F = 4$ is the desired fractal dimension.

## 6.5 Scaling limit: Painlevé again!

A continuum limit may be reached by letting $g$ tend to its critical value $g_c = 1/12$. More precisely, we write

$$g = \frac{1}{12}(1 - \epsilon^4) \qquad \Rightarrow \qquad gR = \frac{1}{6}(1 - \epsilon^2) \tag{141}$$

from eq.(63). In turn, the characteristic equation (125) yields

$$x = e^{-a\epsilon} + O(\epsilon^3) \qquad a = \sqrt{6} \tag{142}$$

As seen from eq.(129), a sensible limit is obtained by writing

$$n = \frac{r}{\epsilon} \tag{143}$$

and letting $\epsilon \to 0$. Writing the scaling variable $r$ as $r = n/\xi$, we see that $\epsilon$ plays the role of the inverse of the correlation length $\xi$. As we approach the critical point, we have $\xi = \epsilon^{-1} = \left((g_c - g)/g_c\right)^{-\nu}$ with a critical exponent $\nu = 1/4$, in agreement with $\nu = 1/d_F$, as expected from general principles. Performing this limit explicitly on the solution (129) yields an explicit formula for the continuum partition function $\mathcal{F}(r)$ of surfaces with two marked points at a geodesic distance *larger or equal to* $r$:

$$\mathcal{F}(r) \equiv \lim_{\epsilon \to 0} \frac{R - R_n}{\epsilon^2 R} = -2 \frac{d^2}{dr^2} \text{Log} \sinh\left(\sqrt{\tfrac{3}{2}}\, r\right) = \frac{3}{\sinh^2\left(\sqrt{\frac{3}{2}}\, r\right)} \tag{144}$$

Upon differentiating w.r.t. $r$, we obtain the continuum partition function for surfaces with two marked points at a geodesic distance *equal to* $r$:

$$\mathcal{G}(r) = -\mathcal{F}'(r) = 3\sqrt{6}\, \frac{\cosh\left(\sqrt{\frac{3}{2}}\, r\right)}{\sinh^3\left(\sqrt{\frac{3}{2}}\, r\right)} \tag{145}$$

This reproduces a conjecture [25] obtained in a transfer matrix formalism of 2D quantum gravity.

Note that the precise form of the scaling function $\mathcal{F}(r)$ may alternatively be obtained by solving the continuum counterpart of eq.(122). Indeed, writing

$$R_n = R(1 - \epsilon^2 \mathcal{F}(n\epsilon)) \tag{146}$$

and expanding eq.(122) up to order 4 in $\epsilon$, we obtain the following differential equation

$$\mathcal{F}''(r) - 3\mathcal{F}^2(r) - 6\mathcal{F}(r) = 0 \tag{147}$$

It is easy to check that $\mathcal{F}(r)$ as given by (144) is the unique solution of (147) with boundary conditions $\mathcal{F}(r) \to \infty$ when $r \to 0$ and $\mathcal{F}(r) \to 0$ when $r \to \infty$. Writing $\mathcal{F}(r) = u(r) - 1$, we note that eq.(147) turns into

$$u^2 - u''/3 = 1 \tag{148}$$

strikingly reminiscent of the Painleve I equation governing the model's all-genus double-scaling limit (102), except for the r.h.s. which is now a constant. The function $u$ leading to $\mathcal{F}$ is simply the unique solution to (148) such that $u(0^+) = \infty$ and $u(+\infty) = 1$.

## 6.6 Generalizations

The results of Sects.6.1-6.5 generalize straightforwardly to the case of arbitrary even valences. Using again the bijection of Sect.4.2, we still have to keep track of excess black leaves. Introducing similarly the generating function $R_n$ for planar graphs with even valences and with two legs at geodesic distance less or equal to $n$, we get a recursion relation by inspecting all configurations of the vertex attached to the root of the corresponding blossom trees. We may use the same rules as those found in the 4-valent case (122). Going clockwise around the vertex and starting from the root, we may encounter blossom trees with up to $p$ excess black leaves or single black leaves. Encountering a black leaf decreases the index $p$ of the objects following it clockwise, while encountering a blossom subtree increases it by 1. Using the "Q-operator" formalism of Sect.6.1, namely that $Q|n\rangle = |n+1\rangle + R_n|n-1\rangle$, we get the general recursion relation

$$R_n = 1 + \sum_{k\geq 1} g_{2k} \langle n-1|Q^{2k-1}|n\rangle \tag{149}$$

to be supplemented with $d/2-1$ initial conditions $R_{-1} = R_{-2} = ...R_{d/2-1} = 0$ ($d = \deg(V)$), and the usual convergence condition $\lim_{n\to\infty} R_n = R$, to the solution $R$ of (70). The explicit solution to (149) with these boundary conditions was derived in [14], and involves soliton-like expressions. It allows for investigating the fractal dimension for multicritical planar graphs, found to be $d_F = 2(m+1)$ for the case of Sect.3.6 (91), and to derive continuum scaling functions for multicritical matter on surfaces with two marked points at a fixed geodesic distance $r$. Writing (149) as $1 = \langle n-1|V'(Q)|n\rangle$, we use again the trick of adding a weight $t$ per face of the graph, which amounts to replacing $V'(Q) \to V'(tQ)$, and multiplying by $t$ leaves us with $t = \varphi(t^2R_n, t^2R_{n\pm1}, ...)$. Taking the multicritical values for $g_{2k}$, and writing $t = t_c(1-\epsilon^{2(m+1)})$, we look for solutions of (149) of the form $R_n = R(1-\epsilon^2\mathcal{F}(r=n\epsilon))$. This gives at order $2(m+1)$ in $\epsilon$ a differential equation for $\mathcal{F}$. Noting that our scaling Ansatz for $R_n$ is the same as that for the double-scaling limit ($r_n = r_c(1-au(y))$) except for the prefactor $R = R_c(1-\epsilon^2)$ we see that $u(r) = 1+\mathcal{F}(r)$ satisfies the generalized Painleve equation (119), but with a constant r.h.s. In differentiated form, this corresponds to writing the commutation relation $[P,Q]=0$ between two differential operators $P$ and $Q$ of the variable $r$, with respective orders $2m+1$ and 2, with $Q = d^2-u$.

The generalization to graphs with arbitrary (even and odd) valences is straightforward, as we simply have to use the "Q-operator" formalism in the combinatorial setting. The functions $S_n$ (resp. $R_n$) generate planar graphs with one leg (resp. two legs), with the leg (resp. second leg) at distance at most $n$ from the external face. The operator $Q$ now acts as $Q|n\rangle = |n+1\rangle + S_n|n\rangle + R_n|n-1\rangle$, where the new contribution corresponds to subtrees of charge 0, that do not affect the numbers of allowed excess black leaves of their followers. We obtain the system of equations

$$0 = \langle n|V'(Q)|n\rangle \qquad 1 = \langle n-1|V'(Q)|n\rangle \tag{150}$$

This generalizes presumably to all planar graph enumeration problems for which a matrix model treatment is available, using orthogonal polynomials involving a natural $Q$ operator, interpreted in the combinatorial setting as describing objects of various charges attached to the root vertex of the corresponding blossom trees. We may infer that in the general multicritical case of a CFT with central charge $c(p,q) < 1$, the scaling function for surfaces with two marked points at geodesic distance at least $r$ is governed by a differential system of the form $[P,Q]=0$, $P$ and $Q$ two differential operators of the variable $r$ of respective degrees $p$ and $q$.

## 7. Planar graphs as spatial branching processes

This last section is devoted to a dual approach to that followed so far, in which we consider the graphs dual to those contributing to the matrix model free energy, namely with prescribed face valences rather than vertex valences. On such a graph, the geodesic distance between vertices is the minimal number of edges visited in a path from one to the other. We will present bijections between classes of such graphs with a specified origin vertex and with a marked vertex at geodesic distance $\leq n$, and labeled trees of arbitrary valences obeying some specific labeling rules.

This allows to make the contact with an active field of probability theory dealing with spatially branching processes. The following is largely based on refs. [13] [16]EWALL [27] [28].

We first concentrate on the quadrangulations, namely the duals of 4-valent graphs.

## 7.1 The dual bijections: labeled trees for planar quadrangulations

We start with a rooted planar quadrangulation, namely a graph with only 4-valent faces (squares), with a marked oriented "root" edge. Let us pick as origin vertex the vertex at which the root edge starts. This choice induces a natural labeling of the vertices of the graph by their geodesic distance to this origin, itself labeled 0 (see Fig.(7.1) (a) for an example). We then note that only two situations may occur for the labeling of vertices around a face, namely

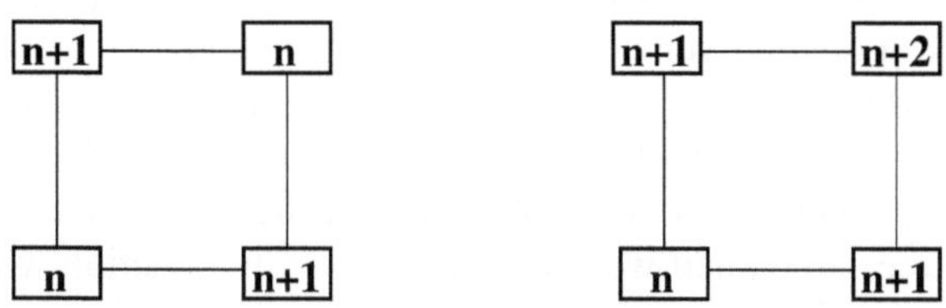


(151)

in which cases the faces are respectively called confluent and normal. The confluent faces have been shaded in the example of Fig.(7.1) (a). We now

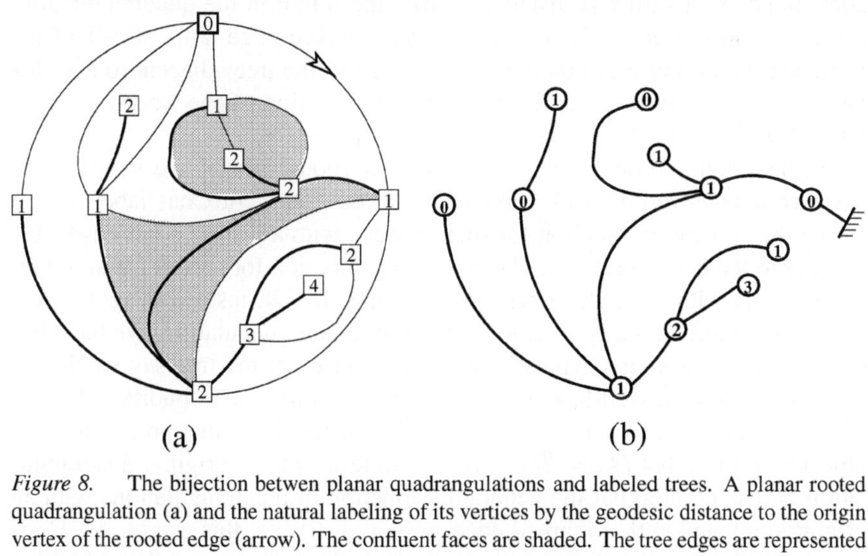


*Figure 8.* The bijection betwen planar quadrangulations and labeled trees. A planar rooted quadrangulation (a) and the natural labeling of its vertices by the geodesic distance to the origin vertex of the rooted edge (arrow). The confluent faces are shaded. The tree edges are represented in thick black lines, and connect all vertices with positive labels. Erasing all but these new edges and the vertices they connect leaves us with a labeled tree (b), which we root at the vertex corresponding to the end of the rooted edge of the initial quadrangulation. Finally, all labels of the tree are shifted by $-1$.

construct new edges as follows:

(152)

in each face of the quadrangulation (including the external face, for which the rules are reversed). This rule may be summarized by saying that we connect via a new edge all the vertices immediately followed clockwise by a vertex with a label one less. These edges are readily seen to connect all vertices of the quadrangulation but the origin. Thus, erasing all but the new edges and the vertices they connect leaves us with a connected labeled tree (see Fig.(7.1) (b)), which we root at the end vertex of the original rooted edge of the quadrangulation, and in which we *subtract* $1$ *from all vertices*[1]. In particular, the vertex attached to the root has label $0$, and all labels are non-negative. Moreover, by the construction rules (152), adjacent labels of the tree may differ only by $0$ or $\pm 1$. Such trees are called well-labeled, and are in bijection with the rooted planar quadrandulations.

The construction rules (152) allow for interpreting the features of the tree in terms of the original quadrangulation. Any vertex labeled $n-1$ in the tree corresponds to a vertex at distance $n$ from the origin in the quadrangulation. From the rules of eq.(152), we see that any marked edge $n \to n+1$ of the quadrangulation corresponds marking an edge of the tree adjacent to a vertex labeled $n$. This in turn may be viewed as the rooting of the tree at a vertex labeled $n$ (the above bijection uses this fact for $n = 0$).

We next define rooted well-labeled trees as rooted labeled trees, with non-negative integer vertex labels, and such that the root vertex has label $n$. Let $R_n$ be the generating function for such objects, with a weight $g$ per edge. According to the above bijection, the generating function for rooted planar quadrangulations with a weight $g$ per face is simply $R_0$. If, instead of rooting the well-labeled tree at the end vertex of the initial quadrangulation, we had chosen to root it elsewhere, typically at another vertex of the tree say labeled $n$, the resulting rooted well-labeled tree would satisfy the extra condition that the label $0$ occurs at least once in the tree. The generating function for such an object is nothing but $G_n = R_n - R_{n-1}$. In terms of the original quadrangulation, this is nothing but the generating function of quadrangulations with an origin vertex and with a marked edge $n \to n+1$ w.r.t. this origin. So $R_n$ is the generating function for planar quadrangulations with an origin and with a marked edge $m \to m+1$, $m \leq n$, and a weight $g$ per face.

The definition of $R_n$ allows to derive a recursion relation of the form

$$R_n = \frac{1}{1 - g(R_{n+1} + R_n + R_{n-1})} \tag{153}$$

where we simply express the labeling rule that the root vertex labeled $n$ may be adjacent to any number of vertices labeled $n, n+1$ or $n-1$, themselves roots of other well-labeled trees. Moreover, for (153) to also make sense at $n = 0$ we must set $R_{-1} = 0$. Removing the constraint that $m \leq n$ by sending $n \to \infty$ leaves us with the generating function $R$ for quadrangulations with an origin and a marked edge, wich also generates the rooted quadrangulations with a marked vertex, and should satisfy the relation

$$R = \frac{1}{1 - 3gR} \tag{154}$$

with $R = 1 + O(g)$. We conclude that the functions $R$ and $R_n$ coincide with those introduced in Sects.4.1 and 6.1.

So we have found another (dual) combinatorial interpretation for the exact solutions (129).

## 7.2 Application I: average numbers of edges and vertices at distance $n$ from a vertex in quadrangulations

A direct application of this new interpretation of $R_n$ concerns properties of large random quadrangulations viewed from their origin. For instance, the average $\langle e_n \rangle_A$ of the number of edges $n \to n+1$ in a quadrangulation with an origin and with say $A$ faces is given by

$$\frac{\langle e_n \rangle}{\langle e_0 \rangle} = \frac{R_{n,A} - R_{n-1,A}}{R_{0,A}} \tag{155}$$

with $R_{n,A}$ as in (135). Again, this is readily computed in the limit $A \to \infty$, where we first note $\langle e_0 \rangle \to 4$ by Euler's relation, and then use a saddle point method just like in (139), resulting in

$$\langle e_n \rangle = \frac{6}{35} \frac{(n^2 + 4n + 2)(5n^4 + 40n^3 + 117n^2 + 148n + 70)}{(n+1)(n+2)(n+3)} \tag{156}$$

This goes as $6n^3/7$ for large $n$, which confirms the value $d_H = 4$ for the fractal dimension, as $\langle e_n \rangle$ sin $d/dn\, n^{d_F} \sim n^{d_F - 1}$.

We may also obtain the average number of vertices at geodesic distance $n$ from the origin, by noting that the corresponding generating function is that of unrooted well-labeled trees with at least a label 0 and a marked vertex with label $n-1$. Abandoning the condition that a label 0 should occur, and decomposing the tree according to the environment of the marked vertex with label

$n-1$ results in the generating function

$$K_{n-1}=\sum_{k=1}^{\infty}\frac{g^k}{k}(R_n+R_{n-1}+R_{n-2})^k=-\text{Log}(1-g(R_n+R_{n-1}+R_{n-2}))=Log(R_{n-1}) \tag{157}$$

where we have incorporated the symmetry factor $1/k$ when the vertex has valence $k$. Finally, the generating function for quadrangulations with an origin and a marked vertex at distance $n$ is

$$V_n = K_{n-1} - K_{n-2} = \text{Log}\left(\frac{R_{n-1}}{R_{n-2}}\right) \tag{158}$$

for $n \geq 2$ and $\text{Log}R_0$ for $n = 1$, while of course $V_0 = 1$. Therefore the average number of vertices at distance $n$ from the origin in a quadrangulation of area $A$ is given by

$$\langle v_n \rangle_A = \text{Log}\left(\frac{R_{n-1,A}}{R_{n-2,A}}\right) \tag{159}$$

easily derived in the large $A$ limit:

$$\langle v_n \rangle = \frac{3}{35}\left((n+1)(5n^2+10n+2)+\delta_{n,1}\right) \tag{160}$$

This goes as $3n^3/7$ for large $n$, also in agreement with $d_F = 4$.

Note that eqs.(157)-(158) also allow to interpret $\text{Log}R_{n-1}$ as the generating function for quadrangulations with an origin and a marked vertex at distance $m \leq n$. In the limit $n \to \infty$, the function $\text{Log}R$ therefore generates the quadrangulations with two marked vertices. In the dual formulation, this corresponds to 4-valent planar graphs with two marked faces: this gives a purely combinatorial derivation in the 4-valent case of the formula (89) obtained above in the matrix model language.

## 7.3 Application II: local environment of a vertex in quadrangulations

Another application of this new graph interpretation of $R_n$ concerns the local environment of the origin. Assume we wish to keep track of the numbers of vertices at some finite distances $p+1$ from the origin, and edges labeled $q \to q+1$ for some specific $p$'sand $q$'s, both less or equal to some given $k$. Then a way to do it is to add extra weights, say $\rho_p$ per vertex labeled $p$ in the corresponding well-labeled tree and $\sigma_p$ per edge adjacent to a vertex labeled $p$ of the well-labeled tree. Indeed, as explained in the previous section, this amounts to adding a weight $\rho_p$ per vertex labeled $p+1$ in the quadrangulation, and a weight $\sigma_p$ per edge $p \to p+1$ in the quadrangulation. This turns the equation (153) into a new set of equations

$$R_n = \frac{\rho_n}{1 - g\sigma_n R_n(\sigma_{n+1}R_{n+1} + \sigma_n R_n + \sigma_{n-1}R_{n-1})}, \qquad n = 0, 1, 2, ..., k+1 \tag{161}$$

with $\rho_{k+1} = \sigma_{k+1} = 1$, while $R_n$ satisfies (153) for all $n \geq k+2$. This is slightly simplified by introducing $Z_n = \sigma_n R_n$ (with $\sigma_n = 1$ for $n \geq k+1$), as we are left with

$$\begin{aligned} Z_n &= \frac{\sigma_n \rho_n}{1 - g\sigma_n(Z_{n+1} + Z_n + Z_{n-1})}, \qquad n = 0, 1, 2, ..., k \\ Z_n &= \frac{1}{1 - g(Z_{n+1} + Z_n + Z_{n-1})}, \qquad n = k+1, k+2, ... \end{aligned} \tag{162}$$

Solving such a system seems quite difficult in general, but we may use the integral of motion (131) to replace the infinite set of equations on the second line of (162) (and the convergence condition of $Z_n$ to $R$), by simply the conserved quantity

$$f(Z_k, Z_{k+1}) = f(R, R) \tag{163}$$

Together with the first line of (162), this gives a system of $k+2$ algebraic relations for the functions $Z_0, Z_1, ..., Z_{k+1}$, which completely determines them order by order in $g$. As an example, let us compute in the case $k = 0$ the generating function including a weight $\rho_0 = \rho$ per vertex labeled 0 in the trees and $\sigma_0 = \sigma$ per edge incident to a vertex labeled 0 in the trees. (This in turn corresponds in the quadrangulations to a weight $\rho$ per vertex labeled 1, i.e. per nearest neighbor of the origin, and a weight $\sigma$ per edge $0 \to 1$.) We get the system:

$$Z_0 = \frac{\rho\sigma}{1 - g\sigma(Z_0 + Z_1)}, \qquad Z_0 Z_1(1 - g(Z_0 + Z_1) - Z_0 - Z_1 = f(R, R) = gR^3 - R \tag{164}$$

which upon eliminating $Z_1$ and reinstating $R_0 = Z_0/\sigma$, boils down to

$$(R_0 - \rho)(1 + R_0 - g\sigma^2 R_0^2 - \rho) - \sigma R_0(R_0 - \rho + gR(1 - gR^2)) + g\sigma^3 R_0^3 = 0 \tag{165}$$

for the generating function $R_0$ for rooted quadrangulations with weights $\rho$ per neighboring vertex of the origin and $\sigma$ per edge adjacent to the origin. $R_0 \equiv R_0(g|\rho, \sigma)$ is the unique solution to (165) such that $R_0 = \rho + O(g)$. Note that we recover $R_0 = R - gR^3$ of (134) when $\rho = \sigma = 1$. As the rooting of the quadrangulation is itself a choice of an edge adjacent to the origin, we may express the corresponding generating function for "unrooted " quadrangulations, namely with just an origin vertex, as

$$\Gamma_0(g|\rho, \sigma) = \int_0^\sigma \frac{ds}{s} R_0(g|\rho, s) \tag{166}$$

simply expressing the rooting of the quadrangulation as $\sigma\partial_\sigma\Gamma_0 = R_0$. The statistical average over quadrangulations of area $A$ of $\rho^{N_1}\sigma^{N_{01}}$ ($N_1$ the number of neighboring vertices of the origin, $N_{01}$ the number edges adjacent to the origin) finally reads

$$\langle\rho^{N_1}\sigma^{N_{01}}\rangle_A = \frac{\Gamma_{0,A}(\rho,\sigma)}{\Gamma_{0,A}(1,1)} = \frac{\int_0^\sigma \frac{ds}{s} R_{0,A}(\rho,s)}{\int_0^1 \frac{ds}{s} R_{0,A}(1,s)} \tag{167}$$

where as usual $\Gamma_{0,A}(\rho,\sigma)$ (resp. $R_{0,A}(\rho,s)$) denotes the coefficient of $g^A$ in $\Gamma_0(g|\rho,\sigma)$ (resp. $R_0(g|\rho,s)$). The limit $\lim_{A\to\infty}\langle\rho^{N_1}\sigma^{N_{01}}\rangle_A = \Gamma$ may again be extracted by a saddle-point expansion. After some algebra, we find

$$6\Gamma(\Gamma+1)(\Gamma+3) - \sigma\big(2\Gamma(1+4\Gamma+\Gamma^2) + 3\rho(\Gamma+1)^2(\Gamma+2)\big) = 0 \tag{168}$$

and $\Gamma$ is uniquely determined by the condition $\Gamma = 1$ for $\sigma = \rho = 1$. For instance, when $\sigma = 1$, we get

$$\Gamma(\rho,1) = \frac{2}{\sqrt{4-3\rho}} - 1 = \sum_{n\geq 1} \rho^n \left(\frac{3}{16}\right)^n \binom{2n}{n} \tag{169}$$

in which we read the probability $P(n) = (3/16)^n\binom{2n}{n}$ for a vertex to have $n$ neighboring vertices in an infinite quadrangulation. Similarly, taking $\rho = 1$, we get

$$\Gamma(1,\sigma) = \frac{1}{2}\left(\sqrt{\frac{6+3\sigma}{6-5\sigma}} - 1\right) \tag{170}$$

which generates the probabilities to have $n$ edges adjacent to a vertex in an infinite quadrangulation. We may also derive the generating function for the conditional probabilities of having $n$ nearest neighboring vertices, given that there is no multiple edge connecting them to the origin, by simply taking $\Gamma(\rho = t/\sigma, \sigma)$ and letting $\sigma \to 0$, which indeed suppresses all contributions from multiply connected vertices. This gives

$$\Pi(t) = \lim_{\sigma\to 0} \Gamma\left(\frac{t}{\sigma}, \sigma\right) = \sqrt{\frac{8-t}{2-t}} - 2 \tag{171}$$

For instance, the probability that a given vertex have no multiple neighbors in an infinite quadrangulation is

$$\Pi(1) = \sqrt{7} - 2 \tag{172}$$

## 7.4 Spatial branching processes

We have seen so far how the information on the geodesic distance from the origin in a rooted planar quadrangulation may be coded by rooted well-labeled trees. The latter give rise to natural examples of so-called spatially branching processes, in the context of which quantities like $R_n$ correspond to certain probabilities.

A spatial branching process consists of two data. First we have a mono-parental population, whose genealogy is described by a rooted tree, the root corresponding to the common ancestor. A standard measure on these trees attaches the probability $(1-p)p^k$ for any vertex to have $k$ descendents. The second data is a labeling of the vertices of the tree by positions say on the integer line $n \in \mathbb{Z}$. Here, we add the rule of the "possessive ancestor" that his children must be at close enough positions from his (namely differing by 0 or $\pm 1$). Let $E(T)$ denote the probability of extinction of the population at generation $T$, then we have the recursion relation

$$E_n(T) = \frac{1-p}{1-\frac{p}{3}(E_{n+1}(T-1)+E_n(T-1)+E_{n-1}(T-1))} \tag{173}$$

Letting $T \to \infty$, we see that the extinction probability $E_n = \lim_{T\to\infty} E_n(T)$ obeys the same equation as $R_n$ (153) upon some rescaling, and we find that $E_n = (1-p)R_n\left(g = \frac{p(1-p)}{3}\right)$, in the case of positions restricted to lie in a half-line (with a "wall" at the origin). Without this restriction, the problem becomes translationally invariant and $E_n = E = (1-p)R\left(g = \frac{p(1-p)}{3}\right)$. Note that the critical point $g = g_c = 1/12$ corresponds here to the critical probability $p = p_c = 1/2$.

In this new setting, we may ask different questions, such as what is the probability for the process to escape from a given interval, say $[0, L]$. Once translated back into $R_n$ terms, this amounts to still imposing the recursion relation (153), but changing boundary conditions into

$$R_{-1} = 0 \qquad \text{and} \qquad R_{L+1} = 0 \tag{174}$$

The escape probability from the interval reads then

$$S_n = 1 - (1-p)R_n\left(g = \frac{p(1-p)}{3}\right) = (1-p)(R - R_n) \tag{175}$$

The equation (153) with the boundary conditions (174) still admits an exact solution expressed by means of the Jacobi $\theta_1$ function

$$\theta_1(z) = 2i\sin(\pi z)\prod_{j\geq 1}(1-2q^j\cos(2\pi z)+q^{2j}) \tag{176}$$

The solution reads $R_n = R_n^{(L)}$, with

$$R_n^{(L)} = R\frac{u_n u_{n+3}}{u_{n+1}u_{n+2}}$$
$$u_n = \theta_1\left(\frac{n+1}{L+5}\right) \tag{177}$$

guaranteeing that the boundary conditions (174) are satisfied, and where the nome $q$ still has to be fixed. The main recursion relation (153) reduces to a quartic equation for the $u_n$'s:

$$u_n u_{n+1} u_{n+2} u_{n+3} = \frac{1}{R} u_{n+1}^2 u_{n+2}^2 + gR(u_{n-1}u_{n+2}^2 u_{n+3} + u_n^2 u_{n+3}^2 + u_n u_{n+1}^2 u_{n+4}) \tag{178}$$

and the latter is satisfied by (177) provided we take

$$R = 4\frac{\theta_1(\alpha)\theta_1(2\alpha)}{\theta_1'(0)\theta_1(3\alpha)}\left(\frac{\theta_1'(\alpha)}{\theta_1(\alpha)} - \frac{1}{2}\frac{\theta_1'(2\alpha)}{\theta_1(2\alpha)}\right)$$
$$g = \frac{\theta_1'(0)^2\theta_1(3\alpha)}{16\theta_1(\alpha)^2\theta_1(2\alpha)\left(\frac{\theta_1'(\alpha)}{\theta_1(\alpha)} - \frac{1}{2}\frac{\theta_1'(2\alpha)}{\theta_1(2\alpha)}\right)^2} \tag{179}$$

for $\alpha = 1/(L+5)$. The identity (178) is proved typically by showing that both sides have the same transformations under $n \to n + L + 5$ and $n \to n + (L+5)/(2i\pi)\mathrm{Log}q$, and that moreover they have the same zeros, this latter condition amounting to (179).

The elliptic solution $R_n$ may be interpreted terms of *bounded* graphs as follows. The quantity $G_n^{(L)} = R_n^{(L)} - R_{n-1}^{(L-1)}$ is the generating function for quadrangulations with an origin and a marked edge $n \to n+1$, which are moreover bounded in the sense that all vertices are distant by at most $L+1$ from the origin.

Taking again the continuum scaling limit of the model leads to the probabilists' Integrated SuperBrownian Excursions (ISE), here in one dimension [29]. The scaling function $\mathcal{U}$ obtained from $R_n = R_c(1-\epsilon^2\mathcal{U})$ in the limit (141), while moreover $r = n\epsilon$ and $\lambda = (L+5)\epsilon$ are kept fixed, reads:

$$\mathcal{U}(r) = 2\wp(z|\omega,\omega') \tag{180}$$

where $\wp$ is the Weierstrass function ($\wp = -\partial_r^2 \mathrm{Log}\,\theta_1$), with half-periods $\omega = \lambda/2$ and $\omega'$, related via the condition that the second invariant $g_2(\omega,\omega') = 3$.

## 7.5 Generalizations

We have so far only discussed quadrangulations and their relations to spatial branching processes (see also [30] [31]). All of the above generalizes to rooted planar graphs with arbitrary even face valences. These are in bijection with rooted well-labeled trees with more involved labeling rules, also called well-labeled mobiles [28]. This allows for a generalization of spatial branching processes, possessing these labeling rules. As we already know that these objects have an interesting variety of multicritical behaviors, this should turn into multicritical generalizations of the ISE.

In [28], the general case covered by two-matrix models is treated as well, and seen to generate Eulerian (i.e. vertex-bicolored) planar graphs. The latter contain as a particular case the gravitational Ising model, and in principle allow for reaching any $c(p,q)$ CFT coupled to 2D quantum gravity. These will lead presumably to interesting generalizations of the ISE.

## 8. Conclusion

In these lectures we have tried to cover various aspects of discrete 2D quantum gravity, namely of statistical matter models defined on random graphs of given topology.

The matrix model approach, when solvable, gives exact recursion relations between quantities eventually leading to compact expressions for the genus expansion of the free energy of the models. We have further investigated the so-called double scaling limit in which both matter and space degrees of freedom become critical, allowing for instance to define and compute a scaling function summarizing the leading singularities of the free energy at all genera, as a function of the renormalized cosmological constant $x$. The final general result takes the form

$$[P,Q]=1, \qquad P=d^p+v_2d^{p-2}+\ldots+v_p, \quad Q=d^q+u_2d^{q-2}+\ldots+u_q \tag{181}$$

with $d=d/dx$, all $v$'s and $u$'s functions of $x$, and $u_2$ proportional to $F''$, the second derivative of the singular part of the all-genus free energy w.r.t $x$.

The combinatorial approach, when bijections with trees are available, also gives exact recursion relations between basic generating functions which can be interpreted in terms of planar graph counting, while keeping track of the geodesic distance between marked points. The expressions for the solutions are completely explicit, allowing for taking a scaling limit, describing the free energy for random surfaces with marked points at a renormalized geodesic distance $r$. We may write the general result for this scaling free energy in the form

$$[P,Q]=0, \qquad P=d^p+v_2d^{p-2}+\ldots+v_p, \quad Q=d^q+u_2d^{q-2}+\ldots+u_q \tag{182}$$

with $d=d/dr$, all $u$'s and $v$'s functions of $r$, and $u_2$ proportional to the scaling two-point function for surfaces with two marked points at geodesic distance $\geq r$.

Remarkably, in all cases solved so far, the exactly solvable geodesic distance problems for planar graphs all correspond to cases where a matrix model solvable by orthogonal polynomials is available. It seems therefore that the bijections with trees exactly parallel the orthogonal polynomial solutions. More precisely, we have observed that a similar abstract "Q-operator" could be introduced in both cases, one of them describing the possible subtrees one can encounter when going counterclockwise around a vertex of a blossom tree, the other describing the multiplication by an eigenvalue $\lambda$ on the basis of orthogonal polynomials.

The two apparently unrelated results (181) and (182) show that something deeper happens here, that deserves to be better understood. One may imagine that there must exist a more general structure which would unify and combine the notions of genus and geodesic distance, and give for instance closed equations for scaling functions of both $x$ and $r$. To reach this, one should first be able to control geodesic distances in higher genus as well, by generalizing the tree bijection techniques explained here only in the planar case. Another possibility could be that matrix models as we know them today may still be only part of a more general setting. Some generalizations of matrix models involving integration of eigenvalues over contours (or linear combinations thereof) in the complex plane may be the correct answer, and relate to the intrinsic geometry of graphs once interpreted combinatorially.

Finally, it is interesting to notice that no continuum field theoretical representation of geodesic distance dependence of random surfaces has been found yet, although 2D quantum gravity is now well understood in terms of the coupling of CFT to the Liouville field theory [2]. The simplicity of the results found here for the various scaling functions comes as a surprise in that respect. Field theory probably still has some way to go before explaining the purely combinatorial results shown here.

## Notes

1. This is just a technical trick to make the precise contact with the generating function $R_n$ of Sects.4 and 6. The reader will have to remember to add up one to each vertex label of the tree to recover its geodesic distance from the origin in the quadrangulation.

## References


[1] P. Di Francesco, P. Mathieu and D. Sénéchal, Conformal Field Theory, Graduate Texts in Contemporary Physics, springer (1996).

[2] V.G. Knizhnik, A.M. Polyakov and A.B. Zamolodchikov, *Mod. Phys. Lett.* A3:819 (1988). J. Distler and H. Kawai, *Nucl. Phys.* B321:509 (1989) F. David, *Mod. Phys. Lett.* A3:1651 (1988)

[3] W. Tutte, A Census of Planar Maps, *Canad. Jour. of Math.* 15:249-271 (1963) A Census of planar triangulations, *Canad. Jour. of Math.* 14:21-38 (1962) A Census of Hamiltonian polygons, *Canad. Jour. of Math.* 14:402417 (1962)

[4] G. 't Hooft, *Nucl. Phys.* B72:461 (1974).

[5] E. Brézin, C. Itzykson, G. Parisi and J.-B. Zuber, Planar Diagrams, *Comm. Math. Phys.* 59:35-51 (1978).

[6] V. Kazakov, *Phys. Lett.* B150:282 (1985); F. David, *Nucl. Phys.* B257:45 (1985); J. Ambjorn, B. Durhuus and J. Frohlich, *Nucl. Phys.* B257:433 (1985); V. Kazakov, I. Kostov and A. Migdal, *Phys. Lett.* 157B:295 (1985).

[7] E. Brézin and V. Kazakov, Exactly solvable field theories of closed strings, *Phys. Lett.* B236:144-150 (1990); M. Douglas and S. Shenker, Strings in less than 1 dimension, *Nucl. Phys.* B335:635 (1990); D. Gross and A. Migdal, Non-perturbative two-dimensional gravity, newblock*Phys. Rev. Lett.* 64 (1990) 127-130.

[8] see for instance the review by P. Di Francesco, P. Ginsparg and J. Zinn-Justin, 2D Gravity and Random Matrices, *Physics Reports* 254:1-131 (1995).

[9] E. Witten, On the topological phase of two-dimensional gravity, *Nucl. Phys.* B340:281-332 (1990); Two dimensional gravity and intersection theory on moduli space, *Surv. in Diff. Geom.* 1:243-310 (1991); M. Kontsevich, Intersection theory on the moduli space of curves and the matrix Airy function, *Comm. Math. Phys.* 147:1-23 (1992).

[10] E. Bender and E. Canfield, The number of degree-restricted rooted maps on the sphere, *SIAM J. Discrete Math.* 7:9-15 (1994).

[11] G. Schaeffer, Conjugaison d'arbreset cartes combinatoires aléatoires PhD Thesis, Université Bordeaux I (1998) Bijective census and random generation of Eulerian planar maps, *Electronic Journal of Combinatorics,* 4:R20 (1997).

[12] J. Bouttier, P. Di Francesco and E. Guitter, Census of planar maps: from the one-matrix model solution to a combinatorial proof, *Nucl. Phys.* B645:477-499 [PM] (2002), arXiv:cond-mat/0207682.

[13] R. Cori, Un code pour les graphes planaires et ses applications, *Société Mathématique de France, Paris 1975;* D. Arquès, Les hypercartes planaires sont des arbres très bien étiquetés, *Discr. Math.* 58:11-24 (1986); M. Marcus and G. Schaeffer, Une bijection simple pour les cartes orientables, http://www.loria.fr/~schaeffe/pub/orientable/directe.ps; P. Chassaing and G. Schaeffer, Random Planar Lattices and Integrated SuperBrownian Excursion, to appear in Probability Theory and Related Fields, arXiv:math.CO/0205226.

[14] J. Bouttier, P. Di Francesco and E. Guitter, Geodesic distance in planar graphs, *Nucl. Phys. B 663[FS]:535-567 (2003), arXiv:cond-mat/0303272.*

[15] M. Staudacher, The Yang-Lee Edge Singularity on a Dynamical Planar Random Surface, *Nucl. Phys.* B336:349-362 (1990).

[16] I. Kostov, *Mod. Phys. Lett.* A4:217 (1989); M. Gaudin and I. Kostov, *Phys. Lett.* B220:200 (1989); I. Kostov and M. Staudacher, *Nucl.Phys.* B384:459 (1992); B. Eynard and J. Zinn-Justin, *Nucl. Phys.* B386:558 (1992); B. Eynard and C. Kristjansen, *Nucl. Phys.* B455:577 (1995) and *Nucl. Phys.* B466:463-487 (1996).

[17] D. Boulatov and V. Kazakov, The Ising model on a random planar lattice: the structure of the phase transition and the exact critical exponents, *Phys. Lett.* B186:379-384 (1987); V. Kazakov, *Nucl. Phys.* B4:93 (Proc. Suppl.) (1998); J.-M. Daul, Q-states Potts model on a random planar lattice, *arXiv:hep-th/9502014;* B. Eynard and G. Bonnet, The Potts-q random matrix model:loop equations, critical exponents, and rational case, *Phys. Lett.* B 463:273-279 (1999), arXiv:cond-mat/9906130; P. Zinn-Justin, The dilute Potts model on random surfaces, *J. Stat. Phys.* 98:245-264 (2001), arXiv:cond-mat/9903385.

[18] I. Kostov, Strings with discrete target space, *Nucl. Phys.* B376:539-598 (1992); Gauge invariant matrix models for A-D-E closed strings, *Phys. Lett.* B297:74-81 (1992).

[19] P. Di Francesco, Rectangular Matrix Models and Combinatorics of Colored Graphs, *Nucl. Phys.* B648:461-496 (2002), arXiv:cond-mat/0207682.

[20] D. Bessis, Comm. Math. Phys. 69:147-163 (1979); D. Bessis, C.Itzykson and J.-B. Zuber, Adv. in Appl. Math. 1:109 (1980).

[21] P. Di Francesco and D. Kutasov, Unitary minimal models coupled to gravity, *Nucl. Phys.* B342:589 (1990) and Integrable models of 2D quantum gravity NATO ASI Series B315:73 (1993).

[22] V. Drinfeld and V. Sokolov, *J. Sov. Math.* 30:1975 (1985); *and Sov. Math. Dokl.* 23:457 No.3 (1981).

[23] M. Bousquet-Mélou and G. Schaeffer, The degree distribution in bipartite planar maps: application to the Ising model, *arXiv:math.CO/0211070.*

[24] J. Bouttier, P. Di Francesco and E. Guitter, Combinatorics of hard particles on planar maps, *Nucl. Phys.* B655:313:341 (2002), arXiv:cond-mat/0211168.

[25] J. Ambjfirn and Y. Watabiki, Scaling in quantum gravity, *Nucl.Phys.* B445:129-144 (1995).

[26] J. Bouttier, P. Di Francesco and E. Guitter, Statistics of planar maps viewed from a vertex: a study via labeled trees, *Nucl. Phys.* B675:631-660[FS] (2003), arXiv:cond-mat/0307606.

[27] J. Bouttier, P. Di Francesco and E. Guitter, Random trees between two walls: Exact partition function, *J. Phys. A: Math. Gen.* 36:12349-12366 (2003), arXiv:cond-mat/0306602.

[28] J. Bouttier, P. Di Francesco and E. Guitter, Planar maps as labeled mobiles, *arXiv:math.CO/0405099 (2004)*, submitted to Elec. Jour. of Combinatorics.

[29] D.J. Aldous, Tree-based models for random distribution of mass, *J. Stat. Phys.* 73:625-641 (1993); E. Derbez and G. Slade, The scaling limit of lattice trees in high dimensions, *Comm. Math. Phys.* 193:69-104 (1998); J.-F. Le Gall, Spatial branching processes, random snakes and partial differential equations, Lectures in Mathematics ETH Zurich, Birkhauser, Basel, 1999.

[30] P. Chassaing and B. Durhuus, Statistical Hausdorff dimension of labelled trees and quadrangulations, *arXiv:math.PR/0311532.*

[31] J. F. Marckert and A. Mokkadem, Limit of normalized quadrangulations: the Brownian map, arXiv:math.PR/0403398.

# EIGENVALUE DYNAMICS, FOLLYTONS AND LARGE $N$ LIMITS OF MATRICES*

Joakim Arnlind
*Royal Institute of Technology*
*Department of Mathematics*
*100 44 Stockholm*
*Sweden*
jarnlind@math.kth.se

Jens Hoppe
*Royal Institute of Technology*
*Department of Mathematics*
*100 44 Stockholm*
*Sweden*
hoppe@math.kth.se

How do the eigenvalues of a "free" hermitian $N \times N$ matrix $X(t)$ evolve in time? The answer is provided by the rational Calogero-Moser systems [5, 13] if (!) the initial conditions are chosen such that $i[X(0), \dot{X}(0)]$ has a non-zero eigenvalue of multiplicity $N-1$; for generic $X(0), \dot{X}(0)$ the question remained unanswered for 30 years.

While it is easy to see that $H = \frac{1}{2} \operatorname{Tr} P^2$ (let's restrict to the real-symmetric case, i.e. $X(t) = R(t)Q(t)R^{-1}(t)$, with $R$ real orthogonal, $Q$ diagonal, $P = \dot{X} = R(\dot{Q}+[R^{-1}\dot{R}, Q])R^{-1}$, $F := \Big[[R^{-1}\dot{R}, Q], Q\Big]$) yields the known Euler-Calogero-Moser system(s),

$$H = \frac{1}{2}\sum_{i=1}^{N} p_i^2 + \frac{1}{2}\sum_{i \neq j} \frac{f_{ij}^2}{(q_i - q_j)^2} \tag{1}$$

*Talk given by the second author.



*E. Brezin et al. (eds.), Applications of Random Matrices in Physics*, 89–94.

with equations of motion [9]

$$\dot{q}_i = p_i, \quad \dot{p}_i = -2\sum{}' \frac{f_{ij}^2}{(q_i - q_j)^3}, \tag{2}$$

$$\dot{f}_{ij} = -\sum{}'' f_{ik} f_{kj} \left[ \frac{1}{(q_i - q_k)^2} - \frac{1}{(q_k - q_j)^2} \right], \tag{3}$$

$\Big($ a correct derivation of the so$(N)$ Poissonbrackets $\{f_{ij}, f_{kl}\} = -\frac{1}{2}(\delta_{jk} f_{il} \pm$ 3 more$)$ being slightly less trivial: with $dA := R^{-1}dR$ the canonical symplectic form for real symmetric matrices can be shown to be(come) $\mathrm{Tr}\,\big( - dQ \wedge dP - dF \wedge dA + F dA \wedge dA \big)$, i.e.

$$\begin{aligned} &-dq_i \wedge dp_i + 2 \sum_{i<j} df_{ij} \wedge da_{ij} \\ &\quad - 2 \sum_{i<j<k} \Big( f_{ij} da_{jk} \wedge da_{ik} + f_{ik} da_{ij} \wedge da_{jk} + f_{jk} da_{ik} \wedge da_{ij} \Big); \end{aligned} \tag{4}$$

as

$$\left( \begin{array}{cc|cc} 0 & -\mathbb{1} & \multicolumn{2}{c}{0} \\ \mathbb{1} & 0 & & \\ \hline \multicolumn{2}{c|}{0} & 0 & 2\cdot\mathbb{1} \\ & & -2\cdot\mathbb{1} & \mathcal{A} \end{array} \right)^{-1} = \left( \begin{array}{cc|cc} 0 & \mathbb{1} & \multicolumn{2}{c}{0} \\ -\mathbb{1} & 0 & & \\ \hline \multicolumn{2}{c|}{0} & \frac{1}{4}\mathcal{A} & -\frac{1}{2}\cdot\mathbb{1} \\ & & +\frac{1}{2}\cdot\mathbb{1} & 0 \end{array} \right) \tag{5}$$

where rows and columns refer to the ordering $(dq_i, dp_i, df_{i<j}, da_{i<j})$, (4) implies

$$\{f_{i<j}, f_{j<k}\} = \frac{1}{4}\mathcal{A}_{i<j,j<k} = -\frac{1}{2} f_{ik} \tag{6}$$

as well as

$$\{f_{i<j}, a_{k<l}\} = -\frac{1}{2}\delta_{ik}\delta_{jl},$$

resp.

$$\text{“}\{r_{ij}, f_{kl}\} = -\frac{1}{2}(\delta_{jk} r_{il} - \delta_{jl} r_{ik})\text{”} \qquad \Big),$$

the originally asked question may be rephrased as:
Is it possible to eliminate the $f_{ij}$ in (2), i.e. express $f_{ij}^2$ in terms of the eigenvalues $q_i(t)$ (and, possibly, their time derivatives)? The key observation in this respect is the existence of conserved quantities

$$\begin{aligned} J_{\alpha\beta} &:= \sum_{k=0}^{\alpha+\beta} (-)^k \mathscr{L}_{\alpha+\beta-k} \mathscr{R}_k \\ \mathscr{R}_{k-1} &:= \dot{\mathscr{R}}_k, \quad \mathscr{L}_{k-1} := \dot{\mathscr{L}}_k, \end{aligned} \tag{7}$$

involving, apart from $\mathscr{R}_{\alpha+\beta} := \operatorname{Tr} Q^{\alpha+\beta} = \operatorname{Tr} X^{\alpha+\beta}$ certain quantities (and their time derivatives)

$$E_{\alpha\beta} := \operatorname{Tr} Q^\alpha F Q^\beta F =: \mathscr{L}_{\alpha+\beta} = -\sum q_i^\alpha q_j^\beta f_{ij}^2, \tag{8}$$

which contain the $f_{ij}^2$ linearly and (for $0 \leq \alpha < \beta \leq N-2$) are independent (for non-coinciding eigenvalues) – thus allowing to express the $f_{ij}^2$ as functions of $s_k = \operatorname{Tr} Q^k$, and (many) conserved quantities.

Interestingly, this general construction [1] yields, if $\dot{X}(0)$ is chosen to be of rank 1 (cp [6]), a first-order formulation of the rational Ruijsenaars-Schneider system.

## Follytons [12]

Through Schrodinger operators the following 4 topics are well known to be related:

A. Spectral theory of linear differential operators

B. Darboux-transformations

C. Integrable non-linear PDE's

D. Lieb-Thirring inequalities

What can one say about higher-order operators in this respect? While the connection between A and C is fairly standard, and structurally "independent" of the degree $l$ (see e.g. [7] or [10]), namely (let's choose $l = 4$):
to every self-adjoint (quartic) linear differential operator on the line,

$$L = \partial^4 + \partial u \partial + v \tag{9}$$

there exists (uniquely) $L^{1/4} = \partial + \sum_{i=1}^{\infty} l_i(x)\partial^{-i}$, such that

$$\dot{L} = [L, M], \tag{10}$$

for every $M^{(m)} := \left(\left(L^{1/4}\right)^m\right)_+$ (and therefore, in particular, for $M :=$ the differential operator part of $L^{3/4}$) consistently defines *partial* differential equations, for $m = 3$ reading ($u = u(x,t)$, $v = v(x,t)$)

$$\begin{aligned} \dot{u} &= 10u''' + 6uu' - 24v' \\ \dot{v} &= 3\left(u''' + uu''\right)' - 8v''' - 6uv', \end{aligned} \tag{11}$$

very little is known about spectral properties of (9). Assuming $u$ and $v$ to go to zero at $\infty$, solutions of $L\psi = E\psi$, with $E = -4\kappa^4 < 0$, go like

$e^{\pm\kappa x}\begin{cases}\cos\kappa x\\ \sin\kappa x\end{cases}$, as $x\rightarrow\infty$ (and therefore have infinitely many zeroes!), but apparently "nothing" is known about degeneracies (not even for the ground state). A crucial assumption later will be that eigenvalues of $L$ should be doubly degenerate; like the lowest eigenvalue ($E_0=-4$) of

$$L=\partial^4+6\partial\frac{1}{\cosh^2 x}\partial-\frac{4}{\cosh^2 x}, \tag{12}$$
$$\psi_\pm=\frac{e^{\pm ix}}{\cosh x}.$$

In order to employ Darboux-transformations (i.e., loosely speaking, relating operators $BA$ to – isospectral, apart from $\ker A$ and $\ker B$ possibly differing – operators $AB$) to derive inequalities for weighted sums of eigenvalues of $L$ it is crucial to be able to *factorize* (9). The authors of [2] tried to do so by writing $L-E_0$ as $(\partial^3+\cdots)(\partial+h)$. The problem with that factorization is that $(L-E_0)\psi=0$ forces $h=-\frac{\psi'}{\psi}$ to be singular (due to the zeroes of $\psi$). The Ansatz

$$L-E_0=A^\dagger A=\big(-\partial^2-f\partial+g-f'\big)\big(-\partial^2+f\partial+g\big) \tag{13}$$

overcomes that problem (s.b.) but leads to the formidable ODE system

$$\begin{aligned}v-E_0&=(g-f')g-fg'-g''\\ u&=-2g-f^2-f'\end{aligned} \tag{14}$$

to be solved for $f$ and $g$. Even for the simplest possible choice for $u$ and $v$ (namely, $u\equiv 0\equiv v$) this leads to a rather non-trivial ODE,

$$2f'''+6ff''+7f'^2+8f'f^2+f^4=1; \tag{15}$$

$\kappa$ is now scaled to $1/2$. on top of this, in order not to run into the same factorization singularities as in the 3+1 split, one has to prove that the (non-constant, if $f'\neq 0$) Wronskian formed out of 2 groundstate wavefunctions,

$$W(x):=\psi_+\psi_-'-\psi_+'\psi_- \tag{16}$$

*does not* vanish; as $\big(-\partial^2+f\partial+g\big)\psi_\pm=0$ implies

$$f=\frac{W'}{W},\quad g=-\frac{1}{W}(\psi_+'\psi_-''-\psi_+''\psi_-'). \tag{17}$$

Both "problems" *can* be (nicely) resolved:
W *is* non-vanishing, and (15), resp.

$$2\big(W''''W-W'''W'\big)+\big(W''^2-W^2\big)=0 \tag{18}$$

has an explicit 4-parameter family of exact solutions; $\hat{W} = a + be^x + ce^{-x} + d\cos x + e\sin x$ solves (18), provided $4bc + d^2 + e^2 = \frac{1}{2}a^2$. $f = -\frac{\hat{W}'}{\hat{W}}$ and $g = -\frac{1}{2}(f^2 + f')$ then leads to a 4-parameter class of potentials $u$ and $v$ (via $L = \partial^4 + \partial u \partial + v$) making $u(x + 4t)$, $v(x + 4t)$ exact solutions of (11), "Follytons".

## gl($N \to \infty$)

For each 2-dimensional surface $\Sigma_2$ there exists a basis $\{Y_\alpha\}_{\alpha=0}^{\infty}$ of the Poisson-algebra of (real) functions (on $\Sigma_2$), and a basis $\{T_a^{(N)}\}_{a=0}^{N^2-1}$ of $\mathrm{u}(N)$ such that (for all $a, b, c$)

$$\lim_{N\to\infty} \mathrm{Tr}\left(\left[T_a^{(N)}, T_b^{(N)}\right] T_c^{(N)}\right) = \int_{\Sigma_2} \{Y_a, Y_b\} Y_c. \tag{19}$$

This correspondence was originally discovered for $S^2$ [11], 10 years later for $T^2$ [8], conjectured to hold for higher genus surfaces [3], and then proven for (almost general, compact) Kahler manifolds [4].

## Acknowledgements

One of us (JH) would like to thank M. Bordemann, A. Laptev and J. Ostensson for collaboration(s) on related subjects, as well as Didina Serban and all organizers of the Les Houches School for the kind invitation.

## References

[1] Arnlind J. and Hoppe J. Eigenvalue-Dynamics off the Calogero-Moser System. *Lett. Math. Phys.* 68:121-129 (2004).
Eigenvalue dynamics off Calogero-Moser. *IHES preprint P/03/41 (July 2003).*

[2] Beals R., Deift P. and Tomei C. Direct and inverse scattering on the line. *Mathematical surveys and monographs (AMS)* 28:1–209 (1988).

[3] Bordemann M., Hoppe J., Schaller P. and Schlichenmaier M. gl($\infty$) and geometric quantization. *Comm. Math. Phys.* 138:209–244 (1991)

[4] Bordemann M., Meinrenken E. and Schlichenmaier M. Toeplitz quantization of Kahler manifolds and gl($N$), $N \to \infty$ limits. *Comm. Math. Phys.* 165:281–296 (1994).

[5] Calogero F. Exactly solvable one-dimensional many-body problems. *Lett. Nouvo Cim.*, 13:411–416 (1975).

[6] Calogero F. A technique to identify solvable dynamical systems, and a solvable generalization of the goldfish many-body problem. *J. Math. Phys* 45 (2004).

[7] Drinfeld V., Sokolov V. *Journal of Sov. Math.* 30 (1985).

[8] Fairlie D., Fletcher P., Zachos C. *Phys. Lett. B* 218:203 (1989).
Hoppe, J. *Phys. Lett. B* 215:706 (1988)

[9] Gibbons J. and Hermsen T. A generalization of the Calogero-Moser system. *Physica D*, 11:337–348 (1984).
Wojciechowski S. An integrable marriage of the Euler equations with the Calogero-Moser system. *Physics Letters*, 111A no 3:101–103 (1985).

[10] Hoppe J. Lectures on Integrable systems m10. Springer Verlag (1992).

[11] Hoppe J. Quantum Theory of Massless Relativistic Surface. *MIT Ph.D. Thesis* (1982). Available at http://www.aei.mpg.de/~hoppe.

[12] Hoppe J., Laptev A. and Ostensson J. Follytons and the Removal of Eigenvalues for Fourth Order Differential Operators. *math-ph/0311011* (2003).

[13] Moser J. Three integrable hamiltonian systems connected with isospectral deformations. *Adv. Math.*, 16:197–220 (1975).

# RANDOM MATRICES AND SUPERSYMMETRY IN DISORDERED SYSTEMS

K.B. Efetov
*Theoretische Physik III, Ruhr-Universität Bochum, 44780 Bochum, Germany, and L.D. Landau Institute for Theoretical Physics, 117940 Moscow, Russia*
efetov@tp3.rub.de

**Abstract** It is described how one comes to the Wigner-Dyson random matrix theory (RMT) starting from a model of a disordered metal. The lectures start with a historical introduction where basic ideas of the RMT and theory of disordered metals are reviewed. This part is followed by an introduction into supermathematics (mathematics operating with both commuting and anticommuting variables). The main ideas of the supersymmetry method are given and basic formulae are derived. Both level-level correlations and fluctuations of amplitudes of wave functions are discussed. It is shown how one can both obtain known formulae of the RMT and go beyond. In the last part some recent progress in the further development of the method and possible perspectives are discussed.

**Keywords:** Random matrices, disordered systems, supersymmetry, non-linear sigma-model.

## Introduction

### 0.1 Wigner-Dyson Theory

According to basic principles of quantum mechanics the energy spectrum of a particle in a limited volume is discrete. The precise values of the energy depend on the boundary conditions and the interactions in system. In many cases these quantities can be calculated with a certain accuracy. However, often the interactions are so complicated that calculations for the single levels become impossible. On the other hand, the complexity of the interactions can lead to the idea of a statistical description in which information about separate levels is neglected and only averaged quantities are studied. Density of states, energy level and wave functions correlations are quantities that can be studied in the statistical approach. Sometimes, it is sufficient to study the average of, e.g., the density of states or its variance. In other cases one may be interested in a full statistical description that can be achieved calculating distribution func-



*E. Brezin et al. (eds.), Applications of Random Matrices in Physics,* 95–137.

tions. Studying the level statistics is, to some extent, analogous to the statistical study of the motion of atoms and molecules, which is the subject of statistical physics.

The idea of the statistical description of the energy levels was first proposed by Wigner [1] for study of highly excited nuclear levels in complex nuclei. In such nuclei a large number of particles interact in an unknown way and the main assumption was that the interactions were equally probable. Of course, in order to specify the meaning of the words "equally probable" one had to formulate a statistical hypothesis in terms of a probability distribution that would play the role of the Gibbs distribution.

This was done in Ref. [1] in the following way. Choosing a complete set of eigenfunctions as a basis, one represents the Hamiltonian $H$ as a matrix with matrix elements $H_{mn}$. The matrix elements $H_{mn}$ are assumed random with a certain probability distribution. It is clear that the distribution should not depend of the basis chosen, which implies an invariant form of the distribution function. In the language of the random matrices $H_{mn}$ the corresponding distribution function can contain only $Trf\left(H\right)$, where $f$ is a function.

The first statistical theory [1] was based on a Gaussian distribution. According to the Gaussian statistical hypothesis a physical system having $N$ quantum states has the statistical weight $D\left(H\right)$

$$D\left(H\right) = A\exp\left[-\sum_{m,n=1}^{N}\frac{\left|H_{mn}\right|^{2}}{2a^{2}}\right] = A\exp\left[\frac{-TrH^{2}}{2a^{2}}\right] \tag{1}$$

In Eq. (1) the parameter $a$ is a cutoff excluding strong interaction, $H$ is a random $N \times N$ matrix, and $A$ is a normalization coefficient.

It is important to emphasize that the weight $D\left(H\right)$ is rather arbitrary and other forms for the distribution functions can be suggested. Of course, the dependence of the mean energy level spacing $\Delta\left(\varepsilon\right)$

$$\Delta^{-1}\left(\varepsilon\right) = \left\langle tr\delta\left(\varepsilon - H\right)\right\rangle_{D} \tag{2}$$

on the energy $\varepsilon$ is different for different distributions $D\left(H\right)$. (In Eq. (2) the symbol $\langle ...\rangle_D$ stands for the averaging with the distribution $D\left(H\right)$). For example, the distribution function $D\left(H\right)$ for the Gaussian distribution, Eq. (1), has the form of a semicircle (Wigner semicircle law) (for a review, see, e.g. Ref.[2]).

What is more interesting, level correlations described, for example, by the level-level correlation function $R\left(\omega\right)$

$$R\left(\omega\right) = \Delta^{2}\left(\varepsilon\right)\left\langle Tr\delta\left(\varepsilon - H\right)Tr\delta\left(\varepsilon - \omega - H\right)\right\rangle \tag{3}$$

prove to be universal in the limit $N \to \infty$ provided the energy $\omega$ is much smaller than the characteristic scale of the variation of $\Delta\left(\varepsilon\right)$. For the Wigner semicircle law, the latter condition means that the energy $\varepsilon$ is not close to the points $\varepsilon_0$, $-\varepsilon_0$, where the quantity $\Delta^{-1}(\varepsilon)$ proportional to the average density of states turns to zero.

In the absence of magnetic interactions violating the time reversal symmetry, the wave functions and matrix elements $H_{mn}$ in Eq. (1) can be chosen real. In this case the statistical properties of the systems are described by real symmetric random matrices. The ensemble of real symmetric matrices with the Gaussian distribution is often called the *Gaussian orthogonal ensemble* (GOE).

If the magnetic interactions are present in the system, the time reversal and spin-rotation symmetry is violated and the wave functions are no longer real. This means that one should deal with general Hermitian matrices without any additional symmetry and integrate over the matrix elements $H_{mn}$ using only the constraint $H_{mn} = (H_{nm})^*$. This system is called the *Gaussian unitary ensemble* (GUE).

The third possible type of the symmetry arises when the system is time-reversal invariant but does not have central symmetry. In this case it is also impossible to make all the matrix elements real. Nevertheless an additional symmetry exists in this case. According to the Kramers theorem all levels of the system remain doubly degenerate and every eigenvalue of the matrix $H$ must appear twice. Matrices consisting of real quaternions $H_{mn}$ of the form

$$\begin{pmatrix} p_{mn} & q_{mn} \\ -q^*_{mn} & p^*_{mn} \end{pmatrix}$$

and satisfying the condition $H_{mn} = (H_{nm})^+$ have this property. The corresponding ensemble is called the *Gaussian symplectic ensemble* (GSE).

Somewhat different distribution functions were introduced later by Dyson [3] who suggested characterizing the system not by its Hamiltonian but by an unitary $N \times N$ matrix $S$ whose elements give the transition probabilities between states, where, again, $N$ is the number of the levels. This matrix is related to the Hamiltonian $H$ of the system in a complicated way that is not specified in the theory. According to the Dyson hypothesis the correlation properties of $n$ successive energy levels of the system $(n \ll N)$ are statistically equivalent to those of $n$ successive angles provided all the unitary matrices $S$ have equal probabilities. Again, depending on the symmetry, one can distinguish among the three different ensembles. The corresponding ensembles are called *Circular ensembles.*

As concerns the complex nuclei, the orthogonal ensembles are most relevant for their description because in order to change the level statistics one needs, e.g. huge magnetic fields that hardly exist. However, the other two ensembles have been under intensive discussions for problems of mesoscopic physics (for a review, see [4]). Moreover, it has been realized that one might formulate additional ensembles of e.g. chiral matrices relevant for studying properties of models for QCD (for a review, see the lecture by Jac Verbaarschot[5]). It has been proven later that in total 10 different symmetry classes exist [6]. Most of the new ensembles may be relevant to different disordered mesoscopic systems due the presence of, e.g., superconductivity or additional symmetries of the lattice.

I do not plan to discuss in these lectures the non-standard symmetry classes and restrict myself by the Wigner-Dyson (WD) statistics. It is relevant to say that calculation of the level-level correlation function, Eq. (3), starting from the Gaussian or Circular ensembles is not a simple task. The conventional method of the evaluation is using orthogonal polynomials[2]. The procedure is not difficult for the unitary ensembles but one has to put a considerable effort to perform the calculations for the orthogonal and symplectic ones. As it has been mentioned, the final results are universal in the limit $N \to \infty$ and can be written for the orthogonal, unitary and symplectic ensembles in the form

$$R_{orth}(\omega) = 1 - \frac{\sin^2 x}{x^2} - \frac{d}{dx}\left(\frac{\sin x}{x}\right)\int_1^\infty \frac{\sin xt}{t}dt \tag{4}$$

$$R_{unit}(\omega) = 1 - \frac{\sin^2 x}{x^2} \tag{5}$$

$$R_{sympl}(\omega) = 1 - \frac{\sin^2 x}{x^2} + \frac{d}{dx}\left(\frac{\sin x}{x}\right)\int_0^1 \frac{\sin xt}{t}dt \tag{6}$$

where $x = \pi\omega/\Delta$, and $\Delta$ is the mean level spacing.

The functions $R(\omega)$, Eqs. (4, 5, 6), tend to 1 in the limit $\omega \to \infty$, which means that the correlations are lost in this limit. In the opposite limit $x \to 0$ they turn to zero as $x^\beta$, where $\beta = 1, 2, 3$ for the orthogonal, unitary and symplectic ensembles, respectively. This means that the probability of finding a level at the distance $\omega$ from another level decays at small $\omega$. The effect is know as "level repulsion". It is important that Eqs. (4-6) describe the level-level correlation function, Eq. (3), for both the Gaussian and Circular ensembles.

## 0.2 Small disordered particles

Nowadays the relevance of the random matrix theory (RMT) to mesoscopic physics is almost evident and it is the starting point of many works on transport in quantum dots, electromagnetic response of metallic grains, etc. However, it took quite a long time before the ideas of the RMT penetrated from nuclear to condensed matter physics.

There were several reasons for this slow development. First, until the end of 60's of the last century most of the objects studied in condensed matter physics were macroscopic and the discreteness of the energy levels could be neglected.

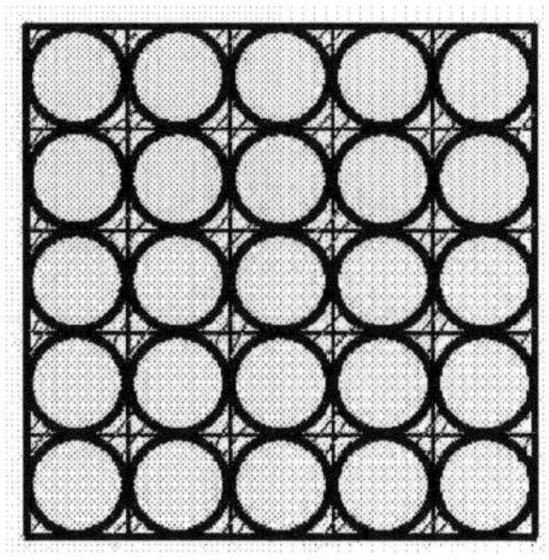

*Figure 1.* Structure of a granular metal.

Therefore the question about the level statistics was not so interesting. Second, from the theoretical point of view it was not clear at all how one could come to the Wigner-Dyson level statistics starting from the Schrødinger equation. It was clear that making perturbation theory in interaction or disorder did not lead to the anything that would resemble Eqs. (4-6).

As soon as experimentalists started investigation of granular materials (which happened in 60's), a theory that would describe small metal systems became nevessary. Such materials consist of small metal particles (grains) with the diameter down to $10 - 100A$. These grains can be covered by an insulator and therefore be well isolated from each other. A schematic view of the pattern can be found in Fig.1

It is clear that one can speak now about discrete levels and study their statistics. Of course, one needs low temperatures in order to prevent inelastic processes smearing the levels but this is not so difficult (temperatures $< 1K$ can be sufficient).

In practice, the form of the grains can be not very regular, they can contain defects and impurities and therefore the energy spectrum strongly fluctuates from grain to grain. All this true even if one neglects the electron-electron interaction. So, one naturally comes to the idea to describe the levels statistically.

The first work on the application of the RMT to small metallic grains was done by Gorkov and Eliashberg (GE) [7]. These authors studied the electromagnetic response of the system of the grains and therefore they needed an information about the level-level correlation in a single grain. As the RMT is purely phenomenological and nothing is assumed about the origin of the randomness, this theory was taken by GE to describe the correlations. Starting from the explicit form of the level-level correlation function $R(\omega)$, Eqs. (4-6), they calculated the desired physical quantity.

GE identified correctly physical situations when the three symmetry classes might be used. In the absence of any magnetic and spin orbit interactions they

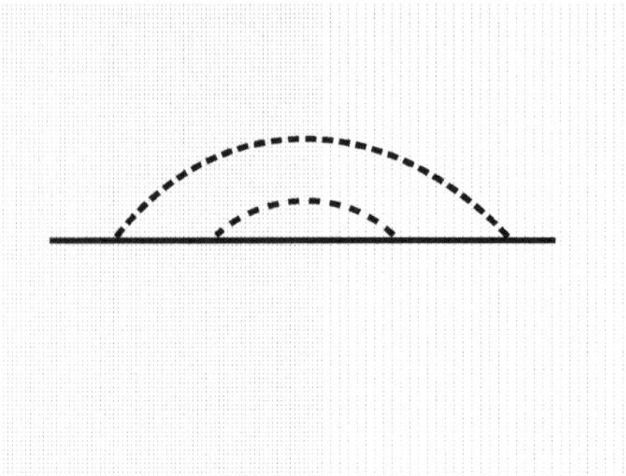

*Figure 2.* Typical diagrams for the Green function in a disordered metal.

suggested to use the orthogonal ensemble. If a magnetic field is applied or there are magnetic impurities in the grains, the unitary ensemble should be applicable. If there are no magnetic interactions but spin-orbital impurities are present, the grains should be described by the symplectic ensemble.

Being the first application of the RMT in solid state physics, the paper[7] remained the only application during the next 17 years. This is not surprising because using calculational schemes existed in that time no indication in non-trivial level correlations could be seen. Let us discuss this point in more details.

Studying a disordered system and neglecting electron-electron interactions one can start with the following Hamiltonian

$$H = \varepsilon\left(\hat{\mathbf{p}}\right) + U\left(\mathbf{r}\right) \tag{7}$$

where $\varepsilon\left(\hat{\mathbf{p}}\right)$ is the operator of the kinetic energy and calculate the Green function $G_\varepsilon = \left(\varepsilon - H\right)^{-1}$ performing an expansion in the disorder potential $U\left(\mathbf{r}\right)$. Usually, it is assumed that the $U\left(\mathbf{r}\right)$ is random and its fluctuations are Gaussian with

$$\langle U\left(\mathbf{r}\right)\rangle = 0, \ \langle U\left(\mathbf{r}\right) U\left(\mathbf{r}'\right)\rangle = \frac{1}{2\pi\nu\tau}\delta\left(\mathbf{r}-\mathbf{r}'\right) \tag{8}$$

where $\nu$ is the density of states and $\tau$ is the mean scattering time.

Making the perturbation theory in the random potential $U\left(\mathbf{r}\right)$ and averaging over this potential can be done using the "cross technique" [8]. Only diagrams without intersections of the impurity lines are important in the standard approximation of the weak disorder. For the one particle Green function the typical diagrams are represented in Fig.2.

As a result, one obtains for the averaged Green function $\langle G\rangle$ the following expression

$$G = \frac{1}{\varepsilon - \varepsilon\left(\mathbf{p}\right) \pm i/2\tau} \tag{9}$$

where "+" corresponds to the retarded and "−" to the advanced Green functions.

Standard calculations based on a summation of "ladder diagrams" lead in this case to the classical Drude formula for, e.g., the conductivity $\sigma(\omega)$

$$\sigma(\omega) = \frac{\sigma_0}{1 - i\omega\tau}, \quad \sigma_0 = 2e^2\nu D_0 \tag{10}$$

where $D_0 = v_0^2\tau/3$ is the classical diffusion coefficient, $\omega$ is the frequency and $e$ is the electron charge.

So long as the grain size remained larger than the atomic distances, no deviations from this formula could be found and therefore the question about the applicability of the RMT remained open. It is relevant to mention works by Kubo performed in approximately the same time. In Ref. [9]he argued that even very small irregularities (with size of the order of atomic distances) of the shape of the metallic grains must lead to lifting of all degeneracies of eigenstates that are present in ideally spherical particles. This lead him to the conclusion that the mean energy level spacing $\Delta$ is inversely proportional to the volume $V$ of the particle

$$\Delta = (\nu V)^{-1} \tag{11}$$

where $\nu$ is density of states at the Fermi surface of the metal. Later Kubo suggested [10] that the spacing distribution had to follow the Poisson law. The latter differs essentially from the Wigner-Dyson RMT by the absence of the level repulsion. The second work by Kubo, Ref.[10], was published several years after the work by Gorkov and Eliashberg, Ref. [7], and this shows that the applicability of the Wigner-Dyson statistics to the small metal particles was far from being established in that time.

The situation started to change only at the end of 70's with the new developments in the theory of Anderson localization. In the publication[11] a new scaling idea was put forward for description of disordered samples of an arbitrary dimensionality. The most unusual was a prediction that two dimensional initially metallic samples could not remain metals in the presence of an arbitrary weak disorder and had to acquire insulating properties. Again, using Eq. (10) it was not clear why something had to happen in two dimension (the localization in one-dimensional chains had been proven before and it was clear that the ladder diagrams summation leading to Eq. (10) was not sufficient for that case).

Trying to understand how something unusual could happen in two dimensions Gorkov, Larkin and Khmelnitskii [12] investigated more complicated diagrams and found that a certain class of diagrams could lead to a divergence in any dimensionality $d \leqslant 2$. These are "fan" diagrams with the maximal number of crossings. They are represented in Fig. 3 and their sum is a new effective

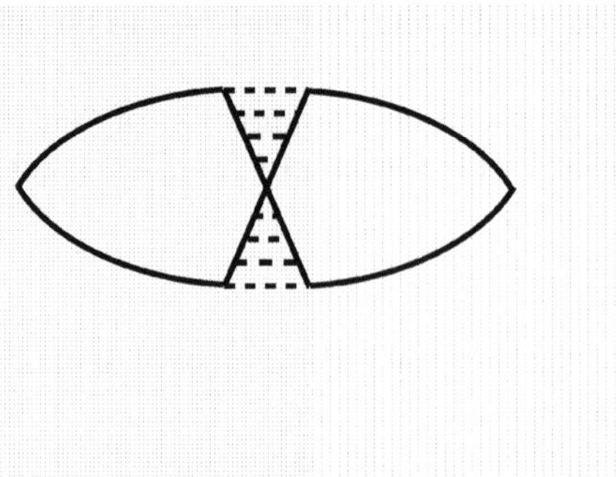

*Figure 3.* Cooperon: a singular correction to conductivity.

mode that is usually called "cooperon". This mode has a form of a diffusion propagator and its contribution to the conductivity can be written in the form

$$\sigma\left(\omega\right)=\sigma_{0}\left(1-\frac{1}{\pi\nu}\int\frac{1}{D_{0}\mathbf{k}^{2}-i\omega}\frac{d^{d}\mathbf{k}}{\left(2\pi\right)^{d}}\right) \tag{12}$$

It is clear from Eq. (12) that in the limit of low frequencies $\omega\rightarrow 0$ the second term in the brackets in Eq. (12) diverges in any dimension $d\leqslant 2$. This means that in a disordered film the quantum correction to the classical conductivity diverges and this signals (but, of course, does not prove) the possibility of the localization.

The scaling theory of the localization and the discovery of the new diffusion modes was revolutionary in the theory of disordered systems but how can these findings be related to the Wigner-Dyson theory?

Actually, Eq. (12) is the key to understanding that there can be something beyond the classical Eq. (10) in small metal particles. One needs only realizing that the case of small metal grains corresponds to the zero dimensionality of the integral in Eq. (12). Due to the finite size the values of the momentum $\mathbf{k}$ are quantized such that $k_a = 2\pi n_a/L_a$, where $n_a = 0, \pm 1, \pm 2, \pm 3......$ , $a = (x, y, z)$, and $L_a$ is the size of the grain in the $a$ -direction.

At low frequencies $\omega \ll D_0/L^2$ the most important contribution in the integral in Eq. (12) comes from the zero harmonics with $\mathbf{k} = 0$ and one can see that this contribution strongly diverges when $\omega\rightarrow 0$. Moreover, the quantum correction is proportional to $\Delta/\omega$, where $\Delta$ is given by Eq. (11) and this is what one can expect from Eqs. (4-6).

So, Eq. (12) really signals that something nontrivial can happen in metal grains and the WD theory is not excluded. The diffusion modes play a prominent role and it seems, at first glance, that one should merely write proper diagrams and sum their contribution. However, even if this were possible this would hardly correspond to Eq. (4-6). The problem is that the expansion in terms of $\Delta/\omega$ cannot take into account the oscillating part in Eqs. (4-6) even in principle. Summing the diagrams one can hope to reproduce only non-oscillating asymptotics of these equations. This means that another approach has to be developed.

The possibility to demonstrate that the energy level and wave functions statistics can really be described by the RMT came first with the development of the supersymmetry approach [13, 14]. This method is based on a representation of Green functions of a disordered metal in terms of an integral over both commuting and anticommuting variables. Singling out excitations with the lowest energy (diffusion modes) one can reduce calculations to a supermatrix non-linear model. The supersymmetry method allowed to prove for the first time that the level-level correlation function for disordered particles is really described by Eqs. (4-6). Later, using the supersymmetry technique Verbaarschot, Weidenmuller and Zirnbauer [15] have derived Eqs. (4-6) starting directly from the Gaussian ensembles, Eq. (1).

When deriving the non-linear $\sigma$-model for metallic particles it was very important that they contained disorder. However, it is not the necessary condition. Several years later Bohigas, Gianonni and Schmidt [16] conjectured that the RMT should describe correctly spectral properties of quantum systems which are chaotic in their classical limit. In particular, the Wigner-Dyson statistics had to be observed in clean metal particles (quantum billiards) provided their shape was such that classical motion would be chaotic. Their hypothesis was made on the basis of extensive numerics. For a review of the subsequent activity in these fields the book [17] is a good reference.

Historically, the description of disordered systems with a non-linear $\sigma$-model has been suggested by Wegner [18] using the replica method and integration over conventional complex variables. In the first work there were problems with convergence of functional integrals and therefore the replica approach was further developed in the publications [19] and [20]. The $\sigma$-model of Ref. [20] was obtained by the integration over conventional variables and, as a result, the group of the matrices $Q$ was non-compact. In contrast, the starting point of Ref. [19] was a representation of Green functions in terms of integrals over anticommuting (Grassmann) variables and this lead to a compact group of the matrices $Q$.

Although both the replica and supersymmetry approach are equivalent when doing the perturbation theory in the diffusion modes, the latter method is much more efficient for non-perturbative calculations like the study of the level-level correlations. This had become clear shortly after the works [19, 20] were finished and this drawback of the replica approach motivated the development of the supersymmetry one. It should be noticed that recently the oscillating behavior of the level-level correlation function $R(\omega)$ has been obtained [21–23] using the compact replica $\sigma$-model of Ref. [19]. However, the procedures used in these references are considerably more complicated than the calculations by the supersymmetry method and the limit of low frequencies $\omega \lesssim \Delta$ is still hardly achievable.

It is fair to say that the replica approach allows including electron-electron interactions in a comparatively easy way [24, 25] and this has been the main motivation in the attempts [21–23] to obtain non-perturbative results within the replica technique. At the same time, it was believed for a long time that an inclusion of the electron-electron interaction into the supersymmetry scheme was impossible. However, this is not quite so and, at least, not very strong interaction can be incorporated in the supermatrix $\sigma$-model [26].

It follows from this discussion that the supersymmetry approach is better suitable for making connections with the RMT and therefore the present lectures contain discussions based on this method only. It is possible neither review here all works made in this direction nor present all details of the calculations. For a more detailed information see the book [4] and more recent reviews [27–31]. It is relevant to mention here that the word *supersymmetry* has appeared in the condensed matter physics in the publication by Parisi and Sourlas [32], who discovered a complex symmetry in a model describing ferromagnets in a random magnetic field. They used a concept of superspace including both commuting and anticommuting variables.

Two other related directions of the use of the RMT are reviewed at this school by Boris Altshuler (Quantum Chaos) and Jac Verbaarschot (QCD).

In the next sections I want to present the main ideas of the supersymmetry approach and show how it can be used for both the level correlations and wave functions statistics. It will be shown how to obtain the Wigner-Dyson statistics and how to go beyond it. A new development concerning a generalization of the supersymmetric $\sigma$-model to more complicated situations will be outlined in the last section.

# 1. Supersymmetry method

## 1.1 Supermathematics

The supersymmetry method is based on the use of the so called Grassmann variables $\chi_i$, $i = 1, 2, ..., n$ (the elements of the Grassmann algebra) that are introduced in a completely formal way. These are abstract objects but in many cases abstract mathematical constructions drastically influence the development of physics. For example, nobody can dispute the usefulness of complex numbers for physics, but what is the physical meaning of $\sqrt{-1}$? Here I want to remind the reader basic formulae concerning definitions and operations with objects containing combinations of the Grassmann variables and conventional numbers (supermathematics).

The Grassmann variables are some mathematical objects obeying the following anticommutation rules [33]

$$\{\chi_i, \chi_j\} = \chi_i \chi_j + \chi_j \chi_i = 0 \tag{13}$$

for any $1 \leq i, j \leq n$.

The anticommutation rules, Eq. (13) hold in particular for $i = j$ and we see that the square of an arbitrary variable $\chi_i$ is zero

$$\chi_i^2 = 0 \tag{14}$$

For any anticommuting variable $\chi$ one can introduce its "complex conjugate) $\chi^*$. It is assumed by the definition that $(\chi^*)^* = -\chi$, such that the "square of the modulus is "real"

$$(\chi_i^* \chi_i)^* = -\chi_i \chi_i^* = \chi_i^* \chi_i \tag{15}$$

The anticommuting variables $\chi_i$, Eq. (13-15), remained not very useful until Berezin introduced integrals over these variables. The integrals are nothing more than formal symbols introduced as follows

$$\int d\chi_i = \int d\chi_i^* = 0, \int \chi_i d\chi_i = \int \chi_i^* d\chi_i^* = 1 \tag{16}$$

It is implied that the "differentials" $d\chi_i$, $d\chi_i^*$ anticommute with each other and with the variables $\chi_i$, $\chi_i^*$

$$\begin{gathered} \{d\chi_i, d\chi_j\} = \{d\chi_i, d\chi_j^*\} = \{d\chi_i^*, d\chi_j^*\} = 0 \\ \{d\chi_i, \chi_j\} = \{d\chi_i, \chi_j^*\} = \{d\chi_i^*, \chi_j\} = \{d\chi_i^*, \chi_j^*\} = 0 \end{gathered} \tag{17}$$

The definition, Eq. (16), is sufficient for introducing integrals of an arbitrary function. If such a functions depends only on one variable $\chi_i$ it must be linear in $\chi_i$ because already $\chi_i^2 = 0$. Assuming that the integral of a sum of two functions equals the sum of the integrals we calculate the integral of the sum with Eq. (16). The repeated integrals are implied by integrals over several variables. This enables us to calculate the integral of a function of an arbitrary number of variables.

The most important for the development of the supersymmetry method are Gaussian integrals. The direct integration shows that the following relation is fulfilled

$$I = \int \exp\left(-\chi^+ A \chi\right) \prod_{i=1}^{n} d\chi_i^* d\chi_i = DetA \tag{18}$$

where $A$ is an $n \times n$ Hermitian matrix and

$$\chi = \begin{pmatrix} \chi_1 \\ \chi_2 \\ \cdot \\ \cdot \\ \chi_n \end{pmatrix}, \chi^+ = \begin{pmatrix} \chi_1^* & \chi_2^* & \cdot & \cdot & \chi_n^* \end{pmatrix} \tag{19}$$

Eq. (18) differs from the corresponding equation for the commuting variables by giving $\det A$ instead $(\det A)^{-1}$. This remarkable difference is the basis of the supersymmetry method presented in these lectures. In addition to Eq. (18) one can write one more useful integral

$$I_2 = \frac{\int \chi_i \chi_k^* \exp\left(-\chi^+ A \chi\right) \prod_{l=1}^n d\chi_l^* d\chi_l}{\int \exp\left(-\chi^+ A \chi\right) \prod_{l=1}^n d\chi_l^* d\chi_l} = \left(A^{-1}\right)_{ik} \tag{20}$$

In contrast to the integral $I$, Eq. (18), the integral $I_2$, Eq. (20), is completely similar to the corresponding integral over conventional numbers. Eq. (20) can be proven by the differentiation of $\ln I$ in $A_{ki}$.

The next step is the introduction of supervectors and supermatrices. An $n+m$ component supervector is introduced as

$$\Phi = \begin{pmatrix} \chi \\ S \end{pmatrix} \tag{21}$$

where the $n$- component vector $\chi$ is defined in Eq. (19). The $m$-component vector $S$ has a similar form

$$\begin{pmatrix} S_1 \\ S_2 \\ \cdot \\ \cdot \\ S_m \end{pmatrix} \tag{22}$$

but its components are conventional complex numbers.

In analogy with conventional vectors one can introduce the Hermitian conjugation

$$\Phi^+ = \left(\Phi^T\right)^* \tag{23}$$

and the scalar product

$$\Phi^{i+}\Phi^j = \sum_{\alpha=1}^n \chi_\alpha^{i*} \chi_\alpha^j + \sum_{\alpha=1}^m S_\alpha^{i*} S_\alpha^j \tag{24}$$

A linear transformation $F$ in the space of the supervectors converts a supervector $\Phi$ into another supervector $\tilde{\Phi}$

$$\tilde{\Phi} = F\Phi \tag{25}$$

Of course, the supervector $\tilde{\Phi}$ must have the same structure, Eq. (21), as the supervector $\Phi$. This imposes a restriction on the structure of the supermatrix $F$ corresponding to the linear transformation $F$ : it has to be of the form

$$F = \begin{pmatrix} a & \sigma \\ \rho & b \end{pmatrix} \tag{26}$$

In Eq. (26) $a$ and $b$ are $n \times n$ and $m \times m$ matrices containing only commuting variables, $\sigma$ and $\rho$ are $n \times m$ and $m \times n$ matrices consisting of anticommuting ones. Matrices having the structure, Eq. (26) can be called supermatrices.

Two supermatrices $F$ and $G$ of the rank $(m+n) \times (n+m)$ are assumed to multiply according to the conventional rules

$$(FG)_{ik} = \sum_{l=1}^{m+n} F_{il} G_{lk} \tag{27}$$

and one can see that $FG$ is a supermatrix of the same form. In order to define the supertranspose $F^T$ of the supermatrix $F$ one should use the notion of the scalar product of two supervectors, Eq. (24). Again, by analogy with the conventional definition the supermatrix $F^T$ is introduced as

$$\Phi_1^T F^T \Phi_2 = (F\Phi_1)^T \Phi_2 \tag{28}$$

The transpose of a conventional matrix is obtained by transposing its indexes. This is not as simple for the supermatrices. Writing out the scalar product on both sides of Eq. (28) explicitly and using the anticommutation relation, Eq. (13), one can see that the supermatrix $F^T$ is equal to

$$F^T = \begin{pmatrix} a^T & -\rho^T \\ \sigma^T & b^T \end{pmatrix} \tag{29}$$

where $a^T$, $b^T$, $\sigma^T$, and $\rho^T$ stand for the conventional transposition of the matrices $a$, $b$, $\sigma$, and $\rho$.

Using the scalar product, Eq. (28) one obtains immediately

$$(F_1 F_2)^T = F_2^T F_1^T \tag{30}$$

The Hermitian conjugate $F^+$ of the matrix $F$ can be defined in a standard way

$$F^+ = \left(F^T\right)^* \tag{31}$$

Combining Eqs. (15, 30, 31) one can obtain

$$(F_1 F_2)^+ = F_2^+ F_1^+ \text{ and } \left(F^+\right)^+ = F \tag{32}$$

The latter equality shows that the operation of the Hermitian conjugation is inverse to itself. The same is not generally true for the transposition

$$\left(F^T\right)^T \neq F \tag{33}$$

A very important operation in the theory of conventional matrices is taking the trace of a matrix. If one takes the trace of a product of several matrices it is invariant under cyclic permutations of the matrices. However, due to the presence of anticommuting elements a proper operation for the supermatrices should be defined in a different way. The supertrace $STrF$ of matrix of the form, Eq.(26) is defined as

$$STrF = Tra - Trb \tag{34}$$

where the symbol $Tr$ stands for the conventional trace.

Although somewhat strange, the definition, Eq. (34) is very useful because it is this operation that provides the invariance under the cyclic permutations. We obtain for arbitrary supermatrices $F_i$ of the form, Eq. (26)

$$STrF_1F_2 = STrF_2F_1 \tag{35}$$

and

$$STr\left(F_1F_2...F_n\right) = STr\left(F_nF_1F_2...F_{n-1}\right) \tag{36}$$

In addition to the supertrace it is convenient to introduce a superdeterminant of the supermatrix $F$

$$\ln SDetF = STr \ln F \tag{37}$$

The superdeterminant $SDetF$ can also be written as

$$SDetF = Det\left(a - \sigma b^{-1}\rho\right) Detb^{-1} \tag{38}$$

The connection between the superdeterminant and supertrace enables us to prove immediately the multipliticity of the superdeterminant

$$SDet\left(F_1F_2\right) = \left(SDetF_1\right)\left(SDetF_2\right) \tag{39}$$

The rules of the operations with the supervectors and supermatrices are very convenient because they are similar to those of conventional linear algebra. In fact, one can manipulate superobjects in exactly the same way as conventional objects. This simplifies calculations with quantities containing both types of variables considerably.

Now we can write Gaussian integrals over supervectors that generalize the integrals over conventional complex numbers or Grassmann variable. A direct calculation shows that

$$I^s = \int \exp\left(-\Phi^+ F\Phi\right) d\Phi^* d\Phi = SDetF, \tag{40}$$

$$d\Phi^* d\Phi = \pi^{-m} \prod_{i=1}^{n} d\chi_i^* d\chi_i \prod_{k=1}^{m} dS_i^* dS_i$$

and

$$I_2^s = \frac{\int \Phi_i \Phi_k^* \exp\left(-\Phi^+ F \Phi\right) d\Phi^* d\Phi}{\int \exp\left(-\Phi^+ F \Phi\right) d\Phi^* d\Phi} = \left(F^{-1}\right)_{ik} \tag{41}$$

The formulae written in this subsection give complete information about integrals over the Grassmann variables, supervectors, and supermatrices. This information will be directly used for constructing the supersymmetry method.

## 1.2 Physical quantities as integrals over supervectors. Averaging over disorder

Eq. (41) enables us to express physical quantities in terms of functional integrals over supervectors. The form of the integrals that will be obtained is such that averaging over disorder can be performed the beginning of all calculations.

I start with the Schrødinger equation for electrons without any electron-electron interactions but in a presence of an external potential containing both regular and irregular parts. The regular part can describe potential walls and other features of the system whereas the irregular part $H_1$ of the Hamiltonian stands for disorder. The Schrødinger equation takes the form

$$H\phi_k = \varepsilon_k \phi_k, \; H = H_0 + H_1, \; \langle H_1 \rangle = 0 \tag{42}$$

where $\phi_k$ and $\varepsilon_k$ are eigenfunctions and eigenvalues, respectively. The angular brackets $\langle ... \rangle$ stand for the averaging over disorder.

The most important physical quantities can be expressed in terms of retarded $G_\varepsilon^R$ and advanced $G_\varepsilon^A$ Green functions of the Schrødinger equation. Using the spectral expansion the Green functions $G_\varepsilon^{R,A}$ can be written in the form

$$G_\varepsilon^{R,A}\left(\mathbf{r}, \mathbf{r}'\right) = \sum_k \frac{\phi_k\left(\mathbf{r}\right) \phi_k^*\left(\mathbf{r}'\right)}{\varepsilon - \varepsilon_k \pm i\delta} = \sum_k G_{\varepsilon k}^{R,A} \phi_k\left(\mathbf{r}\right) \phi_k^*\left(\mathbf{r}'\right) \tag{43}$$

These functions satisfy the equation

$$\left(\varepsilon - H\right) G_\varepsilon^{R,A}\left(\mathbf{r}, \mathbf{r}'\right) = \delta\left(\mathbf{r} - \mathbf{r}'\right) \tag{44}$$

The average density of states $\rho\left(\varepsilon\right)$ (this quantity is proportional to $\Delta^{-1}$, Eq. (2)) can be written as

$$\langle \rho\left(\varepsilon, \mathbf{r}\right) \rangle = \left\langle \sum_k \phi_k^*\left(\mathbf{r}\right) \phi_k\left(\mathbf{r}\right) \delta\left(\varepsilon - \varepsilon_k\right) \right\rangle \tag{45}$$

$$= \frac{1}{\pi} \left\langle \mathrm{Im}\, G_\varepsilon^A\left(\mathbf{r}, \mathbf{r}\right) \right\rangle \tag{46}$$

whereas the level-level correlation function $R(\omega)$ takes the form

$$R(\omega) = \left(\frac{\Delta}{\pi}\right)^2 \left\langle \sum_{k,m} \operatorname{Im} G^A_{k,\varepsilon-\omega} \operatorname{Im} G^A_{m,\varepsilon} \right\rangle \tag{47}$$

We see from Eqs. (45, 47) that, as soon as we are able to average the Green functions or their products over the disorder, the quantities of interest are found. However, this cannot be done directly using Eqs. (45, 47) and we need another representation for the Green functions. Of course, one can do perturbation theory in the disorder potential but, as it has been already discussed, such an approach can hardly help in obtaining the Wigner-Dyson statistics.

What will be done now is writing the Green functions in a form that would be suitable for averaging over the disorder in the very beginning. This can be conveniently done with the integrals over the supervectors. I want to present here the main scheme only. All necessary details can be found in the book [4]. As the main interest is to calculate the level-level correlation function $R(\omega)$, Eq. (47), all formulae will be written for this case. Calculating the density of states, Eq. (45), is a simpler and less interesting task.

Let us introduce 8-component supervectors $\psi$ consisting of 4-component supervectors $\psi^1$ and $\psi^2$ such that

$$\psi^m = \begin{pmatrix} \theta^m \\ v^m \end{pmatrix}, \ \theta^m = \frac{1}{\sqrt{2}} \begin{pmatrix} \chi^{m*} \\ \chi^m \end{pmatrix}, \ v^m = \frac{1}{\sqrt{2}} \begin{pmatrix} S^{m*} \\ S^m \end{pmatrix}, \tag{48}$$

$m = 1, 2.$

For the supervectors $\psi$ of the form of Eq. (48) one can define, in addition to transposition and Hermitian conjugation, the operation of the "charge conjugation"

$$\bar{\psi} = (C\psi)^T, \ \bar{\psi}^m = \begin{pmatrix} \bar{\theta}^m, & v^m \end{pmatrix} \tag{49}$$

In Eq. (49), $T$ stands for transposition, and $C$ is the supermatrix of the form

$$C^{mn} = \Lambda^{mn} \begin{pmatrix} c_1 & 0 \\ 0 & c_2 \end{pmatrix}$$

where $\Lambda$ is the diagonal supermatrix

$$\Lambda = \begin{pmatrix} \mathbf{1} & 0 \\ 0 & -\mathbf{1} \end{pmatrix} \tag{50}$$

with $\mathbf{1}$ the unity $4 \times 4$ unity matrix.

The matrices $c_1$ and $c_2$ have the form

$$c_1 = \begin{pmatrix} 0 & -1 \\ 1 & 0 \end{pmatrix}, \ c_2 = \begin{pmatrix} 0 & 1 \\ 1 & 0 \end{pmatrix} \tag{51}$$

The function $R(\omega)$ is determined by products $G^A G^R$, $G^R G^R$ and $G^A G^A$. The last two products are not interesting because the average of their products is equal for weak disorder to the product of their average. When calculating these quantities the diffusion modes discussed in the Introduction do not appear. Therefore let us concentrate on the calculation of the product $G^A G^R$. Using Eqs. 40, 41) we can write this quantity as

$$G^A_{\varepsilon-\omega}(\mathbf{r},\mathbf{r})\, G^R_{\varepsilon}(\mathbf{r}',\mathbf{r}') = \left(\frac{\Delta}{\pi}\right)^2 \int \psi^1(\mathbf{r})\, \bar{\psi}^1(\mathbf{r})\, \psi^2(\mathbf{r}')\, \bar{\psi}^2(\mathbf{r}') \exp(-L)\, D\psi \tag{52}$$

where the Lagrangian $L$ has the form

$$L = i \int \bar{\psi}(\mathbf{r}) \left(-\tilde{H}_0 - U(\mathbf{r}) - \frac{1}{2}(\omega + i\delta)\Lambda\right) \psi(\mathbf{r})\, d\mathbf{r} \tag{53}$$

where $U(\mathbf{r})$ the impurity potential. The operator $\tilde{H}_0$ equals

$$\tilde{H}_0 = \varepsilon(-i\nabla_r) - \varepsilon + \frac{\omega}{2} \tag{54}$$

where $\varepsilon(-i\nabla_r)$ is the spectrum.

I would like to draw attention at this point that the weight denominator is absent in Eq. (52), which contrasts the analogous integrals, Eqs. (20,41). Actually, this is a consequence of the fact that $F$ in Eq. (41) is taken as unity in the space of the $2 \times 2$ supermatrices when writing Eq. (52). In other words, the weight denominator is absent due to the difference of the results of the Gaussian integration over the anticommuting variables $\chi$, Eq. (18), (one obtains $Det A$) and the corresponding formula for the integration over conventional numbers that gives $(Det A)^{-1}$.

The possibility to write the Green function in the form of Eq. (52) without the weight denominator is the reason why the integration over the supervectors is used. As the weight denominator is absent one can immediately average over impurities in Eq. (52).

Let us assume for simplicity that the distribution of the random potential $U(\mathbf{r})$ in Eq. (53) is specified by Eq. (8). Then, the averaging over the random potential is simple and one obtains again Eq. (52) but now the Lagrangian $L$ should be written as

$$L = \int \left[-i\bar{\psi}\tilde{H}_0\psi + \frac{1}{4\pi\nu\tau}\left(\bar{\psi}\psi\right)^2 - \frac{i(\omega + i\delta)}{2}\bar{\psi}\Lambda\psi\right] d\mathbf{r} \tag{55}$$

Eq. (55) shows that we have reduced the initial disordered problem to a regular model with an $\psi^4$ interaction. Of course, it does not help to solve it exactly but now we can use approximations well developed in theory of interacting particles.

The Lagrangian $L$ in Eq. (55) is similar to those studied in field theory. Let us remark that at $\omega = 0$ the Lagrangian is invariant under rotations of the supervectors in the superspace because it depends on the square of "length" only. The frequency $\omega$ violates this symmetry and, if one uses an analogy with spin models, plays the role of an "external field". The violation of supersymmetry by the frequency is due to the fact that Eq. (55) is written for the product $G^R G^A$ (the presence of the matrix $\Lambda$ is a direct consequence of this). The symmetry would not be violated in the corresponding integrals for $\langle G^R G^R \rangle$ and $\langle G^A G^A \rangle$. The averaging of the simpler Lagrangian corresponding to the density of states (one averaged Green function) results also in a model with the interaction $\psi^4$, but the supersymmetry in this case is not violated. The diffusion modes discussed previously exist only as a result of the violation of the supersymmetry (Goldstone modes).

## 1.3 Spontaneous breaking of the symmetry and Goldstone modes. Non-linear $\sigma$-model

It is clearly not possible to calculate any correlation function with the Lagrangian $L$, Eq. (55), exactly (except for the case of the one-dimensional chain or the Bethe lattice, where one can write recurrence equations. Further calculations will be performed in the limit of large mean free times $\tau$, which correspond to a weak interaction in the Lagrangian $L$. However, the use of the standard perturbation theory as we have seen is impossible even in this limit because of the existence of the diffusion modes and so one should try to use non-perturbative approaches.

One of the standard approaches used for such a type of theories is the mean field approximation. According to this scheme one simplifies the $\psi^4$ interaction replacing pairs $\psi\psi$ by their averages. For the interaction $\psi^4$ there can be six different pairings, which can be written as follows

$$L_{int} = \frac{1}{4\pi\nu\tau} \int \left(\bar{\psi}\psi\right)^2 d\mathbf{r} \rightarrow L_1 + L_2 + L_3, \tag{56}$$

$$L_1 = \frac{1}{4\pi\nu\tau} \sum_{\alpha,\beta} \int 2 \left\langle \bar{\psi}_\alpha \psi_\alpha \right\rangle_{eff} \bar{\psi}_\beta \psi_\beta d\mathbf{r},$$

$$L_2 = \frac{1}{4\pi\nu\tau} \sum_{\alpha,\beta} \int 2\bar{\psi}_\alpha \left\langle \psi_\alpha \bar{\psi}_\beta \right\rangle_{eff} \psi_\beta d\mathbf{r},$$

$$L_3 = \frac{1}{4\pi\nu\tau} \sum_{\alpha,\beta} \int \left( \left\langle \bar{\psi}_\alpha \bar{\psi}_\beta \right\rangle_{eff} \psi_\beta \psi_\alpha + \bar{\psi}_\alpha \bar{\psi}_\beta \left\langle \psi_\beta \psi_\alpha \right\rangle \right) d\mathbf{r}$$

In Eqs.(56), the symbol

$$\langle ... \rangle_{eff} = \int (...) \exp\left(-L_{eff}\right) D\psi$$

stands for the functional averaging with the effective Lagrangian $L_{eff} = L_0 + L_1 + L_2 + L_3$, where $L_0$ is the quadratic part of the Lagrangian $L$ in Eq. (55), and $\alpha$ and $\beta$ stand for the components of the supervectors $\bar{\psi}$ and $\psi$.

In fact the terms in $L_2$ and $L_3$ are equal to each other. The average $\langle \bar{\psi}_\alpha \psi_\alpha \rangle$ renormalizes the energy $\varepsilon$ and is not important. The averages in $L_2$ and $L_3$ can be both commuting and anticommuting variables, depending on the subscripts $\alpha$ and $\beta$. The final Lagrangian $L_{eff}$ takes the form

$$L_{eff} = \int \left[ -i\bar{\psi} \left( \tilde{H} + \frac{1}{2}(\omega + i\delta)\Lambda + \frac{iQ}{2\tau} \right) \psi d\mathbf{r} \right] \tag{57}$$

with the $8 \times 8$ supermatrix $Q$ satisfying the following self-consistency equation

$$Q = \frac{2}{\pi\nu} \langle \psi\bar{\psi} \rangle_{eff} \tag{58}$$

Calculating the Gaussian integral in Eq. (58) with the help of Eq.(41) we obtain

$$Q = \frac{1}{\pi\nu} \int g_0(\mathbf{p}) \frac{d^d\mathbf{p}}{(2\pi)^d} \tag{59}$$

$$g_0(\mathbf{p}) = i \left( \varepsilon(\mathbf{p}) + \frac{\omega + i\delta}{2} - \varepsilon + \frac{1}{2}(\omega + i\delta)\Lambda + \frac{iQ}{2\tau} \right)^{-1} \tag{60}$$

The integral over the momenta $\mathbf{p}$ in Eq. (60) has both a real and an imaginary part. As concerns the imaginary part the main contribution comes from the region $|\varepsilon(\mathbf{p}) - \varepsilon| \gg \tau^{-1}, \omega$ and, therefore, is proportional to the unit matrix $\mathbf{1}$. This contribution leads to a small renormalization of the energy $\varepsilon$. Assuming that this energy has already been renormalized we can forget about the imaginary part of $Q$ and concentrate on the real part. The main contribution to the real part of Eq. (60) comes from the region $|\varepsilon(\mathbf{p}) - \varepsilon| \sim \tau^{-1} \ll \varepsilon$. Introducing the variable $\xi = \varepsilon(\mathbf{p}) - \varepsilon$ we can rewrite Eq. (60) in the form

$$Q = \frac{i}{\pi} \int_{-\infty}^{\infty} \left( \xi + \frac{1}{2}(\omega + i\delta)\Lambda + \frac{iQ}{2\tau} \right)^{-1} d\xi \tag{61}$$

Eq. (61) determines the contribution to the real part of $Q$ only and has at $\omega \neq 0$ one evident solution

$$Q = \Lambda, \ \omega \neq 0 \tag{62}$$

However, putting $\omega = 0$ we see that any $Q$ of the form

$$Q = V\Lambda\bar{V} \tag{63}$$

where $V$ is an arbitrary unitary supermatrix, $V\bar{V} = 1$, satisfies Eq. (61). Here the symbol of conjugation " $-$ " for an arbitrary matrix $A$ means the following

$$\bar{A} = CA^T C^T \tag{64}$$

with $C$ from Eqs. (49, 51). The supermatrix $Q$ is self conjugate, $Q = \bar{Q}$.

The degeneracy of the ground state, Eq. (63), leads to the existence of the low-lying Goldstone modes, and their contribution to physical quantities should be taken into account properly. These Goldstone modes are just the diffusion modes discussed in the Introduction. In the language of spin models the degeneracy of the solution, Eq. (63) is equivalent to the degeneracy due to an arbitrary spin direction.

Substituting the mean field solution, Eq. (63), into the effective Lagrangian into Eq. (57) we obtain zero, which shows that this approximation is not sufficient. In order to describe a contribution of the diffusion modes we have to take into account slow variations of the supermatrix $Q$ in space. This can be done assuming that the supermatrix $V$ is a slow function of the coordinates $\mathbf{r}$ and expanding in the gradients of $V$ (or, equivalently, $Q$).

As a result of the expansion in the gradients and small frequencies one can obtain a functional $F\left[Q\right]$ describing slow variations of the supermatrix $Q$ in space

$$F\left[Q\right] = \frac{\pi\nu}{8} STr \int \left[D_0 \left(\nabla Q\right)^2 + 2i\left(\omega + i\delta\right)\Lambda Q\right] d\mathbf{r} \tag{65}$$

where $D_0$ is the classical diffusion coefficient, Eq. (7), and the supermatrix $Q$ is described by Eq. (63). The free energy functional, Eq. (65), has the form of a non-linear $\sigma$-model.

In order to calculate, e.g. the level-level correlation function $R\left(\omega\right)$, Eq. (47), one should express it in terms of a functional integral over the supermatrix $Q$. As a result, this function takes the form

$$R\left(\omega\right) = \frac{1}{2} - \frac{1}{2V^2} \operatorname{Re} \int Q_{11}^{11}\left(\mathbf{r}\right) Q_{11}^{22}\left(\mathbf{r}'\right) \exp\left(-F\left[Q\right]\right) DQ d\mathbf{r} d\mathbf{r}' \tag{66}$$

where $V$ is the volume of the system. In Eq. (66) the superscripts of $Q$ enumerate "retarded-advanced blocks" , the subscripts relate to the matrix elements within these blocks.

Eqs. (65, 66) show that the calculation of physical quantities for disordered systems can be reduced to study of a supermatrix non-linear $\sigma$-model. There is a variety of physical problems that can be solved by considering this model in different dimensionalities. To some extent, the problem of the level statistics in a limited volume is the simplest one because it corresponds to the zero-dimensional $\sigma$-model (no dependence of $Q$ on the coordinates).

Adding magnetic or spin orbit interactions results in additional "external fields" partially breaking the symmetry. If these interactions are not very weak they simply cut some degrees of freedom. As a result, one comes again to Eq. (65) but with a reduced symmetry of the supermatrix $Q$.

There can be three different types of the symmetries:

1. In the absence of both magnetic and spin-orbital interactions the system is invariant under time reversal and spatial inversion. It will be called "orthogonal" because, as we will see, one comes in small particles to the Wigner-Dyson statistic for the orthogonal ensemble.

2. In the presence of magnetic interactions the time reversal symmetry is violated and this will be called unitary ensemble.

3. In the absence of magnetic interactions but in the presence of spin-orbital ones, one obtains the symplectic ensemble.

These three ensembles lead to quite different results in different situations and each of them should be considered separately.

## 1.4 Level statistics in a limited volume

Eq. (63) specifies a general form for the $8 \times 8$ supermatrix $Q$. In other words, it obeys the constraint

$$Q^2 = 1 \tag{67}$$

However, this constraint is not sufficient to determine unambiguously the precise structure of $Q$ because Eq. (67) could correspond both to rotations on a sphere and on a hyperboloid. It turns out that the symmetry of the supermatrix $Q$ is more complex than those corresponding to these two possibilities. The supermatrix $Q$ consists of two parts: one describing rotations on the sphere and the other on the hyperboloid, such that the group of rotations of $Q$ is a mixture of compact and non-compact groups of rotations [4]. These two parts are glued by anticommuting elements. We can describe the explicit form of the supermatrix $Q$ writing it in the form

$$Q = UQ_0\bar{U}, \tag{68}$$

where

$$U = \begin{pmatrix} u & 0 \\ 0 & v \end{pmatrix} \tag{69}$$

and $\bar{u}u = 1$, $v\bar{v} = 1$. All Grassmann variables are included in the $4 \times 4$ supermatrices $u$ and $v$. These matrices contain also some phases. Their form is simple and not very important for our discussion. More details can be found in Ref. [4].

The form of $Q_0$ is more interesting and can be written as

$$Q_0 = \begin{pmatrix} \cos\hat{\theta} & i\sin\hat{\theta} \\ -i\sin\hat{\theta} & -\cos\hat{\theta} \end{pmatrix}, \ \hat{\theta} = \begin{pmatrix} \hat{\theta}_{11} & 0 \\ 0 & \hat{\theta}_{22} \end{pmatrix} \tag{70}$$

The $2 \times 2$ matrices are different for different classes of the symmetry.

We see that the symmetry of $Q_0$ corresponds to a group rotations on both the sphere and the hyperboloid. The explicit form of the matrices $\hat{\theta}_{11}$ and $\hat{\theta}_{22}$ can be expressed as follows

$$\hat{\theta}_{11} = \begin{pmatrix} \theta & 0 \\ 0 & \theta \end{pmatrix}, \; \hat{\theta}_{22} = i \begin{pmatrix} \theta_1 & \theta_2 \\ \theta_2 & \theta_1 \end{pmatrix} \tag{71}$$

with $0 < \theta < \pi$, $\theta_1 > 0$, $\theta_2 > 0$ for the orthogonal ensemble,

$$\hat{\theta}_{11} = \begin{pmatrix} \theta & 0 \\ 0 & \theta \end{pmatrix}, \; \hat{\theta}_{22} = i \begin{pmatrix} \theta_1 & 0 \\ 0 & \theta_1 \end{pmatrix} \tag{72}$$

with $0 < \theta < \pi$, $\theta_1 > 0$ for the unitary ensemble, and

$$\hat{\theta}_{11} = \begin{pmatrix} \theta_1 & \theta_2 \\ \theta_2 & \theta_1 \end{pmatrix}, \; \hat{\theta}_{22} = i \begin{pmatrix} \theta & 0 \\ 0 & \theta \end{pmatrix} \tag{73}$$

for the symplectic one.

Eqs. (65,66, 70-73) specify the non-linear supermatrix $\sigma$-model. What remains to do for the level-level correlation function, Eq. (66), is to calculate the functional integral over $Q$. Many of the physical quantities can also be expressed in a form of a correlation function of the supermatrices $Q$ with the free energy functional $F[Q]$, Eq. (65).

Let us consider now the level-level correlation function in a limited volume. The functional integral, Eqs. (65, 66) can be considerably simplified in the limit of small frequencies. In a finite volume one can expand the supermatrix $Q$ in Fourier series. Then, it is not difficult to understand from Eq. (65) that for $\omega \ll D_0/L^2$, where $L$ is the sample size, only the zero space harmonics is essential. In this case one can integrate over the supermatrices $Q$ not varying in space and the integral for the function $R(\omega)$, Eq. (66), becomes just a definite integral over several variables. This integral takes the following form

$$R(\omega) = \frac{1}{2} - \frac{1}{2} \operatorname{Re} \int Q_{11}^{11} Q_{11}^{22} \exp\left(-F_0[Q]\right) dQ \tag{74}$$

where $F_0[Q]$ takes the form

$$F_0[Q] = \frac{i\pi(\omega + i\delta)}{4\Delta} STr(\Lambda Q) \tag{75}$$

One can say that Eqs. (74, 75) determine a zero dimensional non-linear $\sigma$-model. In general, it is clear that the dimensionality of the $\sigma$-model is determined at not very high temperatures by the geometry of the sample. For example, the one dimensional $\sigma$-model describes a long wire of a finite thickness.

In order to calculate the integral over $Q$ in Eq. (74) it is very convenient to use Eqs. (70, 73). We see immediately that the supermatrix $U$, Eq. (69), drops out from $F_0[Q]$ entering the pre-exponential in Eq. (74) only. This allows us to integrate first over the elements of $U$, and reduce the integral to the variables $\hat{\theta}_{11}$, $\hat{\theta}_{22}$ only. The integration over the Grassmann variables is, according to Eq. (16), a very simple task. Actually, one has to calculate also Jacobians arising when changing from the integration over $Q$ to the integration over the "eigenvalues" $\lambda$, $\lambda_1$, $\lambda_2$. As a result, one comes to the following integrals for all three ensembles

$$R_{orth}(\omega) = 1 + \mathrm{Re} \int_1^\infty \int_1^\infty \int_{-1}^1 \frac{(\lambda_1\lambda_2 - \lambda)^2 (1 - \lambda^2)}{\left(\lambda_1^2 + \lambda_2^2 + \lambda^2 - 2\lambda\lambda_1\lambda_2 - 1\right)^2} \qquad (76)$$

$$\times \exp\left[i(x + i\delta)(\lambda_1\lambda_2 - \lambda)\right] d\lambda_1 d\lambda_2 d\lambda$$

$$R_{unit}(\omega) = 1 + \frac{1}{2}\mathrm{Re} \int_1^\infty \int_{-1}^1 \exp\left[i(x + i\delta)(\lambda_1 - \lambda)\right] d\lambda d\lambda_1 \qquad (77)$$

$$R_{sympl}(\omega) = 1 + \mathrm{Re} \int_1^\infty \int_0^1 \int_{-1}^1 \frac{(\lambda - \lambda_1\lambda_2)^2 (\lambda^2 - 1)}{\left(\lambda^2 + \lambda_1^2 + \lambda_2^2 - 2\lambda_1\lambda_2\lambda - 1\right)^2} \qquad (78)$$

$$\times \exp\left[i(x + i\delta)(\lambda - \lambda_1\lambda_2)\right] d\lambda_1 d\lambda_2 d\lambda$$

We see that the integrals over the supermatrix $Q$ can be reduced to integrals over the "eigenvalues" $\lambda$, $\lambda_1$, $\lambda_2$. Depending on the ensemble one obtains twofold or threefold integrals. Calculation of such integrals is a much simpler task than calculation of integrals over a large number $N$ of variables encountered in RMT[2]. Although the reduction to the integrals over the eigenvalues has been carried out for the level-level correlation function only, the corresponding manipulations for studing other physical quantities are the same. The only thing that remains to be done when calculating different physical quantities is to write a proper pre-exponential and carry out integration over the elements of supermatrices $u$ and $v$ entering the pre-exponential only. Then one obtains integrals over the variables $\lambda$, $\lambda_1$, $\lambda_2$ analogous to those in Eqs. (76, 77, 78).

The integration over $\lambda$ and $\lambda_1$ in Eq. (77) for the unitary ensemble is simple. At first glance, the integrals for the orthogonal and symplectic ensembles look terrifying. However, they can be calculated by Fourier transforming the integrals from the frequencies $\omega$ to the real time $t$. As a result, one comes to the Wigner-Dyson formulae, Eqs. (4-6), which demonstrates that the level-level correlation function for a small metal particle is really the same as that given by the RMT. This is how one proves the relevance of the RMT for disordered systems [13]. Actually, the model of the disordered metal was the first microscopic model for which the Wigner-Dyson statistics had been proven.

Of course, this is possible for not very high frequencies $\omega \ll D_0/L^2$. In the opposite limit $\omega \gg D_0/L^2$, the situation is no longer zero dimensional but one can do perturbation theory in diffusion modes. This calculation was done in Ref. [34].

One can also come to the zero dimensional $\sigma$-model starting from the Gaussian distribution for the random matrices, Eq. (1), in the limit of large $N$. This was done in the review, Ref. [15]. Therefore, the supersymmetry method can be considered as an alternative to the method of the orthogonal polynomials[2] in RMT.

Using the non-linear $\sigma$-model, Eq. (65), one can consider thick disordered wires. This corresponds to the one dimensional $\sigma$-model. In this case one can use a transfer matrix technique that allows to reduce calculation of a one dimensional functional integral to solving of an effective "Schrødinger equation". The exact solution found for this model[35] proves localization of all states (vanishing of the conductivity) for any arbitrarily weak disorder. In the language of random matrices this quasi-one-dimensional model corresponds to a model of random banded matrices[36].

The exact solution can also be found for a model with disordered grains connected in a such a way that they constitute a Bethe lattice. For this model, using recursion relations one can write a non-linear integral equation[37]. It was demonstrated that within the model on the Bethe lattice one could have a metal-insulator transition with a very unusual critical behavior. The model on the Bethe lattice has been shown to be equivalent to models of certain sparse random matrices [38]. Sparse matrices are relevant for description of diluted spin models, some combinatorial optimization problems[39] and other interesting systems.

Properties of two dimensional disordered metals can be studied using a renormalization group scheme [18–20, 13, 14]

# 2. Wave functions Fluctuations in a Finite volume. Multifractality

## 2.1 General definitions

In this Section statistics of wave functions is discussed. Investigation of wave functions is complimentary to study of energy levels. In the language of random matrices, energy levels correspond to eigenvalues of the matrices whereas the wave functions relate to eigenvectors.

Study of wave functions has become popular in condensed matter physics not long ago with the development of mesoscopic physics. One can study, e.g., electron tunneling through so called quantum dots, which are actually small quantum wells. At low temperatures one can reach a resonance tunnelling regime when the electron tunnel via one resonance level. In this case the tunnelling amplitude is very sensitive to the wave function of the resonance state.

Wave functions can also be measured in optical and acoustic resonators where they are electromagnetic or sound waves, respectively.

We start with the standard Schrødinger equation

$$H\phi_\alpha(r) = \varepsilon_\alpha \phi_\alpha(r) \tag{79}$$

that determines the eigenenergies $\varepsilon_\alpha$ and eigenfunctions $\phi_\alpha(r)$.

We assume that a finite volume $V$ is considered, such that the spectrum of the energies $\varepsilon_\alpha$ is discrete.

The full statistics of the wave functions $\phi_\alpha(r)$ at a given point $r$ can be described by the following distribution function $f$

$$f(t) = \Delta \left\langle \sum_\alpha \delta\left(t - |\phi_\alpha(r)|^2\right) \delta(\varepsilon - \varepsilon_\alpha) \right\rangle \tag{80}$$

The function $f(t)$, Eq. (80), gives the probability that the square of the absolute value of the wave function (intensity) at the point $r$ and energy $\varepsilon$ is equal to $t$. The distribution function $f(t)$ and the wave functions $\phi_\alpha(r)$ are assumed to be properly normalized such that

$$t_0 = 1, \; t_1 = V^{-1} \tag{81}$$

where $t_n(V)$ are coefficients of the so called inverse participation ratio

$$t_n = \int_0^\infty t^n f(t)\, dt = \Delta \left\langle \sum_\alpha |\phi_\alpha(r)|^{2n} \delta(\varepsilon - \varepsilon_\alpha) \right\rangle \tag{82}$$

These coefficients indicate very sensitively the degree of localization of states through their dependence $t_n(V)$ on the volume of the system. In a pure metal or a ballistic chaotic box where the wave functions extend over the whole system one has

$$t_n \propto V^{-n} \tag{83}$$

If disorder makes the localization length $L_c$, at which the typical wave functions decay, is much shorter than the sample size $L \sim V^{1/d}$, the coefficients $t_n$ are insensitive to $L$. However, a very interesting information about the development of localization can be gained through an analysis of $t_n(V)$ for small samples with $L < L_c$.

As soon as the localization length $L_c$ exceeds the sample size, any length scale disappears and, in the language of the coefficients $t_n$, this is described as

$$Vt_n \propto L^{-\tau(n)}, \; \tau(n) = (n-1)\, d^*(n) \tag{84}$$

where $d^* (n)$ may differ from the physical dimension $d$ of the system and be a function of $n$. This function gives the values of the *fractal dimensions* $d^* (n)$ for each $n$.

If the behavior of the wave function is described by Eq. (83), the fractal dimension $d^*$ coincides with the physical dimension $d$. We will see that the fractal dimension of a system obeying the RMT coincides with the physical dimension. In such a situation, although the amplitude fluctuations are possible, they are not very strong.

Once we assume that the envelope of a typical wave function at a length scale shorter than $L_c$ obeys a power law $\phi (r) \propto r^{-\mu}$ with a single fixed exponent $\mu < d/2$, the set of the coefficients $t_n$ reveals $d^* = d - 2\mu$ different from $d$ but the same for all $n > d/ (2\mu)$. This is when one speaks of fractal behavior with the *fractal dimension* $d^*$.

If $d^* (n)$ is not a constant, that signals a more sophisticated structure of the wave functions. They can be imagined as splashes of multiply interfering waves at different scales and with various amplitudes, and possibly, self-similarity characterized by a relation between the amplitude $t$ of the local splash of the wave function and the exponent $\mu (t)$ of the envelope of its extended power law tail.

We will see below that the multifractality of the wave functions of two dimensional weakly disordered conductors is the most general property of these systems as soon as the sample size $L$ does not exceed the localization length $L_c$ [40].

## 2.2 Porter-Thomas distribution

Before starting more complicated calculations I would like to present here what is known about the distribution of wave functions from nuclear physics (see, e.g. Refs. [41, 42]) where it was studied for description of level width fluctuations in neutron scattering. The wave function fluctuations are obtained there again from the RMT.

To start the calculation one should choose and arbitrary basis of eigenfunctions $\rho_m (\mathbf{r})$ and expand the function $\phi_n (\mathbf{r})$ in this basis

$$\phi_n (\mathbf{r}) = V^{-1/2} \sum_{m=1}^{\infty} a_{nm} \rho_m (\mathbf{r}) \tag{85}$$

where $V$ is the volume. It is convenient to truncate the basis to a finite $N$-dimensional set and take the limit $N \to \infty$, as is usually done in the RMT. The main statistical hypothesis is that all coefficients $a_{mn}$ are uniformly distributed. The only restriction on the coefficients $\{a_{mn}\}$ is imposed by normalization of the wave functions, and the probability density $P(\{a_{mn}\})$ can be written as

$$\tilde{P}(\{a_{mn}\}) = \frac{2}{\Omega_N}\delta\left(\sum_{m=1}^{N}|a_{mn}|^2 - 1\right) \tag{86}$$

where $\Omega_n$ is the solid angle in $N$ dimensions. Because of the truncation of the basis the condition of completeness of the basis $\{\rho_m\}$ should be written in the form

$$\sum_{m=1}^{\infty}\rho_m^2(\mathbf{r}) \equiv |\vec{\eta}|^2 = N \tag{87}$$

where $\vec{\eta}$ is an $N$-dimensional vector with components $\{\rho_m\}$. The distribution function of the intensities $W(v)$ at the point $\mathbf{r}$ is introduced as

$$W(v) = \frac{2}{\Omega_N}\int \delta\left(v - |\vec{a}\vec{\eta}|^2\right)\delta\left(|\vec{a}|^2 - 1\right)d\vec{a} \tag{88}$$

where the vector $\vec{a}$ is an $N$-dimensional vector with components $\{a_{mn}\}$. The distribution function $W(v)$, Eq. (88) is defined in such a way that

$$\int W(v)\,dv = 1 \tag{89}$$

In the unitary ensemble, one should integrate over complex vectors $\vec{a}$. Integrating first over the component of the vector $\vec{a}$ parallel to the vector $\vec{\eta}$ and using Eq. (87) one obtains

$$W(v) = \frac{2\pi}{N\Omega_N}\int \delta\left(|\vec{a}_\perp|^2 - \left(1 - \frac{v}{N}\right)\right)d\vec{a} \tag{90}$$

where $\vec{a}_\perp$ is the component perpendicular to $\vec{\eta}$. The remaining integration in Eq. (90) can be carried out easily. Taking the limit $N \to \infty$ one obtains for the unitary ensemble a simple formula

$$W(v) = \exp(-v) \tag{91}$$

Computing the integral in Eq. (88) for real vectors $\vec{a}$ and $\vec{\eta}$ one can obtain the distribution function for the orthogonal ensemble

$$W(v) = \frac{1}{\sqrt{2\pi v}}\exp\left(-\frac{v}{2}\right) \tag{92}$$

The functions $W(v)$, Eqs. (91,92) satisfy the normalization conditions, Eq. (89). The distribution functions $W(v)$ are universal and do not depend on details of models for disorder. The amplitudes $v$ are related to $t$ from the previous section as $v = Vt$. The functions $W$ and $f$ are related to each other accordingly. Eqs. (91, 92) are usually referred to as the Porter-Thomas distribution.

## 2.3 Non-linear $\sigma$-model and the statistics of wave functions

Now we concentrate on the calculation of the distribution function $f(t)$, Eq. (80). At first glance, this task does not seem easy. In the previous section we were able to reduce the level-level correlation function to a functional integral over $8 \times 8$ supermatrices $Q$. It became possible because the level-level correlation function $R(\omega)$, Eq. (3), could be expressed in terms of the product of two Green functions, Eq. (47). Actually, the size $8 \times 8$ of the supermatrices is determined by the fact that only two Green functions are needed.

So, in order to calculate the distribution function $f(t)$, Eq. (80), we have to make two necessary steps:

1. To express Eq. (80) in terms of the Green functions.

2. To express products of the Green functions in terms of an integral over supervectors $\psi$. If we really want to make explicit calculations the supervectors $\psi$ should not have too many components.

It turns out that both the steps are possible and the necessary number of the components of the supervector $\psi$ is just $8$ for the orthogonal ensemble and $4$ for the unitary one.

The step 1 is done introducing Green functions $G^{R,A}_{\varepsilon\gamma}$ for a system with smeared levels

$$G^{R,A}_{\varepsilon,\gamma}\left(\mathbf{r},\mathbf{r}'\right) = \sum_{\alpha} \frac{\phi_\alpha\left(\mathbf{r}\right)\phi^*_\alpha\left(\mathbf{r}\right)}{\varepsilon - \varepsilon_\alpha \pm \frac{i\gamma}{2}} \tag{93}$$

Then, Eq. (80) can be written as

$$f(t) = \Delta\left\langle \lim_{\gamma\to 0}\sum_\alpha \delta\left(t - \frac{i\gamma}{2}G^R_{\varepsilon\gamma}\left(\mathbf{r},\mathbf{r}\right)\right)\delta\left(\varepsilon - \varepsilon_\alpha\right)\right\rangle \tag{94}$$

$$= \frac{\Delta}{2\pi}\lim_{\gamma\to 0}\lim_{\beta\to 0}\left\langle\int \delta\left(t - \frac{i\gamma}{2}G^R_{\varepsilon\gamma}\left(\mathbf{r},\mathbf{r}\right)\right)\left(G^A_{\varepsilon\beta}\left(\mathbf{r}',\mathbf{r}'\right) - G^R_{\varepsilon\beta}\left(\mathbf{r}',\mathbf{r}'\right)\right)\right\rangle d\mathbf{r}'$$

In Eq. (94) one should first take the limit $\beta \to 0$ and then $\gamma \to 0$. Since the distribution function $f(t)$ is represented in terms of a function of only two Green functions at two points $\mathbf{r}$ and $\mathbf{r}'$ one can express it in terms of an integral over 8-component supervectors $\psi$ using the Wick theorem. The corresponding Lagrangian is the same as the one for the level-level correlation function, Eq. (55), provided one replaces the frequency $\omega$ by the level width $\gamma$.

The derivation of the corresponding $\sigma$-model is standard and one comes to the following expression for the distribution function $f(t)$

$$f(t) = \lim_{\gamma \to 0} \int \int STr\left(\pi_b^{(1)} Q(\mathbf{r})\right) \delta\left(t - \frac{\pi\nu\gamma}{4} STr\left(\pi_b^{(2)} Q(\mathbf{r}_o)\right)\right) \times (-F[Q]) DQ \frac{d\mathbf{r}}{4V} \tag{95}$$

where the free energy functional $F[Q]$ has the form

$$F = \frac{\pi\nu}{8} \int STr\left[D_0 (\nabla Q)^2 - \gamma\Lambda Q(\mathbf{r})\right] \tag{96}$$

and $\mathbf{r}_o$ is the "observation point". When the system can be described by the zero dimensional $\sigma$- model the distribution function $f(t)$ does not depend on $\mathbf{r}_o$. However, beyond the $0D$ approximation, this function can also be a function of the coordinates. The matrices $\pi_b^{(1,2)}$ in Eq. (95) select from the supermatrix $Q$ its boson-boson sector and have the form

$$\pi_b^{(1)} = \begin{pmatrix} \pi_b & 0 \\ 0 & 0 \end{pmatrix}, \ \pi_b^{(2)} = \begin{pmatrix} 0 & 0 \\ 0 & \pi_b \end{pmatrix}, \ \pi_b = \begin{pmatrix} 0 & 0 \\ 0 & 1 \end{pmatrix} \tag{97}$$

As we have discussed previously, the $\sigma$-model is noncompact, Eqs. (70-73). Therefore, in order to avoid divergent integrals over the variables $\hat{\theta}_{22}$, we must calculate the integrals keeping $\gamma$ finite. The limit $\gamma \to 0$ can be taken only at the end of the calculations. However, it is not very convenient to keep an additional free parameter, and it is better to get rid of the parameter $\gamma$ at an earlier stage. This can be done by integrating over the zero space harmonics of $Q$ in the very beginning of the calculations.

To carry out this procedure one should represent the supermatrix $Q(\mathbf{r})$ in the form of Eq. (63) and change the variables of integration $V(\mathbf{r})$ to $\tilde{V}(\mathbf{r})$ as $V(\mathbf{r}) = V(\mathbf{r}_o)\tilde{V}(\mathbf{r})$. This leads to supermatrices $\tilde{Q}: Q(\mathbf{r}) = V(\mathbf{r}_o)\tilde{Q}(\mathbf{r})\bar{V}(\mathbf{r}_o)$. In terms of the new variables $\tilde{V}(\mathbf{r})$ and $\tilde{Q}(\mathbf{r})$ the gradient term in Eq. (96) preserves its form, but now the condition

$$V(\mathbf{r}_o) = 1, \ \tilde{Q}(\mathbf{r}_o) = \Lambda \tag{98}$$

has to be fulfilled. Changing the variables of the integration for all points $\mathbf{r} \neq \mathbf{r}_o$ from $Q(\mathbf{r})$ to $\tilde{Q}(\mathbf{r})$ one obtains a new free energy functional that does not contain $V(\mathbf{r}_o)$ or $Q(\mathbf{r}_o)$. These variables enter only the pre-exponential and the term with $\gamma$, and, hence, the integral over $V(\mathbf{r}_o)$ can be computed without making approximations. The result of the integration contains only the variables $\tilde{Q}(\mathbf{r})$ with the boundary condition, Eq. (98). This means that the reduced $\sigma$-model obtained in this way operates only with relative variations of the field $Q$ with respect to its value at the observation point.

The limit $\gamma \to 0$ simplifies the computation because the main contribution to the integral over the variable $\theta_{1o}$ entering the parametrization, Eqs. (70-73), is from $\cosh\theta_{1o} \sim 1/\gamma$ (for simplicity we are considering the unitary ensemble but the final results are similar for all ensembles). After standard manipulations one can express the distribution function $f(t)$ in the form

$$f(t) = \frac{1}{V}\frac{d^2\Phi(t)}{dt^2}, \; \Phi(t) = \int_{\tilde{Q}(\mathbf{r}_o)=\Lambda} \exp\left(-\tilde{F}\left[\tilde{Q}, t\right]\right) D\tilde{Q}(\mathbf{r}) \tag{99}$$

where the free energy $\tilde{F}[Q,t]$ has the following form

$$\tilde{F}[Q,t] = \frac{1}{8}\int STr\left[\pi\nu D_0\left(\nabla\tilde{Q}\right)^2 - 2t\Lambda\Pi\tilde{Q}\right] d\mathbf{r} \tag{100}$$

The matrix $\Pi$ selects from $\tilde{Q}$ its noncompact "boson-boson" sector.

If $t$ is not very large one can take into account the zero space harmonics of $Q$ only. Taking into account Eq. (98) we can just put everywhere $Q = \Lambda$, which leads us immediately to the Porter-Thomas distribution, Eq. (91) (and Eq. (92) for the orthogonal ensemble).

Nonetheless, this would be only an approximate procedure because the value $\tilde{Q}(\mathbf{r}) = \Lambda$ does not correspond to the minimum of the functional $\tilde{F}$ when $t \neq 0$. An equation for the minimum can be found by taking into account the noncompact variable $\theta_1$ under the condition at the boundary and at the observation point

$$\mathbf{n}\nabla\theta_1 = 0, \; \theta_1(\mathbf{r}_o) = 0 \tag{101}$$

where $\mathbf{n}$ is the unit vector at the boundary and perpendicular to it.

Using Eq. (101) one writes the equation for the extremum solution $\theta_t$ in the form

$$\Delta_{\mathbf{r}}\theta_t = -\frac{t}{\pi\nu D_0}\exp\left(-\theta_t(\mathbf{r})\right) \tag{102}$$

where $\Delta_{\mathbf{r}}$ is the Laplacian. The solution $\theta_t(\mathbf{r})$ of Eq. (102) has to be substituted into the energy functional $\tilde{F}$, Eq. (100), which takes the form

$$F_t = \frac{1}{2}\int\left[\pi\nu D_0\left(\nabla\theta_t\right)^2 + 2t\exp\left(-\theta_t\right)\right] d\mathbf{r} \tag{103}$$

It is remarkable that Eq. (102) for the non-trivial vacuum of the reduced $\sigma$-model in two dimensions is exactly the Liouville equation known in the conformal theory of $2D$ quantum gravity [43, 44]. Within this model one has to calculate the functional integral over all $\theta$ with the free energy functional determined by Eq. (103). Although this can lead to helpful analogies [45], results that might be anticipated in this way can be used only as intermediate asymptotics.

Eq. (102) for the minimum looks very similar to a saddle point equation derived in Ref. [46] when considering the problem of long-living current relaxation. However, the non-linear term is different, which leads to different solutions.

The most interesting is the solution of Eq. (102) in two dimensions where it can be found exactly. However, the exact solution is somewhat cumbersome and I write here its asymptotics at distances $r$ much smaller than the sample size $L$ (but exceeding the mean free path $l$)

$$\exp(-\theta_t) \approx (l/r)^{2\mu} \tag{104}$$

where $\mu$ is a parameter depending on disorder.

With the same accuracy, the free energy of the vacuum state can be approximated by

$$F_t \approx 4\pi^2 \nu D_0 \left\{ \mu + \mu^2 \ln(L/l) \right\} \tag{105}$$

The parameter $\mu$ can be determined from the following equation

$$\mu \approx \frac{z(T)}{2\ln(L/l)}, \text{ where } ze^z = T \equiv \frac{tV\ln(L/l)}{2\pi^2 \nu D_0} \tag{106}$$

In principle, Eqs. (104-106) determine the distribution function $f(t)$ for arbitrary $t$ ($\mu$ must remain small, though). However, analytical expressions can be written only in the limiting cases $T \ll 1$ and $T \gg 1$. In these limits, the distribution function $f(t)$, Eq. (80), can be written as[40]

$$f(t) = AV \begin{cases} \exp\left(-Vt\left[1 - \frac{T}{2} + \ldots\right]\right), & T \ll 1 \\ \exp\left(-\frac{\pi^2 \nu D_0}{\ln(L/l)} \ln^2 T\right), & T \gg 1 \end{cases} \tag{107}$$

where $A$ is a normalization constant.

We see from Eq. (107) that at small values of the amplitudes, such that $T \ll 1$, the distribution function $f(t)$ agrees with the Porter-Thomas distribution, Eq. (91), thus proving the latter for disordered systems. In this limit one can make expansion in $T$[47].

At large $t$ ($T \gg 1$) the function $f(t)$ has log-normal asymptotics that is strikingly similar to the asymptotics of the distribution function of the local density of states or conductances discovered by Altshuler, Kravtsov and Lerner [48] who came to this result considering renormalization of terms high gradients in the $\sigma$-model. Even the numerical coefficients in the exponentials are the same, although, of course, the logarithms contain different variables. It appears that the log-normal form is really universal. The slower decay of the distribution function $f(t)$ at large $t$ is due to localization effect. Unfortunately, until now it is not clear how the growth of the high gradient terms is related to the existence of the non-trivial vacuum considered in this Section.

As concerns the coefficients of the inverse participation ratio $t_n$, Eq. (82), they show the multifractal behavior, Eq. (84). Using Eqs. (102,106) one comes to the following expression for the fractal dimension $d^*(n)$

$$d^*(n) = 2 - n\left(4\pi^2 \nu D_0\right)^{-1} \tag{108}$$

We see that even for a weak disorder the fractal dimension $d^*(n)$ strongly deviates from 2. Of course, it cannot become negative and this determines the region of the applicability of Eq. (108).

## 3. Recent and possible future developments

In the preceding sections it was demonstrated how one can come to the random matrix theory starting from a model of a disordered metallic particle. This became possible using the supersymmetry method. Actually, to the best of my knowledge, the model of a disordered metal was the first microscopic model for which the Wigner-Dyson statistics was confirmed.

Starting from the first works [7, 13] where the relevance of the Wigner-Dyson theory to the disordered systems was suggested and proven, a huge number of problems have been attacked using these ideas. Calculations were performed either assuming that RMT was applicable for the description of small particles (they are often called "quantum dots") and using methods of the RMT [2] or making direct computation starting from a disordered metal and applying the non-linear supermatrix $\sigma$-model. Reviewing all these application within several lectures is impossible even though several related topics are considered in this volume by Boris Altshuler and Jac Verbaarschot. At this point I can only refer again to recent reviews [4, 6, 17, 27–31, 49] and apologize in case if some references are missing here.

The selection of topics of the present lectures was motivated mainly by the desire to give a feeling of how to calculate within the supersymmetry method both level and wave function correlations. We have seen that one could obtain results that agreed in a certain region of parameters with the predictions of the RMT and, at the same time, go beyond the RMT.

Essential conditions for the derivation of the $\sigma$-model were the absence of the electron-electron interaction and a sufficiently high concentration of impurities. For the problem of the level statistics, the latter condition corresponds to the case when the mean free path $l$ is much smaller than the sample size. In other words, an electron can scatter many times on the impurities in the bulk before it reaches a boundary of the sample.

At the same time, the RMT was initially suggested for description of complex nuclei, where disorder is absent but interactions are strong. A natural question that can be asked is: Can one prove the relevance of the RMT for clean or/and interacting systems? Clearly, the supermatrix $\sigma$-model discussed in the previous sections is not applicable in these situations and one needs a generalization of this method.

It seems that really new ideas are necessary in order to achieve this goal. Nevertheless, first steps towards constructing more general schemes have been done and I want to present here the main ideas of the new approaches.

## 3.1 Supersymmetry with interaction

From the beginning of the use of the supersymmetry method it was clear that the method could be applicable for non-interacting particles only. The method is based on the result of the Gaussian integration, Eq. (18), that gives $DetA$ instead of the usual $(DetA)^{-1}$. Introducing an interaction results in non-quadratic terms in the Lagrangian. Therefore the trick with writing Green functions in terms of a Gaussian integral without a weight denominator does not work anymore. This is the reason why, in contrast to the replica approach where a proper $\sigma$-model has been derived long ago [24, 25], introducing an interaction into the supersymmetry scheme was believed to be impossible.

To some extent, it is true and it is not clear how to include the interaction into the supersymmetry exactly. However, a weak interaction can really be included without considerable difficulties [26]. The initial electron model with the interaction is not supersymmetric and cannot be made supersymmetric by a transformation. This is why one cannot get rid off the weight denominator.

The main idea of Ref. [26] is to replace approximately the initial electron model by an effective supersymmetric model. This is possible for any disorder in the limit of a weak interaction. The effective model takes into account the most important (Hartree-Fock type) diagrams for any fixed configuration of impurities. Then, the derivation of the proper $\sigma$-model is quite standard.

The resulting non-linear $\sigma$-model resembles very much the replica model of Finkelstein[24, 25] but contains supermatrices and does not have replica indices. The supermatrices $Q$ contain, in addition to those for the non-interacting systems, indices for Matsubara frequencies. One should also write properly spin indices. As a result, the supermatrix non-linear $\sigma$-model for electron systems with interaction takes the form (unitary ensemble)

$$
\begin{aligned}
F = & \frac{\pi\nu}{4}\int d\mathbf{r}\, \mathrm{Str}\left[D(\nabla Q)^2 - 4EQ\right] \\
& + \frac{\pi\nu}{4}\int d\mathbf{r}\, \left[\Gamma_2 Q\gamma_2 Q - \Gamma_1 Q\gamma_1 Q\right]
\end{aligned}
\tag{109}
$$

where $\gamma_1$ and $\gamma_2$ are certain operators acting on the supermatrices $Q$, and $\Gamma_1$ and $\Gamma_2$ are scattering amplitudes characterizing the interaction (they are different from those obtained for the replica $\sigma$-model [24, 25]. As usual, one has the constraint $Q^2 = 1$ but now the product of two supermatrices includes summation also summation over the Matsubara frequencies.

Using the sigma-model Eq. (109) renormalization group equations of Refs. ([24, 25]) have been reproduced in the first order in the interaction constants and there is a hope to use this model for non-perturbative calculations.

## 3.2 Method of quasiclassical green functions

Although the interaction is included in Eq. (109), it is written for systems with considerably high concentration of short range impurities. Studying problems for clean systems or systems with a long range disorder one needs a different scheme. Statistical properties of clean chaotic systems are covered in this school by Boris Altshuler and I do not review them here. Instead, I want to concentrate on calculational schemes.

The saddle point equation (61) is not a good approximation for clean systems and systems with long range disorder and we cannot follow the same way as the one used for diffusive models. The method that I want to present now is based on using quasiclassical Green functions.

Introducing an $8 \times 8$ matrix function $G(\mathbf{r}, \mathbf{r}')$ as

$$G(\mathbf{r}, \mathbf{r}') = 2\langle\psi(\mathbf{r})\bar{\psi}(\mathbf{r}')\rangle_\psi \tag{110}$$

we can write in a standard way the following equation for this function

$$\left[H_{0\mathbf{r}} + U(\mathbf{r}) + \Lambda\frac{\omega + i\delta}{2} + iJ(\mathbf{r})\right] G(\mathbf{r}, \mathbf{r}') = i\delta(\mathbf{r} - \mathbf{r}') \tag{111}$$

where the subscript $\mathbf{r}$ of $H_{0\mathbf{r}}$ means that the operator acts on $\mathbf{r}$. The notations are the same as in Eqs. (53, 54), and $J(\mathbf{r})$ is a source term that allows to extract more complicated correlation functions.

Conjugating Eq.(111)we obtain another equation for the matrix $G(\mathbf{r}, \mathbf{r}')$ with the operator $H_{0\mathbf{r}'}$ acting on its second variable

$$G(\mathbf{r}; \mathbf{r}')\left[H_{0\mathbf{r}'} + U(\mathbf{r}') + \Lambda\frac{\omega + i\delta}{2} + iJ(\mathbf{r}')\right] = i\delta(\mathbf{r} - \mathbf{r}') \tag{112}$$

Until now no approximations have been done and Eqs. (111, 112) are exact. Now we can use the assumption that the potential $U(\mathbf{r})$ changes slowly on the wavelength $\lambda_F$. If the mean free path $l$ for the scattering on the random potential exceeds $\lambda_F$ the Green function varies as a function of $\mathbf{r} - \mathbf{r}'$ at distances of the order of $\lambda_F$ but, at the same time, is a slow function of $(\mathbf{r} + \mathbf{r}')/2$. The Fourier transform $G_\mathbf{p}((\mathbf{r} + \mathbf{r}')/2)$ of $G(\mathbf{r}, \mathbf{r}')$ respective to $\mathbf{r} - \mathbf{r}'$ has a sharp maximum near the Fermi surface. In order to cancel large terms we subtract Eq. (112) from Eq. (111). Using the assumption that the potential $U(\mathbf{r})$ is smooth and expanding it in gradients we obtain in the lowest order

$$\left[-\frac{i\mathbf{p}\nabla_{\mathbf{R}}}{m}+i\nabla_{\mathbf{R}}U(\mathbf{R})\frac{\partial}{\partial\mathbf{p}}\right]G_{\mathbf{p}}(\mathbf{R})+\frac{\omega+i\delta}{2}[\Lambda,G_{\mathbf{p}}(\mathbf{R})]+i[J(\mathbf{R}),G_{\mathbf{p}}(\mathbf{R})]=0 \tag{113}$$

where $\mathbf{R}=(\mathbf{r}+\mathbf{r}')/2$ and $[,]$ stands for the commutator. When deriving Eq. (113), not only the potential $U(\mathbf{r})$ but also the function $J(\mathbf{r})$ was assumed to be smooth.

The dependence of the Green function $G_{\mathbf{p}}(\mathbf{R})$ on $|\mathbf{p}|$ is more sharp than on other variables. In order to avoid this sharp dependence we integrate Eq. (113) over $|\mathbf{p}|$. Of course, this procedure makes a sense for very large samples when the level discreteness can be neglected.

The most interesting contribution in the integral over $|\mathbf{p}|$ comes from the vicinity of the Fermi-surface. A contribution given by momenta considerably different from $p_F$ is proportional to the unity matrix an drops out from Eq. (113).

Introducing the function $g_{\mathbf{n}}(\mathbf{r})$

$$g_{\mathbf{n}}(\mathbf{r})=\frac{1}{\pi}\int G_{p\mathbf{n}}(\mathbf{r})\,d\xi,\ \xi=\frac{p^2-p_F^2}{2m} \tag{114}$$

where $\mathbf{n}$ is a unite vector pointing a direction on the Fermi surface, we obtain the final quasiclassical equation

$$\left(v_F\mathbf{n}\nabla-p_F^{-1}\nabla_{\mathbf{r}}U(\mathbf{r})\,\partial_{\mathbf{n}}\right)g_{\mathbf{n}}(\mathbf{r})+\frac{i(\omega+i\delta)}{2}[\Lambda,g_{\mathbf{n}}(\mathbf{r})]-[J,g_{\mathbf{n}}]=0 \tag{115}$$

where

$$\partial_{\mathbf{n}}=\nabla_{\mathbf{n}}-\mathbf{n},\ \nabla_{\mathbf{n}}=-[\mathbf{n}\times[\mathbf{n}\times\frac{\partial}{\partial\mathbf{n}}]]$$

Eq. (115) should be complemented by a boundary condition at the surface of the sample. Considering a closed sample we assume that the current across the border is equal to zero. This leads the boundary condition at the surface

$$g_{\mathbf{n}_\perp}(\mathbf{r})|_{surface}=g_{-\mathbf{n}_\perp}(\mathbf{r})|_{surface} \tag{116}$$

where $\mathbf{n}_\perp$ is the component of the vector $\mathbf{n}$ perpendicular to the surface.

Eq. (115) is similar to an Eilenberger equation written long ago in superconductivity theory [50]. As in the theory of superconductivity, the solution for the Eq.(115) satisfies the condition $g_{\mathbf{n}}^2(\mathbf{r})=\hat{1}$. Eq. (115) is written for a non-averaged potential $U(\mathbf{r})$ and it is valid also in the absence of the long range potential.

The quasiclassical equation, Eq. (115), has been written first by Muzykantskii and Khmelnitskii, Ref. [51] who guessed a functional for which Eq. (115) is just a condition for an extremum. Then, they proceeded to work with this functional without estimating fluctuations near this minimum.

It came later as a surprise that, actually, one could write the proper solution of Eq. (115) in terms of a functional integral over supermatrices exactly[52]. The *exact* solution for Eq. (115) can be written as

$$\begin{aligned} g_{\mathbf{n}}(\mathbf{r}) &= Z^{-1}[J]\int_{Q_{\mathbf{n}}^2=1} Q_{\mathbf{n}}(\mathbf{r})\exp\left(-\frac{\pi\nu}{2}\Phi_J[Q_{\mathbf{n}}(\mathbf{r})]\right)DQ_{\mathbf{n}}, \\ \Phi[Q_{\mathbf{n}}(\mathbf{r})] &= Str\int d\mathbf{r}d\mathbf{n}[\Lambda\bar{T}_{\mathbf{n}}(\mathbf{r})(v_F\mathbf{n}\nabla_{\mathbf{r}} \\ &- p_F^{-1}\nabla_{\mathbf{r}}U(\mathbf{r})\nabla_{\mathbf{n}})T_{\mathbf{n}}(\mathbf{r}) + \left(\frac{i\,(\omega+i\delta)}{2}\Lambda - J(\mathbf{r})\right)Q_{\mathbf{n}}(\mathbf{r})], \\ Q_{\mathbf{n}}(\mathbf{r}) &= T_{\mathbf{n}}(\mathbf{r})\Lambda\bar{T}_{\mathbf{n}}(\mathbf{r}),\ \ \bar{T}_{\mathbf{n}}(\mathbf{r})T_{\mathbf{n}}(\mathbf{r}) = 1 \end{aligned} \tag{117}$$

In Eq. (117), the partition function $Z[J(\mathbf{r})]$ is

$$Z[J] = \int_{Q_{\mathbf{n}}^2=1}\exp\left(-\frac{\pi\nu}{2}\Phi[Q_{\mathbf{n}}(\mathbf{r})]\right)DQ_{\mathbf{n}} \tag{118}$$

and the integration is performed over the self-conjugate supermatrices $Q_{\mathbf{n}} = \bar{Q}_{\mathbf{n}}(\mathbf{r})$ satisfying the following relation

$$Q_{\mathbf{n}}^2(\mathbf{r}) = 1 \tag{119}$$

We see that the quasiclassical Green function $g_{\mathbf{n}}(r)$ can be written in the form of a functional integral over supermatrices $Q_{\mathbf{n}}(\mathbf{r})$ depending both on the coordinates $\mathbf{r}$ and the direction of the momentum $\mathbf{n}$ and satisfying the constraint, Eq. (119). The first term in the free energy functional is written in terms of the supermatrices $T_{\mathbf{n}}$ rather than $Q_{\mathbf{n}}$.. However, it can be written in a form of a Wess-Zumino-Novikov-Witten term containing the supermatrices $Q$ only[51]. Writing this term one should introduce an additional coordinate varying at the interval $[0,1]$.

The model described by the functional $\Phi$, Eqs. (117), is usually referred to as a "ballistic $\sigma$-model". The partition function $Z[J]$, Eq. (118), is unity at $J=0$ due to the supersymmetry and this allows us to average (if necessary) over the smooth potential $U(\mathbf{r})$.

The method of quasiclassical Green functions suggested here for a static external potential does not seem to be restricted by the non-interacting case. There are indications that it can be generalized to describe clean interacting systems. Of course, in this case one should write the quasiclassical equations in time representation because the interaction mixes states with different energies. Study of interacting systems with this method may be a very interesting direction of research.

As concerns attempts to prove the Wigner-Dyson statistics for clean non-interacting systems one can try to start with the functional $\Phi$, Eq. (117). At first glance, we should simply restrict ourselves with the integration over $Q$ depending neither on the coordinates, nor on the momenta. Then, the functional $\Phi$ would contain only the last term and we would have the zero dimensional $\sigma$-model, which leads immediately to the WD statistics.

However, a very important question is whether one averages over the potential $U(\mathbf{r})$ or puts $U(\mathbf{r}) = 0$ and averages over the spectrum. In the former case one gets after averaging over $U(\mathbf{r})$ an additional term in functional $\Phi$ quadratic in gradients. This term leads eventually to a suppression of non-zero harmonics and one can really obtain the zero-dimensional $\sigma$-model (see [52] and references therein).

The situation with $U(\mathbf{r}) = 0$ and averaging over the energy is more interesting. Everything depends on whether the system is classically integrable or chaotic. It is just the situation for which the authors of Ref. [16] made their hypothesis.

It turns out that within the model with the functional $\Phi$, Eq. (117), and $U(\mathbf{r}) = 0$ one cannot come to the zero-dimensional $\sigma$-model. There is a common consensus that a "regularizer" (see e.g. [53]) containing something like square of gradients in coordinates or momenta is necessary in the correct ballistic $\sigma$-model. Aleiner and Larkin [54] argued that in order to come to the zero dimensional $\sigma$-model one had to take into account diffraction, which is clearly absent in the ballistic $\sigma$-model, Eq. (117). They did not manage to include the diffraction in their calculational scheme microscopically and mimiced it by introducing artificial quantum impurities that would correspond to the potential $U(\mathbf{r})$ in Eq. (117). This allowed them to come to the zero dimensional $\sigma$-model, confirm the WD statistics and calculate corrections to it. It is worth emphasizing that the effective potential $U(\mathbf{r})$ was very weak such that the computation was done in the ballistic regime.

As concerns the real physical diffraction, it cannot be directly included in the $\sigma$-model using the quasiclassical scheme and a more sophisticated approach is necessary.

## 3.3 Beyond the quasiclassics

We have seen that the solution of the equations for the Green functions, Eqs. (111, 112), can be written in the quasiclassical approximation in terms of the functional integral over $8 \times 8$ supermatrices $Q_{\mathrm{n}}(\mathbf{r})$. For certain problems this approximation is not sufficient and the natural question is: can we do better than that and find a solution for the Green functions in terms of a functional integral over supermatrices valid at all distances including those of the order of the electron wavelength $\lambda_F$ ?

This attempt has been undertaken recently in Ref. [55], where an integral of such a type was suggested for a solution of Eq. (111). The idea is rather close to the one known in field theory where it is called bosonization[56]. The final expressions obtained in Ref. ([56]) are rather complicated and this method, to the best of my knowledge, has not evolved into an efficient calculational tool.

However, the supersymmetric form of the Green functions considered here seems to promise more and the derivation is rather simple. I follow here a simpler derivation of Ref. [57].

What I want to show now is that the Green function, Eq. (111), can be represented exactly as an integral over supermatrices $Q(\mathbf{r},\mathbf{r}')$ depending on two coordinates $\mathbf{r}$ and $\mathbf{r}\prime$

$$G(\mathbf{r},\mathbf{r}') = Z^{-1}[J]\int Q(\mathbf{r},\mathbf{r}')\exp(-\Phi[Q])DQ \tag{120}$$

where $Z[J]$ is a new partition function

$$Z[J] = \int \exp(-\Phi[Q])DQ \tag{121}$$

and the functional $\Phi[Q]$ has the form

$$\begin{aligned}\Phi[Q] = \frac{i}{2}\mathrm{Str}\int \left(H_{0\mathbf{r}} + U(\mathbf{r}) + \frac{\omega + i\delta}{2}\Lambda\right) \\ \times\, \delta(\mathbf{r}-\mathbf{r}')Q(\mathbf{r},\mathbf{r}')d\mathbf{r}d\mathbf{r}' \qquad (122) \\ +\frac{1}{2}\mathrm{Str}\ln Q - \frac{1}{2}\mathrm{Str}\int J(\mathbf{r},\mathbf{r}')Q(\mathbf{r}',\mathbf{r})d\mathbf{r}d\mathbf{r}'\end{aligned}$$

where $J(\mathbf{r},\mathbf{r}')$ is a source term. The structure of the supermatrix $Q(\mathbf{r},\mathbf{r}')$ in the integral Eq.(120) should be the same as that of the Green function $G(\mathbf{r},\mathbf{r}')$, i.e. be the same as of the product $\psi(\mathbf{r})\bar{\psi}(\mathbf{r}')$. In particular, this means that $Q(\mathbf{r},\mathbf{r}')$ is self-conjugated

$$\bar{Q}(\mathbf{r},\mathbf{r}') \equiv C\, Q^T(\mathbf{r}',\mathbf{r})C^T = Q(\mathbf{r},\mathbf{r}') \tag{123}$$

In order to prove Eq.(120) we write the following identity

$$\begin{aligned}&-2iZ^{-1}[J]\int\left[\int \frac{\delta\exp\left(-\frac{1}{2}\mathrm{Str}\ln \mathrm{Q}\right)}{\delta Q(\mathbf{r}'',\mathbf{r})}Q(\mathbf{r}'',\mathbf{r}')d\mathbf{r}''\right]\times \\ &\exp\left(-\frac{i}{2}\mathrm{Str}\left[\hat{H}_{0\mathbf{r}} + \frac{\omega+i\delta}{2}\Lambda + iJ\right]Q\right)DQ = \\ &= i\delta(\mathbf{r}-\mathbf{r}'),\end{aligned} \tag{124}$$

and integrate over $Q$ by parts. The derivative $\delta/\delta Q$ should act now on both $Q$ and the exponential. At this point, the supersymmetry plays a crucial role. Differentiating first the supermatrix $Q$ we obtain the supermatrix product $(\delta/\delta Q)\, Q$. As the number of the anticommuting variables in the sum over the matrix elements is equal to the number of the boson ones and the derivatives have the opposite signs, this matrix product vanishes. Differentiating the exponential only we come to the following equation

$$Z^{-1}[J]\int d\mathbf{r}''\left(\hat{H}_{0\mathbf{r}}+U\left(\mathbf{r}\right)\frac{\omega+i\delta}{2}\Lambda+iJ\right)(\mathbf{r},\mathbf{r}'')\times$$
$$\int Q(\mathbf{r}'',\mathbf{r}')\exp(-\Phi[Q])DQ=i\delta(\mathbf{r}-\mathbf{r}') \tag{125}$$

Eq. (125) proves immediately that the integral Eq.(120) does satisfy Eq.(111) and we have really the alternative representation of the Green function in terms of an integral over the supermatrices $Q$.

Making the Fourier transform $Q\left(\mathbf{r}',\mathbf{r}''\right)$ in the difference $\mathbf{r}'-\mathbf{r}''$ (Wigner transformation) one can express the functional $\Phi\left[Q\right]$ in terms of the variables $Q_{\mathbf{p}}\left(\mathbf{r}\right)$, where $\mathbf{p}$ is the momentum and $\mathbf{r}$ is the center of mass $\left(\mathbf{r}'+\mathbf{r}''\right)/2$. Then the free energy functional $\Phi$, Eq. (122) can be written as

$$\Phi[Q]=\frac{i}{2}\int \mathrm{Str}\left[\mathcal{H}_J(\mathbf{x})*Q(\mathbf{x})-i\ln Q(\mathbf{x})\right]d\mathbf{x} \tag{126}$$

where $\mathbf{x}=(\mathbf{r},\mathbf{p})$ is the coordinate in the phase space,

$$\mathcal{H}_J(\mathbf{x})=H_0\left(\mathbf{p}\right)+U\left(\mathbf{r}\right)+\frac{\omega+i\delta}{2}+iJ\left(\mathbf{p},\mathbf{r}\right) \tag{127}$$

is classical Hamilton function.

The product $*$ of two matrices $A\left(\mathbf{x}\right)$ and $B\left(\mathbf{x}\right)$ is defined by Moyal formula

$$A(\mathbf{x})*B(\mathbf{x})=A(\mathbf{x})e^{\frac{i}{2}\left(\overleftarrow{\nabla}_{\mathbf{r}}\overrightarrow{\nabla}_{\mathbf{p}}-\overrightarrow{\nabla}_{\mathbf{r}}\overleftarrow{\nabla}_{\mathbf{p}}\right)}B(\mathbf{x}).$$

The scheme of calculations using the Wigner representations and the star product "$*$" is known as Weyl symbol calculus [58]. This method is convenient for quasiclassical expansions.

If the potential $U\left(\mathbf{r}\right)$ is smooth, one can simplify Eq. (126) and come again to Eq. (117). This procedure is described in Ref. [55]. The functional $\Phi$, Eq. (126) has a form of the Lagrangian of a non-commutative field theory [58]. The method suggested here can naturally be called "superbosonization".

At the end I have to warn that Eqs. (120-122, 126) are not complete yet because nothing has been said about the contour of integration over the supermatrices $Q$. This was not very important in Refs. ([55, 57]) where a saddle point approximation was used. However, generally, this question requires a more careful study and this is a subject of a current work. In case if the difficulties are overcome, the superbosonization can become very useful in the theory of random matrices for more general models than the one described by the conventional Eq. (1).

## 4. Summary

In these lectures I tried to achieve two goals:

1. The random matrix theory is to a large extent a phenomenological theory. Therefore, it is very important to have examples when it can be obtained starting from a microscopic model. The model of the disordered metals considered here is the first one for which the relevance of the RMT has been proven. The proof became possible with the help of the supermatrix non-linear $\sigma$-model first derived for other purposes.

2. Having presented the derivation of the $\sigma$-model I demonstrated how one obtains the Wigner-Dyson statistics from its zero-dimensional version. However, in many situations the supersymmetry method allows to go beyond the Wigner-Dyson model and obtain completely different results like the log-normal distribution of the amplitudes of the wave functions of Section 2. Of course, as soon as one has a Hamiltonian one comes to random matrices. However, generally it is not clear how to write the distribution function for these matrices in each particular situation and the supersymmetry method can play an important role for investigation of microscopic models.

I wanted by no means to oppose the RMT and the supersymmetry method to each other. They can be considered as complimentary methods, although with a considerable overlap. We see at this school that the random matrices find more and more applications in many fields of physics, which is an exciting development. I believe, in many cases the supersymmetry technique can also be useful in these new applications and one should keep in mind a possibility of using this scheme.

## Acknowledgements

This work was supported by the Transregio 12 "Symmetries and Universality in Mesoscopic Systems" of German Research Society".

## References

[1] E.P. Wigner, Ann. Math. **53**, 36 (1951)

[2] L.M. Mehta, *Random Matrices,* Academic Press, San Diego (1991)

[3] F.J. Dyson, J. Math. Phys. **3**, 140, 157, 166 (1962)

[4] K.B. Efetov, *Supersymmetry in Disorder and Chaos,* Cambridge University Press, New York (1997)

[5] J.J.M. Verbaarschot (this volume)

[6] A. Altland, and M.R. Zirnbauer, Phys. Rev. B **55**, 1142 (1997)

[7] L.P. Gorkov and G.M. Eliashberg, Zh. Eksp. Teor. Fiz. **48,** 1407 (1965) (Sov. Phys. JETP **21**, 940 (1965))

[8] A.A. Abrikosov, L.P. Gorkov, and I.E. Dzyaloshinskii, *Methods of Quantum Field Theory in Statistical Physics,* Prentice Hall, New York (1963)

[9] R. Kubo, J. Phys. Soc. Jpn. **17**, 975 (1962)

[10] R. Kubo, in *Polarization, Matiere et Rayonnement,* livre jubile en l'honneur du Professeur A. Kastler, Presses Universitaires de France, Paris, p.325 (1969)

[11] E. Abrahams, P.W. Anderson, D.C. Licciardello, and T.V. Ramakrishnan, Phys. Rev. Lett. **42**, 673 (1979)

[12] L.P. Gorkov, A.I. Larkin, and D.E. Khmelnitskii, Pis'ma Zh. Eksp. Teor. Fiz. **30,** 248 (1979) (Sov. Phys. JETP Lett. **30**, 228 (1979))

[13] K.B. Efetov, Zh. Eksp. Teor. Fiz. **82**, 872 (1982) (Sov. Phys. JETP **55,** 514 (1982); J. Phys. C **15**, L909 (1982); Zh. Eksp. Teor. Fiz. **83**, 833 (Sov. Phys. JETP **56**, 467 (1982)

[14] K.B. Efetov, Adv. Phys. **32**, 53 (1983)

[15] J.J.M. Verbaarschot, H.A. Weidenmuller, and M.R. Zirnbauer, Phys. Rep. **129**, 367 (1985)

[16] O. Bohigas, M.J. Gianonni and C. Schmidt, Phys. Rev. Lett. **52**, 1 (1984); J. Physique Lett. **45,** L1615 (1984)

[17] *Supersymmetry and Trace Formula, (*edited by J.P. Keating, D.E. Khmelnitskii, and I.V. Lerner), NATO ASI Series B: Physics Vol. **370**, Kluwer Academic, New York (1999)

[18] F. Wegner, Z. Phys. B **35**, 207(1979)

[19] K.B. Efetov, A.I. Larkin, D.E. Khemelnitskii, Zh. Eksp. Teor. Fiz. **79**, 1120 (1980) (Sov. Phys. JETP, **52**, 568 (1980))

[20] L. Schafer and F. Wegner, Z. Phys. B **38**, 113 (1980)

[21] A. Kamenev and M. Mezard, J. Phys. A: Math. Gen. **32**, 4373 (1999)

[22] I.V. Yurkevich and I.V. Lerner, Phys. Rev. B **60**, 3955 (1999)

[23] E. Kanzieper, Phys. Rev. Lett. **89**, 250201 (2002)

[24] A.M. Finkel'stein, Zh. Eksp. Teor. Fiz. **84**, 168 (1983) (Sov. Phys. JETP **57**, 97 (1983))

[25] A.M. Finkel'stein, in *Electron Liquid in Disordered Conductors,* edited by I.M. Khalatnikov, Soviet Scientific Reviews, Vol. 14 (Harwood, London, 1990)

[26] G. Schwiete and K.B. Efetov, Phys. Rev. B, 2005 (in press), cond-mat/0409546

[27] Y.V. Fyodorov, H.J. Sommers, J. Math. Phys. **38**, 1918 (1997)

[28] T. Guhr, A. Muller-Groeling, H.A.Weidenmuller, Phys. Rep. **299**, 190 (1998)

[29] A.D. Mirlin, Phys. Rep. **326**, 259 (2000)

[30] A. Altland, B.D. Simons, D. Taras-Semchuk, Adv. Phys. **49**, 321 (2000)

[31] A. Altland, B.D. Simons, M.R. Zirnbauer, Phys. Rep. **359**, 283 (2002)

[32] G. Parisi and N. Sourlas, Pys. Rev. Lett. **43**, 744 (1979)

[33] Dokl. Akad. Nauk SSSR, **137**, 31 (1961); *Metod vtorichnogo kvantovaniya, (The method of Second Quantization),* Nauka (1965)*; Introduction to Superanalysis,* MPAP, Vol. 9 D. Reidel, Dordrecht (1987)

[34] B.L. Altshuler and B.I. Shklovskii Zh. Eksp. Teor. Fiz. **91**, 220 (1986) (Sov. Phys. JETP **64**, 127 (1986)

[35] K.B. Efetov and A.I. Larkin, Zh. Eksp. Teor. Fiz. **85**, 764 (1983) (Sov. Phys. JETP, **58**, 444 (1983)

[36] Y.V. Fyodorov and A.D. Mirlin, Phys. Rev. Lett. **67**, 2405 (1991)

[37] K.B. Efetov, Zh. Eksp. Teor. Fiz. **88**, 1032-1052 (1985); (Sov. Phys. JETP. **61**, 606-617, 1985)

[38] A.D. Mirlin and Y.V. Fyodorov, J. Phys. A **24**, 2273 (1991)

[39] M. Mezard and G. Parisi, J. Phys. (Paris) **47**, 1285 (1985); Europhys. Lett. **3**, 1067 (1987)

[40] V.I. Falko and K.B. Efetov, Europhys. Lett. **32**, 627 (1995); Phys. Rev. B**52**, 17413 (1995)

[41] C.E. Porter, *Statistical Theories of Spectra: Fluctuations,* Academic Press, New York (1965)

[42] T.A. Brody, J. Flores, J.B. French, P.A. Mello, A. Panday, and S.S.M. Wong, Rev. Mod. Phys. **53**, 385 (1981)

[43] A.M. Polyakov, Phys. Lett. **103B** 207, 211 (1981)

[44] N. Seiberg, Prog. Theor. Phys. **102** Supl., 319 (1990)

[45] A. Tsvelik, C. Mudry, and Y. Kogan, Phys. Rev. Lett, **77,** 707 (1996)

[46] B.A. Muzykantskii and D.E. Khmelnitskii, Phys. Rev. B **51**, 5480 (1995)

[47] Y.V. Fyodorov and A.D. Mirlin, JETP Lett. **60**, 790 (1994)

[48] B.L. Altshuler, V.E. Kravtsov, and I.V. Lerner, Pis'ma Zh. Eksp. Teor. Fiz. **43**, 342 (1986) (Sov. Phys. JETP Lett. **43,** 441 (1986)); Zh. Eksp. Teor. Fiz. **91**, 2276 (1986) (Sov. Phys. JETP **64**, 1352 (1986)); Phys. Lett. A **134,** 488 (1989)

[49] C.W.J. Beenakker, Rev. Mod. Phys. **69**, 733 (1997)

[50] G. Eilenberger, Z. Phys. **214**, 195 (1968)

[51] B. A. Muzykantskii and D. E. Khmelnitskii, Pis'ma Zh. Eksp. Theo. Fiz. **62** 68 (1995) [JETP Lett. **62** 76 (1995)]; (also see in Ref. [17]).

[52] K. B. Efetov, V. R. Kogan, Phys. Rev. B **67**, 245312 (2003)

[53] A. Altland, C.R. Offer, and B.D. Simons in [17].

[54] I.L. Aleiner and A.I. Larkin, Phys. Rev. B**54**, 14423 (1996); Phys. Rev. E **55** (2): R1243 (1997)

[55] K. B. Efetov, G. Schwiete, K. Takahashi, Phys. Rev. Lett. **92**, 026807 (2004)

[56] P.B. Wiegmann, Nucl. Phys. B **323**, 311 (1989); S.R. Das, A. Dhar, G. Mandal, and S.R. Wadia, Mod. Phys. Lett. A **7**, 71 (1992); A. Dhar, G. Mandal, and S.R. Wadia, *ibid.* **7**, 3129 (1992); D.V. Khveshchenko, Phys. Rev. B **49**, 16893 (1994); *ibid.* **52**, 4833 (1995).

[57] K. B. Efetov, V. R. Kogan, Rev. B **70**, 195326 (2004)

[58] S.W. Mc Donald, Phys. Rep. **158**, 337 (1988)

# HYDRODYNAMICS OF CORRELATED SYSTEMS

## *Emptiness Formation Probability and Random Matrices**

Alexander G. Abanov
*Department of Physics and Astronomy, Stony Brook University, Stony Brook, NY 11794-3800, U.S.A.*
alexandre.abanov@sunysb.edu

**Abstract** A hydrodynamic approach is used to calculate an asymptotics of the *Emptiness Formation Probability* – the probability of a formation of an empty space in the ground state of a quantum one-dimensional many body system. Quantum hydrodynamics of a system is represented as a Euclidian path integral over configurations of hydrodynamic variables. In the limit of a large size of the empty space, the probability is dominated by an instanton configuration, and the problem is reduced to the finding of an instanton solution of classical hydrodynamic equations. After establishing a general formalism, we carry out this calculation for several simple systems – free fermions with an arbitrary dispersion and Calogero-Sutherland model. For these systems we confirm the obtained results by comparison with exact results known in Random Matrix theory. We argue that the nonlinear hydrodynamic approach might be useful even in cases where the linearized hydrodynamics fails.

**Keywords:** Quantum Hydrodynamics, Random Matrices, Instanton, Rare Fluctuation, Emptiness Formation Probability, Calogero-Sutherland Model.

## 1. Introduction

The hydrodynamic approach to correlation functions in quantum many body systems has a long history [1, 2]. Generally, hydrodynamic equations are nonlinear and dispersive. However, usually, it is the linearized hydrodynamics which is used to extract long distance asymptotics of correlation functions [2]. In particular, in one spatial dimension the linearized quantum hydrodynamics (or bosonization) is especially useful [3]. It was argued that corrections to the bosonization due to the nonlinearities of the hydrodynamics (curvature of an

*This is an extended version of the seminar given at the School "Applications of Random Matrices in Physics", Les Houches, June 2004



*E. Brezin et al. (eds.), Applications of Random Matrices in Physics,* 139–161.

underlying fermionic spectrum) are irrelevant in calculation of leading terms of correlation functions [4].

Although the linearized hydrodynamics turned out to be a very powerful tool in studies of correlation functions, there are many phenomena for which nonlinearities are essential. The goal of this lecture is to consider a particular example where the linear approximation fails but nonlinear hydrodynamics can still be used to extract non-perturbative results for correlation functions.

As such an example we consider a particular quantity which plays an important role in the theory of quantum, one-dimensional, integrable systems [5] – the Emptiness Formation Probability (EFP). This correlation function was argued to be the simplest of correlators in some integrable models [5]. The EFP $P(R)$ is essentially, a probability of formation of an empty region of a size $2R$ in the ground state of the many body system. In integrable models the EFP has an exact representation in terms of determinants of Fredholm operators [5, 6] or multiple integrals [7]. These expressions are exact but are very complex, and extracting, e.g., long distance asymptotics $R \to \infty$ from the exact expressions is a non-trivial problem (see Appendix A and references therein for some known results). However, the limit of large $R$ is precisely the limit where hydrodynamic description is applicable, and we illustrate that it, indeed, correctly produces the leading term of the asymptotics of the EFP. The hydrodynamic approach can also be used to calculate more complicated quantities than EFP (which is, essentially, not a dynamic but a ground state property). Moreover, the applicability of hydrodynamics is not limited to integrable systems.

We start with some definitions as well as with making a connection to Random Matrices.

**Emptiness Formation Probability.** Consider a one-dimensional quantum liquid at zero temperature. The wave function of the ground state of the liquid $\Psi_G(x_1, x_2, \ldots, x_N)$ gives the probability distribution $|\Psi_G|^2$ of having all $N$ particles at given positions $x_j$, where $j = 1, \ldots, N$. We introduce the *Emptiness Formation Probability* (EFP) $P(R)$ as a probability of having no particles with coordinates $-R < x_j < R$. Formally we define

$$P(R) = \frac{1}{\langle \Psi_G | \Psi_G \rangle} \int_{|x_j|>R} dx_1 \ldots dx_N \, |\Psi_G(x_1, \ldots, x_N)|^2, \tag{1}$$

or following Ref.[5]

$$P(R) = \lim_{\alpha \to +\infty} \langle \Psi_G | \, e^{-\alpha \int_{-R}^{R} \rho(x)\, dx} \, | \Psi_G \rangle \,, \tag{2}$$

where $\rho(x)$ is a particle density operator

$$\rho(x) \equiv \sum_{j=1}^{N} \delta(x - x_j). \tag{3}$$

We are interested in an asymptotic behavior of $P(R)$ as $\rho_0 R \to \infty$, where $\rho_0$ is the average density of particles in the system. EFP $P(R)$ gives us the probability that the one-dimensional river parts to make a ford of a macroscopic size $2R$.

**Random Matrices.** The EFP (1) introduced for a general one-dimensional quantum liquid is a well-known quantity in the context of spectra of random matrices [8]. Namely, it is the probability of having no eigenvalues in some range. Consider e.g., the joint eigenvalue distribution for the Circular Unitary Ensemble (CUE). The CUE is defined as an ensemble of $N \times N$ unitary matrices with the measure given by de Haar measure. Diagonalizing matrices and integrating out unitary rotations, one obtains [9]

$$\int DU \to \int \prod_{j=1}^{N} d\theta_j \prod_{1\leq j<k\leq N} \left| e^{i\theta_j} - e^{i\theta_k} \right|^{\beta}, \tag{4}$$

where $\beta = 2$ for CUE and $e^{i\theta_j}$ with $j = 1, \ldots, N$ are the eigenvalues of a unitary matrix. One can read the joint eigenvalue distribution

$$\begin{aligned} P_N(\theta_1, \ldots, \theta_N) &= const. \prod_{1\leq j<k\leq N} \left| e^{i\theta_j} - e^{i\theta_k} \right|^{\beta} \\ &\sim \exp\left\{ \frac{\beta}{2} N^2 \int \frac{d\theta\, d\theta'}{(2\pi)^2} \rho(\theta) \ln\left| e^{i\theta_j} - e^{i\theta_k} \right| \rho(\theta') \right\}. \end{aligned} \tag{5}$$

Here we replaced the sums over particles (eigenvalues) to integrals with particle densities. We left only terms which are dominant in the limit of large $N$. [1] Now we introduce the probability of having no eigenvalues on the arc $-\alpha < \theta < \alpha$ as

$$P(\alpha) = \frac{1}{N} \int_{\theta_j \notin [-\alpha,\alpha]} \prod_{j=1}^{N} d\theta_j \prod_{1\leq j<k\leq N} \left| e^{i\theta_j} - e^{i\theta_k} \right|^{\beta}. \tag{6}$$

This quantity is known as $E_\beta(0, \alpha)$ in notations of [9]. For orthogonal, unitary, and symplectic circular ensembles the joint eigenvalue distribution is given by (5) with $\beta = 1, 2, 4$ respectively.

**Spin chains and lattice fermions.** The EFP can also be defined for spin chains where we are interested in the probability of having a ferromagnetic string of the length $n$ in the ground state of the spin chain. The Jordan-Wigner transformation maps spin $1/2$ chain to a one-dimensional lattice gas of spinless fermions. Under this mapping the ferromagnetic string corresponds to a string of empty lattice sites and one can write

$$P(R) = \left\langle \prod_{j=-R}^{R} \psi_j \psi_j^\dagger \right\rangle, \tag{7}$$

where $\psi_j, \psi_j^\dagger$ are annihilation and creation operators of spinless fermions on the lattice site $j$. Therefore, the probability of the formation of a ferromagnetic string in spin chains corresponds to the Emptiness Formation Probability of Jordan-Wigner fermions. We are going to use a language of particles in this paper but all results are also valid for corresponding one-dimensional spin systems.

## 2. Instanton or rare Ductuation method

In the limit of large $R$ ($\rho_0 R \gg 1$) we use a collective description instead of dealing with individual particles. We assume that $u$ is some collective field describing the hydrodynamic motion of a one-dimensional liquid. [2] The dynamics of the liquid is defined by a Euclidian partition function

$$Z = \int Du \; e^{-S[u]}, \tag{8}$$

where $S[u]$ is the Euclidian action

$$S[u] = \int dx \int_{-1/2T}^{1/2T} d\tau \; L(u, \dot{u}) \tag{9}$$

and the inverse temperature $1/T$ defines periodic boundary conditions in the imaginary time $\tau$.

The asymptotic behavior of $P(R)$ in the limit of large $R$ is defined by a rare fluctuation when all particles move away so that at some time $t = 0$ we have no particles in the spatial interval $[-R, R]$. Then with an exponential accuracy

$$P(R) \sim e^{-S_{opt}}. \tag{10}$$

Here $S_{opt}$ is the value of the action (9) on the trajectory $u(x,t)$ which minimizes (9) and is subject to EFP boundary conditions. These are: the EFP boundary condition

$$\rho(t = 0; -R < x < R) = 0, \tag{11}$$

and standard boundary conditions at infinity

$$\begin{aligned} \rho &\to \rho_0, \quad x, \tau \to \infty, \\ v &\to 0, \quad x, \tau \to \infty, \end{aligned} \tag{12}$$

where $\rho, v$ are the density and velocity of particles related somehow to the collective field $u$.

Let us estimate the probability of such rare fluctuation[3]. Without the boundary condition (11) the minimum of the action subject to (12) is obviously given by the constant solution $\rho(x,t) = \rho_0$, $v(x,t) = 0$. The condition (11), however, disturbs this constant solution in some area of space-time around the origin. The spatial extent of this disturbance is of the order of $R$. If we assume that the quantum system we are dealing with is some *compressible* liquid we expect that the typical temporal scale of the disturbance is of the order of $R/v_s$, where $v_s$ is a sound velocity at $\rho = \rho_0$. We conclude that the (space-time) "area" of the disturbance scales as $R^2$ and the action $S_{opt} \sim R^2$. Therefore, we expect a Gaussian decay for (10)

$$P(R) \sim e^{-\alpha R^2}, \tag{13}$$

where $\alpha$ is some (non-universal) constant depending on the details of (9).

This argument can be extended for the case of low but finite temperature. Namely, while the temporal extent of the instanton $R/v_s$ is smaller than the inverse temperature $R/v_s \ll 1/T$ instanton "does not know" that the temperature is not zero and one obtains an intermediate Gaussian decay of EFP (13). However, for long enough $R$ it is the $1/T$ scale that defines the temporal size of the instanton, the space-time area of the disturbance scales as $R$ and one obtains

$$P(R) \sim e^{-\gamma R}. \tag{14}$$

One expects a crossover between the Gaussian and the exponential behavior taking place at $R \sim v_s/T$. The result (14) is very familiar from statistical physics with $\gamma$ measuring the difference (per unit length) of free energies between the true ground state of the liquid and the empty state. The Gaussian decay (13) is a manifestation of an effective increase of the dimensionality from one to two in quantum systems at $T = 0$.

In the above argument we used the assumption of compressibility of the quantum liquid. For incompressible liquids one has a finite correlation length of density fluctuations. The argument can be extended then to obtain an exponential form (14) with the correlation length playing the role of an inverse temperature [11, 12].

## 3. Hydrodynamic approach

The collective description we are looking for is nothing else but a hydrodynamic description in terms of density $\rho$ and current $j = \rho v$ (or velocity $v$) of a one-dimensional liquid. For simplicity, let us consider first the case of a system with Galilean invariance. Then we write the Euclidian action of a liquid as

$$S = \int d^2x \left[ \frac{j^2}{2\rho} + \rho\epsilon(\rho) + \dots \right], \tag{15}$$

where the first term is fixed by the Galilean invariance and is the kinetic energy of the liquid moving as a whole. The second term is the internal energy of the fluid which is determined by the equation of state of the liquid. $\epsilon(\rho)$ is the internal energy per particle at given density $\rho$. The terms denoted by dots are the terms which depend on density and its spatial derivatives. These terms will be small in the problems, where density and velocity gradients are small compared to $\rho_0$. Let us now remember that due to the particle conservation the density and current are not independent variables but are related by a constraint – the continuity equation

$$\partial_t \rho + \partial_x j = 0. \tag{16}$$

One can easily solve the constraint (16) introducing a particle displacement field $u$ such that

$$\begin{aligned} \rho &= \rho_0 + \partial_x u, \\ j &= -\partial_t u. \end{aligned} \tag{17}$$

Microscopic definition of the displacement field is $u(x)+\rho_0 x = \sum_j \theta(x-x_j)$, where $\theta(x)$ is a step function and $x_j$-s are coordinates of particles. It is easy to check that the configuration $u(x,t)$ minimizing $S[u]$ from (15) is given by $\delta S = 0$ or after simple algebra

$$\partial_t v + v\partial_x v = \partial_x \partial_\rho [\rho\epsilon(\rho)], \tag{18}$$

which is the Euler equation of a one-dimensional hydrodynamics. The sign of the r.h.s. of (18) differs from the conventional minus sign because we work in the Euclidian formulation.

The action (15) with the parameterization (17) provide us with a variational formulation of one-dimensional classical hydrodynamics. To calculate the probability of hydrodynamic fluctuations at zero temperature we have to quantize this hydrodynamics. To the best of my knowledge, the first "quantization" of hydrodynamics was done by L. D. Landau in Ref.[1], where he used essentially (15) as a quantum Hamiltonian of the liquid with density and velocity fields satisfying commutation relations $[\rho(x), v(y)] = -i\partial_x \delta(x-y)$. For the purpose of evaluating a rare fluctuation the path integral approach is more useful and we use the partition function (7), where $u(x,t)$ is the displacement field and the action $S[u]$ is given by (15) and (17). We notice here that we did not specify the measure of integration $Du$ in (7). Finding this measure requires a derivation of an effective hydrodynamic formulation from the underlying microscopic physics. However, the "non-flatness" of the measure gives only gradient corrections similar to the already omitted terms denoted by dots in (15). Those corrections to the measure and the action will produce subdominant contributions to the asymptotics of EFP and will be neglected in this work.

Let us now summarize our strategy for the calculation of the leading term of the EFP. We solve *classical* equations of motion (16,18) with EFP boundary conditions (11,12) and then find the value $S_{opt}$ of (15) on the obtained solution. Finally, (10) will give us the dominant contribution to the EFP at $R \to \infty$.

We conclude this section with two remarks. First, it will be convenient for us to generalize the problem and replace the EFP boundary condition (11) by a slightly more general depletion formation probability (DFP) boundary condition [10]

$$\rho(t = 0; -R < x < R) = \bar{\rho}, \tag{19}$$

where $\bar{\rho}$ is some constant density. In the case $\bar{\rho} = 0$ we obtain the EFP problem while for $\bar{\rho}$ close to $\rho_0$ one can use the bosonization technique to calculate $P(R; \bar{\rho})$.

Second remark is that one can easily obtain the functional dependence of $P(R)$ even without solving hydrodynamic equations (16,18) using very simple scaling arguments. Indeed, the equations (16,18) are uniform in space and time derivatives so that if $\rho(x, t)$ and $v(x, t)$ are solutions, then $\rho(\lambda x, \lambda t)$, $v(\lambda x, \lambda t)$ are also solutions of hydrodynamic equations. Choosing $\lambda = R$ we obtain the boundary condition (19) as $\rho(t = 0; -1 < x < 1) = \bar{\rho}$ and the only dependence on $R$ is left over in the integration measure of (15), which gives $S_{opt} \sim R^2$. Thus, we obtain a Gaussian decay of DFP (or EFP in particular) as a function of $R$. Corrections to this behavior come from the terms of higher order in gradients in the hydrodynamic action (denoted by dots in (15)) as well as from fluctuations around the saddle point (classical) trajectory in the partition function (7).

## 4. Linearized hydrodynamics or bosonization

Before proceeding to the general case, let us consider the DFP for

$$\frac{\rho_0 - \bar{\rho}}{\rho_0} \ll 1, \tag{20}$$

i.e., the probability of formation of a small constant density depletion along the long string $-R < x < R$. In this case the deviation of the density from $\rho_0$ is small almost everywhere (see below) and one can use a linearized version of generally nonlinear hydrodynamic equations (16,18). Expanding the classical action (15) in gradients of the displacement field $u$ (17) we obtain in harmonic approximation[4]

$$S_{bos} = \frac{v_{s0}}{\rho_0} \int d^2x \, \frac{1}{2} (\partial_\mu u)^2, \tag{21}$$

where we scaled $v_{s0}t \to t$. The sound velocity $v_{s0}$ at $\rho = \rho_0$ is defined as

$$v_{s0}^2 = \rho \partial_\rho^2 (\rho \epsilon(\rho)) \Big|_{\rho=\rho_0}. \tag{22}$$

The corresponding equation of motion is the Laplace equation

$$\Delta u = 0 \tag{23}$$

with the DFP boundary condition

$$u(x, t = 0) = -(\rho_0 - \bar{\rho})x, \quad \text{for } -R < x < R. \tag{24}$$

It is easy to see that the solution of (23,24) decaying sufficiently fast at infinity is given by

$$u(x,t) = -(\rho_0 - \bar{\rho}) \mathscr{R} \left( z_0 - \sqrt{z_0^2 - R^2} \right), \tag{25}$$

where we introduced the complex notation $z_0 = x + iv_{s0}t$ (where $t$ is the original imaginary time). Indeed, at $t = 0$, $-R < x < R$ the complex coordinate $z_0$ is real and square root in (25) is purely imaginary so that (24) is satisfied.

At space-time infinity $z_0 \to \infty$ we have

$$u(x,t) \approx -\frac{\alpha}{z_0} - \frac{\bar{\alpha}}{\bar{z}_0} \tag{26}$$

with

$$\alpha = \bar{\alpha} = \frac{1}{4} (\rho_0 - \bar{\rho}) R^2. \tag{27}$$

We obtain from (26) that at $z_0 \to \infty$

$$\begin{aligned} \rho &\approx \rho_0 + \frac{\alpha}{z_0^2} + \frac{\bar{\alpha}}{\bar{z}_0^2}, \\ v &\approx -i \frac{v_{s0}}{\rho_0} \left( \frac{\alpha}{z_0^2} - \frac{\bar{\alpha}}{\bar{z}_0^2} \right), \end{aligned} \tag{28}$$

which obviously satisfy the boundary conditions at infinity (12).

Now that we obtained the solution of hydrodynamic equations it is a straightforward problem to calculate the value of the action (21) on this solution[5] and obtain

$$S_{DFP} = \frac{1}{2} \frac{v_{s0}}{\pi \rho_0} \left[ \pi (\rho_0 - \bar{\rho}) R \right]^2. \tag{29}$$

We note that the linearization of hydrodynamic action (21) is not self-consistent near the ends of the string $t = 0, x = \pm R$ where the solution (25) is singular and gradients of $u$ diverge. However, one can estimate the corrections coming from those areas and find that they change the coefficient $[\pi(\rho_0 - \bar{\rho})]$ in front of the $R^2$ in (29) by terms of the higher order in small parameter (20) [10].

We also note here that although the value of the coefficient $\alpha$ given by (27) is approximate and is valid only in the limit of a very weak depletion, the asymptotic forms (26,28) are very general as they depend only on the linearization of hydrodynamic equations at $x, t \to \infty$, where it is always possible.

## 5. EFP through an asymptotics of the solution

The calculation of the EFP has already been reduced to the calculation of the value of the classical action on the solution of equations of motion satisfying EFP boundary conditions. In this section we are going use a Maupertui principle [13] to obtain a simple expression for $S_{opt}$ in terms of the asymptotics of the EFP solution of hydrodynamic equations.

Let us calculate the variation of the action (15) with respect to the displacement field $u$

$$\begin{aligned} \delta S &= \int d^2x \Big\{ - \partial_t(v\delta u) - \partial_x \left[ \left( \frac{v^2}{2} - \partial_\rho(\epsilon\rho) \right) \delta u \right] \\ &+ \delta u \left[ \partial_t v + v\partial_x v - \partial_x \partial_\rho(\epsilon\rho) \right] \Big\}. \end{aligned} \qquad (30)$$

We kept here surface terms (full derivatives) in addition to the last term which produces the equation of motion (18). Now we assume that the action $S$ is calculated on the EFP (or DFP) solution of equations of motion and consider the derivative of this action with respect to the equilibrium background density $\rho_0$. We have

$$\partial_{\rho_0} S_{opt} = \int d^2x \left\{ -\partial_t(v\partial_{\rho_0} u) - \partial_x \left[ \left( \frac{v^2}{2} - \partial_\rho(\epsilon\rho) \right) \partial_{\rho_0} u \right] \right\}, \qquad (31)$$

where we used the fact that $u$ satisfies the Euler equation and dropped the last term in (30). (31) has only derivative terms and can be re-written as a boundary contribution

$$\partial_{\rho_0} S_{opt} = \oint \left[ v\partial_{\rho_0} u \, dx + \left( \partial_\rho(\epsilon\rho) - \frac{v^2}{2} \right) \partial_{\rho_0} u \, dt \right], \qquad (32)$$

where the integral is taken over the infinitely large contour around $xt$ plane. As the integrand in (32) should be calculated at (infinitely) large $x$ and $t$ we can use in (32) the general asymptotics (26,28). After simple manipulations we obtain our main result

$$\partial_{\rho_0} S_{opt} = 2\pi \frac{v_{s0}}{\rho_0} (\alpha + \bar{\alpha}). \qquad (33)$$

As a simple check of this result we take the value of $\alpha$ obtained in bosonization approach (27) and substitute it into (33). We immediately obtain $\partial_{\rho_0} S_{opt} = \pi \frac{v_{s0}}{\rho_0}(\rho_0 - \bar{\rho})R^2$, which is equivalent to (29) up to the terms of higher order in the small parameter (20).

## 6. Free fermions

In this section we find the EFP for a free Fermi gas in one dimension. First, let us find the internal energy $\epsilon(\rho)$ of the gas. The density of fermions is given in terms of Fermi momentum $k_F$ as

$$\rho = \int_{-k_F}^{k_F} \frac{dk}{2\pi} = \frac{k_F}{\pi}. \tag{34}$$

The energy per unit length is

$$\rho\epsilon(\rho) = \int_{-k_F}^{k_F} \frac{k^2}{2}\frac{dk}{2\pi} = \frac{k_F^3}{6\pi} = \frac{\pi^2}{6}\rho^3 \tag{35}$$

where we put $\hbar = 1$ and fermion mass $m = 1$. The energy per particle in a free Fermi gas with density $\rho$ is

$$\epsilon(\rho) = \frac{\pi^2}{6}\rho^2. \tag{36}$$

We calculate the sound velocity using

$$v_s^2 = \rho\partial_\rho^2(\rho\epsilon) \tag{37}$$

and obtain

$$v_s = \pi\rho = k_F \tag{38}$$

– the well known result that the sound velocity and Fermi velocity are the same (remember that in our notations $m = 1$ and $v_F = k_F$).

Hydrodynamic equations (16) and (18) for free fermions are

$$\begin{aligned} \partial_t\rho + \partial_x(\rho v) &= 0, \\ \partial_t v + v\partial_x v &= \pi^2\rho\partial_x\rho. \end{aligned} \tag{39}$$

Introducing a complex field

$$w = \pi\rho + iv, \tag{40}$$

we re-write both equations (39) as a single complex Hopf equation[6]

$$\partial_t w - iw\partial_x w = 0. \tag{41}$$

The latter equation has a general solution

$$w = F(z), \tag{42}$$

where $F(z)$ is an arbitrary analytic function of a complex variable $z$ defined as

$$z = x + iwt. \tag{43}$$

The boundary conditions will determine a particular analytic function $F(z)$ so that the equations (42,43) will define the solution $w(x,t)$ of (41) implicitly. It is easy to check that the unknown function $F(z)$ for DFP is given by

$$F(z) = \pi\bar{\rho} + \pi(\rho_0 - \bar{\rho})\frac{z}{\sqrt{z^2 - R^2}} \tag{44}$$

and the one for EFP can be obtained by putting $\bar{\rho} = 0$ in (44). The space-time configuration of the density and velocity fields minimizing the hydrodynamic action is given by real and imaginary parts of $w(x,t)$ (see Eq. (40)) which is implicitly defined by

$$w - \pi\bar{\rho} = +\pi(\rho_0 - \bar{\rho})\frac{z}{\sqrt{z^2 - R^2}} \tag{45}$$

with (43). This solution is relatively complicated (see Figs. 1, 2) and a direct calculation of the value of the hydrodynamic action (15) on this solution is cumbersome. However, one can get the result very quickly using (33). Indeed, in the limit $x, t \to \infty$ we have $w \to \pi\rho_0$ and $z = x + iwt \to x + i\pi\rho_0 t = z_0$. Therefore, the asymptotics of (45) is given by

$$w - \pi\rho_0 \approx \pi(\rho_0 - \bar{\rho})\frac{R^2}{2z_0^2} \tag{46}$$

and taking, e.g., its real part

$$\rho - \rho_0 \approx \frac{1}{4}(\rho_0 - \bar{\rho})R^2\left(\frac{1}{z_0^2} + \frac{1}{\bar{z}_0^2}\right). \tag{47}$$

Comparing (47) with (28) we extract

$$\alpha = \frac{1}{4}(\rho_0 - \bar{\rho})R^2 \tag{48}$$

and substituting into (33) we derive $\partial_{\rho_0} S_{opt} = \pi^2(\rho_0 - \bar{\rho})R^2$ and

$$S_{opt} = \frac{1}{2}\left[\pi(\rho_0 - \bar{\rho})R\right]^2. \tag{49}$$

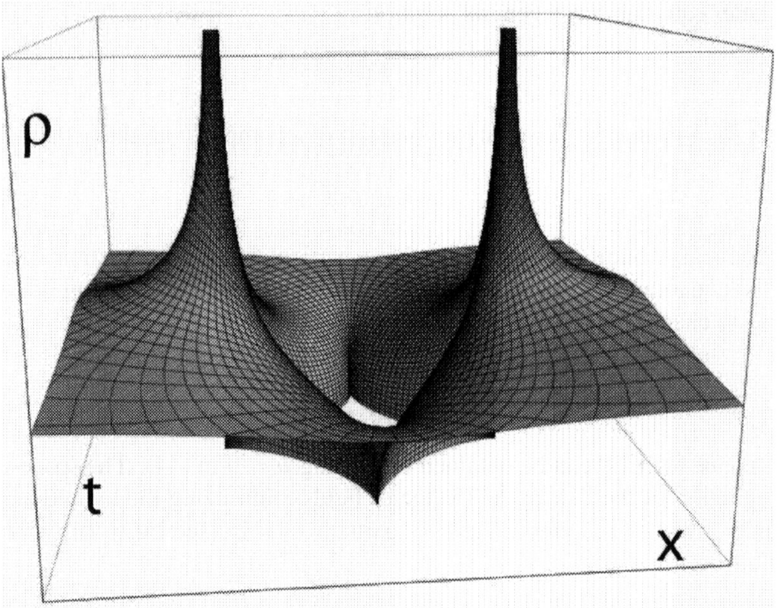


*Figure 1.* The density profile $\rho(x,t)$ is shown for the EFP instanton. The density diverges at points $(x,t)=(\pm R,0)$. The shape of the "Emptiness" is shown in Fig.2.

This gives for DFP and EFP probabilities respectively

$$P_{DFP}(R) \sim \exp\left\{-\frac{1}{2}\left[\pi(\rho_0-\bar{\rho})R\right]^2\right\}, \tag{50}$$

$$P_{EFP}(R) \sim \exp\left\{-\frac{1}{2}\left(\pi\rho_0 R\right)^2\right\}. \tag{51}$$

In the case of free fermions an exact asymptotic expansion of EFP in $1/R$ is known [8] with first few terms given by (B.2). The instanton contribution (51) gives an exact first (Gaussian) term of this expansion.

## 7. Calogero-Sutherland model

Our next example of a one-dimensional liquid is the Calogero-Sutherland model – a model of one-dimensional particles interacting with an inverse square

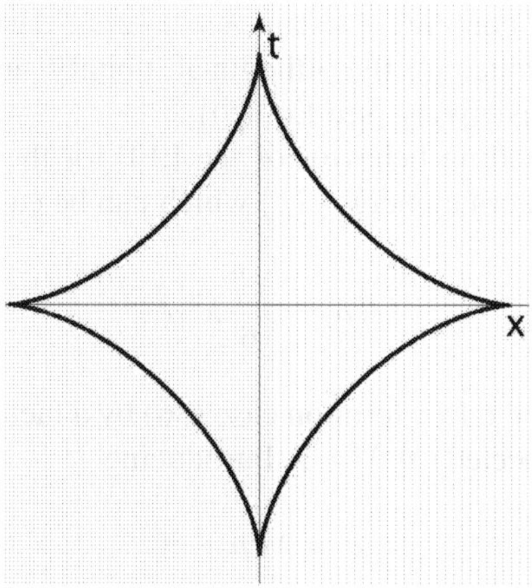


*Figure 2.* The region of the $x - t$ plane in which $\rho(x,t) = 0$ for the EFP instanton for free fermions is shown. The boundary of the region is given by an astroid $x^{2/3} + (\pi\rho_0 t)^{2/3} = R^{2/3}$.

potential. The Hamiltonian of the model is

$$H = -\frac{1}{2}\sum_{j=1}^{N}\frac{\partial^2}{\partial x_j^2} + \frac{1}{2}\sum_{1\le j<k\le N}\frac{\lambda(\lambda-1)}{(x_j-x_k)^2} \tag{52}$$

$$= -\frac{1}{2}\sum_{i=1}^{N}\left(\frac{\partial}{\partial x_i} + \sum_{j=1, j\neq i}^{N}\frac{\lambda}{x_i - x_j}\right)\left(\frac{\partial}{\partial x_i} - \sum_{k=1, k\neq i}^{N}\frac{\lambda}{x_i - x_k}\right).$$

Here we again use the units $\hbar = 1$ and $m = 1$. This model is known to be integrable [14, 15]. We are interested in the thermodynamic limit of (52). The easiest way to go to this limit is to consider (52) in an additional harmonic potential which does not destroy the integrability of the model and then take a limit of number of particles $N \to \infty$ and the strength of the potential going to zero so that the density is kept constant and equal $\rho_0$. We omit all these details which can be found in the original papers [14, 15]. We mention only that the ground state wave function of (52) is

$$\Psi_{GS} = \prod_{j<k}(x_j - x_k)^\lambda \tag{53}$$

and shows an intermediate statistics interpolating between non-interacting bosons ($\lambda = 0$) and non-interacting fermions ($\lambda = 1$). We also notice here that the probability distribution of particle coordinates

$$|\Psi_{GS}|^2 = \prod_{j<k}|x_j - x_k|^{2\lambda} \tag{54}$$

at particular values of coupling constant $\lambda = 1/2,\ 1,\ 2$ coincides with the joint probability of eigenvalues for the orthogonal, unitary, and symplectic random matrix ensembles respectively (see Eq. (5)).

To calculate the leading behavior of the EFP for the Calogero-Sutherland model we need the internal energy $\epsilon(\rho)$ which can be easily found [15]

$$\epsilon(\rho) = \frac{\pi^2}{6}\lambda^2\rho^2. \tag{55}$$

The (55) differs from the free fermion case (36) by a factor of $\lambda^2$ and coincides with the latter (as expected) at $\lambda = 1$. Introducing

$$w = \lambda\pi\rho + iv \tag{56}$$

and repeating the calculations of the previous section we obtain

$$P_{DFP}(R) \sim \exp\left\{-\frac{1}{2}\lambda\left[\pi(\rho_0 - \bar{\rho})R\right]^2\right\}, \tag{57}$$

$$P_{EFP}(R) \sim \exp\left\{-\frac{1}{2}\lambda\left(\pi\rho_0 R\right)^2\right\}. \tag{58}$$

Comparing to the known exact result (B.4) we see that (58), indeed, gives the exact leading asymptotics of the EFP for the Calogero-Sutherland model. Subleading (in $1/R$) corrections to (58) are due to gradient corrections to the hydrodynamic action (55) and to quantum fluctuations around the found instanton.

## 8. Free fermions on the lattice

The goal of this section is to illustrate that the hydrodynamic method we used is not limited to Galilean invariant systems. We use the method to calculate the EFP for a system of non-interacting lattice fermions with an arbitrary dispersion $\varepsilon(k)$. Semiclassically, one can describe the evolution of the degenerate 1D Fermi gas in terms of two smooth and slow functions $k_{R,L}(x,t)$ – right and left Fermi points respectively. The equations of motion are given by

$$\partial_t k_{R,L} + \varepsilon'(k_{R,L})\partial_x k_{R,L} = 0, \tag{59}$$

which is an obvious consequence of the absence of interactions. Indeed, the derivative $\varepsilon'(k_{R,L})$ is the right (or left) Fermi velocity of particles, and (59) is nothing else but the statement that the momentum of a particle does not change in time. The classical action of 1D Fermi gas should reproduce (59) as well as give the correct energy of the system $E = \int dx \int_{k_L(x)}^{k_R(x)} \frac{dk}{2\pi}\,\varepsilon(k)$.

We assume here that the dispersion of fermions is parity invariant $\epsilon(k) = \epsilon(-k)$ and consider the following Euclidian action

$$S = \int d^2x \left\{ \frac{i}{4\pi} \left[ w\partial_x^{-1}\partial_t w - \bar{w}\partial_x^{-1}\partial_t \bar{w} \right] + \int_{-\bar{w}}^{w} \frac{dk}{2\pi}\, \varepsilon(k) \right\}. \qquad (60)$$

Here $\varepsilon(k)$ is the dispersion (including chemical potential) of free fermions and $w, \bar{w}$ are independent variables which are Euclidian versions of $k_R$, $-k_L$. We identify the density of particles as $\rho = \frac{w+\bar{w}}{2\pi}$. The equations of motion corresponding to (60)

$$\begin{aligned} i\partial_t w + \partial_x \varepsilon(w) &= 0, \\ -i\partial_t \bar{w} + \partial_x \varepsilon(\bar{w}) &= 0 \end{aligned} \qquad (61)$$

are, indeed, the Euclidian versions of (59). Equations (61) are complex conjugates of each other ($\overline{\varepsilon(w)} = \varepsilon(\bar{w})$ as $\varepsilon(k)$ is a real function). The difference of these two equations give the continuity equation (16) with the current $j = \frac{\varepsilon - \bar{\varepsilon}}{2\pi i}$. The path integral with the action (60) is the integral over two fields $w, \bar{w}$. We integrate out their difference $w - \bar{w}$ in a saddle point approximation. Namely, we re-write (60) in terms of combinations $w - \bar{w}$ and $w + \bar{w}$. Then, we take a variation of the action with respect to $w - \bar{w}$. This produces the continuity equation relating $w$ and $\bar{w}$. We solve the continuity equation introducing the displacement field. Thus, we have $w$ and $\bar{w}$ in terms of $u$ as the saddle point trajectory. Namely,

$$\begin{aligned} \rho &= \frac{w + \bar{w}}{2\pi} = \rho_0 + \partial_x u, \\ j &= \frac{\varepsilon - \bar{\varepsilon}}{2\pi i} = -\partial_t u. \end{aligned} \qquad (62)$$

We substitute these expressions back into (60) and obtain

$$S[u] = \int d^2x \left\{ -\frac{1}{4\pi}(w - \bar{w}) \left[ \varepsilon(w) - \varepsilon(-\bar{w}) \right] + \int_{-\bar{w}}^{w} \frac{dk}{2\pi}\, \varepsilon(k) \right\}. \qquad (63)$$

We assume now that the integration variable of the path integral is the displacement field $u$. The fields $w, \bar{w}$ in (63) are not independent but are related to $u$ by (62). Deriving (63) we neglected the fluctuations around the saddle point as well as the changes in the measure of path integration. The corrections due to the neglected terms will be of the higher order in field gradients and are not essential for our calculation of the leading term of the EFP.

One can now use (63) in the path integral formulation of the quantum hydrodynamics (7) where the integration is taken over all configurations of the displacement field $u$. We note here that in the Galilean invariant systems the "kinetic term" of the hydrodynamic action (15) is fixed. The Lagrangian of the action (63) is a more complicated function of $\partial_t u$. In this case we derived it using the fact that fermions are free. In a more general problem of interacting particles the derivation of the hydrodynamic action requires the solution of the dynamic many body problem.

The Lagrangian of (63) is given by

$$L = -\frac{1}{4\pi}(w - \bar{w})\left[\varepsilon(w) - \varepsilon(-\bar{w})\right] + \int_{-\bar{w}}^{w} \frac{dk}{2\pi}\,\varepsilon(k). \tag{64}$$

Now, we calculate the EFP for lattice fermions using the the Lagrangian (64) with the Eq. (62), and the results of Appendix A. First, we calculate

$$dL = \frac{\varepsilon + \bar{\varepsilon}}{2} d\rho + \frac{w - \bar{w}}{2i} dj \tag{65}$$

and

$$L_j = v = \frac{w - \bar{w}}{2i}, \tag{66}$$

$$L_\rho = \frac{\varepsilon + \bar{\varepsilon}}{2}, \tag{67}$$

$$L_{jj} = \frac{2\pi}{\varepsilon' + \bar{\varepsilon}'}, \tag{68}$$

$$L_{\rho j} = -\frac{2\pi}{\varepsilon' + \bar{\varepsilon}'}\frac{\varepsilon' - \bar{\varepsilon}'}{2i}, \tag{69}$$

$$L_{\rho\rho} = \frac{2\pi}{\varepsilon' + \bar{\varepsilon}'}|\varepsilon'|^2, \tag{70}$$

where $\varepsilon' = \left|\frac{\partial \varepsilon}{\partial k}\right|_{k=w}$ and $\bar{\varepsilon}' = \left|\frac{\partial \varepsilon}{\partial k}\right|_{k=\bar{w}}$. In terms of the density and velocity

$$w = \pi\rho + iv. \tag{71}$$

We emphasize that because of the absence of the Galilean invariance (the dispersion is *not* $k^2/2m$), the current is *not* $\rho v$ but can be found from (62,66).

We obtain then an interesting relation

$$\kappa \equiv \sqrt{L_{\rho\rho}L_{jj} - L_{\rho j}^2} = \pi. \tag{72}$$

This relation can be considered as a special property of free fermions (with arbitrary dispersion!). Namely, one can trace the origin of (72) to the fact that each fermion occupies a fixed volume in the phase space. [7]

Let us now apply Riemann's trick to equations (61). Namely, we interchange independent and dependent variables so that $x = x(w, \bar{w})$ and $t = t(w, \bar{w})$. Then using (72, 66-70) we re-write (61) as

$$\partial_w(x - i\bar{\varepsilon}'t) = 0 \tag{73}$$

and its complex conjugate. The most general solution of (73) is given implicitly by

$$w = F(z), \tag{74}$$

where $F(z)$ is an arbitrary analytic function and

$$z = x + i\varepsilon'(w)t. \tag{75}$$

The equations (73, 75) reduce to equations for free continuous fermions if $\varepsilon(k) = k^2/2$.

Similarly to free continuous fermions, the EFP boundary conditions specify the form of the function $F(z)$ in (74). We define

$$\tilde{w} = \frac{w - \pi\bar{\rho}}{2(1-\bar{\rho})}, \tag{76}$$

$$\tilde{\theta} = \frac{\pi(\rho_0 - \bar{\rho})}{2(1-\bar{\rho})}, \tag{77}$$

where we assume lattice spacing to be 1 so that the maximal density of fermions on the lattice is 1. Then the EFP solution is given by $F(w)$ such that

$$\sin\tilde{w} = \sin\tilde{\theta}\frac{z}{\sqrt{z^2 - R^2}}. \tag{78}$$

We immediately find at large distances ($z \to z_0 = x + i\varepsilon_0' t \to \infty$)

$$w - \pi\rho_0 \approx 2(1-\bar{\rho})\frac{R^2}{2z_0^2}\tan\tilde{\theta} \tag{79}$$

and using (28) and (71) we extract

$$\alpha = \frac{1}{2\pi}(1-\bar{\rho})R^2\tan\tilde{\theta}. \tag{80}$$

We substitute this expression in (A.7) and obtain

$$\partial_{\rho_0} S = 2\pi(1-\bar{\rho})R^2\tan\frac{\pi(\rho_0 - \bar{\rho})}{2(1-\bar{\rho})}, \tag{81}$$

so that

$$S_{DFP} = -4(1-\bar{\rho})^2R^2\ln\cos\frac{\pi(\rho_0-\bar{\rho})}{2(1-\bar{\rho})}. \tag{82}$$

In particular for $\bar{\rho} = 0$ we obtain for the Emptiness Formation Probability

$$S_{EFP} = -4R^2\ln\cos\frac{\pi\rho_0}{2} = -4R^2\ln\cos\frac{k_F}{2}. \tag{83}$$

This result is the exact first term in $1/R$ expansion (B.10). The next term is $\frac{1}{4}\ln R$. It is interesting to note that as it should be, in the limit $\rho_0 - \bar{\rho} \ll \rho_0$ the result (82) reproduces the bosonization result (29). Also, in the limit $\rho_0, \bar{\rho} \ll 1$ the results (82,83) reproduce the corresponding results (50,51) for free continuous fermions.

# 9. Conclusion

We showed on the example of the Emptiness Formation Probability that one can obtain some exact results for correlation functions using a collective hydrodynamic description. Moreover, the nonlinear hydrodynamic description might work and produce non-perturbative results even in cases where the linearized hydrodynamics fails. We obtained the leading term of the asymptotics of the EFP using the instanton approach. Although the EFP is a property of the ground state of a quantum many body system, the hydrodynamic approach can also be used to study dynamics of quantum systems. It is also not limited to integrable systems.

We considered few simple systems which are in the Luttinger liquid phase at zero temperature. The effects of finite temperature and finite gap in the spectrum of excitations can also be considered using the hydrodynamic approach in the limit when temperature is very low and the gap is very small [10, 12]. In these limits one obtains a crossover between the Gaussian decay of the EFP at intermediate $R$ to the exponential decay at very large $R$.

There are also many questions which are left open. First, it would be nice to obtain the results for the EFP in other integrable systems such as bosons with delta-repulsion and XXZ spin chains. Some results for the EFP have already been obtained using other methods (see Appendix B), while, e.g., the EFP for a XXZ spin chain in the presence of magnetic field is still not known. The application of the hydrodynamic approach developed here to these systems is straightforward and reduces the problem to the problem of finding asymptotics of the solution of a system of classical equations with proper boundary conditions. However, these classical equations are complicated and we haven't yet obtained analytical results for their asymptotics.

The second open question involves the corrections to the leading hydrodynamic approximation that we used in this lecture. There are two sources of such corrections: the gradient corrections to the "classical" hydrodynamic action and quantum fluctuations around the classical saddle point. E.g., if one is interested in the next to leading terms of the asymptotic expansion of the EFP in $1/R$ one needs to include these gradient corrections (which make hydrodynamics dispersive) and quantum fluctuations. Especially interesting would be to obtain the power law pre-factor of the EFP (or $\ln R$ term in the asymptotic expansion). The hydrodynamic calculation might shed some light on a possible universality of the exponent of this pre-factor.

Needless to say, that although we focused in this lecture on the particular correlation function (the EFP), the use of hydrodynamic approach is much broader. In particular, the nonlinear hydrodynamics is important to capture a lot of important nonlinear phenomena which disappear in the linear approximation.

## Acknowledgements

I would like to thank the organizers of the 2004 Les Houches Summer School on Applications of Random Matrices in Physics for a very nice atmosphere of the School. I have greatly benefited from discussions with F. Essler, F. Franchini, V. E. Korepin, S. L. Lukyanov, A. M. Tsvelik, and P. B. Wiegmann. This work was supported by the NSF grant DMR-0348358, and the Theory Institute of Strongly Correlated and Complex Systems at Brookhaven.


## Appendix: Hydrodynamic approach to non-Galilean invariant systems

The hydrodynamic description of Sec. 3 and the approach of Sec. 5 can be easily generalized to non-Galilean invariant systems. We briefly list here relevant formulas leaving the details of (straightforward) calculations to the interested reader. We assume that a one-dimensional compressible liquid can be described by the partition function (7) where the functional integration is taken over all configurations of the displacement field $u$ which defines the hydrodynamic density and current by (17) and where the Euclidian action (9) can be written as

$$S = \int d^2x\, L(\rho, j). \tag{A.1}$$

The Lagrangian density $L(\rho, j)$ is a function of $\rho$ and $j$. As in the main text we neglect the gradient corrections to the action (A.1). In the case of non-Galilean invariant systems these gradient corrections will depend on gradients of current in addition to gradients of density. The case of Galilean invariant systems considered in the main text (15) can be obtained from the general one assuming the particular Galilean invariant form of the Lagrangian $L(\rho, j) = j^2/2\rho + \rho\epsilon(\rho)$.

Variation of (A.1) with respect to $u$ gives a generalized Euler equation which together with the continuity equation reads

$$\partial_t\rho + \partial_x j = 0, \tag{A.2}$$

$$\partial_t L_j - \partial_x L_\rho = 0, \tag{A.3}$$

where $L_j$ means $\partial L/\partial j$ etc.

If the deviation from the ground state is small, we can expand the action (A.1) around $\rho = \rho_0$, $j = 0$. We assume that $L_{\rho j}^{(0)} = 0$, where superscript (0) means that the derivative is calculated at the equilibrium values $\rho = \rho_0$ and $j = 0$. We obtain for the action in harmonic approximation

$$S = \int d^2x \left[L_{jj}^{(0)} u_t^2 + L_{\rho\rho}^{(0)} u_x^2\right]. \tag{A.4}$$

We introduce a complex space-time coordinate

$$z_0 = x + iv_{s0}t, \tag{A.5}$$

where

$$v_{s0} = \sqrt{\frac{L_{\rho\rho}^{(0)}}{L_{jj}^{(0)}}} \tag{A.6}$$

is the sound velocity at equilibrium. Let us define the coefficient $\alpha$ through the asymptotics of an EFP solution at $z_0 \to \infty$ by (26). The asymptotics of the current and density are given by (28).

Our main formula (33) for the variation of the EFP instanton action with respect to the background density becomes

$$\partial_{\rho_0} S = 2\pi\kappa_0 \,(\alpha + \bar{\alpha}), \tag{A.7}$$

where $\kappa_0$ is given by

$$\kappa_0 = \sqrt{L^{(0)}_{\rho\rho} L^{(0)}_{jj}}. \tag{A.8}$$

The coefficient $\kappa_0$ is related to the compactification radius $R_{\rm comp}$ of bosons in the bosonization procedure $\kappa_0 = (2\pi R_{\rm comp})^2$. For free fermions $\kappa_0 = \pi$ which corresponds to $R_{\rm comp} = 1/\sqrt{4\pi}$.

# Appendix: Exact results for EFP in some integrable models

For the sake of reader's convenience in this appendix we list some results obtained for the Emptiness Formation Probability $P(R)$ in integrable one-dimensional systems. We present the results for $S \equiv -\ln P(R)$ which should be compared with the instanton action $S_{opt}$ used in the main text.

**Free continuous fermions** Let us denote

$$s \equiv \pi\rho_0 R. \tag{B.1}$$

We use the fact that the ground state wave function (more precisely $|\Psi|^2$) of free fermions coincides with the joint eigenvalue distribution of unitary random ensemble. For the latter the probability of having no eigenvalues in the range $2R$ of the spectrum was obtained in [8] (see also [9]). First few terms of the expansion in $1/R$ are

$$S = \frac{1}{2}s^2 + \frac{1}{4}\ln s - \left(\frac{1}{12}\ln 2 + 3\zeta'(-1)\right) + O(s^{-2}). \tag{B.2}$$

**Calogero-Sutherland model** The Calogero-Sutherland model [14, 15] (rational version, known also as Calogero model) with $N$-particles is defined as[8]

$$\mathcal{H} = \frac{1}{2}\sum_{j=1}^{N} p_j^2 + \frac{1}{2}\sum_{j,k=1;j\neq k}^{N} \frac{\lambda(\lambda-1)}{(x_j - x_k)^2}, \tag{B.3}$$

where $p_j = -i\partial/\partial x_j$ is the momentum operator of $j$-th particle and $\lambda$ is a dimensionless coupling constant. The wave function of the ground state is proportional to $\prod_{j<k}(x_j - x_k)^\lambda$. At $\lambda = 1$ we have free fermions, while in the case of general $\lambda$ the model (B.3) describes particles with fractional statistics. Using the form of the ground state wave function and thermodynamic arguments [9] one obtains

$$S = \frac{\lambda}{2}s^2 + (1-\lambda)s + O(\ln s). \tag{B.4}$$

or defining

$$s \equiv \sqrt{\lambda}\pi\rho_0 R. \tag{B.5}$$

and

$$\alpha_0 \equiv \frac{\lambda^{1/2} - \lambda^{-1/2}}{2} \tag{B.6}$$

we have

$$S = \frac{1}{2}s^2 - 2\alpha_0 s + O(\ln s). \tag{B.7}$$

The notation $\alpha_0$ originates from the conformal field theory with central charge $c = 1 - 24\alpha_0^2$ which is known to be related to the Calogero-Sutherland model [16]. As far as I know the coefficient in front of $\ln s$ term of the expansion is not known for the general $\lambda$. However, $\lambda = 1/2, 1, 2$ correspond to random matrix ensembles where the full asymptotic expansion is known (see below). In those cases the coefficient of $\ln s$ is $1/8, 1/4, 1/8$ respectively. The natural guess is that

$$S = \frac{1}{2}s^2 - 2\alpha_0 s + \left(\frac{1}{4} - \alpha_0^2\right) \ln s + O(1). \tag{B.8}$$

**Random matrices** For Random Matrix ensembles with $\beta = 1, 2, 4$ the joint eigenvalue distribution is proportional to $\prod_{i<j} |z_i - z_j|^\beta$. The full asymptotic expansion of the quantity $E_\beta(0, 2R)$ corresponding to the EFP $P(R)$ was obtained using properties of Toeplitz determinants [8, 9]. The first few terms of these expansions are given by

$$\begin{aligned}
S_{\lambda=1/2} &= \frac{1}{4}(\pi\rho_0 R)^2 + \frac{1}{2}(\pi\rho_0 R) + \frac{1}{8}\ln(\pi\rho_0 R) - \frac{7}{24}\ln 2 - \frac{3}{2}\zeta'(-1) + O(1/s), \\
S_{\lambda=1} &= \frac{1}{2}(\pi\rho_0 R)^2 + \frac{1}{4}\ln(\pi\rho_0 R) - \frac{1}{12}\ln 2 - 3\zeta'(-1) + O(1/s^2), \\
S_{\lambda=2} &= (\pi\rho_0 R)^2 - (\pi\rho_0 R) + \frac{1}{8}\ln(\pi\rho_0 R) + \frac{4}{3}\ln 2 - \frac{3}{2}\zeta'(-1) + O(1/s).
\end{aligned} \tag{B.9}$$

Here we used $\lambda = \beta/2 = 1/2, 1, 2$ instead of $\beta$. Using notations (B.5,B.6) we can summarize the first three terms of (B.9) in a compact form (B.8).

**Free fermions on the lattice** For non-interacting one-dimensional fermions on the lattice (and the corresponding XY spin chain) the asymptotic behavior of EFP was derived in [17] using the Widom's theorem on the asymptotic behavior of Toeplitz determinants. Introducing the Fermi momentum $k_F = \pi\rho_0$ and using units in which the lattice spacing is 1 we have

$$S = -4R^2 \ln\cos\frac{k_F}{2} + \frac{1}{4}\ln\left[2R\sin\frac{k_F}{2}\right] - \left(\frac{1}{12}\ln 2 + 3\zeta'(-1)\right) + O(R^{-2}). \tag{B.10}$$

In the continuous limit $k_F \to 0$ the (B.10) goes to its continuous version (B.2).

**Bosons with delta repulsion** The model of bosons with short range repulsion is described by

$$\mathcal{H} = \frac{1}{2}\sum_{j=1}^{N} p_j^2 + g \sum_{1\le j<k\le N} \delta(x_j - x_k), \tag{B.11}$$

where $g$ is a coupling constant. It is integrable by Bethe Ansatz [18]. It was derived (conjectured) in Ref. [19] that the leading term of the EFP is

$$S = \frac{1}{2}(KR)^2 \left[1 + I(g/K)\right], \tag{B.12}$$

where $K$ is the Fermi momentum in the Lieb-Liniger solution [18] and

$$I(x) = \frac{1}{2\pi^2}\int_{-1}^{1}\frac{y\,dy}{\sqrt{1-y^2}}\int_{-1}^{1}\frac{z\,dz}{\sqrt{1-z^2}}\log\left(\frac{x^2+(y+z)^2}{x^2+(y-z)^2}\right). \tag{B.13}$$

The limit $I(x \to \infty) = 0$ corresponds to the free fermion result (B.2) (Tonks-Girardeau gas), while the limit $I(x \to 0) = 1$ is the result for the EFP of free bosons.

**XXZ model** The Hamiltonian of an XXZ model is given by

$$H = J \sum_{k} \left[ \sigma_k^x \sigma_{k+1}^x + \sigma_k^y \sigma_{k+1}^y + \Delta \sigma_k^z \sigma_{k+1}^z \right],$$

where the sum is taken over the sites of a one-dimensional lattice and $\sigma^{x,y,z}$ are Pauli matrices. Let us parametrize the anisotropy as $\Delta = \cos \pi\nu$. Then for the EFP we have [20, 21]

$$P(n) \sim A n^{-\gamma} C^{-n^2}, \tag{B.14}$$

as $n = 2R \to \infty$, where

$$C = \frac{\Gamma^2(1/4)}{\pi\sqrt{2\pi}} \exp\left\{ -\int_0^\infty \frac{dt}{t} \frac{\sinh^2(t\nu) e^{-t}}{\cosh(2t\nu)\sinh(t)} \right\} \tag{B.15}$$

and the exponent $\gamma$ was conjectured in [21] to be

$$\gamma = \frac{1}{12} + \frac{\nu^2}{3(1-\nu)}. \tag{B.16}$$

## Notes

1. In fact, one can do a better job including subdominant corrections. See Ref.[9] for details.
2. Later we will use the conventional displacement field as $u$. See Eq. (17).
3. In this section we follow closely the qualitative argument of Ref.[10]
4. The harmonic approximation to the nonlinear hydrodynamics of quantum liquid is equivalent to a linear bosonization approach to interacting one-dimensional particles.
5. See the next section on how to avoid doing this calculation.
6. In real time formalism, instead of $w, \bar{w}$ one introduces "right and left Fermi momenta" $k_{R,L} = \pi\rho \pm v$ which satisfy the Euler-Hopf equations $\partial_t k + k\partial_x k = 0$ reflecting the absence of interactions between fermions.
7. For Calogero-Sutherland model (55) we have $\kappa = \pi\lambda$. It means that the volume of the phase space per particle is changed by the factor of $\lambda$ which reflects the fractional statistics of particles.
8. To prevent particles running to infinity we either add a harmonic potential to (B.3) or put particles on a circle of a large radius.

## References


[1] L. D. Landau, Sov. Phys. ZhETF, **11**, 592 (1941).
*Theory of Superfluidity of Helium II*

[2] See e.g., D. Forster, *Hydrodynamic Fluctuations, Broken Symmetry, and Correlation Functions*, W. A. Benjamin, Inc., Reading, Massachusetts, 1975 and references therein.

[3] M. Stone, editor, *Bosonization*, World Scientific, Singapore, 1994.

[4] F. D. M. Haldane, Phys. Rev. Lett. **47**, 1840 (1981).
*Effective Harmonic-Fluid Approach to Low-Energy Properties of One-Dimensional Quantum Fluids*

[5] V. E. Korepin, N. M. Bogoliubov, and A. G. Izergin, *Quantum Inverse Scattering Mehod and Correlation Functions*, Cambridge University Press, Cambridge, UK, 1993.

[6] V. E. Korepin, A. G. Izergin, F. H. L. Essler, and D. B. Uglov, Phys. Lett. **A 190**, 182 (1994).
*Correlation functions of the spin-1/2 XXX antiferromagnet.*

[7] M. Jimbo, and T. Miwa, *Algebraic analysis of solvable lattice models.*, no 85, Providence: American Mathematical Society, 1995.

[8] J. des Cloizeaux and M. L. Mehta, J. Math. Phys. **14**, 1648 (1973).
*Asymptotic Behavior of Spacing Distributions for Eigenvalues of Random Matrices.*
F. Dyson, Commun. Math. Phys. **47**, 171 (1976).
*Fredholm Determinants and Inverse Scattering Problems.*

[9] M. L. Mehta, *Random Matrices*, 2nd rev. enl. ed. New York: Academic Press, 1991.

[10] A. G. Abanov and V. E. Korepin, Nucl. Phys. **B 647**, 565 (2002).
*On the probability of ferromagnetic strings in antiferromagnetic spin chains.*

[11] A. G. Abanov, and F. Franchini, Phys. Lett. **A 316**, 342 (2003).
*Emptiness formation probability for the anisotropic XY spin chain in a magnetic field.*

[12] F. Franchini and A. G. Abanov, arXiv:cond-mat/0502015 (2005).
*Asymptotics of Toeplitz Determinants and the Emptiness Formation Probability for the XY Spin Chain.*

[13] L. D. Landau and E. M. Lifshitz, *Mechanics : Volume 1 (Course of Theoretical Physics)*, Pergamon Press.

[14] F. Calogero, J. Math. Phys. **10**, 2197 (1969).
*Ground State of a One-Dimensional $N$-Body System.*

[15] B. Sutherland, J. Math. Phys. **12**, 246 (1971);
*Quantum Many Body Problem in One Dimension: Ground State*
Phys. Rev. A **4**, 2019 (1971).
*Exact Results for a Quantum Many-Body Problem in One Dimension*

[16] H. Awata, Y. Matsuo, S. Odake, J. Shiraishi, Phys. Lett. **B 347**, 49-55 (1995).
*Collective field theory, Calogero-Sutherland model and generalized matrix models*

[17] M. Shiroishi, M. Takahashi, and Y. Nishiyama, J. Phys. Soc. Jap. **70**, 3535 (2001).
*Emptiness Formation Probability for the One-Dimensional Isotropic XY Model.*

[18] E. H. Lieb and W. Liniger, Phys. Rev. **130**, 1605 (1963).
*Exact Analysis of an Interacting Bose Gas, I. The General Solution and the Ground State*

[19] A. R. Its, V. E. Korepin, and A. K. Waldron, arXiv:cond-mat/9510068 (1995).
*Probability of Phase Separation for the Bose Gas with Delta Interaction.*

[20] N. Kitanine, J. M. Maillet, N. A. Slavnov, and V. Terras, J. Phys. A: Math. Gen. **35**, L753-L758 (2002).
*Large distance asymptotic behaviour of the emptiness formation probability of the XXZ spin-$\frac{1}{2}$ Heisenberg chain*

[21] V. E. Korepin, S. Lukyanov, Y. Nishiyama and M. Shiroishi, Phys. Lett. **A 312**, 21-26 (2003).
*Asymptotic Behavior of the Emptiness Formation Probability in the Critical Phase of XXZ Spin Chain.*

# QCD, CHIRAL RANDOM MATRIX THEORY AND INTEGRABILITY

J.J.M. Verbaarschot
*Department of Physics and Astronomy, SUNY at Stony Brook, NY 11794*
jacobus.verbaarschot@stonybrook.edu

## 1. Summary

Random Matrix Theory has been a unifying approach in physics and mathematics. In these lectures we discuss applications of Random Matrix Theory to QCD and emphasize underlying integrable structures. In the first lecture we give an overview of QCD, its low-energy limit and the microscopic limit of the Dirac spectrum which, as we will see in the second lecture, can be described by chiral Random Matrix Theory. The main topic of the third lecture is the recent developments on the relation between the QCD partition function and integrable hierarchies (in our case the Toda lattice hierarchy). This is an efficient way to obtain the QCD Dirac spectrum from the low energy limit of the QCD partition function. Finally, we will discuss the QCD Dirac spectrum at nonzero chemical potential. We will show that the microscopic spectral density is given by of the replica limit of the Toda lattice equation. Recent results by Osborn on the Dirac spectrum of full QCD will be discussed.

## 2. Introduction

Applications of Random Matrix Theories to problems in physics have a long history starting with the idea of Wigner [1] to describe the spacing distribution of nuclear levels by an ensemble of real symmetric matrices. Although this is the first application of Random Matrix Theory (RMT) to strong interactions, applications of RMT to QCD started much later. The first paper that put QCD into the context of RMT was the work of 't Hooft on two-dimensional QCD in the limit of a large number of colors [2]. It was shown [3] that the combinatorial factors that enter in this large $N_c$ expansion could be obtained from matrix integrals. Even today, as we have seen in the lectures by Di Francesco [4], this work attracts a great deal of attention. It greatly stimulated the analysis of a complicated nonlinear theory such as QCD by means of much simpler matrix models. Because of the success of the application of RMT to the pla-



*E. Brezin et al. (eds.), Applications of Random Matrices in Physics*, 163–217.

nar expansion of QCD, the hope was that nontrivial results could be derived this way. I will mention three well-known results that have emerged from this line of thought: the Brezin-Gross-Witten model [5, 6], the Eguchi-Kawai [7] reduction, and induced QCD according to Kazakov and Migdal [8].

The Wilson loop in lattice QCD *without* quarks in 1+1 dimensions can can be reduced to the calculation of the unitary matrix integral

$$z(g^2, N_c) = \int_{U \in SU(N_c)} dU e^{g^{-2}\mathrm{Tr}(U+U^\dagger)}. \tag{1}$$

which is known as the Brezin-Gross-Witten model [9, 5, 6]. This reduction was generalized to an arbitrary Wilson loop amplitude in the large $N_c$ limit of lattice QCD in four dimensions and is known as the Eguchi-Kawai reduction [7]. It was shown that Wilson loop amplitudes do not depend on space-time and can be obtained from a single plaquette integral. However, this reduction is not valid in the weak coupling limit [10]. The idea of induced QCD [8] is to induce the plaquette action by a unitary matrix integral. With a vanishing Wilson line [11] this approach turned out not to be successful as well.

The matrix model (1) also appears in the low-energy limit of QCD *with* quarks. However, in this case the integral over $U(N_c)$ is not over the color degrees of freedom but rather over the flavor degrees of freedom, and $g^{-2}$ is replaced by $mV\Sigma/2$ with $\Sigma$ the chiral condensate, $m$ the quark mass and $V$ the volume of space time. It coincides with the full QCD partition function in a domain where the pion Compton wavelength is much larger than the size of the box [12, 13]. In this limit we have

$$\frac{1}{VN_f}\partial_m\, z(m\Sigma V/2, N_f) = \langle \frac{1}{V}\sum_k \frac{1}{i\lambda_k + m}\prod_k (i\lambda_k + m)^{N_f}\rangle, \tag{2}$$

where the $\lambda_k$ are the eigenvalues of the Dirac operator. By expanding in powers of the inverse mass, one obtains sum rules for the inverse Dirac eigenvalues [13] which put constraints on the Dirac spectrum, but do not determine the average spectral density on the scale of the average level spacing (which is known as the microscopic spectral density) and other spectral correlators.

A Random Matrix Theory that describes the fluctuations of the small eigenvalues of the Dirac operator was introduced in [14, 15]. It was shown that chiral RMT is equivalent to the flavor unitary matrix integral (1). The spectral correlation functions were found to be in good agreement with lattice QCD simulations (see [16, 17] for a review of lattice results). One argument to understand this is that level correlations of complex systems on the scale of the average level spacing are universal, i.e. they do not depend on the details of the system. This could be shown rigorously in the context of RMT [18]. However, it was understood later that the generating function for the Dirac spectrum is completely determined by chiral symmetry [19, 20], and its microscopic limit does not change if the Dirac operator replaced by a random matrix. This is one of the main topics of these lectures.

The question has been raised if the quenched spectral density can be obtained from the limit $N_f \to 0$ of (2). This procedure is known as the replica trick. It has been argued that this limit generally gives the wrong result [21]. However, the family of partition functions for different values of $N_f$ are related by the Toda lattice equation [22]. If we take the replica limit of the Toda lattice equation [23] or of the corresponding Painlevé equation [24] we obtain the correct nonperturbative result. This is a second main topic of these lectures.

A third main topic is the discussion of QCD at nonzero baryon chemical potential. In that case the Dirac operator is non-Hermitian and its eigenvalues are scattered in the complex plane. We will show that also in this case the spectral density can be obtained from the Toda lattice equation [25, 26].

A fourth topic is the discussion of the spectral density for QCD with dynamical quarks at nonzero baryon chemical potential. We will discuss recent analytical results by Osborn for a non-Hermitian random matrix model in the universality class of QCD at nonzero chemical potential [27].

We start these lectures with an elementary introduction to QCD and its symmetries. The Dirac spectrum is discussed in section 4. The low energy limit of QCD and partially quenched QCD (see section 5) is equivalent to a RMT with the symmetries of QCD introduced in section 6. In section 6 we also calculate the microscopic spectral density by means of orthogonal polynomials and the supersymmetric method. In section 7 we show that this spectral density can be obtained from the replica limit of the Toda lattice equation. In the same section we connect these results with ideas from the theory of exactly solvable systems such as Virasoro constraints, Painlevé equations, Backlund transformations and the Toda lattice. QCD at nonzero chemical potential is discussed in sections 8 and 9. In section 8 we show that the microscopic spectral density can be obtained from the replica limit of a Toda lattice equation. Recent results for full QCD at nonzero chemical potential are discussed in section 9 and concluding remarks are made in section 10.

Finally, a note about books and reviews on the subject of these lectures. The classic RMT text is the book by Mehta [1] which emphasizes the orthogonal polynomial method. The third edition of this book appeared recently. In the book by Forester [29] the emphasis is on the relation between RMT and solvable models and mathematical physics. A comprehensive review of RMT is given in [30]. A collection of recent papers on RMT can be found in [31] which also contains several reviews. Applications to mesoscopic physics are discussed in [32, 33], applications to QCD in [16] and applications to Quantum Gravity in [34]. Among the pedagogical reviews we mention a general review of RMT [35] and an introduction to the supersymmetric method [36, 37].

*Table 1.* Quark Masses

| | |
|---|---|
| $m_u$ = 4 MeV | $m_c$ = 1.3 GeV |
| $m_d$ = 8 MeV | $m_b$ = 4.4 GeV |
| $m_s$ = 160 MeV | $m_t$ = 175 GeV |

These lecture notes overlap in part with lecture notes I prepared for the Latin American Summer School in Mexico City [37], where the main emphasis was on the supersymmetric method rather than on applications of RMT to QCD.

# 3. QCD

## 3.1 Introduction

QCD (Quantum Chromo Dynamics) is the theory of strong interactions that describes the world of nucleons, pions and the nuclear force. No experimental deviations from this theory have ever been detected. QCD is a theory of quarks which interact via gauge bosons known as gluons. In nature we have 6 different flavors of quarks which each occur in three colors. Each quark is represented by a 4-component Dirac spinor

$$q^f_{i,\mu}, \qquad f = 1, \cdots, N_f = 6, \qquad i = 1, \cdots, N_c = 3 \tag{3}$$

with Dirac index $\mu$. In total we have 18 quarks (plus an equal number of anti-quarks). The gluon fields are represented by the gauge potentials

$$A^{ij}_\mu, \qquad i, j = 1, \cdots, N_c = 3. \tag{4}$$

which, as is the case in electrodynamics, are spin 1 vector fields. The gauge fields are Hermitian and traceless; they span the algebra of $SU(N_c)$. In total we have 8 massless gluons. The 6 quark flavors are known as up, down, strange, charm, bottom and and top. The quark masses are given in Table 1. Only the two lightest quarks are important for low-energy nuclear physics.

First principle calculations of QCD can be divided into three different groups, perturbative QCD, lattice QCD and chiral perturbation theory. The main domain of applicability of perturbative QCD is for momenta above several GeV. Chiral perturbation theory is an expansion in powers of the momentum and the pion mass and is only reliable below several hundred MeV. Although lattice QCD is an exact reformulation of QCD, in practice both the domain of low momenta and high momenta cannot be accessed, and its main domain of applicability lies somewhere in between the two perturbative schemes.

The reason that perturbative QCD is applicable at high energies is asymptotic freedom: the coupling constant $g \to 0$ for momenta $p \to \infty$. This property was instrumental in gaining broad acceptance for QCD as the theory of strong interactions and its discovers were awarded this years Nobel prize.

A second important property of QCD is confinement, meaning that only color singlets can exist as asymptotic states. This empirically known property has been confirmed by lattice QCD calculations. However, a first principle proof of the existence of a mass gap in QCD is still lacking even in the absence of quarks. Because of confinement the lightest particles of the theory are not quarks or gluons but rather composite mesons.

This brings us to the third important property of QCD: chiral symmetry. At low temperatures chiral symmetry is broken spontaneously which, according to Goldstone's theorem gives rises to massless Goldstone bosons. Because the chiral symmetry is slightly broken by the light quark masses, the Goldstone bosons are not exactly massless, but the mass of the pions of 135-138 $MeV$ is an order of magnitude less than a typical QCD scale of about 1 $GeV$. This justifies a systematic expansion in the pion mass and the momenta known as chiral perturbation theory.

A fourth important property of QCD is that a first principle nonperturbative lattice formulation can be simulated numerically. This allows us to compute nonperturbative observables such as for example the nucleon mass and $\rho$-meson mass. Without lattice QCD we would have had only a small number of first principle nonperturbative results and the validity of QCD in this domain would still have been a big question mark.

## 3.2 The QCD partition function

The QCD partition function in a box of volume $V_3 = L^3$ can be expressed in terms of the eigenvalues of the QCD Hamiltonian $E_k$ as

$$Z^{\rm QCD} = \sum_k e^{-\beta E_k}, \tag{5}$$

where $\beta$ is the inverse temperature. At low temperatures, ($\beta \to \infty$), the partition function is dominated by the lightest states of the theory, namely the vacuum state, with an energy density of $E_0/V_3$ and massless excitations thereof. The partition function $Z^{\rm QCD}$ can be rewritten as a Euclidean functional integral over the nonabelian gauge fields $A_\mu$,

$$Z^{\rm QCD}(M) = \int dA_\mu \prod_{f=1}^{N_f} \det(D + m_f) e^{-S^{\rm YM}}, \tag{6}$$

where $S^{\rm YM}$ is the Yang-Mills action given by

$$S^{\rm YM} = \int d^4x [\frac{1}{4g^2} F_{\mu\nu}^{a\ 2} - i\frac{\theta}{32\pi^2} F_{\mu\nu}^a \tilde{F}_{\mu\nu}^a]. \tag{7}$$

The field strength and its dual are given by

$$F^a_{\mu\nu} = \partial_\mu A^a_\nu - \partial_\nu A^a_\mu + f_{abc} A^b_\mu A^c_\nu, \qquad \tilde{F}_{\mu\nu} = \frac{1}{2}\epsilon_{\mu\nu\alpha\beta} F^{\alpha\beta}. \tag{8}$$

The $f_{abc}$ are the structure constants of the gauge group $SU(N_c)$. The gauge fields are denoted by $A_\mu = A_{\mu a}\frac{T^a}{2}$, where $T^a$ are the generators of the gauge group. The integral $\nu \equiv \frac{1}{32\pi^2}\int d^4x F^a_{\mu\nu}\tilde{F}^a_{\mu\nu}$ is a topological invariant, i.e. it does not change under continuous transformations of the gauge fields. An important class of field configurations are instantons. These are topological nontrivial field configurations that minimize the classical action. They are classified according to their topological charge $\nu$. The parameter $\theta$ is known as the $\theta$-angle. Experimentally, its value is consistent with zero. In (6), the mass matrix is diagonal, $M = \text{diag}(m_1, \cdots, m_{N_f})$, but below we will also consider a general mass matrix. The anti-Hermitian Dirac operator in (6) is given by

$$D = \gamma_\mu(\partial_\mu + iA_\mu), \tag{9}$$

where the $\gamma_\mu$ are the Euclidean Dirac matrices with anti-commutation relation $\{\gamma_\mu, \gamma_\nu\} = 2\delta_{\mu\nu}$. In the chiral representation the $\gamma$-matrices are given by

$$\gamma_k = \begin{pmatrix} 0 & i\sigma_k \\ -i\sigma_k & 0 \end{pmatrix}, \quad \gamma_4 = \begin{pmatrix} 0 & 1 \\ 1 & 0 \end{pmatrix}, \quad \gamma_5 = \begin{pmatrix} 1 & 0 \\ 0 & -1 \end{pmatrix}. \tag{10}$$

In this representation the Dirac operator has the structure

$$D = \begin{pmatrix} 0 & id \\ id^\dagger & 0 \end{pmatrix}. \tag{11}$$

The integration measure is defined by discretizing space-time

$$dA^a_\mu = \prod_x dA^a_\mu(x). \tag{12}$$

A particular popular discretization is the lattice discretization where the QCD action is discretized on a hyper-cubic lattice with spacing $a$. The discussion of lattice QCD would be a lecture by itself. For the interested reader we recommended several excellent textbooks on the subject [38–40].

A field theory is obtained by taking the continuum limit, i.e. the limit of zero lattice spacing $a$ for the integration measure discussed above. This limit only exists if we simultaneously adjust the coupling constant, i.e. $g \to g(a)$. If such limit exists the field theory is called renormalizable. For QCD $g(a)$ approaches zero in the continuum limit, a property known as asymptotic freedom.

We will be mainly interested in the eigenvalues of the Dirac operator and how they fluctuate for gauge fields $A_\mu$ distributed according to the QCD action. We will show that below a well-defined scale the fluctuations of the Dirac eigenvalues are given by a RMT with the global symmetries of the QCD.

## 3.3 Symmetries of the QCD partition function

It is well-known that the QCD action is greatly constrained by gauge symmetry, Poincaré invariance and renormalizability. These symmetries determine the structure of the Dirac operator and are essential for its infrared spectral properties. In this section we will discuss the global symmetries of the Euclidean Dirac operator. In particular, the chiral symmetry, the flavor symmetry and the anti-unitary symmetry of the continuum Dirac operator are discussed.

**Axial symmetry.** The axial symmetry, or the $U_A(1)$ symmetry, can be expressed as the anti-commutation relation

$$\{\gamma_5, D\} = 0. \tag{13}$$

This implies that all nonzero eigenvalues occur in pairs $\pm i\lambda_k$ with eigenfunctions given by $\phi_k$ and $\gamma_5\phi_k$. If $\lambda_k = 0$ the possibility exists that $\gamma_5\phi_k \sim \phi_k$, so that $\lambda_k = 0$ is an unpaired eigenvalue. According to the Atiyah-Singer theorem, the total number of such zero eigenvalues is a topological invariant, i.e., it does not change under continuous transformations of the gauge field configuration. Indeed, this possibility is realized by the field of an instanton which is a solution of the classical equations of motion. On the other hand, it cannot be excluded that $\lambda_k = 0$ while $\phi_k$ and $\gamma_5\phi_k$ are linearly independent. However, this imposes additional constraints on the gauge fields that will be violated by infinitesimal deformations. Generically, such situation does not occur.

In a decomposition according to the total number of topological zero modes, the QCD partition function can be written as

$$Z^{\mathrm{QCD}}(M,\theta) = \sum_\nu e^{i\nu\theta} Z_\nu^{\mathrm{QCD}}(M), \tag{14}$$

where

$$Z_\nu^{\mathrm{QCD}}(M) = \langle \prod_f m_f^\nu \prod_k (\lambda_k^2 + m_f^2)\rangle_\nu. \tag{15}$$

Here, $\langle \cdots \rangle_\nu$ denotes the average over gauge-field configurations with topological charge $\nu$ weighted by the Yang-Mills action. If we introduce right-handed and left-handed masses as complex conjugated masses we find that the $\theta$ dependence of the QCD partition function is only through the combination $me^{i\theta/N_f}$. This property can be used to obtain the $\theta$-dependence of the low-energy effective partition function.

**Flavor symmetry.** A second important global symmetry is the flavor symmetry. This symmetry can be best explained by writing the fermion determinant in the QCD partition function as a functional integral over Grassmann variables,

$$\prod_f \det(D + m_f) = \int d\psi d\bar{\psi} e^{\int d^4x \sum_{f=1}^{N_f} \bar{\psi}^f (D+m_f)\psi^f}. \tag{16}$$

In a chiral basis with $\psi_R = \gamma_5 \psi_R$ and $\psi_L = -\gamma_5 \psi_L$, the exponent can be rewritten as

$$\sum_{f=1}^{N_f} \bar{\psi}^f (D + m_f)\psi^f = \bar{\psi}_R^f D \psi_R^f + \bar{\psi}_L^f D \psi_L^f + \bar{\psi}_R^f M_{RL} \psi_L^f + \bar{\psi}_L^f M_{LR} \psi_R^f. \tag{17}$$

To better illuminate the transformation properties of the partition function we have replaced the diagonal mass matrix by $M_{RL}$ and $M_{LR}$.

For $m_f = 0$ we have the symmetry

$$\begin{aligned} \psi_L &\to U_L \psi_L, \quad \bar{\psi}_L \to \bar{\psi}_L U_L^{-1}, \\ \psi_R &\to U_R \psi_R, \quad \bar{\psi}_R \to \bar{\psi}_R U_R^{-1}. \end{aligned} \tag{18}$$

The only condition to be imposed on $U$ and $V$ is that their inverse exists. If the number of left-handed modes is equal to the number of right-handed modes we thus have an invariance under $Gl_R(N_f) \times Gl_L(N_f)$, where $Gl(N_f)$ is the group of complex $N_f \times N_f$ matrices with nonzero determinant. However, if the number of left-handed modes is not equal to the number of right-handed modes, the axial-symmetry group is broken to an $Sl(N_f)$ subgroup whereas the vector symmetry with $U_L = U_R$ remains unbroken. For $m_f = 0$ the flavor symmetry is thus broken explicitly to $Gl_V(N_f) \times Sl_A(N_f)$ by instantons or the anomaly. A $Gl_V(1)$ subgroup of $Gl_V(N_f)$ corresponds to baryon number conservation and is usually not considered when flavor symmetries are discussed.

What is much more important, though, is the spontaneous breaking of the axial flavor symmetry. From lattice QCD simulations and phenomenological arguments we know that the expectation value $\langle \bar{\psi}\psi \rangle = \langle \bar{\psi}_R \psi_R \rangle + \langle \bar{\psi}_L \psi_L \rangle \approx -(240\, MeV)^3$ in the vacuum state of QCD instead of the symmetric possibility $\langle \bar{\psi}\psi \rangle = 0$. Phenomenologically, this is known because the pions are much lighter than the $\sigma$ mesons. The spontaneous breaking of the axial symmetry also follows from the absence of parity doublets. For example, the pion mass and the $a_0$ mass are very different ($m_\pi = 135 MeV$ and $m_\delta = 980 MeV$).

For fermionic quarks there is no need to extend the symmetry group to $Gl_R(N_f) \times Gl_L(N_f)$. In that case we will only consider the usual $SU_R(N_f) \times SU_L(N_f)$ flavor symmetry and it spontaneous breaking to $SU_V(N_f)$. In the case of bosonic quarks, we will see in the next section that it is essential to consider the complex extension of $SU(N_f)$. Notice that the complex extension of the symmetry group does not change the number of conserved currents.

On easily verifies that $\langle\bar{\psi}\psi\rangle$ is only invariant for $U_L = U_R$. The vacuum state thus breaks the chiral symmetry down to $Gl_V(N_f)$. In agreement with the Vafa-Witten theorem [41] only the axial symmetries can be broken spontaneously. We also observe that the *complete* axial group is broken which is known as the maximum breaking [42] of chiral symmetry.

**Flavor symmetry for bosonic quarks.** For bosonic quarks the Goldstone bosons cannot be parameterized by a unitary matrix. The reason is that symmetry transformations have to be consistent with the convergence of the bosonic integrals. Let us consider the case of one bosonic flavor. Then

$$\det{}^{-1}\begin{pmatrix} m & id \\ id^\dagger & m \end{pmatrix} = \frac{1}{\pi^2}\int d^2\phi_1 d^2\phi_2 \exp\left[-\begin{pmatrix} \phi_1^* \\ \phi_2^* \end{pmatrix}\begin{pmatrix} m & id \\ id^\dagger & m \end{pmatrix}\begin{pmatrix} \phi_1 \\ \phi_2 \end{pmatrix}\right], \tag{19}$$

so that the integral is convergent for $\mathrm{Re}(m) > 0$. The most general flavor symmetry group of the action in (19) is $Gl(2)$ that can be parameterized as

$$U = e^H V \quad \text{with} \quad H^\dagger = H \quad \text{and} \quad VV^\dagger = 1. \tag{20}$$

For $U$ to be a symmetry transformation for $m = 0$ we require that

$$U^\dagger \begin{pmatrix} 0 & id \\ id^\dagger & 0 \end{pmatrix} U = \begin{pmatrix} 0 & id \\ id^\dagger & 0 \end{pmatrix}, \tag{21}$$

so that $H$ has to be a multiple of $\sigma_3$, and $V$ has to be a multiple of the identity. The transformations $V$ in (20) are not broken by the mass term and therefore represent the vector symmetry. Only the symmetry transformation $\exp(s\sigma_3)$ is broken by the mass term so that the axial transformations are parameterized by

$$U = \begin{pmatrix} e^s & 0 \\ 0 & e^{-s} \end{pmatrix} \qquad \text{with} \quad s \in \langle -\infty, \infty \rangle. \tag{22}$$

For $N_f$ bosonic flavors the axial transformations are parameterized by

$$U = \begin{pmatrix} e^H & 0 \\ 0 & e^{-H} \end{pmatrix} \qquad \text{with} \qquad H^\dagger = H, \tag{23}$$

which is the coset $Gl(N_f)/U(N_f)$.

## 3.4 Anti-unitary symmetries and the three-fold way

The QCD partition function with three or more colors in the fundamental representations has no anti-unitary symmetries. As will be discussed below, for two colors with fundamental fermions and for adjoint fermions, the Dirac operator has an anti-unitary symmetry. The classification of the QCD Dirac operator according to anti-unitary symmetries was introduced in [15].

**Anti-unitary symmetries and the Dyson index.** The value of the Dyson index is determined by the anti-unitary symmetries of the system. If there are no anti-unitary symmetries the Hamiltonian is Hermitian and the value of the Dyson index is $\beta_D = 2$.

An anti-unitary symmetry operator, which can always be written as $A = UK$ with $U$ unitary and $K$ the complex conjugation operator, commutes with the Hamiltonian of the system

$$[H, UK] = 0. \tag{24}$$

We can distinguish two possibilities

$$(UK)^2 = 1 \quad \text{or} \quad (UK)^2 = -1. \tag{25}$$

corresponding to $\beta_D = 1$ and $\beta_D = 4$, respectively. The argument goes as follows. The symmetry operator $A^2 = (UK)^2 = UU^*$ is unitary, and in an irreducible subspace, it is necessarily a multiple of the identity, $UU^* = \lambda \mathbf{1}$. Because of this relation, $U$ and $U^*$ commute so that $\lambda$ is real. By unitarity we have $|\lambda| = 1$ which yields $\lambda = \pm 1$.

When $\beta_D = 1$ it is always possible to find a basis in which the Hamiltonian is real. Starting with basis vector $\phi_1$ we can construct $\psi_1 = \phi_1 + UK\phi_1$. Then choose $\phi_2$ perpendicular to $\psi_1$ and define $\psi_2 = \phi_2 + UK\phi_2$ with

$$(\phi_2 + UK\phi_2, \psi_1) = (UK\phi_2, \psi_1) = ((UK)^2\phi_2, UK\psi_1)^* = (\phi_2, \psi_1)^* = 0.$$

The next basis vector is found by choosing $\phi_3$ perpendicular to $\psi_1$ and $\psi_2$, etc.. In this basis the Hamiltonian is real

$$\begin{aligned} H_{kl} &= (\psi_k, H\psi_l) = (UK\psi_k, UKH\psi_l)^* = (\psi_k, HUK\psi_l)^* = (\psi_k, H\psi_l)^* \\ &= H^*_{kl}. \end{aligned} \tag{26}$$

The best known anti-unitary operator in this class is the time-reversal operator for which $U$ is the identity matrix.

In the case $(UK)^2 = -1$ all eigenvalues of the Hamiltonian are doubly degenerate. This can be shown as follows. If $\phi_k$ is and eigenvector with eigenvalue $\lambda_k$, then it follows from (24) that also $UK\phi_k$ is an eigenvector of the Hamiltonian with the same eigenvalue. The important thing is that this eigenvector is perpendicular to $\phi_k$ [13],

$$(\phi_k, UK\phi_k) = (UK\phi_k, (UK)^2\phi_k)^* = -(\phi_k, UK\phi_k). \tag{27}$$

In this case it is possible to construct a basis for which the Hamiltonian matrix can be organized into real quaternions [43]. The eigenvalues of a Hermitian quaternion real matrix are quaternion scalars, and the eigenvalues of the original matrix are thus doubly degenerate in agreement with (27). The best known example in this class is the Kramers degeneracy for time reversal invariant systems with half-integer spin but no rotational invariance. For example, for spin $\frac{1}{2}$ the time reversal operator is given by $\sigma_2 K$ with $(K\sigma_2)^2 = -1$.

Next we will discuss the anti-unitary symmetries of the QCD Dirac operator.

**QCD in the fundamental representation.** For three or more colors, QCD in the fundamental representation does not have any anti-unitary symmetries and $\beta_D = 2$. QCD with two colors is exceptional. The reason is the pseudo-reality of $SU(2)$:

$$A_\mu^* = (\sum_k a_k \frac{\tau_k}{2})^* = -\tau_2 A_\mu \tau_2, \tag{28}$$

where the $\tau_k$ are the Pauli matrices acting in color space.

From the explicit representation for the $\gamma$-matrices it follows that

$$\gamma_\mu^* = \gamma_2\gamma_4\gamma_\mu\gamma_2\gamma_4. \tag{29}$$

For the Dirac operator $iD = i\gamma_\mu\partial_\mu + \gamma_\mu A_\mu$ we thus have

$$[KC\gamma_5\tau_2, D] = 0, \tag{30}$$

where $K$ is the complex conjugation operator and $C = \gamma_2\gamma_4$ is the charge conjugation matrix. Because $(KC\gamma_5\tau_2)^2 = 1$ we have that $\beta_D = 1$. Using the argument of Eq. (26) a basis can be constructed such that the Dirac matrix is real for any $A_\mu$.

**QCD in the adjoint representation.** For QCD with gauge fields in the adjoint representation the Dirac operator is given by

$$D = \gamma_\mu\partial_\mu + f^{abc}\gamma_\mu A_{a\mu}, \tag{31}$$

where the $f^{abc}$ denote the structure constants of the gauge group. Because of the complex conjugation property of the $\gamma$-matrices we have that

$$[\gamma_2\gamma_4\gamma_5 K, D] = 0. \tag{32}$$

One easily verifies that in this case

$$(\gamma_2\gamma_4\gamma_5 K)^2 = -1, \tag{33}$$

so that the eigenvalues of $D$ are doubly degenerate (see section 3.4). This corresponds to the case $\beta_D = 4$, so that it is possible to organize the matrix elements of the Dirac operator into real quaternions.

# 4. The Dirac Spectrum in QCD

In this section we show that the smallest eigenvalues of the QCD Dirac operator are related to the chiral condensate by means of the Banks-Casher relation. This result is used to define the microscopic spectral density.

## 4.1 Banks-Casher relation

The order parameter of the chiral phase transition, $\langle\bar{\psi}\psi\rangle$, is nonzero only below a critical temperature or a critical chemical potential. As was shown by Banks and Casher [44], $\langle\bar{\psi}\psi\rangle$ is directly related to the eigenvalue density of the QCD Dirac operator per unit four-volume

$$\Sigma \equiv |\langle\bar{\psi}\psi\rangle| = \lim \frac{\pi\langle\rho(0)\rangle}{V}. \tag{34}$$

For eigenvalues $\{\lambda_k\}$ the average spectral density is given by

$$\rho(\lambda) = \langle\sum_k \delta(\lambda - \lambda_k)\rangle. \tag{35}$$

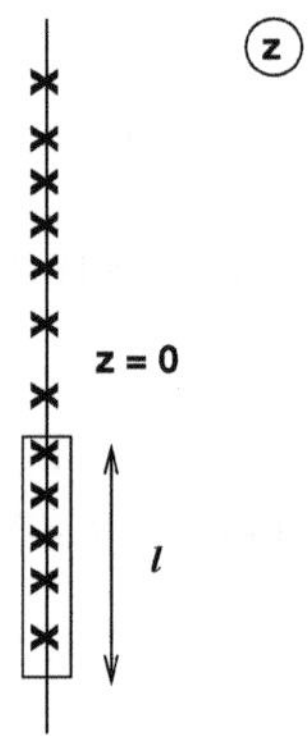


*Figure 1.* A typical Dirac spectrum. To derive the Banks-Casher relation we integrate the resolvent over the rectangular contour in this figure. (Figure taken from [45].)

To show the Banks-Casher relation we study the resolvent defined by

$$G(z) = \sum_k \frac{1}{z + i\lambda_k}. \tag{36}$$

It can be interpreted as the electric field at $z$ of charges at $i\lambda_k$. Using this analogy it is clear that the resolvent changes sign if $z$ crosses the imaginary axis. Let us look at this in more detail. A typical Dirac spectrum is shown in Fig. 1. The average number of eigenvalues in the rectangular contour in this figure is $\rho(\lambda)l$. If we integrate the resolvent along this contour we find

$$\oint G(z) = il(G(i\lambda+\epsilon) - G(i\lambda-\epsilon)) = 2\pi i\rho(\lambda)l, \tag{37}$$

where the second identity follows from Cauchy's theorem. Using the symmetry of the spectrum we obtain

$$\mathrm{Re}G(i\lambda+\epsilon) = \pi\rho(\lambda). \tag{38}$$

Near the center of the spectrum the imaginary part of the resolvent is negligible. Using that the chiral condensate is related to the resolvent by

$$\langle\bar{\psi}\psi\rangle = -\lim_{m\to 0}\lim_{V\to\infty}\frac{1}{V}G(m), \tag{39}$$

immediately results in the Banks-Casher relation (34). The order of the limits in (34) is important. First we take the thermodynamic limit, next the chiral limit and, finally, the field theory limit.

The resolvent of the QCD Dirac spectrum can be obtained from

$$G(z; m_1, \cdots, m_{N_f}) = \frac{\partial}{\partial z}\bigg|_{z=z'} \log Z_\nu^{\mathrm{pq}}(z, z', m_f), \tag{40}$$

with the so called partially quenched QCD partition function given by

$$Z_\nu^{\mathrm{pq}}(z, z', m_f) = \int dA\,\frac{\det(D+z)}{\det(D+z')}\prod_{f=1}^{N_f}\det(D+m_f)\,e^{-S^{\mathrm{YM}}}. \tag{41}$$

For $z = z'$ this partition function coincides with the QCD partition function.

In addition to the regular quarks, the partition function (41) has additional bosonic and fermionic ghost quarks. Our aim is to find the chiral Lagrangian corresponding to (41). If we are successful, we have succeeded in deriving a generating function for the infrared limit of the QCD Dirac spectrum.

## 4.2 Microscopic spectral density

An important consequence of the Bank-Casher formula (34) is that the eigenvalues near zero virtuality are spaced as

$$\Delta\lambda = 1/\rho(0) = \pi/\Sigma V. \tag{42}$$

For the average position of the smallest nonzero eigenvalue we obtain the estimate

$$\lambda_{\min} = \pi/\Sigma V. \tag{43}$$

This should be contrasted with the eigenvalue spectrum of the non-interacting Dirac operator. Then the eigenvalues are those of a free Dirac particle in a box with eigenvalue spacing equal to $\Delta\lambda \sim 1/V^{1/4}$ for the eigenvalues near $\lambda = 0$. Clearly, the presence of gauge fields leads to a strong modification of the spectrum near zero virtuality. Strong interactions result in the coupling of many degrees of freedom leading to extended states and correlated eigenvalues. Because of asymptotic freedom, the spectral density of the Dirac operator for large $\lambda$ behaves as $V\lambda^3$. In Fig. 2 we show a plot of a typical average spectral density of the QCD Dirac operator for $\lambda \geq 0$. The spectral density for negative $\lambda$ is obtained by reflection with respect to the $y$-axis. More discussion of this figure will be given in section 5.3.

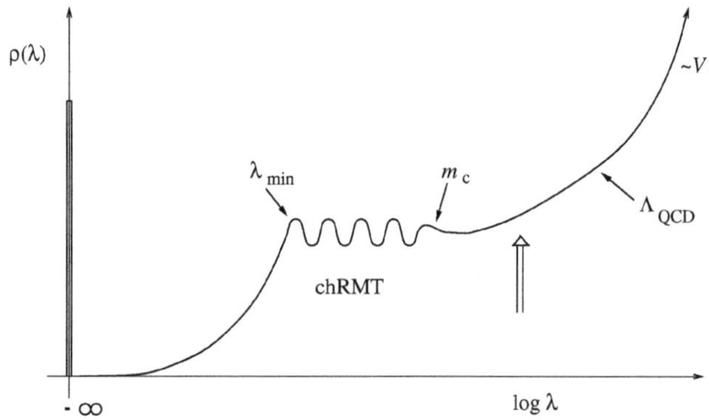


*Figure 2.* Schematic picture of the average spectral density of QCD Dirac operator. (Taken from [46].)

Because the eigenvalues near zero are spaced as $\sim 1/\Sigma V$ it is natural to introduce the microscopic spectral density [14]

$$\rho_s(u) = \lim_{V\to\infty} \frac{1}{V\Sigma}\rho(\frac{u}{V\Sigma}) \quad \text{with} \quad u = \lambda V\Sigma. \tag{44}$$

We expect that this limit exists and converges to a universal function which is determined by the global symmetries of the QCD Dirac operator. In section 6, we will calculate $\rho_s(u)$ both for the simplest theory in this universality class, which is chiral Random Matrix Theory (chRMT), and for the partial quenched chiral Lagrangian which describes the low-energy limit of the QCD partition function. We will find that the two results coincide below the Thouless energy.

## 5. Low energy limit of QCD

In this section we derive the chiral Lagrangian that provides an exact description of QCD at low energies.

## 5.1 The chiral lagrangian

For light quarks the low energy limit of QCD is well understood. It is given by a chiral Lagrangian that describes the interactions of the pseudo-scalar mesons. The reason is that pions are Goldstone bosons which are the only light degrees of freedom in a confining theory such as QCD. To lowest order in the quark masses and the momenta, the chiral Lagrangian is completely determined by chiral symmetry and Lorentz invariance. In the case of $N_f$ light quarks with chiral symmetry breaking according to $SU_L(N_f) \times SU_R(N_f) \to SU_V(N_f)$ the Goldstone fields are given by $U \in SU(N_f)$. Under an $SU_L(N_f) \times SU_R(N_f)$ transformation of the quark fields given in (18), the Goldstone fields $U$ transform in the same way as the chiral condensate

$$U \to U_R U U_L^{-1}. \tag{45}$$

The symmetry (18) is broken the mass term. However, the full symmetry can be restored if we also transform the mass term as

$$M_{RL} \to U_R M_{RL} U_L^{-1}, \qquad M_{LR} \to U_L M_{LR} U_R^{-1}. \tag{46}$$

The low energy effective theory should have the same invariance properties. To second order in the momenta and first order in the quark mass matrix we can write down the following invariant terms:

$$\mathrm{Tr}(\partial_\mu U \, \partial_\mu U^\dagger), \qquad \mathrm{Tr}(M_{RL} U^\dagger), \qquad \mathrm{Tr}(M_{LR} U). \tag{47}$$

Since the QCD partition function is invariant under $M_{RL} \leftrightarrow M_{LR}$, the effective partition function should also have this symmetry. The action of the Goldstone fields is therefore given by the so called Weinberg Lagrangian [47, 48]

$$\mathcal{L}_{\text{eff}}(U) = \frac{F^2}{4} \, \mathrm{Tr}(\partial_\mu U \partial_\mu U^\dagger) - \frac{\Sigma}{2} \, \mathrm{Tr}(M_{RL} U^\dagger + M_{LR} U), \tag{48}$$

where $F$ is the pion decay constant, and $\Sigma$ is the chiral condensate. The Goldstone fields can be parametrized as $U = \exp(i\sqrt{2}\Pi_a t^a / F)$, with the generators of $SU(N_f)$ normalized according to $\mathrm{Tr}\, t^a t^b = \delta^{ab}$. This chiral Lagrangian has been used extensively for the analysis of pion-pion scattering amplitudes [48].

To lowest order in the pion fields we find for equal quark masses $m$

$$\mathcal{L}_{\text{eff}}(U) = \frac{1}{2} \partial_\mu \Pi^a \partial^\mu \Pi^a + N_f \Sigma m + \frac{\Sigma m}{F^2} \Pi^a \Pi^a. \tag{49}$$

This results in the pion propagator $1/(p^2 + m_\pi^2)$ with pion mass given by the Gellmann-Oakes-Renner relation

$$m_\pi^2 = \frac{2m\Sigma}{F^2}. \tag{50}$$

It also illustrates the identification of $\Sigma$ as the chiral condensate.

## 5.2 The low energy limit of $Z_\nu^{\rm pq}$

The low-energy limit of the partially quenched QCD partition function can be derived along the same lines as the derivation of the chiral Lagrangian obtained in previous section. In this case, ignoring convergence questions for the moment, the global flavor symmetry of (41) is given by

$$Gl_R(N_f+1|1) \times Gl_L(N_f+1|1). \tag{51}$$

We already have seen that convergence requirements restrict the axial symmetry for bosonic quarks to $Gl(N_f)/U(N_f)$. Although the axial flavor symmetry group of the fermionic quarks is not a priori determined by convergence requirements we will see in this section that supersymmetry necessarily imposes that this symmetry group is compact, i.e. equal to $U(N_f)$.

Under transformation (51) the quarks fields with $N_f+1$ fermionic components and one bosonic component, transform as

$$\psi_R \to U_R\psi_R, \quad \psi_L \to U_L\psi_L, \quad \bar\psi_R \to \bar\psi_R U_R^{-1}, \quad \bar\psi_L \to \psi_L U_L^{-1}. \tag{52}$$

The subscripts refer to the right-handed (R) or left-handed (L) quarks. For $M=0$ and $\nu=0$ this is a symmetry of the QCD action. For $M\neq 0$ this symmetry can be restored if we also transform the mass term according to

$$M_{RL} \to U_R M_{RL} U_L^{-1}, \qquad M_{LR} \to U_L M_{LR} U_R^{-1}. \tag{53}$$

In the sector of topological charge $\nu$ the partially quenched partition function transforms as

$$Z_\nu^{\rm pq}(M_{RL}, M_{LR}) \to {\rm Sdet}^\nu[U_R U_L]^{-1} Z_\nu^{\rm pq}(M_{RL}, M_{LR}). \tag{54}$$

The Goldstone bosons corresponding to the breaking of of the axial subgroup $Gl_A(N_f+1|1)$ transform as $Q \to U_R Q U_L^{-1}$. If we factorize the Goldstone fields into the zero momentum modes $Q_0$ and the nonzero momentum modes $Q(x)$ as

$$Q = Q_0 Q(x), \tag{55}$$

one can easily show that the low energy effective partition function which the above transformation properties is given by

$$Z_\nu^{\rm pq}(M) = \int_{Q\in Gl(N_f+1|1)} dQ {\rm Sdet}^\nu(Q_0) e^{-\int d^4x \mathcal{L}^{\rm pq}(Q)}, \tag{56}$$

where

$$\mathcal{L}^{\rm pq}(Q) = \frac{F^2}{4}{\rm Str}\partial_\mu Q^{-1}\partial_\mu Q + \frac{\Sigma}{2}{\rm Str}(M_{RL}Q^{-1}) + \frac{\Sigma}{2}{\rm Str}(M_{LR}Q). \tag{57}$$

We already have seen that the boson-boson block of $Gl(N_f+1|1)$ is $Gl(1)/U(1)$. If we parameterize the field $Q$ as

$$Q = e^{\sum_k T_k \pi_k / F}, \tag{58}$$

with $T_k$ the generators of $G(N_f+1|1)$, to second order in the Goldstone fields the mass term is given by $\mathrm{Str}(\Sigma M \sum_k T_k^2 \pi_k^2 / F^2)$. Let us take $M$ diagonal positive definite. Because of the supertrace there is a relative minus sign between the boson-boson and fermion-fermion modes. The boson-boson modes are noncompact and require that the overall minus sign of the mass term is negative. In order to avoid tachyonic fermion-fermion Goldstone modes, we have to compensate the minus sign of the supertrace. This can be done by choosing the parameters that multiply the fermion-fermion generators purely imaginary. This corresponds to a compact parametrization of the fermion-fermion Goldstone manifold. This integration manifold is the maximum Riemannian submanifold [49] of $Gl(N_f+1|1)$ and will be denoted by $\hat{G}l(N_f+1|1)$.

## 5.3 The mesoscopic limit of QCD

In chiral perturbation theory, the different domains of validity where analyzed by Gasser and Leutwyler [12]. A similar analysis applies to partially quenched chiral perturbation theory [51]. The idea is as follows. The $Q$ field can be decomposed as [12]

$$Q = Q_0 e^{i\psi(x)}. \tag{59}$$

where $Q_0$ is a constant (zero-momentum) field. For momenta $p = \pi k/L$ with $k$ integer , the kinetic term of the $\psi$ fields behaves as

$$\frac{1}{2}\partial_\mu \psi^a(x) \partial_\mu \psi^a(x) \sim L^{-2} \psi^2(x). \tag{60}$$

We observe that the magnitude of the fluctuations of the $\psi^a$ fields are of order $1/L$ which justifies a perturbative expansion of $\exp(i\psi(x))$. The fluctuations of the zero modes, on the other hand, are only limited by the mass term

$$\frac{1}{2} V \Sigma \mathrm{Str} M (Q_0 + Q_0^{-1}). \tag{61}$$

For quark masses $m \gg 1/V\Sigma$, the field $Q_0$ fluctuates close to the identity and the $Q_0$ field can be expanded around the identity as well. If $m \ll \Lambda_{\mathrm{QCD}}$ we are in the domain of chiral perturbation theory. For

$$\frac{\Sigma m}{F^2} \ll \frac{1}{\sqrt{V}} \tag{62}$$

the fluctuations of the zero modes dominate the fluctuations of the nonzero modes, and only the contribution from the zero modes has to be taken into account for the calculation of an observable. In this limit the so called finite volume partition function is given by [12, 13]

$$Z_{N_f}^{\rm eff}(M,\theta) \sim \int_{U\in SU(N_f)} dU\, e^{V\Sigma {\rm Re\,Tr}\, MUe^{i\theta/N_f}}, \tag{63}$$

where the $\theta$-dependence follows from the dependence of the QCD partition function on the combination $me^{i\theta/N_f}$ only (see section 3.3). We emphasize that any theory with the same pattern of chiral symmetry breaking as QCD can be reduced to the same extreme infrared limit.

The effective partition function at fixed $\nu$ follows by Fourier inversion

$$Z_\nu^{\rm eff}(M) = \frac{1}{2\pi}\int_0^{2\pi} d\theta e^{-i\nu\theta} Z^{\rm eff}(M,\theta). \tag{64}$$

Combining the integral over $SU(N_f)$ and the integral over $U(1)$ we find that

$$Z_\nu^{\rm eff}(M) = \int_{U(N_f)} {\det}^\nu(U) e^{V\Sigma {\rm Re\,Tr}\, MU^\dagger}. \tag{65}$$

The same arguments apply to the partially quenched chiral Lagrangian. There is an important difference. The mass of the ghost-quarks is an external parameter which can take on any value we wish. The mass of the Goldstone modes containing these quarks is given by

$$M_{zz} = \frac{2z\Sigma}{F^2}. \tag{66}$$

Therefore, independent of the quark masses there is always a domain where the fluctuations of the zero momentum modes dominate the fluctuations of the nonzero momentum modes. This domain is given by [51]

$$z \ll \frac{F^2}{\Sigma L^2} \equiv m_c. \tag{67}$$

In this domain, the Compton wavelength of the Goldstone bosons with mass $M_{zz}$ is much larger than the size of the box. Because the time scale conjugate to $m_c$ is of the order of the diffusion time across the length of the box, this domain is known as the ergodic domain. In order that the non-Goldstone modes do not contribute to the partition function we have to require that $L \gg 1/\Lambda_{\rm QCD}$.

In the Dirac spectrum we can thus distinguish three important scales:

$$\lambda_{\rm min} \ll m_c \ll \Lambda_{\rm QCD}. \tag{68}$$

For $z \ll m_c$ we are in the zero momentum sector of the theory. If $z$ is of the order of $\lambda_{\rm min}$ or less we have to take into account quantum fluctuations to all orders. For $\lambda_{\rm min} \ll z \ll m_c$, the integral over zero modes can be calculated perturbatively by a loop expansion. For $m_c \ll z \ll \Lambda_{QCD}$, chiral perturbation theory still applies, but the zero momentum modes no longer dominate the partition function. For $z \gg \Lambda_{QCD}$, chiral perturbation theory is not applicable to the spectrum of the Dirac operator.

In the ergodic domain the QCD partition function in the sector of topological charge $\nu$ is given by [19]

$$Z_\nu^{\rm pq}(M) = \int_{Q \in \hat{Gl}(N_f+1|1)} dQ\, {\rm Sdet}^\nu Q\, e^{V\frac{\Sigma}{2}\,{\rm Str}(MQ+MQ^{-1})}. \tag{69}$$

The number of QCD Dirac eigenvalues that is described by this partition function is of the order $m_c/\Delta\lambda = F^2L^2$. This number increases linearly in $N_c$ for $N_c \to \infty$ which was recently found in lattice simulations [50].

In section (6.3) we will study this partition function in the quenched limit ($N_f = 0$) and show it coincides with the chRMT result [19, 20].

**Comparison to disordered systems.** In the book by Efetov [32] it is shown that the diffusion of electrons in a disordered medium can be described by the effective action

$$F(Q) = \frac{\pi\nu}{8} D \int d^dx [{\rm Tr}(\nabla Q)^2 - \frac{\pi i \nu\omega}{4} {\rm Tr}\Lambda Q], \tag{70}$$

where $Q$ are the Goldstone fields, $\nu$ is the density of states, $D$ is the diffusion constant and $\omega$ is the energy difference between the advanced and the retarded Green's functions. The matrix $\Lambda$ is a diagonal matrix with matrix elements $\pm 1$ corresponding to the causal character of the Green's functions. The Goldstone bosons arise because of the spontaneous breaking of the symmetry between the advanced and retarded Green's functions.

If we compare this effective action to the chiral Lagrangian (57) we can make the identification

$$\frac{F^2}{4} \leftrightarrow \frac{\pi\nu D}{8}, \qquad \frac{\pi\omega\nu}{4} \leftrightarrow \frac{M\Sigma}{2}, \qquad \nu \leftrightarrow \frac{\rho(E)}{V}, \tag{71}$$

which can be rewritten as

$$M \leftrightarrow \frac{\omega}{2}, \qquad \Sigma \leftrightarrow \pi\nu, \qquad F^2 \leftrightarrow \frac{\pi\nu D}{2}. \tag{72}$$

The domain where the kinetic term factorizes from the partition function is therefore given by

$$L^2 \ll \frac{F^2}{M\Sigma} \leftrightarrow \frac{D}{\omega}. \tag{73}$$

In the theory of disordered mesoscopic systems the corresponding energy scale is known as the Thouless energy. It is defined by [52, 53]

$$E_c = \frac{\hbar D}{L^2}, \tag{74}$$

where $D$ is the diffusion constant for the diffusive motion of electrons in a disordered sample. The time conjugate to $E_c$ is the time scale over which an electron diffuses across the sample. Therefore, the domain where $\hbar\omega \ll E_c$ is known as the ergodic domain. The time scale in mesoscopic physics corresponding to $\Lambda_{QCD}$ is the elastic scattering time $\tau_e$. The domain in between $E_c$ and $\hbar/\tau_e$ is known as the diffusive domain. This domain is characterized by diffusive motion of electrons in the disordered sample described by the Lagrangian (70).

# 6. Chiral RMT and the QCD Dirac spectrum

## 6.1 The chiral ensembles

The chiral ensembles are defined as the ensembles of $N \times N$ Hermitian matrices with block structure [14, 15]

$$D = \begin{pmatrix} 0 & iC \\ iC^\dagger & 0 \end{pmatrix}, \tag{75}$$

and probability distribution given by (for equal quark masses $m$)

$$P(C)dC = \mathcal{N}\det{}^{N_f}(D+m)e^{-\frac{N\beta_D}{4}\mathrm{Tr}C^\dagger C}DC. \tag{76}$$

The integration measure $dC$ is the product of differentials of the independent parts of the matrix elements of $C$, and $N_f$ is a real parameter (corresponding to the number of quark flavors in QCD). The matrix $C$ is a rectangular $n\times(n+\nu)$ matrix. The nonzero eigenvalues of the matrix $D$ occur in pairs $\pm\lambda_k$. This can be seen as follows. If

$$D\begin{pmatrix} a \\ b \end{pmatrix} = \lambda\begin{pmatrix} a \\ b \end{pmatrix} \quad \text{then} \quad D\begin{pmatrix} a \\ -b \end{pmatrix} = -\lambda\begin{pmatrix} a \\ -b \end{pmatrix}. \tag{77}$$

Generically, the matrix $D$ in (75) has exactly $|\nu|$ zero eigenvalues. For this reason, $\nu$ is identified as the topological quantum number. The normalization constant of the probability distribution is denoted by $\mathcal{N}$. We can distinguish ensembles with real, complex, or quaternion real matrix elements. They are denoted by $\beta_D = 1$, $\beta_D = 2$, and $\beta_D = 4$, respectively. In addition to the global symmetries of QCD, this partition function has a large unitary invariance given by

$$C \to UCV^{-1}, \tag{78}$$

where $U$ and $V$ are orthogonal, unitary, or symplectic matrices, respectively. Therefore, the corresponding ensembles are known as the chiral Gaussian Orthogonal Ensemble (chGOE), the chiral Gaussian Unitary Ensemble (chGUE), and the chiral Gaussian Symplectic Ensemble (chGSE), in this order.

Using the invariance (78) it is always possible to decompose $C$ as

$$C = U\Lambda V^{-1}, \tag{79}$$

where $\Lambda$ is a diagonal matrix with $\lambda_k \geq 0$. The joint probability distribution for the eigenvalues is obtained by transforming to $\Lambda$, $U$ and $V$ as new integration variables. The Jacobian is given by

$$J \sim \prod_k \lambda_k^{\nu\beta_D - 1} \prod_{k<l} |(\lambda_k^2 - \lambda_l^2)|^{\beta_D} \tag{80}$$

resulting in the joint eigenvalue distribution

$$P(\{\lambda\})d\{\lambda\} = \mathcal{N}|\Delta(\{\lambda^2\})|^{\beta_D} \prod_k \lambda_k^{\alpha}(\lambda_k^2 + m^2)^{N_f} e^{-N\beta_D\lambda_k^2/4} d\lambda_k, \tag{81}$$

where $\alpha = \beta_D - 1 + \beta_D\nu$. We note that the distribution of the eigenvectors factorizes from the distribution of the eigenvalues factorizes.

## 6.2 Mathematical methods

In this section we will discuss the orthogonal polynomial method, the resolvent expansion method, the replica trick and the supersymmetric method which are widely used in Random Matrix Theory.

**Resolvent expansion methods.** These methods are based on expanding the resolvent in a geometric series

$$G(z) = \langle \mathrm{Tr}\frac{1}{z-H}\rangle = N\frac{1}{z} + \langle \mathrm{Tr}\frac{1}{z}H\frac{1}{z}\rangle + \langle \mathrm{Tr}\frac{1}{z}H\frac{1}{z}H\frac{1}{z}\rangle + \cdots . \tag{82}$$

In the large $N$ limit the averages are given by a sum of planar diagrams. Let us illustrate this for the GUE. In this case the "propagator" is given by

$$\langle H_{ij}H_{kl}\rangle = \frac{1}{N}\delta_{il}\delta_{jk}. \tag{83}$$

For example, as was explained in the course of Di Francesco [4], for $\mathrm{Tr}H^4$ term we have two planar diagrams of order $N^3$ and one diagram of order $N^2$.

**The orthogonal polynomial method.** The oldest method is the orthogonal polynomial method [1]. In principle, one obtains expressions that are exact for finite size matrices. The drawback of this method is that it requires a probability distribution that is invariant under basis change of the random matrix. In general the probability density can be written as

$$P(x_1, \cdots, x_n) = \Delta^{\beta_D}(\{x_k\}) \prod_{k=1}^{n} w(x_k), \tag{84}$$

where $w(x)$ is a weight function and the Vandermonde determinant is given by

$$\Delta(\{x_k\}) = \prod_{k>l}(x_k - x_l). \tag{85}$$

The method is based on the identity

$$\Delta(\{x_k\}) = \begin{vmatrix} 1 & \cdots & 1 \\ x_1 & \cdots & x_n \\ \vdots & & \vdots \\ x_1^{n-1} & \cdots & x_n^{n-1} \end{vmatrix} = \begin{vmatrix} P_0(x_1) & \cdots & P_0(x_n) \\ P_1(x_1) & \cdots & P_n(x_n) \\ \vdots & & \vdots \\ P_{n-1}(x_1) & \cdots & P_{n-1}(x_n) \end{vmatrix}, \tag{86}$$

where the $P_k$ are monic orthogonal polynomials defined by

$$\int dx w(x) P_k(x) P_l(x) = h_k \delta_{kl}. \tag{87}$$

Because of these relations, integrals over the eigenvalues can be performed by means of orthogonality relations. In the next section we illustrate this method by the calculation of the microscopic spectral density for the chGUE.

**The replica trick.** The replica trick is based on the identity

$$G(z) = \frac{1}{V}\langle \mathrm{Tr} \frac{1}{z+iD} \rangle = \lim_{r\to 0} \frac{1}{Vr} \partial_z \langle \det{}^r(iD+z) \rangle. \tag{88}$$

The recipe is to calculation the partition function for positive or negative integer values of $r$ and then analytically continue to $r = 0$. For positive (negative) integer values of $r$ the average determinant can be calculated by rewriting it as a Grassmann (complex) Gaussian integral. Then $D$ appears linear in the exponent which allows us to perform the average for a Gaussian distribution of $D$. The replica trick works without problems for perturbative calculations but usually fails in the nonperturbative calculations. As example consider the following expression for the the modified Bessel function $I_\nu(z)$:

$$I_\nu(z) = \frac{1}{\pi}\int_0^\pi e^{z\cos\theta} \cos\nu\theta d\theta - \frac{\sin\nu\pi}{\pi}\int_0^\infty e^{-z\cosh t - \nu t} dt. \tag{89}$$

We would have missed the second term if we calculate the Bessel function only for integer values of $\nu$. The replica trick can be made to work if we consider a family of partition functions related by a Toda lattice equation. This will be discussed in detail in the next two lectures.

**The supersymmetric method.** The supersymmetric method [74] is based on the identity

$$G(z) = \frac{1}{V}\partial_z \left\langle \frac{\det(iD+z)}{\det(iD+z')} \right\rangle \Big|_{z'=z}. \tag{90}$$

The determinant can be written as a Grassmann integral and inverse determinant as a complex integral. For $z' = z$ this partition function has an exact supersymmetry. The advantage of this method is that it is mathematically rigorous, but it requires a deep understanding of super mathematics. For example, finite expressions can be obtained from singular terms that do not depend on the Grassmann variables (and are zero upon integration). For a discussion of these so-called Efetov-Wegner terms we refer to the original literature [54, 37].

## 6.3 The microscopic spectral density of the chGUE

In this subsection we calculate the microscopic spectral density by the orthogonal polynomials method [55] and the supersymmetric method [19, 20].

**Orthogonal polynomials.** For the chGUE the joint probability distributions only depends on the square of the eigenvalues and we use $x_k = \lambda_k^2$ as new variables. In terms of these variables the weight function is given by

$$w(x) = (xn\Sigma^2)^a e^{-n\Sigma^2 x}, \tag{91}$$

with $a = N_f + |\nu|$. The monic orthogonal polynomials corresponding to this weight function can be expressed in terms Laguerre polynomials

$$P_k(x) = \frac{(-1)^k k!}{(\Sigma^2 n)^k} L_k^a(x\Sigma^2 n) \tag{92}$$

with normalization constants $h_k$ given by $h_k = k!(k+a)!/(n\Sigma^2)^{2k+1}$. The eigenvalue density is given by (with c a constant)

$$\begin{aligned}
\rho(x_1) &= c\int \prod_{k=2}^{n}[w(x_k)dx_k] \begin{vmatrix} P_0(x_1) & \cdots & P_0(x_n) \\ \vdots & & \vdots \\ P_{n-1}(x_1) & \cdots & P_{n-1}(x_n) \end{vmatrix}^2, \\
&= \sum_{\sigma\pi} \mathrm{sg}(\sigma\pi) \prod_{k=2}^{n}[w(x_k)dx_k] P_{\sigma(0)}(x_1)\cdots P_{\sigma(n-1)}(x_n) \\
&\quad \times P_{\pi(0)}(x_1)\cdots P_{\pi(n-1)}(x_n), \\
&= (n-1)!\prod_l h_l \sum_{k=0}^{n-1}\frac{1}{h_k}P_k^2(x_1)w(x_1).
\end{aligned} \tag{93}$$

The microscopic spectral density is obtained by taking the limit $n \to \infty$ at fixed $z = 2n\Sigma\lambda = 2n\Sigma\sqrt{x}$. In this limit the weight function is given by $w(x) = (x\Sigma^2 n)^a$ and the Laguerre polynomials behave as

$$L_k^a(x\Sigma^2 n) \to k^a(x\Sigma^2 nk)^{-a/2} J_a(2\Sigma\sqrt{xnk}). \tag{94}$$

In the limit $n \to \infty$, the sum can be replaced by an integral resulting in the microscopic spectral density [55]

$$\begin{aligned} \rho(z)dz &\sim z\sum_{k=0}^{n-1} J_a^2\left(z\sqrt{\frac{k}{n}}\right) \approx z\int_0^1 tdtJ_a^2(zt), \\ &= 2z(J_a^2(z) - J_{a+1}(z)J_{a-1}(z)). \end{aligned} \tag{95}$$

**Supersymmetric method.** In this section we evaluate the resolvent of QCD for the simplest case of $N_f = 0$ and $\nu = 0$ in the domain $z \ll F^2/\Sigma L^2$. In this domain the partition function is given by

$$Z(J) = \int_{Q\in\hat{Gl}(1|1)} dU \exp\left[\frac{\Sigma V}{2}\mathrm{Str}\begin{pmatrix} z+J & 0 \\ 0 & z \end{pmatrix}(Q+Q^{-1})\right], \tag{96}$$

where the integration is over the maximum super-Riemannian sub-manifold of $Gl(1|1)$. This manifold is parametrized by

$$Q = \exp\begin{pmatrix} 0 & \alpha \\ \beta & 0 \end{pmatrix}\begin{pmatrix} e^{i\phi} & 0 \\ 0 & e^s \end{pmatrix}. \tag{97}$$

The integration measure is the Haar measure which in terms of this parameterization and where $\delta Q \equiv Q^{-1}dQ$ is given by

$$\mathrm{Sdet}\frac{\delta Q_{kl}}{\delta\phi\,\delta s\,\delta\alpha\,\delta\beta}\; d\alpha d\beta d\phi ds. \tag{98}$$

It is straightforward to calculate the Berezinian going from the variables $\{\delta Q_{11},\, \delta Q_{22},\, \delta Q_{12},\, \delta Q_{21}\}$ to the variables $\{\delta\phi,\, \delta s,\, \delta\alpha,\, \delta\beta\}$. The derivative matrix is given by

$$B = \frac{\delta Q_{kl}}{\delta\phi\,\delta s\,\delta\alpha\,\delta\beta} = \begin{pmatrix} i & 0 & \frac{\beta}{2} & \frac{\alpha}{2} \\ 0 & 1 & \frac{\beta}{2} & \frac{\alpha}{2} \\ 0 & 0 & e^{s-i\phi} & 0 \\ 0 & 0 & 0 & e^{-s+i\phi} \end{pmatrix}. \tag{99}$$

Using the definition of the graded determinant one simply finds that $\mathrm{Sdet}B = i$. Up to a constant, the integration measure is thus given by $d\phi ds d\alpha d\beta$.

We also need

$$\frac{1}{2}(Q+Q^{-1}) = \begin{pmatrix} \cos\phi(1+\frac{\alpha\beta}{2}) & \alpha(e^s - e^{-i\phi}) \\ \beta(e^{i\phi} - e^{-s}) & \cosh s(1-\frac{\alpha\beta}{2}) \end{pmatrix}. \tag{100}$$

After differentiating with respect to the source term $(G(z) = \partial_J \log Z(J)|_{J=0})$ this results in (with $x = V\Sigma z$)

$$\frac{G(z)}{V\Sigma} = \int \frac{d\phi ds d\alpha d\beta}{2\pi} \cos\phi(1+\frac{\alpha\beta}{2}) e^{x\cos\phi(1+\frac{\alpha\beta}{2}) - x\cosh s(1-\frac{\alpha\beta}{2})}. \tag{101}$$

With the Grassmann integral given by the coefficient of $\alpha\beta$ we obtain

$$\frac{G(z)}{V\Sigma} = \int \frac{ds d\phi}{4\pi}[\cos\phi + x(\cos\phi + \cosh s)\cos\phi] e^{x(\cos\phi - \cosh s)}.$$

All integrals can be expressed in terms of modified Bessel functions. We find

$$\frac{G(z)}{V\Sigma} = I_1(x)K_0(x) + \frac{x}{2}(I_2(x)K_0(x) + I_0(x)K_0(x) + 2I_1(x)K_1(x)), \tag{102}$$

which can be further simplified by the recursion relation $I_2(x) = I_0(x) - 2I_1(x)/x$. As final result we obtain [51, 19, 20]

$$\frac{G(z)}{V\Sigma} = x(I_0(x)K_0(x) + I_1(x)K_1(x)). \tag{103}$$

This calculation can be generalized to arbitrary $N_f$ and arbitrary $\nu$. The calculation for arbitrary $N_f$ is much more complicated, but with a natural generalization of the factorized parameterization, and using some known integrals over the unitary group, one arrives at the following expression in terms of modified Bessel functions

$$\frac{G(z)}{V\Sigma} = \frac{\nu}{x} + x(I_a(x)K_a(x) + I_{a+1}(x)K_{a-1}(x)), \tag{104}$$

where $a = N_f + |\nu|$. This result is in complete agreement with the resolvent obtained [51] from integrating microscopic spectral density (95).

For $a = 0$ this result is plotted in Fig. 3. We observe that, below some scale, lattice QCD data obtained by the Columbia group [56] closely follow this curve. The predictions of chRMT or of the partially quenched chiral Lagrangian have been studied by numerous lattice simulations [16, 17, 57, 50]. In all cases, agreement has been found in the expected domain of applicability.

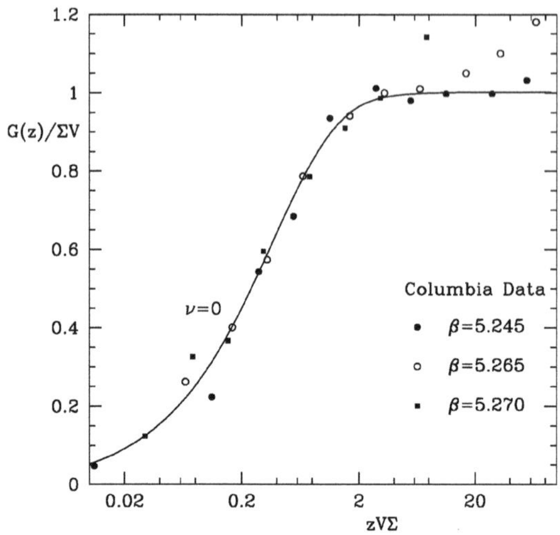


*Figure 3.* The resolvent of quenched QCD. The points represent lattice data obtained by the Columbia group, and the theoretical prediction (103) is given by the solid curve. (Taken from ref. [45].)

# 7. Integrability and the QCD partition function

## 7.1 Virasoro constraints

In this section we derive the small mass expansion of the QCD partition function by means of recursion relations for the partition function known as Virasoro constraints. The starting point is the QCD partition function in the ergodic regime given in (65). The quantities

$$G_\nu(t_k) = \det^{-\nu}(M) Z_\nu(M) \qquad (105)$$

are invariant under the $U(N_f) \times U(N_f)$ transformations $M \to V_1 M V_2^{-1}$. Therefore, $G_\nu(t_k)$ only depends on the eigenvalues of $M^\dagger M$ which can be parameterized in terms of the moments

$$t_k \equiv \frac{1}{k}\mathrm{Tr}\left(\frac{MM^\dagger}{4}\right)^k . \qquad (106)$$

A differential equation for the $G(t_k)$ is obtained from the unitarity relation

$$\frac{1}{Z_\nu}\sum_{a=1}^{N_f}\frac{\partial^2 Z_\nu}{\partial M_{ba}\partial M^\dagger_{ac}} = \frac{1}{4}\sum_{a=1}^{N_f}\langle U^\dagger_{ab}U_{ca}\rangle = \frac{1}{4}\delta_{bc}. \tag{107}$$

Notice that the factor $\Sigma V$ is included in $M$. This relation can be rewritten as

$$\left[\frac{\partial^2}{\partial M_{ba}\partial M^\dagger_{ac}} + \nu M^{-1}_{ab}\frac{\partial}{\partial M^\dagger_{ac}}\right] G_\nu(t_k) = \frac{1}{4}G_\nu(t_k)\delta_{bc}. \tag{108}$$

Using the chain rule and the assumption that the matrix elements of $(MM^\dagger)^{s-1}$ are independent for different values of $s$, we obtain [22]

$$[\mathcal{L}_s - \delta_{s,1}]G_\nu(t_k) = 0 \quad s \geq 1. \tag{109}$$

The Virasoro operators defined by

$$\mathcal{L}_s = \sum_{k=1}^{s-1}\frac{\partial}{\partial t_k}\frac{\partial}{\partial t_{s-k}} + \sum_{k\geq 1k} t_k\frac{\partial}{\partial t_{s+k}} + (N_f+\nu)\frac{\partial}{\partial t_s}, \quad s\geq 1 \tag{110}$$

satisfy the Virasoro algebra

$$[\mathcal{L}_r, \mathcal{L}_s] = (r-s)\mathcal{L}_{r+s}. \tag{111}$$

Therefore, if $G_\nu$ satisfies

$$[\mathcal{L}_1 - 1]G_\nu = 0 \qquad \text{and} \qquad \mathcal{L}_2 G_\nu = 0, \tag{112}$$

then all Virasoro constraints are satisfied. This justifies the independence assumption above (109).

**Solution of the Virasoro constraints.** We can expand $G_\nu$ as

$$G_\nu = 1 + a_1 g_1 + a_2 t_2 + a_{11}t_1^2 + \cdots . \tag{113}$$

From the first Virasoro constraint we obtain

$$\begin{aligned} \mathcal{L}_1 G_\nu &= a_2 t_1 + (N_f+\nu)(a_1 + 2a_{11}t_1) + \cdots , \\ &= 1 + a_1 t_1 + a_2 t_2 + a_{11}t_1^2 + \cdots . \end{aligned} \tag{114}$$

By equating the coefficients of the $t_k$ we find

$$a_1 = \frac{1}{N_f+\nu}, \qquad a_2 + 2(N_f+\nu)a_{11} = a_1. \tag{115}$$

From the second Virasoro constraint, $\mathcal{L}_2 G_\nu = a_{11} + (N_f+\nu)a_2 = 0$, we obtain $a_2 = -a_{11}/(N_f+\nu)$. Continuing this way we can obtain all coefficients in the expansion of the $G(t_k)$. This results in the small mass expansion of the partition function [58]

$$\frac{Z_\nu^{\rm QCD}(M)}{\det^\nu M} = [1 + \frac{{\rm Tr} M M^\dagger}{4(N_f+\nu)} + \frac{1}{32}\frac{{\rm Tr}(MM^\dagger)^2}{(N_f+\nu)((N_f+\nu)^2-1)} + \cdots]. \quad (116)$$

An extension of this expansion to all three Dyson classes can be found in [59].

The small mass expansion can be used to obtain sum rules for the inverse eigenvalues of the Dirac operator [13, 58]. The QCD partition function can be expanded as (the prime indicates that $\lambda_k \neq 0$)

$$\begin{aligned} Z_\nu^{\rm QCD}(M) &= m^{\nu N_f} \left\langle {\prod_k}' \lambda_k^{2N_f} (1+\frac{m^2}{\lambda_k^2})^{N_f}) \right\rangle_\nu \quad (117)\\ &= m^{\nu N_f} (\langle {\prod_k}' \lambda_k^{2N_f} \rangle_\nu + m^2 N_f \langle {\prod_k}' \lambda_k^{2N_f} {\sum_k}' \frac{1}{\lambda_k^2} \rangle_\nu + \cdots) \end{aligned}$$

This results in the expansion

$$\frac{Z_\nu^{\rm QCD}(M)}{\lim_{m\to 0} m^{-\nu N_f} Z_\nu^{\rm QCD}(M)} = 1 + m^2 N_f \langle {\sum_k}' \frac{1}{\lambda_k^2} \rangle_\nu^{\rm QCD}. \quad (118)$$

By equating this expansion to the expansion (116) for equal masses given by

$$1 + \frac{N_f (V\Sigma)^2}{4(N_f+\nu)} m^2, \quad (119)$$

we obtain the Leutwyler-Smilga sum rule [13] for the inverse Dirac eigenvalues

$$\frac{1}{V^2} {\sum_k}' \frac{1}{\lambda_k^2} = \frac{\Sigma^2}{4(N_f+\nu)}. \quad (120)$$

**Flavor-topology duality.** If we construct an $(N_f+\nu)\times(N_f+\nu)$ matrix $\bar{M}$ with $\bar{M}_{ij} = M_{ij}$ for $i,j \leq N_f$ and $\bar{M}ij = 0$ otherwise, we have that ${\rm Tr}(MM^\dagger)^k = {\rm Tr}(\bar{M}\bar{M}^\dagger)^k$. Since the Virasoro constraints only depend on the combination $N_f+\nu$ we have

$$\det{}^{-\nu} M Z_{\nu,N_f}(M) = Z_{\nu=0,N_f+\nu}(\bar{M}, \bar{M}^\dagger). \quad (121)$$

This relation is known at the flavor-topology duality [60].

## 7.2 $\tau$-function

The unitary integral in the QCD partition function can actually be evaluated analytically for an arbitrary number of flavors. We will show that it can be rewritten as a $\tau$-function. The unitary matrix integrals can then be evaluated by means of a Harish-Chandra-Itzykson-Zuber type integral and the use of flavor-topology duality.

**Itzykson-Zuber integral.** We consider the integral

$$I = \int dU dV e^{\frac{1}{2}\mathrm{Tr}(U^\dagger RVS + SV^\dagger RU)}, \tag{122}$$

where $U \in U(N_1)$ and $V \in U(N_2)/U^{N_2}(1)$ and the integral is over the Haar measure of these groups. The matrices $R$ and $S$ are arbitrary rectangular complex matrices. Without loss of generality, they can be taken diagonal with $R_{kk} = r_k > 0$ and $S_{kk} = s_k > 0$ and all other matrix element equal to zero. Using the diffusion equation method one can derive the result [61]

$$I = c\prod_k (r_k s_k)^\nu \frac{\det I_\nu(r_k s_l)}{\Delta(\{s_k^2\})\Delta(\{r_k^2\})}. \tag{123}$$

This result first appeared in the Russian literature [62] as a solution of the Laplace equation. It was proved independently in [61].

**The QCD partition function is a $\tau$-function.** In this subsection we show that the finite volume QCD partition function is a $\tau$-function. Using the flavor-topology duality (121) with $\bar{M}_{kk} = x_k$ for $k \le N_f$ we can write

$$Z_{\nu,N_f}(M) = \det{}^\nu M \int_{U\in U(N_f+\nu)} dU e^{\frac{1}{2}(\bar{M}U^\dagger + \bar{M}^\dagger U)}. \tag{124}$$

This integral is given by (122) with $R$ equal to the $N_f \times (N_f + \nu)$ matrix with $r_k = x_k$, and the $N_f$ diagonal matrix elements of $S$ are expanded as $s_k = 1 + \delta s_k$. For $\delta s_k \to 0$ the matrix elements $\det I_\nu(x_k s_l)$ can be expanded as

$$I_\nu(x_k s_l) = \sum_{j=1}^{N_f} \frac{x_j}{(j-1)!} I_\nu^{(j-1)}(x_k)(\delta s_l)^{j-1} \equiv \sum_{j=1}^{N_f} A_{kj}B_{jl}. \tag{125}$$

with (the upper index between brackets such as $(k)$ denotes the $k$'th derivative)

$$A_{kj} = x_k^{j-1} I_\nu^{(j-1)}(x_k), \qquad B_{jl} = \frac{(\delta s_l)^{j-1}}{(j-1)!}. \tag{126}$$

Up to a constant, the determinant of $B$ is given by

$$\det B \sim \Delta(\{(1+\delta s_k)^2\}), \tag{127}$$

and cancels against the denominator in (123). We finally obtain the result

$$Z_{\nu,N_f}(M) = \frac{\det[x_k^{j-1} I_\nu^{(j-1)}(x_k)]}{\Delta(\{x_k^2\})}. \tag{128}$$

This results was first obtained in [63] and independently for equal masses in [9, 64]. Using the identities such as

$$x^2\partial_x^2 = (x\partial_x)^2 - x\partial_x, \quad x^3\partial_x^3 = (x\partial_x)^3 - 3(\partial_x)^2 + 2x\partial_x, \tag{129}$$

we can rewrite this partition function in terms of derivatives $\delta_k \equiv x_k \partial_{x_k}$ as

$$Z_{\nu,N_f}(M) = \frac{\det[\delta_k^{j-1} I_\nu(x_k)]}{\Delta(\{x_k^2\})}. \tag{130}$$

This form of the partition function is also known as a $\tau$-function [22, 65].

**QCD partition function for equal masses.** The limit of equal masses in the partition function (130) can be obtained by writing

$$x_k = x(1 + \delta x_k), \tag{131}$$

and taking the limit $\delta x_k \to 0$. Because all columns are the same for $\delta x_k = 0$ we have to expand the matrix elements to order $(\delta x_k)^{N_f - 1}$. The expansion to this order can be combined into

$$\det[\delta_k^{j-1} I_\nu(x_k)] = \det[x^{l-1}(\delta_x^{j-1} I_\nu(x))^{(l-1)}] \det[\frac{(\delta x_k)^{l-1}}{(l-1)!}]. \tag{132}$$

Using that

$$\det[(\delta x_k)^{l-1}] = \Delta(\{\delta x_k\}), \quad \Delta(\{x_k^2\}) = x^{N_f(N_f-1)}\Delta(\{\delta x_k\}), \tag{133}$$

we obtain the partition function

$$Z_\nu(x) = c x^{-N_f(N_f-1)} \det[x^{l-1}((x\partial_x)^{j-1} I_\nu(x))^{(l-1)}]. \tag{134}$$

With the help of the identities (129) the derivatives can be combined into derivatives $(x\partial_x)^p$ resulting in

$$Z_\nu(x) = c x^{-N_f(N_f-1)} \det[(x\partial_x)^{l+j-2} I_\nu(x)]. \tag{135}$$

## 7.3 Toda lattice equation

In this section we show that the QCD partition function for equal masses satisfies a Toda lattice equation. This result and generalizations thereof were first obtained in [66]. The Toda lattice was originally introduced as a one dimensional lattice in which neighboring atoms interact via a potential that depends exponentially on the distance. The Hamiltonian equations of motion of this system can be written in the form of the Toda lattice equation discussed below. Because of the existence of a Lax pair, they have infinitely many constants of motion. For a more elaborate discussion of the Toda lattice equation and the relation to integrable systems, we refer to [29, 67, 68]. Several subsections below are based on the paper by Forrester and Witte [69].

**The Sylvester identity.** We all know how to expand a the determinant matrix with respect to its co-factors given by

$$C_{ij} = \frac{\partial}{\partial_{A_{ij}}} \det A. \tag{136}$$

What is less known is that there exists a remarkable identity that relates co-factors to the double co-factors defined by

$$C_{ij;pq} = \frac{\partial^2}{\partial_{A_{ij}} \partial_{A_{pq}}} \det A. \tag{137}$$

This identity, which is known as the Sylvester identity [70], is given by

$$C_{ij} C_{pq} - C_{iq} C_{pj} = \det A C_{ij;pq}. \tag{138}$$

For example, it holds for a $2 \times 2$ matrix with $i = j = 1$ and $p = q = 2$.

**Toda lattice equation.** We apply the Sylvester identity to the determinant that appears in the partition function (135). For $i = j = N_f - 1$ and $p = q = N_f$ we obtain

$$C_{N_f-1,N_f-1} C_{N_f,N_f} - C_{N_f-1,N_f} C_{N_f,N_f-1} = \det A C_{N_f-1,N_f-1;N_f,N_f}, \tag{139}$$

with matrix $A$ given by

$$A_{jk} \equiv (x\partial_x)^{l+j-2} I_\nu(x). \tag{140}$$

The derivative of a determinant is equal to the sum of determinants with one of the rows replaced by its derivatives, or is equal to sum of the determinants with one of the columns replaced by its derivative (in both cases we have in total $N_f$ terms). For the matrix $A$ only differentiating the last row or column gives a nonzero result. This allows us to rewrite the co-factors as derivatives of $\det A$. In particular, we find

$$\begin{aligned} C_{N_f-1,N_f} &= -x\partial_x \det A_{N_f-1}, \quad C_{N_f,N_f-1} = -x\partial_x \det A_{N_f-1}, \\ C_{N_f,N_f} &= \det A_{N_f-1}, \qquad C_{N_f-1,N_f-1} = (x\partial_x)^2 \det A_{N_f-1}. \\ C_{N_f-1,N_f-1;N_f,N_f} &= \det A_{N_f-2}. \end{aligned} \tag{141}$$

To obtain the second last identity we first differentiate the columns and then the rows. Inserting this in (139) we find

$$(x\partial_x)^2 \log \det A_{N_f-1} = \frac{\det A_{N_f} \det A_{N_f-2}}{\det^2 A_{N_f-1}}. \tag{142}$$

Next we substitute the relation between $\det A_{N_f}$ and the partition function

$$\det A_{N_f} = \frac{1}{c} x^{N_f(N_f-1)} Z_\nu^{N_f}(x). \tag{143}$$

The prefactor contributes a factor $x^2$ to the r.h.s. of (142) but does not contribute to its l.h.s.. After raising $N_f$ by 1 we obtain the celebrated Toda lattice equation [22, 65]

$$(x\partial x)^2 \log Z_\nu^{N_f}(x) = cx^2 \frac{Z_\nu^{N_f+1} Z_\nu^{N_f-1}}{[Z_\nu^{N_f}(x)]^2}. \tag{144}$$

By performing the $U$-integral in the partition function for equal masses

$$Z_{N_f,\nu}(x) = \int_{U\in U(N_f)} dU \det{}^\nu U e^{\frac{x}{2}\mathrm{Tr}(U+U^{-1})}, \tag{145}$$

by a saddle point approximation including the Gaussian fluctuations we obtain the large $x$ limit

$$Z_\nu^{N_f}(x) \sim \frac{e^{N_f x}}{x^{N_f^2/2}}. \tag{146}$$

Using this result to normalize the partition function we find that $c = N_f$.

## 7.4 Painlevé system

The partition function (145) can be obtained from the "double" scaling limit of the random matrix partition function

$$Z_{N_f,\nu}^{\mathrm{RMT}}(x) = \int \prod_{k=1}^N [d\lambda_k \lambda_k^{2\nu+1}(\lambda_k^2+m^2)^{N_f}] m^{\nu N_f} |\Delta(\{\lambda_l^2\})|^2 e^{-\frac{N\Sigma^2}{2}\sum_l \lambda_l^2}. \tag{147}$$

where $x = mN\Sigma$ is kept fixed for $N \to \infty$. Using $x_k^2 = \lambda_k^2\Sigma^2 N^2 + x^2$ as new integration variables, we obtain in this limit

$$\begin{aligned} Z_{N_f,\nu}^{\mathrm{RMT}}(x) &= \prod_k [\int_{x^2}^\infty dx_k^2 (x_k^2 - x^2)^\nu x_k^{2N_f}] x^{\nu N_f} e^{x^2/4} e^{-\sum_k x_k^2/2N} \\ &\equiv x^{\nu N_f} e^{x^2/4} E_N([0, s = x^2], N_f, \nu). \end{aligned} \tag{148}$$

$E_N([0, s = x^2], N_f, \nu)$ can be interpreted as the probability that there are no eigenvalues in the interval $[0, x^2]$ for the joint probability distribution given by the integrand of (148). For $\nu = 0$ this is the probability of the partition function with topological charge $N_f$ and no flavors. If we introduce $\sigma(t)$ by

$$E_N([0, s], N_f, \nu) = e^{-\int_0^s \frac{dt}{t}(\sigma(t) + \frac{1}{2}\nu(N_f+\nu))}, \tag{149}$$

then the function $\sigma(t)$ satisfies the Painlevé equation [69]

$$(t\sigma'')^2 - (N_f^2 - \nu^2)(\sigma')^2 + \sigma'(4\sigma' - 1)(\sigma - t\sigma') - \frac{\nu^2}{16} = 0. \tag{150}$$

The boundary conditions for this differential equation follow from the asymptotic behavior of the partition function (145). Using (146) we find for the large $x$ behavior of $E_N([0, s = x^2, N_f, \nu)$

$$E_N([0, s = x^2, N_f, \nu) \sim x^{-N_f^2/2 - \nu N_f} e^{N_f x - x^2/4}, \tag{151}$$

so that the large-$s$ behavior of $\sigma(s)$ is given by

$$\sigma(s) \sim \frac{N_f}{2}\sqrt{s} + \frac{s}{4} - \frac{\nu^2}{2} + \frac{N_f^2}{4}. \tag{152}$$

This Painlevé equation can be derived from the equations of motion of the Hamiltonian [69]

$$tH = q^2p^2 - (q^2 + (N_f + \nu)q - t)p + N_f q \tag{153}$$

with the identification

$$\sigma(t) = -\left.(tH)\right|_{t \to \frac{t}{4}} - \frac{1}{2}(N_f + \nu)\nu + \frac{t}{4}. \tag{154}$$

For example, we have the equations of motion

$$(tH)' \;=\; p, \quad (tH)'' = p' = -\frac{\partial H}{\partial q}, \quad tq' = \frac{\partial tH}{\partial p}. \tag{155}$$

Such Hamiltonians play an important role in the theory of exactly solvable models. Hamiltonians with different values of $\nu$ and $N_f$ are connected by a Backlund transformation. This is a canonical transformation together with $(\nu, N_f) \to (\bar{\nu}, \bar{N}_f)$ such that, in the new variables, the same Painlevé equation is satisfied. In our case we have the Backlund transformation (at fixed $\nu$)

$$\begin{aligned} T: \qquad N_f &\to N_f + 1, \\ H_{N_f} &\to H_{N_f+1} = H_{N_f} + q_{N_f} - q_{N_f}p_{N_f}, \\ q_{N_f} &\to q_{N_f+1}, \\ p_{N_f} &\to p_{N_f+1}. \end{aligned} \tag{156}$$

Below we do not need the explicit transformation rules of $q_{N_f}$ and $p_{N_f}$, but of the inverse transformation of $q_{N_f}(p_{N_f}-1)$ which is given by

$$T^{-1}q_{N_f}(p_{N_f}-1) \to -q_{N_f}(p_{N_f}-1) + N_f + \nu - \frac{N_f}{p_{N_f}}. \tag{157}$$

If we define the $\tau$-function by

$$\tau_{N_f} = e^{\int_0^t H_{N_f} dt}, \tag{158}$$

we can easily derive the equalities

$$\begin{aligned} t\partial_t \log \frac{\tau_{N_f-1}\tau_{N_f+1}}{\tau^2_{N_f}} &= tH_{N_f-1} - tH_{N_f} + tH_{N_f+1} - tH_{N_f}, \\ &= -T^{-1}(q_{N_f}(1-p_{N_f}) + (q_{N_f}(1-p_{N_f}), \\ &= t\partial_t \log \partial_t [tH_{N_f}], \\ &= t\partial_t \log \partial_t [t\partial_t \log \tau_{N_f}]. \end{aligned} \tag{159}$$

To derive the second last equality we have used the inverse Backlund transformation and the Hamilton equations (155). Integrating this equation once and putting the integration constant equal to zero we find the Toda lattice equation

$$(t\partial_t)^2 \log \tau_{N_f} = t\frac{\tau_{N_f+1}\tau_{N_f-1}}{\tau^2_{N_f}}. \tag{160}$$

**Solutions of the Painlevé equation.** The probability $E([0,s],N_f,\nu)$ is related to the partition function (145) by

$$E([0,s],N_f,\nu) = s^{-N_f\nu/2}e^{-s/4}Z_{N_f\nu}(\sqrt{s}). \tag{161}$$

For $N_f = 0$ the partition function is normalized to 1 so that

$$\sigma_{N_f=0}(s) = \frac{s}{4} - \frac{\nu^2}{2}. \tag{162}$$

Indeed this is a solution of the $PIII$ Painlevé equation (150).

For $N_f = 1$ we have that

$$E([0,s],N_f=1,\nu) = s^{-\nu/2}e^{-s/4}I_\nu(\sqrt{s}) = e^{-\int_0^s \frac{dt}{t}(\sigma(t)+\frac{1}{2}\nu(1+\nu))}, \tag{163}$$

resulting in another solution of (150)

$$\sigma_{N_f=1}(s) = \frac{s}{4} - \frac{\nu^2}{2} - s\frac{d}{ds}\log I_\nu(\sqrt{s}). \tag{164}$$

For $x \to 0$ the modified Bessel function behaves as $I_\nu(x) \sim x^\nu$ so that the $t$-integral in (163) is well-behaved for $t \to 0$. Only recursion relations of Bessel functions are required to show that (163) is a solution of the Painlevé equation. Since $(-1)^\nu K_\nu(x)$ satisfies the same recursion relations as $I_\nu(x)$ this provides us with another solution of the the Painlevé equation. This solution corresponds to the partition function with $N_f = -1$ where

$$E([0,s], N_f = -1, \nu) = s^{\nu/2} e^{-s/4} K_\nu(\sqrt{s}), = e^{-\int_0^s \frac{dt}{t}(\sigma(t)+\frac{1}{2}\nu(-1+\nu))}, \quad (165)$$

and satisfies the boundary condition with $N_f = -1$.

**The bosonic partition function.** The natural interpretation of $N_f = -1$ is as a bosonic flavor. In this section we will derive the low energy limit of the QCD partition function for $N_f$ bosonic flavors with equal masses. We already have seen in section 3.3 that the Goldstone manifold for $n$ bosonic quarks is given by $Gl(n)/U(n)$. Using the same invariance arguments as before one obtains the low-energy effective partition function

$$Z^\nu_{-n} = \int_{Q \in Gl(n)/U(n)} \det{}^\nu(Q) e^{\frac{1}{2} V \Sigma \mathrm{Tr}\, M(Q+Q^{-1})}. \quad (166)$$

In this case $Q$ can be diagonalized as $Q = U \mathrm{diag}(e^{s_k}) U^{-1}$, so that an eigenvalue representation of this partition function is given by [71]

$$\int \prod_k ds_k \prod_k e^{\nu s_k} \prod_{k<l} (e^{s_k} - e^{s_l})(e^{-s_k} - e^{-s_l}) e^{x \sum_k \cosh s_k}. \quad (167)$$

The Vandermonde determinant can be written as

$$\prod_{k<l} (e^{s_k} - e^{s_l}) = \det[e^{p s_q}]_{0 \le p \le n-1,\, 1 \le q \le n} \quad (168)$$

and a similar expression for $s_k \to -s_k$. By expanding the two determinants the integrals can be written as modified Bessel functions which can be combined into a determinant as follows [71]

$$Z_{-n}(x) = c_{-n} \det[K_{\nu+k+l}(x)]_{0 \le k,l \le n-1}. \quad (169)$$

From the observation that $(-1)^\nu K_\nu(x)$ and $I_\nu(x)$ satisfy the same recursion relations, and that the factor $(-1)^\nu$ does not affect the determinant, (169) can be rewritten as the $\tau$-function

$$Z_{-n}(x) = \frac{c_{-n}}{x^{n(n-1)}} \det[(x\partial_x)^{k+1} Z_{-1}(x)]_{0 \le k,l \le n-1}. \quad (170)$$

with $Z_{-1}(x) = K_\nu(x)$. The bosonic partition function can also be analyzed along the same lines as the fermionic partition function. On the other hand this derivation can be simply modified to obtain the partition function for $N_f$ fermionic flavors with equal mass.

The bosonic partition function thus satisfies the same Toda lattice equation as the fermionic partition function. The derivative of the resolvent The semi-infinite hierarchies are connected by

$$\lim_{n\to 0} \frac{1}{n}(x\partial_x)^2 \log Z_n^\nu(x). \tag{171}$$

which is related to a derivative of the resolvent.

## 7.5 Replica limit of the Toda lattice equation

The resolvent can be obtained from the replica limit of the fermionic partition function

$$G(z) = \lim_{n\to 0} \frac{1}{n} \log Z_n^\nu(z). \tag{172}$$

If we take the replica limit of the fermionic ($n < 0$) or bosonic ($n > 0$) partition functions directly, we will obtain a result that differs from the supersymmetric calculation. These problems can be avoided if the take the replica limit of the Toda lattice equation. With the normalization $Z_0(x) = 1$ we obtain the relation

$$x\partial_x x G(x) = 2x^2 Z_1^\nu(x) Z_{-1}^\nu(x). \tag{173}$$

Inserting the expressions for $Z_1$ and $Z_{-1}$ we find [23]

$$G(x) = \frac{\nu}{x} + x(K_\nu(x)I_\nu(x) + K_{\nu-1}(x)I_{\nu+1}(x)), \tag{174}$$

which agrees with the result obtained by the supersymmetric method (104). This result has also been derived from the solution of the Painlevé equation (150) for $n \to 0$ [24].

The validity of the replica limit of the Toda lattice equation can be proved by extending to Toda lattice hierarchy to include an additional spectator boson with mass $y$ and using the identity [72]

$$\lim_{n\to 0} \frac{1}{n}(x\partial_x)^2 \log Z_n^\nu(x) = \lim_{y\to x} x\partial_x(x\partial_x + y\partial_y) \log Z_{1,-1}(x, y). \tag{175}$$

## 7.6 Replica limit for the GUE two-point function

We have two possibilities for the generating function of the two-point function of the Gaussian Unitary Ensemble: a fermionic generating function or a bosonic generating function. The fermionic (bosonic) generating function for the two-point function is defined by

$$Z_n(x, y) = \int dHP(H)\det{}^n(x + i\epsilon + H)\det{}^n(y - i\epsilon + H), \qquad (176)$$

with $n > 0$ ($n < 0$). We will consider the microscopic limit where $\pi(x - y)N\rho(x) \equiv r$ is kept fixed for $N \to \infty$. In that case the two-point function in the center of the spectrum only depends on $r$ and is given by

$$R_2(r) = -\lim_{n\to 0}\frac{1}{n^2}\partial_r^2 Z_n(ir), \qquad (177)$$

both in the fermionic and the bosonic case. In an eigenvalue representation of the Goldstone fields the microscopic limit of the generating function $Z_n(ir)$ can be written as [21, 37]

$$Z_n(r) = \int_{-1}^{1}\prod_k du_k \prod_{k<l}(u_k - u_l)^2 e^{ir\sum_k u_k}. \qquad (178)$$

This partition (178) can be written as a $\tau$-function. The first step is to expand the Vandermonde determinant

$$Z_n(r) = \int_{-1}^{1}\prod_k du_k \sum_{\sigma\pi} sg(\sigma\pi) u_1^{\sigma(1)+\pi(1)} \cdots u_n^{\sigma(n)+\pi(n)} e^{ir\sum_k u_k}. \qquad (179)$$

Next we use that

$$\int_{-1}^{1} du_k u_k^a e^{iru_k} = (\partial_{ir})^a Z_1(r), \qquad (180)$$

which results in [73]

$$Z_n(r) = n![\det(\partial_{ir})^{i+j} Z_1(x)]_{0\le i,j\le n-1}. \qquad (181)$$

The partition function $Z_1(x)$ is given by

$$Z_1(r) = \int_{-1}^{1} du e^{iru}. \qquad (182)$$

The microscopic limit of bosonic partition function can be rewritten similarly. The main difference is the convergence requirements of the bosonic integrals which are essential for the structure of the Goldstone manifold. In an eigenvalue representation of the Goldstone fields we find [21]

$$Z_{-n}(r) = \int_{1}^{\infty}\prod_k du_k \prod_{k<l}(u_k - u_l)^2 e^{ir\sum_k u_k}. \qquad (183)$$

This partition function can also be written as a $\tau$-function. By expanding the Vandermonde determinant we can express this generating function as a determinant of derivatives

$$Z_{-n}(r) = n![\det(\partial_{ir})^{i+j} Z_{-1}]_{0\le i,j\le n-1}, \tag{184}$$

with $Z_{-1}(r)$ given by

$$Z_{-1}(r) = \int_1^\infty du e^{iru}. \tag{185}$$

Because of the derivative structure of the partition function, we can again use the Sylvester identity to derive a Toda lattice equation. In this case we find

$$\partial_{ir}^2 \log Z_n(x) = n^2 \frac{Z_{n+1}(r) Z_{n-1}(r)}{[Z_n(r)]^2}, \tag{186}$$

where the factor $n^2$ follows from the choice of the normalization constants. We have made this choice because the left hand side is proportional to $n^2$. The two-point correlation is given by the replica limit of (186)

$$\begin{aligned} R_2(r) &= -\lim_{n\to 0} \frac{1}{n^2} \partial_r^2 \log Z_n(r) = Z_1(r) Z_{-1}(r) \\ &= \int_{-1}^1 du e^{iux} \int_1^\infty e^{iux} = 2i \frac{\sin x}{x} \frac{e^{ix}}{x}, \end{aligned} \tag{187}$$

which is the correct analytical result for the two-point function. This derivation explains the factorization of the two-point function into a compact and a non-compact integral which characterizes the result obtained by a supersymmetric calculation [74]. The fermionic partition functions, the bosonic partition functions and the super-symmetric partition function form a single integrable hierarchy which are related by the Toda lattice equation [23]. A closely related way to derive the two-point function of the GUE is to take the replica limit of the corresponding Painlevé equation. For a discussion of this approach we refer to [24] which preceded our work [23] on the Toda lattice.

## 8. QCD at finite baryon density

In this Chapter we study the quenched microscopic spectrum of the QCD Dirac operator at nonzero chemical potential when the Dirac operator is non-Hermitian with eigenvalues scattered in the complex plane. Using the replica limit of the Toda lattice equation we obtain the exact analytical result for the microscopic spectral density [25].

## 8.1 General remarks

The average spectral density of a non-Hermitian operator is given by

$$\rho(\lambda) = \langle \sum_k \delta^2(\lambda - \lambda_k) \rangle, \tag{188}$$

and the average resolvent is defined by

$$G(z) = \left\langle \sum_k \frac{1}{i\lambda_k + z} \right\rangle. \tag{189}$$

Using that $\partial_{z^*}(1/z) = \pi\delta^2(z)$ we easily derive

$$\partial_{z^*} G(z)|_{z=\lambda} = \pi\rho(\lambda). \tag{190}$$

The resolvent can be interpreted as the electric field in the plane at point $z$ from charges located at the position of the eigenvalues. For example, Gauss law is given by

$$\oint_C G(z)dz = 2\pi i Q, \tag{191}$$

where $Q$ is the number of eigenvalues enclosed by $C$.

## 8.2 The Ginibre ensemble

The obtain a better understanding of the resolvent for a non-Hermitian random matrix ensemble, we first consider the Ginibre ensemble [75] defined by the probability distribution

$$\rho(C) = e^{-N \mathrm{Tr} CC^\dagger}, \tag{192}$$

with $C$ a complex $N \times N$ matrix. The eigenvalues of $C$ are given by the solutions of the secular equation $\det(C - \lambda_k) = 0$. If all eigenvalues are different, the matrix $C$ can be decomposed as

$$C = V \Lambda V^{-1}, \tag{193}$$

where $V$ is a similarity transformation and $\Lambda = \mathrm{diag}(\lambda_1, \cdots, \lambda_N)$. The joint eigenvalue distribution of the Ginibre ensemble is obtained by using the decomposition

$$C = UTU^{-1}, \tag{194}$$

with $U$ a unitary matrix and $T$ an triangular matrix with the eigenvalues of $C$ on the diagonal. After integrating out the upper triangular matrix elements we obtain (see the lectures of Zabrodin [76] for a derivation),

$$\rho(\{\lambda_k\} = |\Delta(\{\lambda_k\})|^2 e^{-N\sum_k |\lambda_k|^2}. \tag{195}$$

This distribution can be interpreted as repulsive charges in the plane balanced by an external force $N|z|$. The resolvent is equal to the electric field of the eigenvalues. For an equilibrium distribution we have $G(z) = N|z|$. Using that the eigenvalue density is spherically symmetric, we find from Gauss law

$$2\pi r|G| = 2\pi \int_0^r \rho(r')dr', \tag{196}$$

so that $\rho(r) = N/\pi$. Because the total number of eigenvalues is equal to $N$, they are located inside the circle $|z| = 1$. The resolvent is thus given by

$$G(z) \quad = \quad Nz^*\theta(1-|z|) + \frac{N}{z}\theta(|z|-1). \tag{197}$$

## 8.3 QCD at nonzero chemical potential

The QCD partition function at nonzero chemical potential $\mu$ is given by

$$Z_{\text{QCD}} = \sum_k e^{-\beta(E_k - \mu N_k)}, \tag{198}$$

where $E_k$ is the energy of the state, and $N_k$ is the quark number of the state. At zero temperature ($\beta \to \infty$) the partition function does not depend on $\mu$ for $\mu < m_N/N_N$, where $N$ is the particle with the smallest value of $m_N/N_N$. For QCD $N$ is the nucleon with quark number $N_N = 3$. This implies that the chiral condensate does not depend on $\mu$ for $\mu < m_N/N_N$.

The QCD partition function can be written as a Euclidean path integral with the fermionic part of the Lagrangian density defined by

$$\mathcal{L} = \bar{\psi}D\psi + m\bar{\psi}\psi + \mu\bar{\psi}\gamma_0\psi. \tag{199}$$

with $D$ the anti-Hermitian Dirac operator. Since $\mu\gamma_0$ is Hermitian, the Dirac operator as a whole is non-Hermitian. As a consequence, the eigenvalues are scattered in the complex plane [77]. The fermion determinant is in general complex. This means that it is not possible to study the QCD partition function by means stochastic methods which severely limits our knowledge of QCD at nonzero chemical potential.

The question we wish to address is if there is a domain where the fluctuations of the Dirac eigenvalues are universal and can be obtained from a random matrix partition function with the global symmetries QCD, or equivalently from a chiral Lagrangian. In this domain we will calculate the resolvent and the spectral density from the replica limit of the Toda lattice equation [25, 26].

**Generating function for the quenched spectral density.** The quenched spectral density is given by the replica limit [20, 79, 80]

$$\rho^{\text{quen}}(z, z^*) = \lim_{n\to 0} \frac{1}{\pi n}\partial_z\partial_{z^*} \log Z_n(z, z^*), \tag{200}$$

with generating function given by

$$Z_n(z, z^*) = \langle \det{}^n(D + \mu\gamma_0 + z)\det{}^n(-D + \mu\gamma_0 + z^*)\rangle. \tag{201}$$

The product of the determinants in (201) can be written as the determinant of [81, 82]

$$\begin{pmatrix} id+\mu & 0 & z & 0 \\ 0 & id-\mu & 0 & z^* \\ z & 0 & id^\dagger+\mu & 0 \\ 0 & z^* & 0 & id^\dagger-\mu \end{pmatrix} \equiv \begin{pmatrix} id+\mu_1 & M_{RL} \\ M_{LR} & id^\dagger+\mu_2 \end{pmatrix} \tag{202}$$

where we have used the decomposition of the Dirac operator given in (11). We observe that the $U(2n) \times U(2n)$ flavor symmetry is broken by the chemical potential term and the mass term. Invariance is recovered by transforming the mass term as in the case of zero chemical potential (see (46)) and the chemical potential term by a local gauge transformation [83]. For the chemical potential matrices the latter transformation is simply given by

$$\mu_1 \to U_R \mu_1 U_R^{-1}, \qquad \mu_2 \to U_L \mu_2 U_L^{-1}. \tag{203}$$

The low-energy limit of quenched QCD should have the same transformation properties. In the domain $\mu \ll 1/L$ and $z\Sigma \ll F^2/L^2$ we only have to consider the zero momentum modes. Using that the Goldstone fields transform as $U \to U_R U U_L^{-1}$ we can write down the following invariants to first order in the quark mass and to second order in the chemical potential

$$\mathrm{Tr}\mu_k^2, \quad \mathrm{Tr}U\mu_1 U^{-1}\mu_2, \quad \mathrm{Tr}M_{RL}U, \qquad \mathrm{Tr}M_{LR}U. \tag{204}$$

The low energy effective partition function is therefore given by

$$Z_{n,\nu}(z, z^*) = \int_{U(2n)} dU \det{}^\nu U \, e^{-\frac{F^2\mu^2 V}{4}\mathrm{Tr}[U,B][U^{-1},B] + \frac{\Sigma V}{2}\mathrm{Tr}M(U+U^{-1})}, \tag{205}$$

where

$$B = \begin{pmatrix} 1_n & 0 \\ 0 & -1_n \end{pmatrix}, \qquad M = \begin{pmatrix} z1_n & 0 \\ 0 & z^*1_n \end{pmatrix}. \tag{206}$$

**Random matrix model.** The partition function (205) can be obtained from the large $N$ limit a random matrix model with the global symmetries of the QCD partition function. For $\nu = 0$ he model is defined by an integral over $N/2 \times N/2$ complex matrices [14, 84, 80],

$$Z(m_f, \mu) = \int dW \prod_{f=1}^{N_f} \det(D(\mu) + m_f) e^{-N\Sigma^2 \mathrm{Tr}WW^\dagger}. \tag{207}$$

The Dirac matrix has the structure

$$D(\mu) = \begin{pmatrix} 0 & iW + \mu \\ iW^\dagger + \mu & 0 \end{pmatrix}. \tag{208}$$

For QCD with three or more colors in the fundamental representation, the matrix $W$ is complex ($\beta_D = 2$). One can also introduce random matrix ensembles with $\beta_D = 1$ or $\beta_D = 4$ by choosing the matrix elements of $W$ real or quaternion real, respectively [85].

An alternative random matrix model [27] is obtained by replacing the identity matrix that multiplies $\mu$ by a complex matrix with the same distribution as $W$. This random matrix model is in the same universality class but turns out to be mathematically simpler. In particular, the joint eigenvalue distribution has been derived [27] which makes it possible to calculate correlation functions by the orthogonal polynomial method.

**Mean field analysis.** The macroscopic spectral density of the partition function (205) can be easily obtained by means of a saddle point approximation [82]. Using an Ansatz that is diagonal in replica space,

$$U = \begin{pmatrix} \cos\theta & e^{i\phi}\sin\theta \\ -e^{-i\phi}\sin\theta & \cos\theta \end{pmatrix}, \tag{209}$$

the partition function is given by

$$Z_n = e^{nV[2\mu^2F^2\sin^2\theta + \Sigma(z+z^*)\cos\theta]}. \tag{210}$$

The extrema are at

$$\cos\theta = 1, \quad \text{or} \quad \cos\theta = \frac{\Sigma(z+z^*)}{4F^2\mu^2}. \tag{211}$$

The critical value of $\mu$ is at the point where the two saddle points coincide

$$\mu_c^2 = \frac{\Sigma|z+z^*|}{4F^2}. \tag{212}$$

The partition function at the saddle point is given by

$$\begin{aligned} \mu < \mu_c &: \quad Z_n = e^{nV\Sigma(z+z^*)}, \\ \mu > \mu_c &: \quad Z_n = e^{nV(2\mu^2F^2 + \Sigma^2(z+z^*)^2/8F^2\mu^2)}. \end{aligned} \tag{213}$$

For the resolvent and the spectral density we then find

$$\begin{aligned} \mu < \mu_c &: \quad G^{\rm quen}(z) = V\Sigma, \quad \rho^{\rm quen}(z) = 0, \\ \mu > \mu_c &: \quad G^{\rm quen}(z) = \frac{V\Sigma^2(z+z^*)}{4\mu^2F^2}, \quad \rho^{\rm quen}(z) = \frac{\Sigma^2 V}{4\mu^2F^2}. \end{aligned} \tag{214}$$

The eigenvalues are located inside a strip of width $4F^2\mu^2/\Sigma$.

## 8.4 The microscopic spectral density

The microscopic spectral density cannot be obtained from a mean field analysis. The assumption in the derivation in the previous section was that saddle point is proportional to the identity in replica space, so that the replica limit can be obtained from the calculation with one replica. The generating function for the microscopic spectral density depends in a nontrivial way on the number of replicas which, as we have seen before, can be obtained from the replica limit of a Toda lattice equation. In this subsection we closely follow [25]. The starting point is a remarkable integration formula to be discussed next.

**Integration formula.** By decomposing a $U(2n)$ matrix as

$$U = \begin{pmatrix} u_1 & \\ & u_2 \end{pmatrix} \begin{pmatrix} v_1 & \\ & v_2 \end{pmatrix} \begin{pmatrix} \sqrt{1-b^2} & b \\ b & -\sqrt{1-b^2} \end{pmatrix} \begin{pmatrix} v_1^\dagger & \\ & v_2^\dagger \end{pmatrix}, \quad (215)$$

with $u_1$, $u_2$, $v_1 \in U(n)$, $v_2 \in U(n)/U^n(1)$ and $b$ a diagonal matrix, the following integration formula can be proved [25]

$$\begin{aligned} &\int_{U(2n)} dU \det{}^\nu U e^{\frac{1}{2}\mathrm{Tr}[M(U+U^{-1}]+\sum_p a_p \mathrm{Tr}[(UBU^{-1}B)^p]} \\ &= \frac{c_n}{(xy)^{n(n-1)}} \det[(x\partial_x)^k (y\partial_y)^l Z_{1,\nu}(x,y)]_{0\le k,l \le n-1}, \end{aligned} \quad (216)$$

where $c_n$ is an $n$-dependent constant and

$$Z_{1,\nu}(x,y) = \int_0^1 \lambda d\lambda I_\nu(\lambda x) I_\nu(-\lambda y) e^{2\sum_p a_p \cos(2p \cos^{-1}\lambda)}. \quad (217)$$

**Toda lattice equation at nonzero chemical potential.** Using the integration formula (216) for $p = 1$ we find that the zero momentum partition function $Z_n^\nu(z, z^*)$ (see eq. (205)) can be written as

$$Z_{n,\nu}(z,z^*) = \frac{c_n}{(zz^*)^{n(n-1)}} \det[(z\partial_z)^k (z^*\partial_{z^*})^l Z_{1,\nu}(z,z^*)]_{0\le k,l\le n-1}, \quad (218)$$

where

$$Z_{1,\nu}(z,z^*) = \int_0^1 \lambda d\lambda e^{-2VF^2\mu^2(\lambda^2-1)} |I_\nu(\lambda z V\Sigma)|^2. \quad (219)$$

By applying the Sylvester identity to the determinant in (218) for $i = j = n-1$ and $p = q = n$ and expressing the cofactors as derivatives, we find a recursion relation that can be written in the form of the Toda lattice equation

$$z\partial_z z^*\partial_{z^*} \log Z_n^\nu(z,z^*) = \frac{\pi n}{2}(zz^*)^2 \frac{Z_{n+1}^\nu(z,z^*) Z_{n-1}^\nu(z,z^*)}{[Z_n^\nu(z,z^*)]^2}. \quad (220)$$

For the spectral density we find the simple expression ($Z_{0,\nu}(z, z^*) = 1$)

$$\rho^{\text{quen}}(z, z^*) = \lim_{n\to 0} \frac{1}{\pi n} \partial_z \partial_{z^*} \log Z_{n,\nu}(z, z^*) = \frac{zz^*}{2} Z_{1,\nu}(z, z^*) Z_{-1,\nu}(z, z^*). \tag{221}$$

What remains to be done is to calculate the bosonic partition function for $n = -1$ which will be completed in the next subsections.

**The bosonic partition function.** In this subsection we evaluate the low-energy limit of the QCD partition function at nonzero chemical potential for one bosonic quark and one conjugate bosonic quark. We closely follow [25]. Because of convergence requirements, the inverse determinants of nonhermitian operators have to regulated. This is achieved by expressing them as the determinant of a larger Hermitian operator [81]

$$\begin{aligned}
&\det{}^{-1}\begin{pmatrix} z & id+\mu \\ id^\dagger+\mu & z \end{pmatrix} \det{}^{-1}\begin{pmatrix} z^* & -id+\mu \\ -id^\dagger+\mu & z^* \end{pmatrix} \\
&= \lim_{\epsilon\to 0} \det{}^{-1}\begin{pmatrix} \epsilon & 0 & z & id+\mu \\ 0 & \epsilon & id^\dagger+\mu & z \\ z^* & -id+\mu & \epsilon & 0 \\ -id^\dagger+\mu & z^* & 0 & \epsilon \end{pmatrix} \qquad (222)\\
&= \int \exp[i \sum_{j=1}^{N/2} \phi_k^{j*} \begin{pmatrix} \epsilon & z & id+\mu & 0 \\ z^* & \epsilon & 0 & id-\mu \\ -id^\dagger+\mu & 0 & \epsilon & -z^* \\ 0 & -id^\dagger-\mu & -z & \epsilon \end{pmatrix}_{kl} \phi_k^j].
\end{aligned}$$

The mass matrices are given by

$$\zeta_1 = \begin{pmatrix} \epsilon & z \\ z^* & \epsilon \end{pmatrix} \quad \text{and} \quad \zeta_2 = \begin{pmatrix} \epsilon & -z^* \\ -z & \epsilon \end{pmatrix} = -I\zeta_1 I. \tag{223}$$

with $I \equiv i\sigma_2$. For the random matrix model (208) we have that $d = W$. The Gaussian integral over $W$ results in the 4-boson term $\exp[-2\text{Tr}\overline{Q_1}\,\overline{Q_2}/N]$ with

$$\overline{Q_1} \equiv \begin{pmatrix} \phi_1^* \cdot \phi_1 & \phi_1^* \cdot \phi_2 \\ \phi_2^* \cdot \phi_1 & \phi_2^* \cdot \phi_2 \end{pmatrix}, \qquad \overline{Q_2} \equiv \begin{pmatrix} \phi_3^* \cdot \phi_3 & \phi_3^* \cdot \phi_4 \\ \phi_4^* \cdot \phi_3 & \phi_4^* \cdot \phi_4 \end{pmatrix}, \tag{224}$$

and we have used the notation $\phi_k^* \cdot \phi_l = \sum_{i=1}^{N/2} \phi_k^{i*} \phi_l^i$. Instead of the usual Hubbard-Stratonovitch transformation, we linearize the 4-boson interaction term by the Hermitian matrix $\delta$ function

$$\delta(Q_i - \overline{Q_i}) = \frac{1}{(2\pi)^4} \int dF e^{-i\text{Tr}F(Q_i - \overline{Q_i})}, \tag{225}$$

where the integral is over Hermitian matrices $F$. We thus find the identity

$$\exp\left[-\frac{2}{N}\mathrm{Tr}\overline{Q_1}\,\overline{Q_2}\right] \sim \int dQ_1 dQ_2 \int dF dG e^{\mathrm{Tr}[-iF(Q_1-\overline{Q_1})iG(Q_2-\overline{Q_2})-\frac{2}{N}Q_1Q_2]}. \tag{226}$$

The integral over the $\phi_k$ is uniformly convergent in $F$ and $G$ which justifies the interchange of the order of the integrals. This results in the partition function

$$\begin{aligned} Z_{-1} &= \int dQ_1 dQ_2 \int dF dG e^{\mathrm{Tr}[-i\frac{N}{2}FQ_1-i\frac{N}{2}GQ_2+i\frac{N}{2}\zeta_1^T(Q_1-IQ_2I)-\frac{N}{2}Q_1Q_2]} \\ &\quad \times \det{}^{-\frac{N}{2}}\begin{pmatrix} \epsilon+F & \mu\sigma_3 \\ \mu\sigma_3 & \epsilon+G \end{pmatrix}, \end{aligned} \tag{227}$$

where we have used a block notation and the mass matrices (223). We have also simplified this integral by changing integration variables according to $F \to F - \zeta_1^T$ and $G \to G + I\zeta_1^T I$ and $Q_i \to NQ_i/2$, $i = 1, 2$. For reasons of convergence we have kept the infinitesimal increments inside the determinant. In the weak nonhermiticity limit, where $\mu^2 N$ is kept fixed for $N \to \infty$, the determinant can be approximated by

$$\begin{aligned} \det{}^{-\frac{N}{2}}\begin{pmatrix} \epsilon+F & \mu\sigma_3 \\ \mu\sigma_3 & \epsilon+G \end{pmatrix} &= \det{}^{-\frac{N}{2}}(\epsilon+F)\det{}^{-\frac{N}{2}}(\epsilon+G) \\ &\quad \times \exp\left[\frac{N\mu^2}{2}\frac{1}{\epsilon+F}\sigma_3\frac{1}{\epsilon+G}\sigma_3\right](1+\mathcal{O}\left(\frac{1}{N}\right)). \end{aligned} \tag{228}$$

The $F$ and $G$ variables in the $\mu^2 N$ term can be replaced the saddle point values of $F$ and $G$ at $\mu = 0$ given by $(\epsilon+F)Q_1 = i$ and $(\epsilon+G)Q_2 = i$. The remaining integrals over $F$ and $G$ are Ingham-Siegel integrals given by [86]

$$\int dF \det{}^{-n}(\epsilon+F)e^{i\mathrm{Tr}QF} \sim \theta(Q)\det{}^{n-p}(Q)e^{-i\epsilon\mathrm{Tr}Q}, \tag{229}$$

where the integral is over $p\times p$ Hermitian matrices, $\mathrm{Im}\epsilon < 0$, and $\theta(Q)$ denotes that $Q$ is positive definite. These manipulations result in

$$\begin{aligned} Z_{-1}(z, z^*; \mu) &= \int dQ_1 dQ_2 \theta(Q_1)\theta(Q_2)\det{}^{\frac{N}{2}-2}(Q_1Q_2) \\ &\quad \times e^{\mathrm{Tr}[i\frac{N}{2}\zeta_1^T(Q_1-IQ_2I)-\frac{N}{2}Q_1Q_2-\frac{N}{2}\mu^2 Q_1\sigma_3 Q_2\sigma_3]}. \end{aligned} \tag{230}$$

In the limit $N \to \infty$ the integrals over the massive modes can be performed by a saddle point approximation. The saddle point equations are given by

$$Q_1^{-1} - Q_2 = 0, \qquad Q_2^{-1} - Q_1 = 0. \tag{231}$$

Both equations can be rewritten as

$$Q_1 = Q_2^{-1}, \tag{232}$$

and therefore only four of the modes, which we choose to be $Q_2$, can be integrated out by a saddle-point approximation. The quadratic fluctuations give rise to a factor $\pi^2/\det^2 Q_1$. The integral over the remaining modes has to be performed exactly. We thus arrive at the partition function [25]

$$Z_{-1}(z, z^*; \mu) = \int \frac{dQ_1}{\det^2 Q_1} \theta(Q_1) e^{\mathrm{Tr}[i\frac{N}{2}\zeta_1^T(Q_1 - IQ_1^{-1}I) - \frac{N}{2}\mu^2 Q_1\sigma_3 Q_1^{-1}\sigma_3]}. \tag{233}$$

Before evaluating this integral, we rederive this partition function based on the symmetries of the QCD partition function.

**Symmetries of $Z_{-1}(\mu)$.** For $\mu = 0$ and $\zeta_1 = \zeta_2 = 0$ the symmetry of the partition function (222) are the $Gl(2) \times Gl(2)$ transformations,

$$\begin{aligned} &\begin{pmatrix} \phi_1 \\ \phi_2 \end{pmatrix} \to U_1 \begin{pmatrix} \phi_1 \\ \phi_2 \end{pmatrix}, \qquad \begin{pmatrix} \phi_1^* \\ \phi_2^* \end{pmatrix} \to \begin{pmatrix} \phi_1^* \\ \phi_2^* \end{pmatrix} U_2^{-1}, \\ &\begin{pmatrix} \phi_3 \\ \phi_4 \end{pmatrix} \to U_2 \begin{pmatrix} \phi_3 \\ \phi_4 \end{pmatrix}, \qquad \begin{pmatrix} \phi_3^* \\ \phi_4^* \end{pmatrix} \to \begin{pmatrix} \phi_3^* \\ \phi_4^* \end{pmatrix} U_1^{-1}, \end{aligned} \tag{234}$$

where we have disregarded convergence. This symmetry can be extended to nonzero mass or chemical potential if we adopt the transformation rules

$$\begin{aligned} \zeta_1 &\to U_2 \zeta_1 U_1^{-1}, & \zeta_2 &\to U_1 \zeta_2 U_2^{-1}, \\ \mu_1 &\to U_2 \mu_1 U_2^{-1}, & \mu_2 &\to U_1 \mu_2 U_1^{-1}, \end{aligned} \tag{235}$$

where $\mu_1$ is the chemical potential matrix that is added $id$ and $\mu_2$ is the chemical matrix that is added to $-id^\dagger$. These matrices are introduced for the sake of discussing the transformation properties of the partition function (222) and will ultimately be replaced by their original values $\mu_1 = \mu_2 = \mu\sigma_3$. The chiral symmetry is broken spontaneously to $Gl(2)$ by the chiral condensate. Because the bosonic integral has to converge, the Goldstone manifold is not $Gl(2)$ but rather $Gl(2)/U(2)$, i.e. the coset of positive definite matrices as in the case of zero chemical potential. Under a $Gl(2) \times Gl(2)$ transformation the Goldstone fields transform as

$$Q \to U_1 Q U_2^{-1}. \tag{236}$$

The low energy effective partition function should have the same transformation properties as the microscopic partition function (222). To second order in $\mu$ and first order in the mass matrix we can write down the following invariants

$$\mathrm{Tr}\zeta_1 Q^{-1}, \quad \mathrm{Tr}\zeta_2 Q, \quad \mathrm{Tr}Q\mu_1 Q^{-1}\mu_2, \quad \mathrm{Tr}\mu_1\mu_2. \tag{237}$$

We also have the discrete symmetry that the partition function is invariant under the interchange of $\zeta_1$ and $\zeta_2$. This symmetry implies that the coefficients of the two mass terms in the effective partition function are the same. Using that the integration measure on positive definite Hermitian matrices is given by $dQ/\det^2 Q$, we finally arrive at the effective partition function

$$Z_{-1,\nu}(z,z^*) = \int_{Q\in Gl(2)/U(2)} \frac{\det^\nu(Q)dQ}{\det^2(Q)} e^{-\frac{F^2\mu^2 V}{4}\mathrm{Tr}[Q,B][Q^{-1},B]+\frac{i\Sigma V}{2}\mathrm{Tr}(\zeta_1 Q+\zeta_2 Q^{-1})}, \tag{238}$$

The partition function (233) is recovered after making the identification $V \to N$, $\Sigma \to 1$ and $F^2 \to 1$ and $Q \to Q^T$.

**Calculation of the integral over $Q$.** To evaluate the integral (238) we use the parameterization

$$Q = e^t \begin{pmatrix} e^r \cosh s & e^{i\theta}\sinh s \\ e^{-i\theta}\sinh s & e^{-r}\cosh s \end{pmatrix}. \tag{239}$$

where

$$r \in \langle -\infty, \infty \rangle, \quad s \in \langle -\infty, \infty \rangle, \quad t \in \langle -\infty, \infty \rangle, \quad \theta \in \langle 0, \pi \rangle \quad . \tag{240}$$

The Jacobian relating the measures $dQ/\det^2 Q$ and $drdsdtd\theta$ is given by

$$J = 4e^{4t}\cosh s \sinh s. \tag{241}$$

We first perform the integral over $r$, which gives a factor $2K_0(2N\epsilon\cosh s\cosh t)$ with leading singularity given by $\sim -\log\epsilon$. This factor is absorbed in the normalization of the partition function. Then the integral over $\theta$ gives a Bessel function. Introducing $u = \sinh s$ as new integration variable we find [25]

$$\begin{aligned} Z^\nu_{-1}(z,z^*) &= C_{-1}\int_{-\infty}^{\infty} dt \int_0^\infty du e^{2\nu t} J_0(2Vu(x^2\cosh^2 t + y^2\sinh^2 t)^{1/2}) \\ &\times e^{-\mu^2 F^2 V(1+2u^2)}. \end{aligned} \tag{242}$$

To do the integral over $u$ we use the known integral

$$\int_0^\infty dx x^{a+1} e^{-\alpha x^2} J_a(\beta x) = \frac{\beta^a}{(2\alpha)^{a+1}} e^{-\beta^2/4\alpha}. \tag{243}$$

This results in

$$Z_{-1,\nu}(z,z^*) = \frac{C_{-1}e^{-V\mu^2F^2}}{4\mu^2F^2V}\int_{-\infty}^{\infty} dt e^{2\nu t} e^{-\frac{V(x^2\cosh^2 t + y^2\sinh^2 t)}{2\mu^2F^2}}. \tag{244}$$

Using that $\cosh^2 t = \frac{1}{2} + \frac{1}{2}\cosh 2t$ and $\sinh^2 t = -\frac{1}{2} + \frac{1}{2}\cosh 2t$, the integral over $t$ can be rewritten as a modified Bessel function resulting in [25]

$$Z_{-1,\nu}(z = x + iy, z^*) = C_{-1}\, e^{\frac{V\Sigma^2(y^2-x^2)}{4\mu^2 F^2}} K_\nu\left(\frac{V\Sigma^2(x^2+y^2)}{4\mu^2 F^2}\right). \quad (245)$$

**The dirac spectrum at nonzero chemical potential.** The final result for the quenched spectral density is obtained by substituting the partition functions $Z_1^\nu(z, z^*)$ and $Z_{-1,\nu}(z = x + iy, z^*)$ in expression (221) obtained from the replica limit of the Toda lattice equation. We find,

$$\begin{aligned}\rho^{\rm quen}(x,y) &= \frac{V^3\Sigma^4}{2\pi F^2\mu^2}(x^2+y^2)e^{\frac{V\Sigma^2(y^2-x^2)}{4\mu^2F^2}} K_\nu\left(\frac{V\Sigma^2(x^2+y^2)}{4\mu^2F^2}\right) \\ &\quad \times \int_0^1 \lambda d\lambda e^{-2VF^2\mu^2\lambda^2)} |I_\nu(\lambda z V\Sigma)|^2. \end{aligned} \quad (246)$$

The normalization constant has been chosen such that the $\mu \to 0$ limit of $\rho^{\rm quen}(x,y)$ for large $y$ is given by $\Sigma V/\pi$ (see below).

In the limit $\mathrm{Re}(z)\Sigma/\mu^2F^2 \ll 1$ the upper limit of the integral in (246) can be extended to infinity. Using the known integral

$$\int_0^\infty \lambda d\lambda e^{-2VF^2\mu^2\lambda^2}|I_\nu(\lambda z V\Sigma)|^2 = \frac{e^{\frac{(z^2+z^{*2})\Sigma^2 V}{8\mu^2F^2}}}{4\mu^2F^2V} I_\nu\left(\frac{zz^*V\Sigma^2}{4\mu^2F^2}\right), \quad (247)$$

the spectral density can be expressed as

$$\rho^{\rm quen}(x,y) = \frac{2}{\pi}u^2 zz^* K_\nu(zz^*u)I_\nu(zz^*u) \quad \text{with} \quad u = \frac{V\Sigma^2}{4\mu^2F^2}. \quad (248)$$

Therefore, the spectral density becomes a universal function that only depends on a single parameter $u$. This parameter can be rewritten in a more physical way as $u = \pi\rho^{\rm asym}(x,y)$. For the dimensionless ratio we obtain

$$\frac{\rho^{\rm quen}(x,y)}{\rho_{\rm asym}(x,y)} = 2uzz^*K_\nu(zz^*u)I_\nu(zz^*u), \quad (249)$$

which is universal combination that depends only on a single universal combination $zz^*u$. (This result was obtained in collaboration with Tilo Wettig).

In the thermodynamic limit the Bessel functions can be approximated by their asymptotic limit. This results in

$$\rho^{\rm quen}(x,y) = \frac{V^2\Sigma^2}{2\pi F\mu\sqrt{2\pi V}}\int_0^1 d\lambda e^{-2VF^2\mu^2(\lambda-\frac{|x|\Sigma}{2F^2\mu^2})^2}. \quad (250)$$

For $V \to \infty$ the integral over $\lambda$ can be performed by a saddle point approximation. If the saddle point is outside the range $[0, 1]$ the integral vanishes for $V \to \infty$. We thus find for the spectral density

$$\rho^{\rm quen}(x, y) = \frac{V\Sigma^2}{4\pi\mu^2 F^2} \quad \text{for} \quad |x| < \frac{2F^2\mu^2}{\Sigma}. \tag{251}$$

and $\rho^{\rm quen}(x, y) = 0$ outside this strip. This result is in agreement with the mean field analysis [82] of the effective partition function given in section 8.3. For the integrated eigenvalue density we find

$$\int_{-\infty}^{\infty} dx \rho^{\rm quen}(x, y) = \frac{\Sigma V}{\pi} \tag{252}$$

in agreement with the eigenvalue density at $\mu = 0$.

## 9. Full QCD at nonzero chemical potential

In quenched QCD the chiral condensate $G(m) \sim m$ in the region where the eigenvalues are located. We have argued before that in full QCD the chiral condensate does not depend on $m$ for $\mu < m_N/N_N$. We thus have

$$G(m) \equiv \langle \sum_k \frac{1}{m + i\lambda_k} \prod_l (i\lambda_l + m)^{N_f} \rangle = G(m \to 0) \text{ for } \mu < m_N/N_N. \tag{253}$$

The conclusion is that the presence of the fermion determinant completely alters the vacuum structure of the theory. The question we wish to address is how we can understand this based on the spectrum of the QCD Dirac operator. For simplicity we only consider the case of $N_f = 1$.

The average spectral density in full QCD is defined by

$$\rho^{\rm full}(x, y, \mu) = \langle \sum_k \delta^2(x + iy - \lambda_k) \det(D + m + \mu\gamma_0) \rangle. \tag{254}$$

The low-energy limit of the generating function for the spectral density can again be written as a $\tau$-function [26]. The spectral density is then obtained from the replica limit of the corresponding Toda lattice equation. The result is

$$\begin{aligned} \rho^{\rm full}(x, y, \mu) &= \frac{V^3(x^2 + y^2)\Sigma^4}{2\pi\mu^2 F^2} e^{\frac{V(y^2 - x^2)\Sigma^2}{4\mu^2 F^2}} K_0\left(\frac{V(x^2 + y^2)\Sigma^2}{4\mu^2 F^2}\right) \\ &\times \int_0^1 t dt e^{-2V\mu^2 F^2 t} \left(I_0^*(z\Sigma V t) - \frac{I_0^*(z\Sigma V) I_0 m V \Sigma t}{I_0(m V \Sigma)}\right) I_0(z\Sigma V t). \end{aligned} \tag{255}$$

It was first obtained from the random matrix model [27] using the method of complex orthogonal polynomials developed in [87, 88]. To appreciate this result, we considers its asymptotic expansion for $V \to \infty$. For $m = 0$ and $x > 0$ we can derive the asymptotic result for difference

$$\rho^{\rm quen}(x, y, \mu) - \rho^{\rm full}(x, y, \mu) \quad \sim \quad \frac{\sqrt{z\Sigma}}{2\mu^3 F^3} e^{\frac{Vz^{*2}\Sigma^2}{8\mu^2 F^2} - \frac{Vx^2\Sigma^2}{2\mu^2}} e^{z^*\Sigma V}. \quad (256)$$

The behavior near the extremum at $(x, y) = (4\mu^2 F^2/3\Sigma, 0)$ is given by

$$\rho^{\rm quen}(x, y, \mu) - \rho^{\rm full}(x, y, \mu) \sim e^{\frac{2}{3}\mu^2 F^2 V} e^{\frac{2}{3} iy\Sigma V}. \quad (257)$$

We have oscillations on the scale $1/\Sigma V$ with an amplitude that diverges exponentially with the volume [26]. In the thermodynamic limit, these oscillations are visible in a domain where the real part of the exponent in (256) is positive. This region is given by intersection of the the inside of the ellipses

$$3(x \pm \frac{4}{3}\mu^2 F^2 \Sigma)^2 + y^2 = \frac{16}{3}\mu^4 F^2 \Sigma^2. \quad (258)$$

and the strip $|x| < 2F^2\mu^2/\Sigma$. At the mean field level this can be reinterpreted as a region where Kaon condensation takes place [91, 92].

## 10. Conclusions

The existence of two formulations of QCD at low energy, first as a microscopic theory of quarks and gluons and second as an effective theory of weakly interacting Goldstone bosons, imposes powerful constraints on either of the theories. The effective theory is completely determined by the symmetries of the microscopic theory, and the mass dependence of the effective theory imposes sum rules on the inverse Dirac eigenvalues. In particular this means that any theory with the same symmetry breaking pattern and a mass gap will be subject to the same constraints. The simplest microscopic theory is chiral Random Matrix Theory.

However, more can be done than constraining the inverse Dirac eigenvalues by sum rules. The key observation is that the generating function for the resolvent amounts to a partition function with additional flavors with a mass $z$ equal to the value for which the resolvent is calculated. Again we have a microscopic theory and an effective theory with the same low energy limit. Because $z$ is a free parameter, it can always be chosen such that the Compton wavelength of the corresponding Goldstone boson is much larger than the size of the box. In this region the $z$ dependence of the partition function is determined by the mass term of the chiral Lagrangian which is a simple matrix integral.

To obtain the Dirac spectrum we have to quench the determinant corresponding to $z$. This can be done in two ways: by the replica trick or by the supersymmetric method. Although, the supersymmetric method is straightforward, the *naive* replica trick is technically somewhat simpler. The problem is that the *naive* replica trick does not give correct nonperturbative results. One way out is if the dependence on the number of replicas $n$ is known as an analytical function of $n$ around $n = 0$. It was argued by Kanzieper that the $n$-dependence can be obtained from the solution of the Painlevé equation. The other way out, which has been advocated in these lectures, is if the partition function for $n = 0$ is related by a recursion relation to partition functions with a nonzero integer number of flavors. The replica limit of this Toda lattice equation gives us nonperturbative correlation functions. This is very efficient formulation of the problem. The structure of the final answer already has the factorized structure of the Toda lattice formulation. We could also say that the supersymmetric partition function connects two semi-infinite hierarchies.

New results with the Toda lattice method were obtained for QCD at nonzero chemical potential. In this case the low-energy effective partition functions are also related by the Toda lattice equation. This made it possible to express the microscopic spectral density as the product of the partition function with one fermionic flavor and the partition function with one bosonic flavor. This result has later been reproduced by RMT with the method of orthogonal polynomials.

More surprisingly the Toda lattice method also gives the correct result for QCD at nonzero chemical potential with dynamical fermions. Because of the phase of the fermion determinant, a breakdown of this method for this case would not have been a surprise. However, the concept of integrability that also reigns this case, is so powerful that replica limit can be taken in exactly the same way as in the quenched case. The result for the spectral density shows oscillations on the scale of $1/V$ and an amplitude that diverges exponentially with $V$. This structure is necessary to obtain a nonzero chiral condensate in the chiral limit.

The Toda lattice method has been applied to quite a few cases in the symmetry class $\beta = 2$ (see [24, 23, 89, 25, 26, 90]). Our conjecture is that all microscopic correlation functions in this class can be obtained from the replica limit of a Toda lattice equation. A much tougher problem is analysis of the replica limit for the other Dyson classes. We are not aware of any progress on this problem and encourage the reader to confront this challenge.

## Acknowledgements

G. Akemann, D. Dalmazi, P. Damgaard, A. Halasz, J.C. Osborn, M. Sener, K. Splittorff, M. Stephanov, D. Toublan, H.A. Weidenmueller, T. Wettig and M.R. Zirnbauer are thanked for sharing their insights which led to the joint publications on which these lectures are based. Kim Splittorff is thanked for a critical reading of the manuscript.

## References

[1] E.P. Wigner, Ann. Math. **62**, 548 (1955).

[2] G. 't Hooft, Nucl. Phys. B **75**, 461 (1974).

[3] E. Brezin, C. Itzykson, G. Parisi and J. B. Zuber, Commun. Math. Phys. **59**, 35 (1978).

[4] P. Di Francesco, arXiv:math-ph/0406013.

[5] D. J. Gross and E. Witten, Phys. Rev. D **21**, 446 (1980).

[6] E. Brezin and D. J. Gross, Phys. Lett. B **97**, 120 (1980).

[7] T. Eguchi and H. Kawai, Phys. Rev. Lett. **48**, 1063 (1982).

[8] V. A. Kazakov and A. A. Migdal, Nucl. Phys. B **397**, 214 (1993).

[9] I. Bars and F. Green, Phys. Rev. D **20**, 3311 (1979).

[10] G. Bhanot, U. M. Heller and H. Neuberger, Phys. Lett. B **113**, 47 (1982).

[11] I. I. Kogan, G. W. Semenoff and N. Weiss, Phys. Rev. Lett. **69**, 3435 (1992).

[12] J. Gasser and H. Leutwyler, Phys. Lett. **188B**, 477 (1987).

[13] H. Leutwyler and A. Smilga, Phys. Rev. **D46**, 5607 (1992).

[14] E.V. Shuryak and J.J.M. Verbaarschot, Nucl. Phys. **A560**, 306 (1993).

[15] J.J.M. Verbaarschot, Phys. Rev. Lett. **72**, 2531 (1994); Phys. Lett. **B329**, 351 (1994).

[16] J. J. M. Verbaarschot and T. Wettig, Ann. Rev. Nucl. Part. Sci. **50**, 343 (2000).

[17] P. H. Damgaard, Nucl. Phys. Proc. Suppl. **106**, 29 (2002). [arXiv:hep-lat/0110192].

[18] G. Akemann, P. H. Damgaard, U. Magnea and S. Nishigaki, Nucl. Phys. B **487**, 721 (1997).

[19] J. C. Osborn, D. Toublan and J. J. M. Verbaarschot, Nucl. Phys. **B 540**, 317 (1999).

[20] P. H. Damgaard, J. C. Osborn, D. Toublan and J. J. M. Verbaarschot, Nucl. Phys. **B 547**, 305 (1999).

[21] J.J.M. Verbaarschot and M.R. Zirnbauer, J. Phys. **A18**, 1093 (1985).

[22] A. Mironov, A. Morozov and G. W. Semenoff, Int. J. Mod. Phys. A **11**, 5031 (1996).

[23] K. Splittorff and J.J.M. Verbaarschot, Phys. Rev. Lett. **90**, 041601 (2003).

[24] E. Kanzieper, Phys. Rev. Lett. **89**, 250201 (2002).

[25] K. Splittorff and J. J. M. Verbaarschot, Nucl. Phys. B **683**, 467 (2004).

[26] G. Akemann, J. C. Osborn, K. Splittorff and J. J. M. Verbaarschot, arXiv:hep-th/0411030.

[27] J. C. Osborn, Phys. Rev. Lett. **93**, 222001 (2004).

[28] M.L. Mehta, *Random Matrices*, 3rd edition, Elsevier, Amsterdam 2004.

[29] P. Forrester, *Log-gases and Random matrices*, Web Book - available at `http://www.ms.unimelb.edu.au/~matpjf/matpjf.html`.

[30] T. Guhr, A. Mueller-Groeling, and H.A. Weidenmueller, Phys. Rep. **299** (1998) 189.

[31] *Special Issue: Random Matrix Theory*, J. Phys. **A 36** (2003), P.J. Forrester, N.C. Snaith and J.J.M. Verbaarschot, eds..

[32] K.B. Efetov, *Supersymmetry in disorder and chaos*, Cambridge University Press, (1997).

[33] C.W.J. Beenakker, Rev. Mod. Phys. **69**, 731 (1997).

[34] P. Di Francesco, P. H. Ginsparg and J. Zinn-Justin, Phys. Rep. **254**, 1 (1995).

[35] M. Stephanov, J.J.M. Verbaarschot and T. Wettig, in Supplement 1 of *Wiley Encyclopedia of Electrical and Electronics Engineering*, J.G. Webster, Ed., Wiley, 2001.

[36] Y. Fyodorov, in *Mesoscopic Quantum Physics*, Elsevier Science, 1995.

[37] J. J. M. Verbaarschot, arXiv:hep-th/0410211.

[38] M. Creutz, *Quarks, Gluons, and Lattices*, Cambridge University Press, Cambridge 1983.

[39] H.J. Rothe, *Lattice Gauge Theories: an Introduction*, World Scientific, Singapore 1992.

[40] I. Montvay, *Quantum Fields on a Lattice*, Cambridge University Press, Cambridge 1994.

[41] C. Vafa and E. Witten, Nucl. Phys. **B234**, 173 (1984).

[42] M. Peskin, Nucl. Phys. **B175**, 197 (1980); S. Dimopoulos, Nucl. Phys. **B168**, 69 (1980); M. Vysotskii, Y. Kogan and M. Shifman, Sov. J. Nucl. Phys. **42**, 318 (1985).

[43] C.E. Porter, *Statistical Theory of Spectra: Fluctuations*, Academic Press, New York 1965.

[44] T. Banks and A. Casher, Nucl. Phys. **B 169**, 103 (1980).

[45] K. Splittorff and J. J. M. Verbaarschot, arXiv:hep-th/0408107.

[46] D. Toublan and J. J. M. Verbaarschot, arXiv:hep-th/0008086.

[47] S. Weinberg, Physica A **96**, 327 (1979).

[48] J. Gasser and H. Leutwyler, Ann. Phys. **158**, 142 (1984); Nucl. Phys. B **250**, 465 (1985); H. Leutwyler, Ann. Phys. **235**, 165 (1994).

[49] M.R. Zirnbauer, J. Math. Phys. **37**, 4986 (1996).

[50] R. Narayanan and H. Neuberger, arXiv:hep-lat/0405025.

[51] J.J.M. Verbaarschot, Phys. Lett. **B368**, 137 (1996).

[52] B.L. Altshuler, I.Kh. Zharekeshev, S.A. Kotochigova and B.I. Shklovskii, Zh. Eksp. Teor. Fiz. **94**, 343 (1988); B.L. Altshuler and B.I. Shklovskii, Zh. Exp. Teor. Fiz. **91** , 220 (1986).

[53] G. Montambaux, in *Quantum Fluctuations, Les Houches, Session LXIII*, E. Giacobino, S. Reynaud and J. Zinn-Justin, eds., Elsevier Science, 1995, cond-mat/9602071.

[54] M.R. Zirnbauer and F.D.M. Haldane, Phys. Rev. **B52**, 8729 (1995); P.B. Gossiaux, Z. Phuhar and H.A. Weidenmueller, cond-mat/9803362.

[55] J. J. M. Verbaarschot and I. Zahed, Phys. Rev. Lett. **70**, 3852 (1993).

[56] S. Chandrasekharan, Nucl. Phys. Proc. Suppl. **42**, 475 (1995); S. Chandrasekharan and N. Christ, Nucl. Phys. Proc. Suppl. **42**, 527 (1996); N. Christ, Lattice 1996.

[57] M. E. Berbenni-Bitsch, S. Meyer, A. Schafer, J. J. M. Verbaarschot and T. Wettig, Phys. Rev. Lett. **80**, 1146 (1998); R. G. Edwards, U. M. Heller and R. Narayanan, Phys. Rev. D **60**, 034502 (1999); M. Gockeler, H. Hehl, P. E. L. Rakow, A. Schafer and T. Wettig, Phys. Rev. D **59**, 094503 (1999); M. Gockeler, H. Hehl, P. E. L. Rakow, A. Schafer and T. Wettig, Phys. Rev. D **65**, 034503 (2002). P. H. Damgaard, U. M. Heller, R. Niclasen and K. Rummukainen Nucl. Phys. Proc. Suppl. **83**, 197 (2000); P. H. Damgaard, M. C. Diamantini, P. Hernandez and K. Jansen, Nucl. Phys. Proc. Suppl. **106**, 772 (2002); L. Giusti, M. Luscher, P. Weisz and H. Wittig, JHEP **0311**, 023 (2003).

[58] P. H. Damgaard and K. Splittorff, Nucl. Phys. B **572**, 478 (2000).

[59] D. Dalmazi and J. J. M. Verbaarschot, Phys. Rev. D **64**, 054002 (2001).

[60] J. J. M. Verbaarschot, arXiv:hep-th/9710114.

[61] T. Guhr and T. Wettig, J. Math. Phys. **37**, 6395 (1996); A. D. Jackson, M. K. Sener and J. J. M. Verbaarschot, Phys. Lett. B **387**, 355 (1996).

[62] F.A. Berezin and F.I.Karpelevich, Doklady Akad. Nauk. SSSR **118**, 9 (1958).

[63] R.C. Brower and M. Nauenberg, Nucl. Phys. B **180**, 221 (1981); R. Brower, P. Rossi and C. I. Tan, Nucl. Phys. B **190**, 699 (1981).

[64] J. B. Kogut, M. Snow and M. Stone, Nucl. Phys. B **200**, 211 (1982).

[65] G. Akemann and P. H. Damgaard, Phys. Lett. B **432**, 390 (1998). G. Akemann and P. H. Damgaard, Nucl. Phys. B **576**, 597 (2000).

[66] S. Kharchev, A. Marshakov, A. Mironov, A. Morozov and A. Zabrodin, Nucl. Phys. B **380**, 181 (1992).

[67] G. Dunne, *Self-Dual Chern-Simons Theories*, Lecture Notes in Physics 36, Springer, Heidelberg 1995.

[68] C.A. Tracy and H.A. Widom, Comm. Math. Phys. **159**, 151 (1994).

[69] P.J. Forrester and N.S. Witte, math-ph/0201051.

[70] J.J. Sylvester, Compt. Rend. Acad. Sc. **54**, 129 (1862).

[71] D. Dalmazi and J. J. M. Verbaarschot, Nucl. Phys. B **592**, 419 (2001).

[72] K. Splittorff and J. J. M. Verbaarschot, Nucl. Phys. B **695**, 84 (2004).

[73] E. Kanzieper, Nucl. Phys.**B 596**, 5 (2001).

[74] K.B. Efetov, Adv. Phys.**32**, 53 (1983).

[75] J. Ginibre, J. Math. Phys. **6**, 440 (1965).

[76] A. Zabrodin, arXiv:hep-th/0412219.

[77] I.M. Barbour, N.E. Behilil, E. Dagotto, F. Karsch, A. Moreo, M. Stone, and H.W. Wyld, Nucl. Phys. **B 275**, 296 (1986); M.-P. Lombardo, J.B. Kogut, and D.K. Sinclair, Phys. Rev. **D 54**, 2303 (1996); H. Markum, R. Pullirsch and T. Wettig, Phys. Rev. Lett. **83**, 484 (1999); E. Bittner, M. P. Lombardo, H. Markum and R. Pullirsch, Nucl. Phys. Proc. Suppl. **106**, 468 (2002); E. Bittner, S. Hands, H. Markum and R. Pullirsch, Prog. Theor. Phys. Suppl. **153**, 295 (2004); G. Akemann and T. Wettig, Nucl. Phys. Proc. Suppl. **129**, 527 (2004).

[78] V.L. Girko, *Theory of random determinants* Kluwer Academic Publishers, Dordrecht, 1990.

[79] A. Goksch, Phys. Rev. **D 37** , 1014 (1988).

[80] M.A. Stephanov, Phys. Rev. Lett. **76**, 4472 (1996).

[81] J. Feinberg and A. Zee, Nucl. Phys. B 504, 579 (1997); Nucl. Phys. B 501, 643 (1997).

[82] D. Toublan and J.J.M. Verbaarschot, Int. J. Mod. Phys. **B 15**, 1404 (2001).

[83] J. B. Kogut, M. A. Stephanov and D. Toublan, Phys. Lett. B **464**, 183 (1999).

[84] A. D. Jackson and J. J. M. Verbaarschot, Phys. Rev. D **53**, 7223 (1996).

[85] M. A. Halasz, J. C. Osborn and J. J. M. Verbaarschot, Phys. Rev. D **56**, 7059 (1997).

[86] Y. V. Fyodorov, Nucl. Phys. B **621**, 643 (2002).

[87] Y.V. Fyodorov, H.-J. Sommers and B.A. Khoruzhenko, Ann. Inst. H. Poincare - Phys. Theor. **68**, 449 (1998).

[88] G. Akemann, Phys. Rev. Lett. **89**, 072002 (2002); J.Phys. A **36**, 3363 (2003).

[89] E. Kanzieper, arXiv:cond-mat/0312006.

[90] T. Andersson, P. H. Damgaard and K. Splittorff, arXiv:hep-th/0410163.

[91] J.B. Kogut and D. Toublan, Phys. Rev. D **64**, 034007 (2001).

[92] J.C. Osborn, K. Splittorff and J.J.M. Verbaarschot, in preparation.

# EUCLIDEAN RANDOM MATRICES: SOLVED AND OPEN PROBLEMS

Giorgio Parisi
*Dipartimento di Fisica, Sezione INFN, SMC and UdRm1 of INFM,*
*Università di Roma "La Sapienza",*
*Piazzale Aldo Moro 2, I-00185 Rome (Italy)*
giorgio.parisi@roma1.infn.it

**Abstract** In this paper I will describe some results that have been recently obtained in the study of random Euclidean matrices, i.e. matrices that are functions of random points in Euclidean space. In the case of *translation invariant* matrices one generically finds a phase transition between a *phonon* phase and a *saddle* phase. If we apply these considerations to the study of the Hessian of the Hamiltonian of the particles of a fluid, we find that this phonon-saddle transition corresponds to the dynamical phase transition in glasses, that has been studied in the framework of the mode coupling approximation. The Boson peak observed in glasses at low temperature is a remanent of this transition. We finally present some recent results obtained with a new approach where one deeply uses some hidden supersymmetric properties of the problem.

## 1. Introduction

In the last years many people have worked on the problem of analytically computing the properties of Euclidean random matrices [1]-[15]. The problem can be formulated as follows.

We consider a set of $N$ points $(x_i)$ that are randomly distributed with some given distribution.Two extreme examples are:

- The $x$'s are random independent points with a flat probability distribution, with density $\rho$.
- The $x$'s are one of the many minima of a given Hamiltonian.

In the simplest case of the first example, given a function $f(x)$, we consider the $N \times N$ matrix:

$$M_{i,k} = f(x_i - x_k) . \tag{1}$$

The problem consists in computing the properties of the eigenvalues and of the eigenvectors of $M$. Of course, for finite $N$ they will depend on the instance



*E. Brezin et al. (eds.), Applications of Random Matrices in Physics,* 219–260.

of the problem (i.e. the actually choice of the $x$'s), however system to system fluctuations for intensive quantities (e.g. the spectral density $g(z)$) disappear when we consider the limit $N \to \infty$ at fixed particle density $\rho$.

The problem is not new; it has been carefully studied in the case where the positions of the particles are restricted to be on a lattice [16]. The case where the particles can stay in arbitrary positions, that is relevant in the study of fluids, has been less studied, although in the past many papers have been written on the argument [1]-[15]. These off-lattice models present some technical (and also physical) differences with the more studied on-lattice models.

There are many possible physical motivations for studying these models, that may be applied to electronic levels in amorphous systems, very diluted impurities, spectrum of vibrations in glasses (my interest in the subject comes from this last field).

I will concentrate in these lectures on the spectral density and on the Green functions, trying to obtaining both the qualitative features of these quantities and to present reasonable accurate computations (when possible). This task is not trivial because there is no evident case that can be used as starting point for doing a perturbation expansion. Our construction can be considered as a form of a mean field theory (taking care of some corrections to it): more sophisticated problems, like localization or corrections to the naive critical exponents will be only marginally discussed in these notes.

A certain number of variations to equation (1) are interesting. For example we could consider the case where we add a fluctuating term on the diagonal:

$$M_{i,k} = \delta_{i,k} \sum_j f(x_i - x_j) - f(x_i - x_k) . \tag{2}$$

This fluctuating diagonal term has been constructed in such a way that

$$\sum_k M_{i,k} = 0 . \tag{3}$$

Therefore the diagonal and the off-diagonal matrix elements are correlated.

In general it may be convenient to associate a quadratic form to the matrix $M$: the quadratic form ($\mathcal{H}[\phi]$) is defined as:

$$\mathcal{H}[\phi] = \frac{1}{2} \sum_{i,k} \phi_i \phi_k M_{i,k} . \tag{4}$$

In the first case we considered, eq. (1), we have that:

$$\mathcal{H}[\phi] = \sum_{i,k} f(x_i - x_k) \phi_i \phi_k . \tag{5}$$

In this second case, eq. (2), the associated quadratic form is given by

$$\mathcal{H}[\phi] = \sum_{i,k} \phi_i \phi_k M_{i,k} = \frac{1}{2} \sum_{i,k} f(x_i - x_k)(\phi_i - \phi_k)^2 \,. \tag{6}$$

Here the matrix $M$ is non-negative if the function $f$ is non-negative. The matrix $M$ has always a zero eigenvalue as consequence of the invariance of the quadratic form under the symmetry $\phi_i \to \phi_i + \lambda$; the presence of this symmetry has deep consequences on the properties of the spectrum of $M$ and, as we shall see later, *phonons* are present. Many of the tricky points in the analytic study are connected to the preserving and using this symmetry in the computations.

In the same spirit we can consider a two-body potential $V(x)$ and we can introduce the Hamiltonian

$$H[x] = \frac{1}{2} \sum_{i,k} V(x_i - x_k) \,. \tag{7}$$

We can construct the $3N \times 3N$ Hessian matrix

$$M_{i,k} = \frac{\partial^2 H}{\partial x_i x_k} = \delta_{i,k} \sum_j V''(x_i - x_j) - V''(x_i - x_k) \,, \tag{8}$$

where for simplicity we have not indicated space indices. Also here we are interested in the computation of the spectrum of $M$. The translational invariance of the Hamiltonian implies that the associated quadratic form is invariant under the symmetry $\phi_i \to \phi_i + \lambda$ and a phonon-like behavior may be expected.

This tensorial case, especially when the distribution of the $x$'s is related to the potential $V$, is the most interesting from the physical point of view (especially for its mobility edge [16, 15]). Here we stick to the much simpler question of the computation of the spectrum of $M$. We shall develop a field theory for this problem, check it at high and low densities, and use a Hartree type method.

Our aim is to get both a qualitative understanding of the main properties of the spectrum and of the eigenvalues and a quantitate, as accurate as possible, analytic evaluation of these properties. Quantitative accurate results are also needed because the computation of the spectral density is a crucial step in the microscopic computation of the thermodynamic and of the dynamic properties of glass forming systems. In some sense this approach can be considered as an alternative route to obtain mode-coupling like results [17], the main difference being that in the mode coupling approach one uses a coarse grained approach and the hydrodynamical equations, while here we take a fully microscopic point of view.

Many variations can be done on the distribution of points, each one has its distinctive flavor:

- The distribution of the points $x$ is flat and the points are uncorrelated: they are uniformly distributed in a cube of size $L$ and their number is $N = \rho L^3$. Here, as in the following cases, we are interested in the thermodynamic limit where both $L$ and $N$ go to infinity at fixed $\rho$.
- The points $x$ are distributed with a distribution that is proportional to $\exp(-\beta H[x])$ where $H[x]$ is a given function.
- The points $x$ are one of the many solutions of the equation $\partial H/\partial x_i = 0$. This last problem may be generalized by weighting each stationary point of $H$ with an appropriate weight.

The last two cases are particularly interesting when

$$M_{i,k} = \frac{\partial^2 H}{\partial x_i x_k}, \tag{9}$$

and consequently

$$\mathcal{H}[\phi] = \frac{1}{2}\sum_{i,k} \phi_i \phi_k \frac{\partial^2 H}{\partial x_i x_k}. \tag{10}$$

If this happens the distribution of points and the matrix are related and the theoretical challenge is to use this relation in an effective way.

In the second section of these lectures, after the introduction we will present the basic definition of the spectrum, Green functions, structure functions and related quantities. In the third section we will discuss the physical motivation for this study and we will present a first introduction to the replica method. In the fourth section we will give a few example how a field theory can be constructed in such way that it describes the properties of randomly distributed points. In the fifth section we will discuss in details the simples possible model for random Euclidean matrices, presenting both the analytic approach and some numerical simulations. In the sixth section we shall present a similar analysis in a more complex and more physically relevant case, where phonons are present due to translational invariance. Finally, in the last section we present some recent results that have been obtained in the case of correlated points.

## 2. Basic definitions

The basic object that we will calculate are the matrix element of the resolvent

$$G[x](z)_{i,j} = \left(\frac{1}{z - M[x]}\right)_{ij}, \tag{11}$$

(we use the notation $M[x]$ in order to stress that the matrix $M$ depends on all the points $[x]$. Sometimes we will suppress the specification "$[x]$", where it is obvious. We can define the sample dependent Green function

$$G[x](y_1, y_2, z) = \sum_{i,j} \delta(y_1 - x_i)\delta(y_2 - x_j)G[x](z)_{i,j} = \sum_{i,j} \delta(y_1 - x_i)\delta(y_2 - x_j)\left(\frac{1}{z - M[x]}\right)_{ij} . \tag{12}$$

The quantity that we are interested to compute analytically are the sample averages of the Green function, i.e.

$$\overline{G}(y_1, y_2, z) = \overline{G[x](y_1, y_2, z)} , \tag{13}$$

where the overline denotes the average over the random position of the points.

If the problem is translational invariant, after the average on the positions of the points $[x]$ we have

$$\overline{G}(y_1, y_2, z) = \overline{G}(y_1 - y_2, z) , \tag{14}$$

where the function $\overline{G}(x, z)$ is smooth apart from a delta function at $x = 0$. It is convenient to consider the Fourier transform of $\overline{G}(x, z)$, i.e.

$$G(p, z) = \frac{1}{N}\overline{\sum_{ij} e^{ip(x_i - x_j)}\left[\frac{1}{z - M}\right]_{ij}} = \frac{1}{N}\overline{\int dxdy e^{ip(x-y)} G[x](x, y, z)} = \int dx e^{ipx} \tilde{G}(x, z) , \tag{15}$$

The computation of the function $G(p, z)$ will one of the main goals of these lectures. It is convenient to introduce the so called dynamical structure factor (that in many case can be observed experimentally) that here is defined as [1]:

$$S_e(p, E) = -\frac{1}{\pi} \lim_{\eta \to 0^+} G(p, E + i\eta) . \tag{16}$$

One must remember that the structure function is usually defined using as variable $\omega = \sqrt{E}$:

$$S(p, \omega) = 2\omega S_e(p, \omega^2) , \tag{17}$$

The resolvent will also give us access to the density of states of the system:

$$\begin{aligned} g_e(E) &= -\frac{1}{N\pi} \lim_{\eta \to 0^+} \overline{\sum_{i=1}^{N} \left[\frac{1}{E + i\eta - M}\right]_{ii}} , \\ &= -\frac{1}{\pi} \lim_{\eta \to 0^+} \lim_{p \to \infty} G(p, E + i\eta) . \end{aligned} \tag{18}$$

# 3. Physical motivations

There are many physical motivations for studying these problems, apart from the purely mathematical interest. Of course different applications will lead to different forms of the problems. In this notes I will concentrate my attention on the simplest models, skipping the complications (that are not well understood) of more complex and interesting models.

## 3.1 Impurities

We can consider a model where there are some impurities in a solid that are localized at the points $x$'s. There are many physical implementation of the same model; here we only present two cases:

- There may be electrons on the impurities and the amplitude for hopping from an impurity ($i$) to an other impurity ($k$) is proportional to $f(x_i - x_k)$. The electron density is low, so that the electron-electron interaction can be neglected.

- There is a local variable $\phi_i$ associated to the impurity (e.g. the magnetization) and the effective Hamiltonian at small magnetizations is given by eq.(5).

In both cases it is clear that the physical behavior is related to the properties of the matrix $M$ that is of the form discussed in the introduction. If the concentration of the impurities is small (and the impurities are embedded in an amorphous system) the positions of the impurities may be considered randomly distributed in space.

## 3.2 Random walks

Let us assume that there are random points in the space and that there is a particle that hops from one point to the other with probability per unit time given by $f(x_i - x_k)$.

The evolution equation for the probabilities of finding a particle at the point $i$ is given by

$$\frac{dP_i}{dt} = \sum_k f(x_i - x_k)P_k - P_i \sum_k f(x_i - x_k) = -\sum_k M_{i,k} P_k \,, \qquad (19)$$

where the matrix $M$ is given by eq. (2).

Let us call $P(x,t)$ the probability that a particle starting from a random point at time 0 arrives at at time $t$. It is evident that after averaging of the different systems (or in a large system after averaging over the random starting points) we have that

$$\int_0^\infty dt \exp(-tz) \int dx \exp(ipx) P(x,t) = -G(p,-z)\,, \tag{20}$$

so that the function $P(x,t)$ can be reconstructed from the knowledge of the function $G(p,z)$ defined in eq. (15).

## 3.3 Instantaneous modes

We suppose that the points $x$'s are distributed according to the probability distribution

$$\exp(-\beta H[x])\,, \tag{21}$$

where $H[x]$ is for example a two body interaction given by eq.(7) and $\beta = 1/(kT)$.

The eigenvectors of the Hessian of the Hamiltonian (see eq.(8)) are called instantaneous normal modes (INN). The behavior of the INN at low temperature, especially for systems that have a glass transition [20, 19], is very interesting [21]. In particular there have been many conjectures that relate the properties of the spectrum of INN to the dynamic of the system when approaching the glass transition.

However it has been realized in these years that much more interesting quantities are the instantaneous saddle modes (ISN) [22, 23, 25, 24]. Given an equilibrium configuration $x_e$ at temperature $T$ one defined the inherent saddle $x_s$ corresponding to that configuration $x_e$ in the following way. We consider all the saddles of the Hamiltonian, i.e. all the solutions of the equations

$$\frac{\partial H}{\partial x_i} = 0\,. \tag{22}$$

The inherent saddle $x_s$ is the nearest saddle to the configuration $x_e$. The Boltzmann distribution at temperature $T$ induces a probability distribution on the space of all saddles. The physical interesting quantities are the properties of the Hessian around these saddles. It turns out that they have a very interesting behavior near the glass transition in the case of glass forming materials.

It is clear that the computation of the spectrum in these case is much more complex: we have both to be able to compute the correlations among the points in this non-trivial case and to take care of the (often only partially known) correlations.

A possible variation on the same theme consists in consider the ensemble of all saddles of given energy $E = H[x]$: there are indications that also in this case there may be a phase transition in the properties of the eigenvalues and eigenvectors when $E$ changes and this phase transition may be correlated to some experimentally measured properties of glassy systems, i.e. the Boson Peak, as we will see later.

# 4. Field theory

## 4.1 Replicated field theory

The resolvent $G(p, z)$ defined in eq.(15) can be written as a propagator of a Gaussian theory:

$$\begin{aligned} \mathcal{Z}[J] &\equiv \int \prod_k d\phi_k \exp\left(-\frac{1}{2}\sum_{lm} \phi_l\,(z-M)_{lm}\,\phi_m + \sum_l \phi_l J_l\right), \\ G(p,z) &= \frac{1}{N}\sum_{ij} \frac{\delta^2}{\delta J_i \delta J_j}\, \overline{\exp(ip(x_i - x_j)) \ln \mathcal{Z}[J]}\bigg|_{J=0}, \end{aligned} \tag{23}$$

where the overline denotes the average over the distribution $P[x]$ of the positions of the particles. We have been sloppy in writing the Gaussian integrals (that in general are not convergent): one should suppose that the $\phi$ integrals go from $-a\infty$ to $+a\infty$ where $a^2 = -i$ and $z$ has a positive (albeit infinitesimal) imaginary part in order to make the integrals convergent [16, 18]. This choice of the contour is crucial for obtaining the non compact symmetry group for localization [18], but it is not important for the density of the states.

In other words, neglecting the imaginary factors, we can introduce a probability distribution over the field $\phi_i$,

$$P[\phi] \propto \prod_k d\phi_k \exp\left(-\frac{1}{2}\sum_{lm} \phi_l\,(z-M)_{lm}\,\phi_m\right). \tag{24}$$

If we denote by $\langle\cdot\rangle$ the average with respect to this probability (please notice that here everything depends on $[x]$ also if this dependence is not explicitly written) we obtain the simple result:

$$\langle \phi_i \phi_k \rangle = G_{i,k}\,. \tag{25}$$

A problem with this approach is that the normalization factor of the probability depends on $[x]$ and it is not simple to get the $x$-averages of the expectation values. Replicas are very useful in this case. They are an extremely efficient way to put the dirty under a simple carpet (on could also use the supersymmetric approach of Efetov where one puts the same dirty under a more complex carpet [16]).

For example the logarithm in eq.(23) is best dealt with using the replica trick [1, 2, 25]:

$$\ln \mathcal{Z}[J] = \lim_{n\to 0} \frac{1}{n}\left(\mathcal{Z}^n[J] - 1\right). \tag{26}$$

The resolvent can then be computed from the $n$-th power of $\mathcal{Z}$, that can be written using $n$ *replicas*, $\phi_i^{(a)}$ ($a = 1, 2, \ldots, n$) of the Gaussian variables of eq.(23):

$$G(p,z) = \lim_{n\to 0} G^{(n)}(p,z)\,,$$

$$NG^{(n)}(p,z) = \tag{27}$$

$$\sum_{ij} \exp(ip(x_i - x_j)) \int \prod_{k,a} d\phi_k^{(a)}\, \phi_i^{(1)} \phi_j^{(1)} \exp\left( -\frac{1}{2} \sum_{l,m,c} \phi_l^{(c)} (z-M)_{lm} \phi_m^{(c)} \right).$$

Indeed

$$\int \prod_{k,a} d\phi_k^{(a)}\, \phi_i^{(a)} \phi_j^{(b)} \exp\left( -\frac{1}{2} \sum_{l,m,c} \phi_l^{(c)} (z-M)_{lm} \phi_m^{(c)} \right) \propto \delta_{a,b} =$$
$$\left[\frac{1}{z-M}\right]_{i,j} \det(z-M)^{-n/2} \tag{28}$$

and the physically interesting case is obtained only when $n = 0$.

In this way one obtains a $O(n)$ symmetric field theory. It well known that an $O(n)$ symmetric theory is equivalent to an other theory invariant under the $O(n+2|1)$ symmetry ([26]), where this group is defined as the one that leaves invariant the quantity

$$\sum_{a=1,n+2} (\phi^a)^2 + \bar{\psi}\psi\,, \tag{29}$$

where $\psi$ is a Fermionic field. Those who hate analytic continuations, can us the group $O(2|1)$ at the place of the more fancy $O(0)$ that is defined only as the analytic continuation to $n = 0$ of the more familiar $O(n)$ groups.

Up to this point everything is quite general. We still have to do the average over the random positions. This can be done in an explicit or compact way in some of the previous mentioned cases and one obtains a field theory that can be studied using the appropriate approximations. Before doing this it is may be convenient to recall a simpler result, i.e. how to write a field theory for the partition function of a fluid.

## 4.2 The partition function of a Duid

Let us consider a system with $N$ (variable) particles characterized by a classical Hamiltonian $H_N[x]$ where the variable $x$ denote the $N$ positions. In the simplest case the particles can move only in a finite dimensional region (a box $B$ of volume $V_B$) and they have only two body interactions:

$$H_N[x] = \frac{1}{2} \sum_{i,k} V(x_i - x_k)\,. \tag{30}$$

At given $\beta$ the canonical partition function can be written as

$$\mathcal{Q}_N = \int_B dx_1 \dots dx_N \exp(-\beta H_N[x]) \,, \tag{31}$$

while the gran-canonical partition function is given by

$$\mathcal{Q}(\zeta) = \sum_N \frac{\zeta^N}{N!} \mathcal{Q}_N \,, \tag{32}$$

where $\zeta$ is the fugacity.

We aim now to find out a field theoretical representation of the $\mathcal{Q}(\zeta)$. This can be formally done [28] (neglecting the problems related to the convergence of Gaussian integrals) by writing

$$\exp\left(-\frac{\beta}{2}\sum_{i,k} V(x_i - x_k)\right) =$$
$$\mathcal{N}^{-1} \int d\sigma \exp\left(\frac{1}{2\beta}\int dxdy V^{-1}(x-y)\sigma(x)\sigma(y) + \sum_i \sigma(x_i)\right) \,, \tag{33}$$

where $\mathcal{N}$ is an appropriate normalization factor such that

$$\mathcal{N}^{-1} \int d\sigma \exp\left(\frac{1}{2\beta}\int dxdy V^{-1}(x-y)\sigma(x)\sigma(y) + \int J(x)\sigma(x)\right) =$$
$$\exp\left(-\int dxdy \beta V(x-y) J(x) J(y)\right) . \tag{34}$$

The reader should notice that $V^{-1}(x-y)$ is formally defined by the relation

$$\int dz V^{-1}(x-z) V(z-y) = \delta(x-y) \,. \tag{35}$$

The relation eq.(33) trivially follows from the previous equations if we put

$$J(x) = \sum_i \delta(x - x_i) \,. \tag{36}$$

It is convenient to us the notation

$$d\mu[\sigma] \equiv \mathcal{N}^{-1} d\sigma \exp\left(\frac{1}{2\beta}\int dxdy V^{-1}(x-y)\sigma(x)\sigma(y)\right) . \tag{37}$$

With this notation we find

$$\mathcal{Q}_N = \int d\mu[\sigma] \left(\int dx \exp(\sigma(x))\right)^N . \tag{38}$$

We finally get

$$\mathcal{Q}(\zeta) = \int d\mu[\sigma] \exp\left(\zeta \int dx \exp(\sigma(x))\right) . \tag{39}$$

People addict with field theory can use this compact representation in order to derive the virial expansion, i.e. the expansion of the partition in powers of the fugacity at fixed $\beta$.

The same result can be obtained in a more straightforward way [29] by using a functional integral representation for the delta function:

$$\exp\left(-\frac{\beta}{2}\sum_{i,k} V(x_i - x_k)\right) =$$

$$\int d[\rho]\delta_F\left[\rho(x) - \sum_{i=1,N} \delta(x - x_i)\right] \exp\left(-\frac{\beta}{2}\int dxdy\rho(x)\rho(y)V(x-y)\right) , \tag{40}$$

where $\delta_F$ stands for a functional Dirac delta:

$$\delta_F[f] = \int d\sigma \exp\left(i \int dx f(x)\lambda(x)\right) . \tag{41}$$

We thus find

$$\mathcal{Q}_N = \int dx_1 dx_N \int d[\rho]d[\lambda] \tag{42}$$

$$\exp\left(-\frac{\beta}{2}\int dxdy\rho(x)\rho(y)V(x-y) + i\lambda(x)\rho(x) - i\sum_{i=1,N}\lambda(x_i)\right)$$

$$= \int d[\rho]d[\lambda] \exp\left(-\frac{\beta}{2}\int dxdy\rho(x)\rho(y)V(x-y) + i\lambda(x)\rho(x)\right)$$

$$\left(\int dx \exp(i\lambda(x))\right)^N .$$

At the end of the day we get

$$\mathcal{Q}(\zeta) = \int d[\lambda] \exp\left(-\frac{\beta}{2}\int dxdy\lambda(x)\lambda(y)V^{-1}(x-y) + \zeta\int dx \exp(i\lambda(x)\right). \tag{43}$$

where, by doing the Gaussian integral over the $\rho$, we have recover the previous formula eq.(33), if we set $i\lambda(x) = \sigma(x)$.

In order to compute the density and its correlations it is convenient to introduce an external field

$$\sum_i U(x_i) . \tag{44}$$

This field is useful because its derivatives of the partition function give the density correlations. As before the partition function is given by

$$\mathcal{Q}(\zeta|U) = \int d\mu[\sigma] \exp\left(\zeta \int dx \exp(\sigma(x) + \beta U(x))\right) . \tag{45}$$

In this way one finds that the density and the two particle correlations are given by:

$$\rho = \langle \sum_i \delta(x - x_i) \rangle = \zeta \langle \exp(\sigma(0)) \rangle$$
$$C(x) = \langle \sum_{i,k} \delta(x_i - 0)\delta(x_k - x) \rangle = \zeta^2 \langle \exp(\sigma(0) + \sigma(x)) \rangle . \tag{46}$$

In the low density limit (i.e. $\zeta$ near to zero) one finds

$$\rho = \zeta \exp(-\beta V(0)) ,$$
$$C(x) = \rho^2 \exp(-\beta V(x)) . \tag{47}$$

that is the starting point of the virial expansion.

## 5. The simplest case

In this section we shall be concerned with simple case for random Euclidean matrices, where the positions of the particles are completely random, chosen with a probability $P(x) = \rho/V_B$.

We will consider here the simplest case, where

$$M_{i,k} = f(x_i - x_k) \tag{48}$$

and $f(x)$ is bounded, fast decreasing function at infinity (e.g. $\exp(-x^2/2)$).

We shall study the field-theory perturbatively in the inverse density of particles, $1/\rho$. The zero$^{th}$ order of this expansion (to be calculated in subsection 5.1) corresponds to the limit of infinite density, where the system is equivalent to an elastic medium. In this limit the resolvent (neglecting $p$ independent terms) is extremely simple:

$$G(p, z) = \frac{1}{z - \rho\tilde{f}(p)} . \tag{49}$$

In the above expression $\tilde{f}(p)$ is the Fourier transform of the function $f$, that due to its spherical symmetry, is a function of only $(p^2)$ . We see that the dynamical structure function has two delta functions at frequencies $\omega = \pm\sqrt{\rho\tilde{f}(p)}$.

It is then clear that Eq.(49) represents the undamped propagation of plane waves in our harmonic medium, with a dispersion relation controlled by the function $\tilde{f}$. The high order corrections to Eq.(49) (that vanishes as $1/\rho$) will be calculated later. They take the form of a complex self-energy, $\Sigma(p, z)$, yielding

$$G(p,z) = \frac{1}{z - \epsilon(p) - \Sigma(p,z)}, \tag{50}$$

$$S_E(p,E) = -\frac{1}{\pi} \frac{Im\Sigma(p,E)}{(E - \epsilon(p) - Re\Sigma(p,E))^2 + (Im\Sigma(p,E))^2}. \tag{51}$$

The dynamical structure factor, is no longer a delta function, but close to its maxima it has a Lorentzian shape. From eq. (51) we see that the real part of the self-energy renormalizes the dispersion relation. The width of the spectral line is instead controlled by the imaginary part of the self-energy.

Subsection Field theory Let us first consider the spectrum: it can be computed from the imaginary part of the trace of the resolvent:

$$R(z) = \frac{1}{N} \overline{\mathrm{Tr}\frac{1}{z - M}}, \tag{52}$$

where the overline denotes the average over the positions $x_i$.

It is possible to compute the resolvent from a field theory written using a replica approach. We shall compute $\Xi_N \equiv \overline{\det(z - M)^{-n/2}}$, and deduce from it the resolvent by using the replica limit $n \to 0$.

It is easy to show that one can write $\Xi_N$ as a partition function over replicated fields $\phi_i^a$, where $i \in \{1...N\}$, $a \in \{1...n\}$:

$$\Xi_N = \int \prod_{i=1}^{N} \frac{dx_i}{V} \int \prod_{i=1}^{N} \prod_{a=1}^{n} d\phi_i^a$$

$$\exp\left( -\frac{z}{2} \sum_{i,a} (\phi_i^a)^2 + \frac{1}{2} \sum_{i,j,a} f(x_i - x_j) \phi_i^a \phi_j^a \right). \tag{53}$$

In order to simplify the previous equations let us introduce the Bosonic [2] fields $\psi_a(x) = \sum_{i=1}^{N} \phi_i^a \delta(x - x_i)$ together with their respective Lagrange multiplier fields $\hat{\psi}_a(x)$. More precisely we introduce a functional delta function:

$$\delta_F\left(\psi_a(x) - \sum_{i=1}^{N} \phi_i^a \delta(x - x_i)\right) =$$

$$\int d[\hat{\psi}_a] \exp\left( i \int dx \hat{\psi}_a(x) \left( \psi_a(x) - \sum_{i=1}^{N} \phi_i^a \delta(x - x_i) \right) \right). \tag{54}$$

One can integrate out the $\phi$ variables, leading to the following field theory for $\Xi_N$, where we have neglected all the $z$ independent factors that disappear when

$n = 0$:

$$\Xi_N = \int D[\psi_a, \hat{\psi}_a] A^N \exp(S_0) \tag{55}$$

where

$$\begin{aligned} S_0 &= i \sum_a \int dx \, \hat{\psi}_a(x) \psi_a(x) + \frac{1}{2} \sum_a \int dx dy \, \psi^a(x) f(x-y) \psi^a(y) \, , \\ A &= \exp\left[ -\frac{1}{2z} \sum_a \hat{\psi}^a(x)^2 \right] \, . \end{aligned} \tag{56}$$

It is convenient to go to a grand canonical formulation for the disorder: we consider an ensemble of samples with varying number of $(N)$, and compute the grand canonical partition function $\mathcal{Z}(\zeta) \equiv \sum_{N=0}^{\infty} \Xi_N \zeta^N / N!$ that is equal to:

$$\mathcal{Z} = \int D[\psi_a, \hat{\psi}_a] \exp(S_0 + \zeta A). \tag{57}$$

In the $n \to$ limit, one finds that $\rho = \zeta$ . The definition of the field $\psi$ implies that the correlation is given (at $x \neq 0$) by:

$$\overline{G}(x,z) \equiv \sum_{i,k} \delta(x - x_i) \delta(x_j) \left( \frac{1}{z - M} \right)_{i,k} = \langle \psi_a(x) \psi_a(0) \rangle \tag{58}$$

A simple computation (taking care also of the contribution in $x = 0$) gives

$$G(p,z) = \frac{1}{\rho \tilde{f}(p)} - \lim_{n \to 0} \frac{1}{\rho \tilde{f}^2(p)} \int d^d x d^d y \, \exp(ip(x-y)) \, \langle \hat{\psi}^{(1)}(x) \hat{\psi}^{(1)}(y) \rangle \, . \tag{59}$$

so that the average Green function can be recovered from the knowledge of the $\hat{\psi}$ propagator.

Notice that we can also integrate out the $\psi$ field thus replacing $S_0$ by $S_0'$, where

$$S_0' = \frac{1}{2} \sum_a \int dx dy \, \hat{\psi}^a(x) f^{-1}(x-y) \hat{\psi}^a(y), \tag{60}$$

where $f^{-1}$ is the integral operator that is the inverse of $f$.

The expression (57) is our basic field theory representation. We shall denote by brackets the expectation value of any observable with the action $S_0 + S_1$. As usual with the replica method we have traded the disorder for an interacting replicated system.

It may be convenient to present a different derivation of the previous result: we write

$$\exp\left(-\frac{z}{2}\sum_{i,a}(\phi_i^a)^2+\frac{1}{2}\sum_{i,j,a}f(x_i-x_j)\phi_i^a\phi_j^a\right)=$$

$$\int d[\omega]\exp\left(-\frac{1}{2}\sum_a\int dxdy\;\omega(x)_a\omega(y)_a f^{-1}(x-y)+\sum_{i,a}\left(\omega(x_i)_a\phi(x_i)_i^a-\frac{z}{2}(\phi_i^a)^2\right)\right). \tag{61}$$

If we collect all the terms at a given site $i$, we get

$$\int\prod_a d\phi_i^a\exp\left(\sum_a\left(\omega(x_i)_a\phi(x_i)_i^a-\frac{z}{2}(\phi_i^a)^2\right)\right)\,. \tag{62}$$

The $\phi$ integrals are Gaussian and they can be done: for $n=0$ one remains with

$$\exp\left(\sum_a\omega(x_i)_a^2/z\right) \tag{63}$$

and one recovers easily the previous result eq.(56).

The basic properties of the field theory are related to the properties of the original problem in a straightforward way. We have seen that the average number of particles is related to $\zeta$ through $N=\zeta V\langle A\rangle$, so that one gets $\zeta=\rho$ in the $n\longrightarrow 0$ limit because $\langle A\rangle=1$. From the generalized partition function $\mathcal{Z}$, one can get the resolvent $R(z)$ through:

$$R(z)=-\lim_{n\to 0}\frac{2}{nN}\frac{\partial\log\mathcal{Z}}{\partial z}\,. \tag{64}$$

## 5.1 High density expansion

Let us first show how this field theory can be used to derive a high density expansion.

At this end it is convenient to rescale $z$ as $z=\rho\hat{z}$ and to study the limit $\rho\to\infty$ at fixed $\hat{z}$. Neglecting addictive constants, the action, $S$, can be expanded as:

$$\frac{1}{2}\sum_a\int dxdy\;\hat{\psi}^a(x)f^{-1}(x-y)\hat{\psi}^a(y)-\frac{1}{2\hat{z}}\int dx\sum_a\hat{\psi}_a(x)^2+O(1/(\rho\hat{z}^2)\,, \tag{65}$$

Gathering the various contributions and taking care of the addictive constants, one gets for the resolvent:

$$\rho R(z)=\frac{1}{z}+\frac{1}{\rho}\int dk\left(\frac{1}{(\hat{z}-\tilde{f}(k))}-\frac{1}{\hat{z}}\right)\,. \tag{66}$$

One can study with this method the eigenvalue density for eigenvalues $|\lambda| \sim O(\rho)$, by setting $\hat{z} = \lambda + i\epsilon$ and computing the imaginary part of the resolvent in the small $\epsilon$ limit. For $\rho \to \infty$ the leading term gives a trivial result for the eigenvalue density $g(\lambda)$ i.e.

$$g(\lambda) = \delta(\lambda) \,. \tag{67}$$

Including the leading large $\rho$ correction that we have just computed, we find that $g(\lambda)$ develops, away from the peak at $\lambda \sim 0$, a component of the form:

$$g(\lambda) \sim \frac{1}{\rho} \int dk \delta(\lambda - \tilde{f}(k)) \,. \tag{68}$$

The continuos part of the spectrum can also be derived from the following simple argument. We suppose that the eigenvalue $\omega_i$ is a smooth function of $x_i$. We can thus write:

$$\sum_k f(x - x_k)\omega(x_k) \approx \rho \int dy f(x-y)\omega(y) = \rho\lambda\omega(x) \tag{69}$$

and the eigenvalues are the same of the integral operator with kernel $f$.

This argument holds if the discrete sum in eq. (69) samples correctly the continuous integral. This will be the case only when the density $\rho$ is large enough that the function $\omega(x)$ doesn't oscillate too much from one point $x_j$ to a neighboring one. This condition in momentum space imposes that the spatial frequency $|k|$ be small enough: $|k| \ll \rho^{1/d}$ ( $\rho^{1/d}$ is the inverse of the average interparticle distance).

The same condition is present in the field theory derivation. We assume that $\tilde{f}(k)$ decreases at large $k$, and we call $k_M$ the typical range of $k$ below which $\tilde{f}(k)$ can be neglected. Let us consider the corrections of order $\rho^{-1}$ in eq. (66). It is clear that, provided $\hat{z}$ is away from 0, the ratio of the correction term to the leading one is of order $k_M^d \rho^{-1}$, and the condition that the correction be small is just identical to the previous one. The large density corrections near to the peak $z = 0$ cannot be studied with this method.

In conclusion the large $\rho$ expansion gives reasonable results at non-zero $z$ but it does not produce a well behaved spectrum around $z = 0$. In the next sections we shall see how to do it.

The reader may wonder if there is a low density expansion: the answer is positive [13], however it is more complex than the high density expansion and it cannot be discussed in these notes for lack of space.

## 5.2 A direct approach

Let us compute directly what happens in the region where $\rho$ is very large. We first make a very general observation: if the points $x$ are random and uncorrelated we have that

$$\sum_i A(x_i) = \rho \int dx A(x) \,,$$
$$\sum_{i,k} B(x_i, x_k) = \sum_{i,k;i\neq k} B(x_i, x_k) + \sum_i B(x_i, x_i) =$$
$$\rho^2 \int dx dy B(x,y) + \rho \int dx B(x,x) \tag{70}$$

When the density is very high the second term can be neglected and we can approximate multiple sum with multiple integrals, neglecting the contribution coming from coinciding points [2, 3, 7].

We can apply these ideas to the high density expansion of the Green function in momentum space. Using a simple Taylor expansion in $1/z$ we can write:

$$G(p,z) = \frac{1}{z}\sum_R (-z)^{-R}\overline{M^R(p)} \tag{71}$$

where for example

$$M^3(p) = N^{-1} \sum_{k_0,k_1 k_2,k_3} f(x_{k_0} - x_{k_1}) f(x_{k_1} - x_{k_2}) f(x_{k_2} - x_{k_3}) \exp(ip(x_{k_0} - x_{k_3}) \tag{72}$$

The contribution of all different points gives the large $\rho$ limit, and taking care of the contribution of pairs of points that are not equal gives the subleading corrections. In principle everything can be computed with this kind of approach that is more explicit than the field theory, although the combinatorics may become rather difficult for complex diagrams and the number of contributions become quite large when one consider higher order corrections [6, 7].

Generally speaking the leading contribution involve only convolutions and is trivial in momentum space, while at the order $1/rho^m$, there are $m$ loops for diagrams that have $m$ intersections: the high $\rho$ expansion is also a loop expansion.

## 5.3 A variational approximation

In order to elaborate a general approximation for the spectrum, that should produce sensible results for the whole spectrum, we can use a standard Gaussian variational approximation in the field theory representation, that, depending on the context, appears under various names in the literature, like Hartree-Fock approximation the Random Phase Approximation or CPA.

Here we show the implementation in this particular case. By changing $\hat{\psi} \to i\hat{\psi}$ in the representation for the partition function, we obtain:

$$\mathcal{Z} = \int D[\hat{\psi}^a] \exp(S)) \,, \tag{73}$$

where

$$S = -\frac{1}{2}\int dxdy \sum_a \hat{\psi}^a(x) f^{-1}(x,y)\hat{\psi}^a(y)$$
$$+\rho z^{-n/2}\int dx \exp\left(\frac{1}{2z}\sum_a \hat{\psi}^a(x)^2\right) . \tag{74}$$

We look for the best quadratic action

$$S_v = -(1/2)\sum_{ab}\int dxdy G^{-1}_{ab}(x,y)\hat{\psi}^a(x)\hat{\psi}^b(y) \tag{75}$$

that approximates the full interacting problem.

This procedure is well known in the literature and it gives the following result. If we have a field theory with only one field and with action

$$\int dp D(p)\phi(p)^2 + \int dx V(\phi(x)), \tag{76}$$

the propagator is given by

$$G(p) = \frac{1}{D(p)+\Sigma}, \tag{77}$$

where

$$\Sigma = \langle V''(\phi)\rangle_G . \tag{78}$$

In other words $\Sigma$ is a momentum independent self-energy that is equal to the expectation value of $V''(\phi)$ in theory where the field $\phi$ is Gaussian and has a propagator equal to $G$.

In the present case, there are some extra complications due to the presence of indices; the appropriate form of the propagator is $G_{ab}(p) = \delta_{ab}G(p)$. After same easy computations, one finds that $\tilde{G}(p)$ satisfies the self consistency equation:

$$\tilde{G}(p) = \frac{1}{f^{-1}(p)-\Sigma}, \quad \Sigma = -\frac{\rho}{z-\int dk\tilde{G}(k)} \tag{79}$$

and the resolvent is given by:

$$R(z) = \frac{1}{z-\int dk\tilde{G}(k)} . \tag{80}$$

Formulas (79,80) provide a closed set of equations that allow us to compute the Gaussian variational approximation to the spectrum for any values of $z$ and the density; if the integral would be dominated by a single momentum, we would recover the usual Dyson semi-circle distribution for fully random matrices . In sect. 5.4 we shall compare this approximation to some numerical estimates of the spectrum for various functions $f$ and densities. The variational approximation correspond to the computation of the sum of the so called tadpole diagrams, that can be often done a compact way.

The main advantage of the variational approximation is that the singularity of the spectrum at $z = 0$ is removed and the spectrum starts to have a more reasonable shape.

Another partial resummation of the $\rho$ expansion can also be done in the following way: if one neglects the triangular-like correlations between the distances of the points (an approximation that becomes correct in large dimensions, provided the function $f$ is rescaled properly with dimension), the problem maps onto that of a diluted random matrix with independent elements. This problem can be studied explicitly using the methods of [30, 31]. It leads to integral equations that can be solved numerically. The equations one gets correspond to the first order in the virial expansion, where one introduces as variational parameter the local probability distribution of the field $\phi$ [32]. The discussion of this point is interesting and it is connected to the low-density expansion: however it cannot be done here.

An other very interesting point that we will ignore is the computation of the tails in the probability distribution of the spectrum representation leading to compact expressions for the behavior of the tail [33].

## 5.4 Some numerical experiments

In this section we present the result of some numerical experiments and the comparison with the theoretical results (we also slightly change the notation, i.e. we set $z = \lambda + i\epsilon$.

We first remark that for a function $f(x)$ that has a positive Fourier transform $\tilde{f}(k)$, the spectrum is concentrated an the positive axis. Indeed, if we call $\omega_i$ a normalized eigenvector of the matrix $M$ defined in (48), with eigenvalue $\lambda$, one has:

$$\sum_{ij} \omega_i f(x_i - x_j) \omega_j = \lambda \,, \tag{81}$$

and the positivity of the Fourier transform of $f$ implies that $\lambda \geq 0$.

We have studied numerically the problem in dimension $d = 3$ with the Gaussian function $f(x) = (2\pi)^{-3/2} \exp(-x^2/2)$. In this Gaussian case the high density approximation gives a spectrum

$$g(\lambda) \sim \frac{1}{\rho\pi^2} \frac{1}{\lambda} \left( \frac{1}{2} \log \frac{\rho}{\lambda} \right)^{1/2} \theta(\rho - \lambda) + C\delta(\lambda) \tag{82}$$

Notice that this spectrum is supposed to hold away from the small $\lambda$ peak, and in fact it is not normalizable at small $\lambda$.

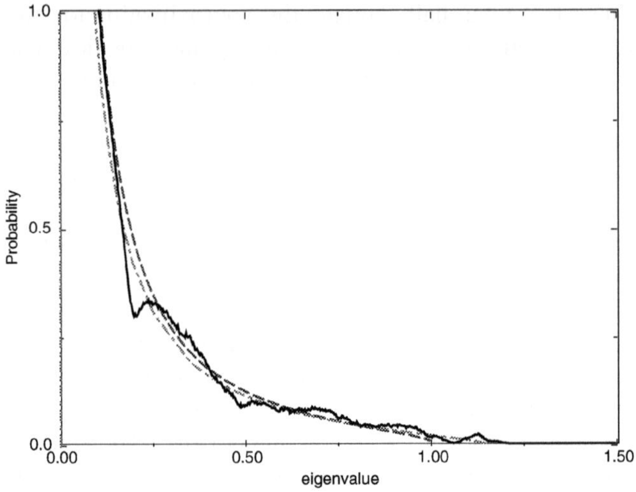


*Figure 1.* Density of eigenvalues of a Euclidean Random Matrix in three dimensions, density $\rho = 1$. The function $f$ is $f(x) = (2\pi)^{-3/2}\exp(-x^2/2)$, and the matrix is defined from eq.(48). The full line is the result of a numerical simulation with $N = 800$ points, averaged over 100 samples. The dashed line is the result from the high density expansion. The dash-dotted line is the result from the Gaussian variational approximation (RPA) to the field theory.

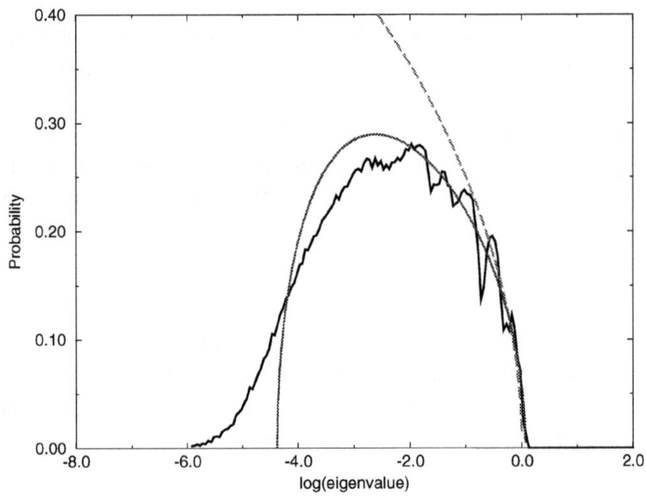


*Figure 2.* Density of the logarithm (in base 10) of the eigenvalues of a Euclidean Random Matrix in three dimensions, density $\rho = 1$ (same data of the previous figure). The dashed line is the result from the high density expansion. The dotted line is the result from the Gaussian variational approximation (RPA) to the field theory.

It is possible to do a variational approximation computation taking care of all the non-linear terms in the action for the fields. This corresponds (as usually) a a resummation of a selected class of diagrams. After some computations, one finds that one needs to solve, given $z = \lambda - i\epsilon$, the following equations for $C(z) \equiv a(z) + ib(z)$:

$$\begin{aligned} \lambda &= \rho \frac{a}{a^2+b^2} + \frac{1}{2\pi^2} \int_0^\infty k^2 dk \frac{e^{k^2/2} - u}{\left(e^{k^2/2} - a\right)^2 + b^2} \\ \epsilon &= \rho \frac{b}{a^2+b^2} - \frac{b}{2\pi^2} \int_0^\infty k^2 dk \frac{1}{\left(e^{k^2/2} - a\right)^2 + b^2} . \end{aligned} \tag{83}$$

One needs to find a solution in the limit where $\epsilon \to 0$.

In fig. (1,2), we plot the obtained spectrum, averaged over 100 realizations, for $N = 800$ points at density $\rho = 1$ (We checked that with a different number of points the spectrum is similar). Also shown are the high density approximation (82), and the result from the variational approximation. We see from fig.(1) that the part of the spectrum $\lambda \in [0.2, 1.5]$ is rather well reproduced from both approximations, although the variational method does a better job at matching the upper edge. On the other hand the probability distribution of the logarithm of the eigenvalues (fig.2) makes it clear that the high density approximation is not valid at small eigenvalues, while the variational approximation gives a sensible result. One drawback of the variational approximation, though, is that it always produces sharp bands with a square root singularity, in contrast to the tails that are seen numerically.

In fig.3, we plot the obtained spectrum, averaged over 200 realizations, for $N = 800$ points at density $\rho = 0.1$. We show also a low density approximation (that we do not describe here [2]), and the result from the variational approximation. We see from fig.(3) that this value of $\rho = 0.1$ is more in the low density regime, and in particular there exists a peak around $\lambda = f(0)$ due to the isolated clusters containing small number of points. The variational approximation gives the main orders of magnitude of the distribution, but it is not able to reproduce the details of the spectrum, in particular the peak due to small clusters. On the other hand the leading term of a low density approximation (introduced in [2]) gives a poor approximation the the overall form of the spectrum. One should use an approach where the advantages of both methods are combined together.

# 6. Phonons

## 6.1 Physical motivations

Inelastic X-ray scattering (IXS) experiments and inelastic neutron scattering on structural glasses and supercooled liquids provided useful information on the dynamics of their amorphous structure, at frequencies larger than 0.1 THz (see for example [34] and references therein). Those experiments show a regime, when the wavelength of the plane wave is comparable with the interparticle distance, where the vibrational spectrum can be understood in terms of propagation of quasi-elastic sound waves, the so-called high frequency sound. This high-frequency sound has also been observed in molecular dynamical simulations of strong and fragile liquids, and it displays several rather universal features. In particular, a peak is observed in the dynamical structure factor at a frequency that depends linearly on the exchanged momentum $p$, in the region $0.1p_0 - 1.0p_0$, $p_0$ being the position of the first maximum in the static structure factor. When extrapolated to zero momentum, this linear dispersion relation yields the macroscopic speed of sound. The width of the spectral line, $\Gamma$ is well fitted by

$$\Gamma(p) = Ap^x \quad , \quad x \approx 2\,, \tag{84}$$

with $A$ displaying a very mild (if any) temperature dependence. Moreover, the same scaling of $\Gamma$ has been found in harmonic Lenhard-Jones glasses [35], and one can safely conclude that the $p^2$ broadening of the high-frequency sound is due to small oscillations in the harmonic approximation. In these context other

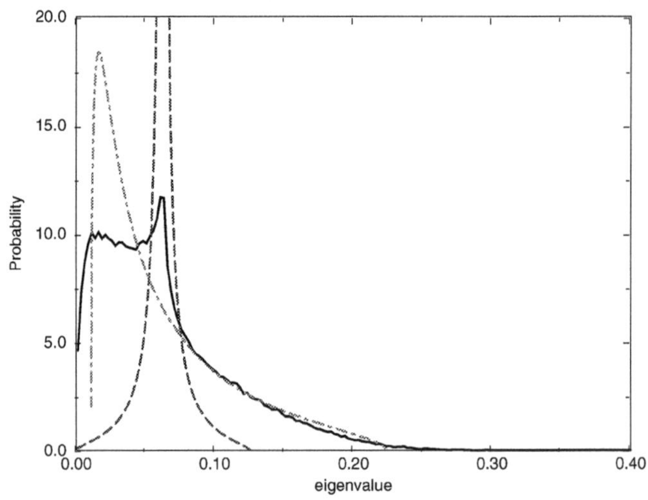


*Figure 3.* Density of eigenvalues of a Euclidean Random Matrix in three dimensions, density $\rho = 0.1$. The function $f$ is $f(x) = (2\pi)^{-3/2}\exp(-x^2/2)$, and the matrix is defined from eq.(48) with $u = 0$. The full line is the result of a numerical simulation with $N = 800$ points, averaged over 200 samples. The dashed line is the result from a low density expansion introduced in [2]. The dash-dotted line is the result from the Gaussian variational approximation (RPA) to the field theory.

interesting problems related with the high-frequency vibrational excitations of these topologically disordered systems [19] regard the origin of the *Boson peak* or the importance of localization properties to understand the dynamics of supercooled liquids [37].

The variety of materials where the $p^2$ broadening appears suggests a straightforward physical motivation. However, the simplest conceivable approximation, a wave propagating on an elastic medium in the presence of random scatterers, yields Rayleigh dispersion: $\Gamma \propto p^4$. This result is very robust: as soon as one assumes the presence of an underlying medium where the sound waves would propagate undisturbed, as in the disordered-solid model [36, 38, 39], the $p^4$ scaling appears even if one studies the interaction with the scatterers nonperturbatively [40]. When the distinction between the propagating medium and the scatterers is meaningless (as it happens for topologically disordered systems), the $p^2$ scaling is recovered.

We want to investigate the problem from the point of view of statistical mechanics of random matrices, by assuming that vibrations are the only motions allowed in the system. The formalism we shall introduce, however, is not limited to the investigation of the high frequency sound and it could be straightforwardly applied in different physical contexts.

Let us look more carefully at the relation between vibrational dynamics in glasses and random matrices. The dynamical structure factor for a system of $N$ identical particles is defined as:

$$S(p,\omega) = \frac{1}{N}\sum_{i,j}\int dt\, \mathrm{e}^{\mathrm{i}\omega t}\left\langle \mathrm{e}^{\mathrm{i}p\cdot(r_j(t)-r_i(0))}\right\rangle , \tag{85}$$

where $\langle \dots \rangle$ denotes the average over the particles positions $r_j$ with the canonical ensemble.

Here it is convenient to consider the normal modes of the glass or the supercooled liquid, usually called instantaneous normal modes (INM) because they are supposed to describe the short time dynamics of the particles in the liquid phase [21]. One studies the displacements $u$ around the random positions $x$, by writing the position of the $i$-th particle as $r_i(t) = x_i + u_i(t)$, and linearizing the equations of motion. Then one is naturally lead to consider the spectrum of the Hessian matrix of the potential, evaluated on the instantaneous position of the particles. Calling $\omega_n^2$ the eigenvalues of the Hessian matrix and $e_n(i)$ the corresponding eigenvectors, the one excitation approximation to the $S(p,\omega)$ at non zero frequency is given in the classical limit by:

$$S^{(1)}(p,\omega) = \frac{k_{\mathrm{B}}T}{m\omega^2}\sum_{n=1}^{N} Q_n(p)\,\delta(\omega-\omega_n)\,, \tag{86}$$

$$Q_n(p) = |\sum_i p\cdot e_n(i)\exp(\mathrm{i}p\cdot x_i)|^2\,. \tag{87}$$

However one cannot always assume that all the normal modes have positive eigenvalues, negative eigenvalues representing the situation where some particles are moving away from the position $x$. Indeed, it has been suggested [37] that diffusion properties in supercooled liquids can be studied considering the localization properties of the normal modes of negative eigenvalues.

A related and better defined problem is the study of normal modes at zero temperature, where the displacements $u$ are taken around the rest positions. By assuming that this structure corresponds to one minimum of the potential energy, one can introduce a harmonic approximation where only the vibrations around these minima are considered, and all the dynamical information is encoded in the spectral properties of the Hessian matrix on the rest positions. The Hessian is in a first approximation a random matrix if these rest positions correspond to a glass phase. It has been shown using molecular dynamics simulation that below the experimental glass transition temperature the thermodynamical properties of typical strong glasses are in a good agreement with such an assumption.

Therefore, the problem of the high-frequency dynamics of the system can be reduced, in its simplest version, to the consideration of random Euclidean matrices, where the entries are deterministic functions (the derivatives of the potential) of the random positions of the particles. As far as the system has momentum conservation in our case, due to translational invariance, all the realizations of the random matrix have a three common normal mode with zero eigenvalue: the uniform translation of the system.

An Euclidean matrix is determined by the particular deterministic function that we are considering, and by the probabilistic distribution function of the particles, that in the INM case is given by the Boltzmann factor. However, for the sake of simplicity we shall concentrate here on the simplest kind of euclidean matrices without spatial correlations and we will neglect the vector indices of the displacement. We consider $N$ particles placed at positions $x_i, i = 1, 2, ..., N$ inside a box, where periodic boundary conditions are applied. Then, the scalar euclidean matrices are given by eq.(2), where $f(x)$ is a scalar function depending on the distance between pairs of particles, and the positions $\{x\}$ of the particles are drawn with a flat probability distribution. Notice that the matrix (2) preserves translation invariance, since the uniform vector $e_0(i) = const$ is an eigenvector of $M$ with zero eigenvalue. Since there are not internal indices (the particle displacements are restricted to be all collinear), we cannot separate longitudinal and transversal excitations.

The dynamical structure factor for a scalar Euclidean matrix is given by

$$S_E(p,E) = \overline{\sum_n Q_n(p)\delta(E-E_n)}\,, \tag{88}$$

$$Q_n(p) = \frac{1}{N}\left|\sum_{i=1}^{N} e_n(i)\exp(ip\cdot x_i)\right|^2, \tag{89}$$

$$S(p,\omega) = 2\omega S_E(p,\omega^2)\,, \tag{90}$$

where the overline stands for the average over the particles position and we have given the definition either in the eigenvalue space ($S_E(p,E)$) and in the frequency space ($S(p,\omega)$).

## 6.2 A more complex Ýeld theory representation

The basic field theory representation is similar to the one of the previous section. The main complication is due to the presence of a diagonal term in the matrix $M$. One way out is to introduce one more pair of Bosonic fields. To perform the spatial integrations it turns out to be convenient to represent the Bosonic fields $\phi$ using new Bosonic fields , i.e.:

$$\chi(x) \equiv \sum_{i,a}(\phi_i^{(a)})^2\delta(x-x_i) \tag{91}$$

$$\psi^{(a)}(x) \equiv \sum_i \phi_i^{(a)}\delta(x-x_i)\,, \tag{92}$$

and using the "Lagrange multipliers" $\hat{\psi^{(a)}}(x), \hat{\chi}(x)$, to enforce the three constraints (92). At this point, the Gaussian variables $\phi_i^{(a)}$ are decoupled and can be integrated out.

Skipping intermediate steps and using the fact that all the replicas are equivalent one can write the Green function as a correlation function:

$$G(p,z) = \lim_{n\to 0}\frac{1}{N}\int d^dx d^dy\; e^{ip(x-y)}\int D[\psi^{(a)},\hat{\psi^{(a)}},\chi,\hat{\chi}]\;\psi^{(1)}(x)\psi^{(1)}(y)A^N\exp(S_0), \tag{93}$$

where we have introduced the following quantities:

$$S_0 \equiv i \sum_{a=1,n} \int d^d x \, \hat{\psi}^{(a)}(x)\psi^{(a)}(x) + \int d^d x \chi(x)\hat{\chi}(x)$$
$$-\frac{1}{2} \sum_{a=1,n} \int d^d x d^d y \psi^{(a)}(x) f(x-y) \psi^{(a)}(y)$$
$$A \equiv \int d^d x \exp \Phi(x)$$
$$\Phi(x) \equiv +i \int d^d y f(y-x)\chi(y) - \frac{1}{2} \frac{\sum_{a=1,n} \left(\hat{\psi}^{(a)}(x)\right)^2}{z + \hat{\chi}(x)} . \tag{94}$$

In order to take easily the limits $N, V \to \infty$ in (93) we shall resort to a grand canonical formulation of the disorder, introducing the partition function $\mathcal{Z}[\rho] \equiv \sum_N A^N \rho^N / N! \, \exp S_0$. Since the average number of particles $\overline{N} = V\rho\langle A \rangle$, in the $n \to 0$ limit we have that $\langle A \rangle = 1$ and the 'activity' $\rho$ is just the density of points, as before. Furthermore, the Gaussian integration over the fields $\psi^{(a)}$ is easily performed, leading to the field theory:

$$\mathcal{Z}[\rho, z] = \int D[\hat{\psi^{(a)}}, \chi, \hat{\chi}] \exp(S_0' + S_1) , \tag{95}$$

where [3]

$$S_0' \equiv -\frac{n}{2} \mathrm{Tr} \ln f + \int d^d x \chi(x)\hat{\chi}(x) -//\frac{1}{2} \sum_{a=1,n} \int d^d x d^d y \, \hat{\psi}^{(a)}(x) f^{(-1)}(x-y) \hat{\psi}^{(a)}(y)$$
$$S_1 \equiv \rho \int d^d x \, (\exp \Phi(x))$$

The resolvent is related to the correlation function by eq.(59). Before computing (59), let us turn to the symmetry due to the translational invariance. Since $e_n(i) = constant$ is an eigenvector of zero eigenvalue of the matrix (2) for every disorder realization, we see that Eq.(15) implies:

$$G(p=0, z) = \frac{1}{z} . \tag{96}$$

Interestingly enough, in the framework of the field theory introduced above, that constraint is automatically satisfied, due to the Ward identity linked to that symmetry. The interaction term $\Phi(x)$ is indeed invariant under the following infinitesimal transformation of order $\eta$:

$$\delta\hat{\psi}^{(1)}(x) = i\eta \left(z + \hat{\chi}(x)\right) , \tag{97}$$
$$\delta\chi(x) = 2i\eta \int d^d y f^{-1}(x-y) \hat{\psi}^{(1)}(y) \left(z + \hat{\chi}(x)\right) , \tag{98}$$

while the whole variation of the action $S = S_0' + S_1$ is due to the non-interacting part $S_0'$:

$$\delta S = -i\eta \frac{z}{\tilde{f}(0)} \int d^d x \hat{\psi}^{(1)}(x) \tag{99}$$

The invariance of $S_1$ hence leads to the following Ward identity:

$$\langle z + \hat{\chi}(x) \rangle = \frac{z}{\tilde{f}(0)} \langle \hat{\psi}^{(1)}(x) \int d^d y \, \hat{\psi}^{(1)}(y) \rangle \tag{100}$$

That is enough to prove that

$$\int d^d y \hat{\psi}^{(1)}(0) \hat{\psi}^{(1)}(y) \tag{101}$$

diverges as $1/z$ at small zeta. If we combine the previous result with the exact relation $\langle \hat{\chi}(x) \rangle = -\rho \tilde{f}(0)$, that can be derived with a different argument, to obtain:

$$(z - \rho \tilde{f}(0)) \frac{\tilde{f}(0)}{z} = \int d^d y \, \langle \hat{\psi}^{(1)}(x) \hat{\psi}^{(1)}(y) \rangle \,, \tag{102}$$

that, together with (59), implies the expected constraint $G(p = 0, z) = 1/z$.

The interacting term in (95) is complicated by the presence of an exponential interaction, meaning an infinite number of vertices. In order to perform the explicit computation of the resolvent $G(p, z)$ one has to introduce some scheme of approximation. We have chosen to deal with the high density limit, where many particles lie inside the range of the interaction $f(r)$. The high density limit ($\rho \gg 1$) of (59) is the typical situation one finds in many interesting physical situations, for example the glassy phase. In order to extract the leading term let us make the expansion $\exp \Phi(x) - 1 \sim \Phi(x)$. In that case the integration over the fields $\chi, \hat{\chi}$ is trivial, because of:

$$\int D[\chi] \exp \left( \int d^d x \chi(x) \left[ \rho \int d^d y f(x - y) + \hat{\chi}(x) \right] \right) = \prod_x \delta(\hat{\chi}(x) + \rho \tilde{f}(0)) \tag{103}$$

and the fields $\hat{\psi}^{(a)}$ are free. In fact, introducing the quantity $a \equiv z - \rho \tilde{f}(0)$, one remains with the Gaussian integration:

$$\mathcal{Z} \propto \int D[\hat{\psi}^{(a)}] \exp -\frac{1}{2} \int d^d x d^d y \sum_a \hat{\psi}^{(a)}(x) K^{-1}(x - y) \hat{\psi}^{(a)}(y) \,, \tag{104}$$

where the free propagator $K$ is defined by:

$$K^{-1}(x - y) \equiv f^{-1}(x - y) + \frac{\rho}{a} \delta(x - y) \,. \tag{105}$$

It is then easy from (59) and (104) to obtain the result:

$$G_0(p,z) = \frac{1}{z-\epsilon(p)}\,, \tag{106}$$

where

$$\epsilon(p) = \sqrt{\rho(\tilde{f}(p) - \tilde{f}(0)} \tag{107}$$

We see that at the leading order, a plane wave with momentum $p$ is actually an eigenstate of the matrix $M$ with eigenvalue $E$, and the disorder does not play any relevant role. In other words, inside a wavelength $2\pi/p$ there is always an infinite number of particles, ruling out the density fluctuations of the particles: the system reacts as an elastic medium.

Let us finally obtain the density of states at this level of accuracy, using Eq. (18) and $G_0$:

$$g_E(E) = \delta\left(E - \rho\tilde{f}(0)\right)\,. \tag{108}$$

We obtain a single delta function at $\rho\tilde{f}(0)$, that is somehow contradictory with our result for the dynamical structure factor: from the density of states one would say that the dispersion relation is Einstein's like, without any momentum dependence! The way out of this contradiction is of course that in the limit of infinite $\rho$ both $\epsilon(p)$ and $\rho\tilde{f}(0)$ diverge. The delta function in eq. (108) is the leading term in $\rho$, while the states that contribute to the dynamical structure factor appear only in the subleading terms in the density of states. The same phenomenon is present in the simpler case discussed in the previous section.

subsectionOne loop

We have seen above that, with the expansion of $\exp\Phi(x)$ up to first order in $\Phi(x)$ the fields $\hat{\psi}$ are non interacting and no self energy is present. Now we shall see that the one-loop correction to that leading term provides the $\frac{1}{\rho}$ contribution to the self energy. In fact by adding the quadratic term to $S_1$ the total action becomes (in the $n \to 0$ limit):

$$\begin{aligned} S = S_0' + S_1 \sim S_0' + \rho \int d^d x \left(\Phi(x) + \frac{1}{2}\Phi^2(x)\right) \\ = -\frac{\rho}{2} \int d^d x d^d y\, \chi(x) f^{(2)}(x-y)\chi(y) + i \int d^d x\, b(x)\chi(x) \\ --\frac{1}{2}\sum_{a=1,n} \int d^d x d^d y\, \hat{\psi}^{(a)}(x) c(x-y)\hat{\psi}^{(a)}(y) \\ +\frac{\rho}{8}\int d^d x \left(\frac{\sum_{a=1,n}\left(\hat{\psi}^{(a)}(x)\right)^2}{z+\hat{\chi}(x)}\right)^2\,, \end{aligned} \tag{109}$$

where

$$\begin{aligned} b(x) &\equiv \hat{\chi}(x) + \rho\tilde{f}(0) - \frac{\rho}{2}\int d^d y \, \frac{\sum_{a=1,n}\left(\hat{\psi}^{(a)}(x)\right)^2}{z+\hat{\chi}(x)} f(y-x) \\ c(x-y) &\equiv f^{(-1)}(x-y) + \frac{\rho}{z+\hat{\chi}(x)}\delta(x-y) \end{aligned} \tag{110}$$

After doing some computations and adding the two contributions coming two different diagrams one gets

$$G(p,z) = \\ G_0(p,z) + G_0^2(p,z)\frac{1}{\rho}\int \frac{d^d q}{(2\pi)^d}\, G_0(q,z)\left(\left[\rho\tilde{f}(p-q) - \rho\tilde{f}(q)\right]\right)^2 \tag{111}$$

The Dyson resummation of all the higher orders terms, that is built by 'decorating' recursively all the external legs $G_0$ with the one loop correction in (111) gives:

$$G(p,z) = \frac{1}{z - \epsilon(p) - \Sigma_1(p,z)}\,, \tag{112}$$

where the self-energy $\Sigma_1(p,z)$ is given by

$$\Sigma_1(p,z) \equiv \frac{1}{\rho}\int \frac{d^d q}{(2\pi)^d}\, G_0(q,z)\left[\rho\tilde{f}(p-q) - \rho\tilde{f}(q)\right]^2 . \tag{113}$$

Let us study in details the low exchanged momentum limit of Eq.(113). It is clear that at $p = 0$ the self-energy vanishes, as required by the Ward identity (100). We need to expand $\tilde{f}(p-q)$ for small $p$, that due to the spherical symmetry of $\tilde{f}$ yields

$$\tilde{f}(p-q) = \tilde{f}(q) - (p\cdot q)\frac{\tilde{f}'(q)}{q} + O(p^2)\,, \tag{114}$$

$$= \tilde{f}(q) + (p\cdot q)\frac{\epsilon'(q)}{q\rho} + O(p^2)\,. \tag{115}$$

Substituting (115) in (113), and performing explicitly the trivial angular integrations in dimensions $d$ we obtain

$$\begin{aligned} \Sigma_1(p,z) &\approx p^2 \frac{2^{1-d}}{\rho d \pi^{d/2}\Gamma(d/2)}\int_0^\infty dq\, q^{d-1}\frac{[\epsilon'(q)]^2}{z-\epsilon(q)} = \\ & p^2 \frac{2^{1-d}}{\rho d \pi^{d/2}\Gamma(d/2)}\int_0^{\epsilon(q=\infty)} d\epsilon\, \frac{[q(\epsilon)]^{d-1}}{q'(\epsilon)(z-\epsilon)}\,. \end{aligned} \tag{116}$$

In the last equation, we have denoted with $q(\epsilon)$ the inverse of the function $\epsilon(q)$. Setting now $z = E + i0^+$, and observing that $\epsilon(p) \approx Ap^2$ for small $p$, we readily obtain

$$Re\Sigma_1(p, E + i0^+) \approx p^2 \frac{2^{1-d}}{\rho d \pi^{d/2} \Gamma(d/2)} \int_0^{\epsilon(q=\infty)} d\epsilon \frac{[q(\epsilon)]^{d-1}}{q'(\epsilon)(E-\epsilon)}, \quad (117)$$

$$Im\Sigma_1(p, E + i0^+) \approx -\frac{\pi 2^{2-d} A}{\rho d \pi^{d/2} \Gamma(d/2)} p^2 [q(E)]^d . \quad (118)$$

Since the principal part is a number of order one, the real part of the self-energy scales like $p^2$ (possibly with logarithmic corrections), and thus the speed of sound of the system renormalizes due to the $1/\rho$ corrections. As a consequence, the function $q(E)$ is proportional to $E^{1/2} \sim p$ at the maximum of the function of $p$ $S_E(p, E)$, and the width of the peak of the $S_E(p, E)$ will scale like $p^{d+2}$. It is then easy to check (see (90)) that in frequency space the width of the spectral line will scale like

$$\Gamma \propto p^{d+1} , \quad (119)$$

as one would expect from Rayleigh scattering considerations.

The result (119) for the asymptotic regime $p << 1$ has been found at the one loop level. In order to predict correctly the spectral properties at very low external momentum $p$, it turns out that one must study the behavior of the two loop contribution, that can be done in details. Nevertheless, the one loop result is already a good starting point to perform detailed comparisons with the numerical simulations. The disadvantage of this approach is that it works near band edge ($\omega = 0$ at high $\rho$ but is not suited for producing the whole spectrum. Due to the complication of the action it is not clear how to do a variational computation and the best it can be done at the present moment is a CPA-like approximation described in the next section.

## 6.3 A CPA like approximation

As in the previous section we consider the resolvent $G(p, z)$ Our aim is to compute $G$ using the appropriate self-consistent equations. A partial resummation of the expansion for the resolvent can be written as

$$G(p, z) = \frac{1}{z - \lambda(p) - \Sigma(p, z)} . \quad (120)$$

The self-energy $\Sigma(p, z)$ is then expanded in powers of $1/\rho$ [2, 7] in the relevant region where $\rho = O(z)$.

If we reformulate the $1/\rho$ expansion in a diagrammatic way we can identify those diagrams with the simple topology of Fig. 4. Topologically, these

*Figure 4.* The diagrams of the $1/\rho$ expansion that are taken into account in our approach. The numbers correspond to the particle-label repetitions.

diagrams are exactly those considered in the usual lattice CPA and in other self consistent approximations. The sum of this infinite subset is given by the solution of the integral equation:

$$\Sigma(p, z) = \frac{1}{\rho} \int \frac{d^3 q}{(2\pi)^3} \left[ \rho \left( \hat{f}(\boldsymbol{q}) - \hat{f}(\boldsymbol{p} - \boldsymbol{q}) \right) \right]^2 G(p, z), \tag{121}$$

where the resolvent is given by Eq. (120). The solution gives us the resolvent, and hence the dynamical structure function and density of states (Eq. 122 below).

We are interested to study the solution of Eq. (121) for different values of $z$ and $\rho$. To be definite, we consider an explicit case where the function $f(r)$ has a simple form, namely $f(r) = \exp[-r^2/(2\sigma^2)]$. This is a reasonable first approximation for the effective interaction [8]. We shall take $\sigma$ as the unit of length and set $p_0 = 1/\sigma$, that is a reasonable choice for $p_0$ for this Gaussian $f(r)$, as discussed in [8]. In this particular case we will solve numerically the self-consistence equation. We will also evaluate by simulation (using the method of moments [41]) the exact dynamical structure function and the density of states by computing the resolvent for concrete realizations of the dynamical matrix, considering a sufficiently high number of particles so that finite volume effects can be neglected. These numerical results will be supplemented by analytic results, that are $f$-independent and can be obtained in the limits $p \to \infty$ and $p \to 0$.

The infinite momentum limit is particularly interesting because of the remarkable result [6, 7] that the density of states $g(\omega)$ can be written as

$$g(\omega) = \lim_{p \to \infty} \frac{\omega^2 S^{(1)}(p, \omega)}{k_{\mathrm{B}} T p^2}. \tag{122}$$

We easily find that in this limit Eq. (121) can be written as:

$$\frac{1}{\rho \mathcal{G}(z)} = \frac{z}{\rho} - \hat{f}(0) - A\mathcal{G}(z) - \int \frac{d^3 q}{(2\pi)^3} \hat{f}^2(\boldsymbol{q}) G(\boldsymbol{q}, z) \,, \tag{123}$$

where

$$\mathcal{G}(z) = \lim_{p \to \infty} G(p, z) \tag{124}$$

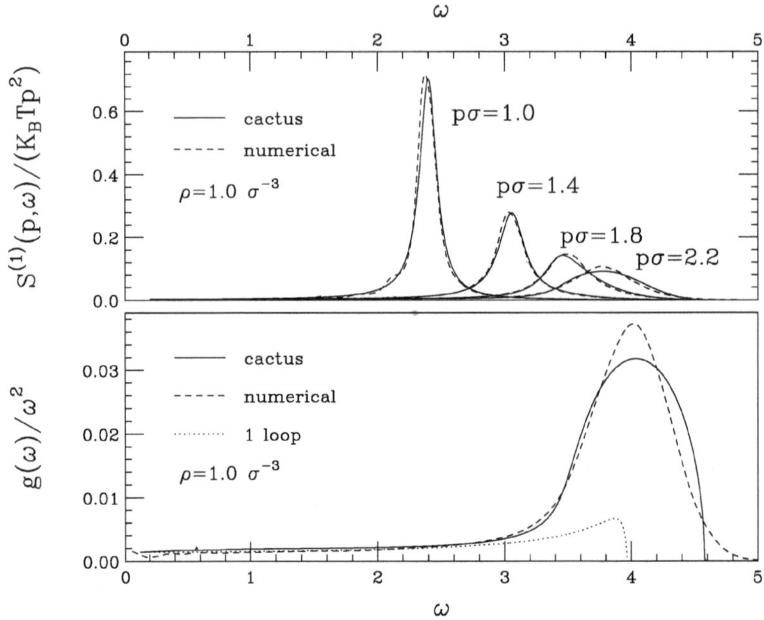


*Figure 5.* Top: dynamic structure factor as obtained from Eq.(121) (full line) and from simulations [7] (dashes). Bottom: the density of states divided by $\omega^2$ (Debye behavior) as obtained from Eq.(123) (full line), simulations (dashes), and first order in the $1/\rho$ expansion (dots).

and

$$\mathcal{A} = (2\pi)^{-3} \int \hat{f}^2(\boldsymbol{q})\, d^3q\,. \tag{125}$$

A simple approximation consists in neglecting the last term in the r.h.s. of (123), that is reasonable at large $z$. This approximation implies a the density of states that is semicircular as a function of $\omega^2$, with width proportional to $\sqrt{\rho}$ and centered at $\omega^2 = \rho\hat{f}(0)$. Translational invariance also requires low-frequency modes. These are given by the neglected term, and in fact it is easy to show that at high density it produces a Debye spectrum that extends between zero frequency and the semicircular part.

In the limit $p \to 0$, the leading contribution to $\Sigma''$ comes from $q \gg p$ in Eq. (121), where $G(q,z) \approx \mathcal{G}(z)$, so we can write for the peak width $\Gamma(p) \approx \Gamma_0(p)$, where

$$\Gamma_0(p) \equiv \pi\rho \frac{g(\omega_p)}{2\omega_p^2} \int \frac{d^3q}{(2\pi)^3} \left[\hat{f}(\boldsymbol{q}) - \hat{f}(\boldsymbol{p}-\boldsymbol{q})\right]^2. \tag{126}$$

The integral is of order $p^2$, so if the spectrum is Debye-like for small frequencies, we get $\Gamma(p) \sim p^2$.

These considerations are verified by the numerical solution of the Gaussian case, that are shown in Fig. 5 for $\rho = 1.0\,\sigma^{-3}$ together with the results for the simulations [7]. Note the good agreement, to be expected for high-densities, and how, for large $p$, $S^{(1)}(p, \omega)$ (Fig. 5, top) tends to the density of states. The density of states from the self-consistent equation (Fig. 5,bottom) also agrees very well with the results from simulations, and is a big improvement over the first term of the expansion in powers of $\rho^{-1}$. The two contributions (Debye and semicircle) mentioned above can be clearly identified. As expected our approximation fails in reproducing the exponential decay of the density of states at high frequencies, that is non perturbative in $1/\rho$ [5] and corresponds to localized states.

Next in Fig. 6 we plot the linewidth as a function of $p$ as obtained from Eq.(121). Notice that we recover the behavior predicted from the first two non-trivial terms in the expansion in powers of $\rho^{-1}$ [7]: the linewidth is proportional to $p^2$ at small $p$ (also predicted by the argument above), then there is a faster growth and finally it approaches to a constant as $S^{(1)}(q, \omega)$ starts to collapse onto the density of states. The inset shows that the contribution Eq. (126) is indeed dominant at small $p$. However, accurate $p^2$ scaling is found only for very small momenta ($p/p_0 < 0.1$), while experiments are done at $0.1 < p/p_0 < 1$. In this crossover region, our approach predicts the existence of non-universal, model dependent small deviations from $p^2$, that are probably hard to measure experimentally. In any case, the effective exponent is certainly less than 4, in contrast with lattice models and consistent with experimental findings. Similar conclusions can be drawn from mode coupling theory (see fig. 8 of [43]).

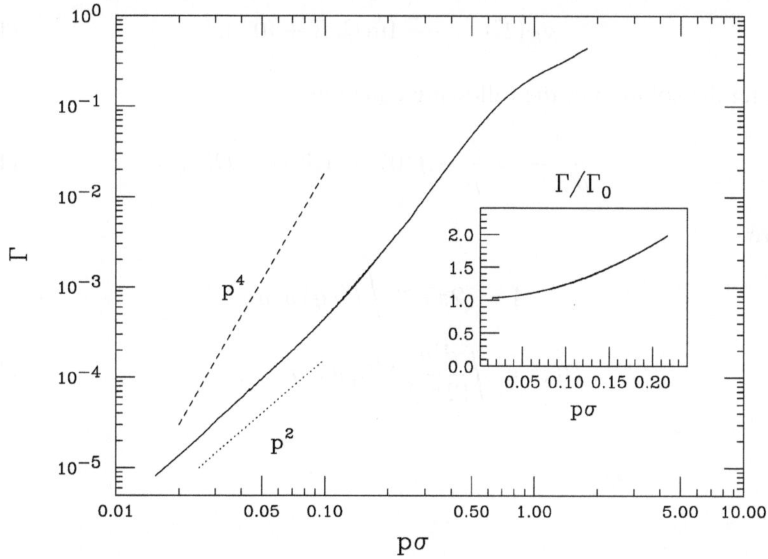


*Figure 6.* Peak width vs. $p$, for $\rho = 0.6\sigma^{-3}$. The inset shows that $\Gamma_0(p)$ is the dominant contribution at small $p$.

## 6.4 The disappearance of phonons

New phenomena are present the function $f$ is no more positive or in the vectorial case. In these case it is possible that the spectrum arrives up to negative values and the density of states $g_E$ does not vanish at $E = 0$. In principle for a given model we should not expect a sharp transition from the situation where the Debye spectrum holds and $g_E(E) \propto E^{1/2}$ because tails are always present. However often tails are small and one can effectively observe a transition from the two regimes.

In the framework where the density of states is computed from a simple integral equation, the tails are neglected and we are in the best situation to observe such a phenomenon.

Let us call $\tau$ a parameter that separate the two regimes. A detailed analysis show that a $\tau = 0$ the density of states behaves as

$$g_E(E) \propto E^{1/4}, \tag{127}$$

while at small $\tau$ and $E$ we have

$$g_e(E, \tau) = E^{1/4} h(\tau E^{1/2}). \tag{128}$$

The detailed argument is a delicate but let us consider an heuristic version of it.

With some work it is possible to obtain from (121) an integral equation even for the density of states. As a matter of fact, defining $\mathcal{G}(z) = G(p = \infty, z)$, the density of states turns out to be

$$g_E(E) = -\frac{2}{\pi} \operatorname{Im} \mathcal{G}(E + i0^+). \tag{129}$$

$\mathcal{G}$ being the solution of the following equation:

$$\frac{1}{\rho \mathcal{G}(z)} = \frac{z}{\rho} - \hat{f}(0) - A\mathcal{G}(z) - B(z) \tag{130}$$

where

$$A = (2\pi)^{-3} \int \hat{f}^2(\boldsymbol{q})\, d^3q\,,$$

$$B(z) = \int \frac{d^3q}{(2\pi)^3} \hat{f}^2(\boldsymbol{q}) G(\boldsymbol{q}, z)\,, \tag{131}$$

With this equation, one needs to know the resolvent at all $q$ to obtain the density of states, due to the last term in the r.h.s. This can be done by solving numerically the self-consistent equation of ref. [11], but here we perform an approximate analysis, that is more illuminating.

The solution of the previous equation is

$$\mathcal{G}(z) = \frac{-\alpha(z) + \sqrt{\alpha(z)^2 - 4A\rho}}{2A\rho}\,, \tag{132}$$

where

$$\alpha(z) = \hat{f}(0) - \frac{z}{\rho} + B(z) \tag{133}$$

The crudest approximation is to neglect the dependence of $B(z)$ from $z$ (we will also assume that $B$ is a smooth function of the other parameters). In this case Eq. 130 is quadratic in $\mathcal{G}$, and one easily finds a semicircular density of states. But the semicircular spectrum misses the Debye part, and a better approximation is needed. So we substitute $G$ in the last term of the r.h.s. by the resolvent of the continuum elastic medium $G_0(z,p) = (z - E(p))^{-1}$. This is reasonable because the $f^2(q)$ factor makes low momenta dominate the integral, and due to translational invariance $G(z,p) \approx G_0(z,p)$ in this region [11]. We shall be looking at small $E$, so to a good approximation

$$B(z) = B_0 + iB_1 z^{1/2} \tag{134}$$

We have two limiting cases. depending on the sign of of $\alpha(0)^2 -\ 4A\rho$.

- When the semicircular part of the density of states does not reach low frequencies, the square root can be Taylor-expanded, and one gets

$$g_e(E) \approx E^{1/2}, \tag{135}$$

  that is precisely Debye's law.

- In the opposite situation, on the other hand, the semicircle arrives also at negative values of $E$ and $g_E(E)$ is different from zero also if $B_1$ where zero.

- Exactly at the critical point we get $g_E(E) \propto \sqrt{\sqrt{E}} = E^{1/4}$ that is the announced result.

Mathematically, the instability arises when $\mathcal{G}(0)$ develops an imaginary part. This can only come from the square root in previous equations. Notice that this instability is a kind of phase transition, where the order parameter is $\mathrm{Im}\,\mathcal{G}(0)$. Doing a detailed computation on finds that this order parameter behaves as $\tau^{\beta}$, with $\beta = 1/2$.

Since the behavior of the propagator at high momentua does not strongly affect the dispersion relation [11], we do not expect deviations from a linear dispersion relation. This has been checked either numerically or solving numerically the self-consistency equation given a particular choice for the function $f(r)$ (see the numerical results section below). It can be argued that this kind of phenomenon is responsible of the Boson peak [12], however we cannot discuss this point for lack of space.

sectionCorrelated points

## 6.5 Various models

When the points are correlated things become more difficult. Already it is difficult to study the statistical properties of correlated points and it is more difficult to study the properties of the matrices that depends on these points. Although some approaches have been developed that allow us to deal with two points correlations [6, 7], it is not evident how to treat the general case where many points correlations are present.

A particular interesting case is when the matrix and the distribution of the points are related. In the best of the possible words there should be extra symmetries that express this relations.

This field is at its infancy, so that I will only describe some general results, without presenting applications, that for the moment do not still exist for the case of Euclidean random matrices.

In the general case we consider an Hamiltonian $H[x]$, such that the stationary equations

$$F_i[x] \equiv \frac{\partial H}{\partial x_i} = 0 \tag{136}$$

have a large number of solutions.

Let us label these solutions with a index $\alpha$. The probability $P[x]$ of a configuration of the points $x$ is assumed to be given by

$$P[x] \propto \sum_\alpha w_\alpha \delta(x - x^\alpha)\,,$$
$$w_\alpha \equiv D[M^\alpha])\exp(-\beta H[x^\alpha])\,, \tag{137}$$

where the matrix $M$ is given by

$$M^\alpha_{i,k} = \frac{\partial^2 H}{\partial x_i x_k}|_{x=x^\alpha}\,. \tag{138}$$

This problem arises for the first time in the framework of spin glasses [44], but its relevance to glasses has been stressed for the first time in [45] The function $F$ may selects the different type of stationary points.

Different interesting possibilities are:

- $D[M] = 1$, i.e. all stationary points have the same weight.
- $D[M] = \text{sign}(\det[M])$, i.e. all stationary points have a weight that can be 1 or -1.
- $D[M] = 1$ only if all the eigenvalues are positive, otherwise it is zero (i.e. minima are selected)

We are eventually interested to study the properties of the matrix

$$\frac{1}{M[x] - z}, \tag{139}$$

when the points $x$ are extracted with the previous probability.

## 6.6 A new supersymmetry

For simplicity I will only restrict myself to the case $z = 0$ where some symmetry are present when $D[M] = \text{sign}(\det[M])$.

Indeed it is evident that

$$\int P[x]d[x]A[x] \propto \int d[x]D[M]\exp(-\beta H[x]) \det M[x]| \prod_i \delta(F_i[x])A[x] \tag{140}$$

The last term simplify to

$$\int d[x] \exp(-\beta H[x]) \det M[x] \prod_i \delta(F_i[x])A[x], \tag{141}$$

when $D[M] = \text{sign}(\det(M))$. For simplicity let us assume that this is the case, without discussing the physical motivations of this choice.

Using usual representations we can write

$$\overline{M^{-1}_{i,k}} = \int d\mu[x, \lambda, \psi, \overline{\psi}]\, \overline{\psi}_i \psi_k, \tag{142}$$

where $d\mu$ is a normalizes measure proportional to

$$d[x]d[\lambda]d[\psi]d[\overline{\psi}] \exp\left( -\beta H[x] + i \sum_k \lambda_k F_k[x] + \sum_{k,j} M_{j,k} \overline{\psi}_j \psi_k \right). \tag{143}$$

Here the $\psi$ are really Fermionic variables (i.e. anticommuting Grassmann variables): they have been introduced for representing the determinant, however they can also be used to compute the matrix elements of the inverse of the matrix $M$.

It was rather unexpected [46, 53] to discover that also for $\beta \neq 0$ the measure $d\mu$ is invariant under a transformation of Fermionic character of BRST type (a supersymmetry in short). (see for example [52, 47]). If $\epsilon$ is an infinitesimal Grassmann parameter, it is straightforward to verify that (143) is invariant under the following transformation,

$$\delta x_i = \epsilon\, \psi_i \qquad \delta\lambda_i = -\epsilon\, \beta\, \psi_i \qquad \delta\bar{\psi}_i = -\epsilon\, x_i \qquad \delta\psi_i = 0 \tag{144}$$

This symmetry is extremely important and it is crucial to use approximation methods that do not break it [48][50]. This has been stressed in the spin glass case [53] in the context of the Tap equations [54].

## 6.7 Physical meaning of the supersymmetry and of is breaking

In this section we will follow a reasoning allowing for an intuitive explanation of the physical meaning of the supersymmetry in terms of a particular behavior of the solutions of the stationary equations [55]-[60].

It may be interesting to concentrate the attention on some of the Ward identities that are generated by the supersymmetry. The simplest one is

$$\langle \overline{\psi}_j \psi_k \rangle = i \langle \lambda_j x_k \rangle \ . \tag{145}$$

Apparently the equation is trivially satisfied. Indeed it is convenient to consider a slightly modified theory where we make the substitution

$$F_j \to F_j - h \ . \tag{146}$$

The final effect is to add an extra term in the exponential equal to $i\lambda_j h$. In other words the r.h.s of equation 145 is

$$\frac{\partial \langle x_k \rangle_h}{\partial h}|_{h=0} \tag{147}$$

The l.h.s can also be computed and one finds that eq. (145) becomes

$$\sum_\alpha w_\alpha \frac{\partial \langle x_k^\alpha \rangle_h}{\partial h}|_{h=0} = \frac{\partial \sum_{\alpha'} w_{\alpha'} \langle x_k^{\alpha'} \rangle}{\partial h}|_{h=0} \ . \tag{148}$$

If we consider solutions where the $\det(M) \neq 0$ we have that

$$\frac{\partial w_\alpha}{\partial h}|_{h=0} = 0 \,, \tag{149}$$

so that the previous equation seems to be always satisfied.

However the we must be careful if most of solutions have a very small $\det(M)$. In this case if we first sent $N$ to infinity and after we do the derivative with respect to $h$ we can find that the set of solutions in not a continuos functions of $h$. Solutions may bifurcate or disappear for any arbitrary small variation of $h$ and the previous relations are no more valid.

This phenomenon has been studied in the framework of infinite range random matrices, where the function that plays the role of $H$ is the free energy as function of the magnetizations (i.e. the TAP free energy) in spin glass type models. One finds that depending on the parameters there are two phase, one where the supersymmetry is exact, the other where the supersymmetry is spontaneously broken. This last phenomenon has been discovered last year and at the present moment one is trying to fully understand its consequences.

In the framework of Euclidean random theory there are two questions that are quite relevant and may be the most interesting for Euclidean Random theory:

- How to construct and to use in a practical way a formalism where the supersymmetry of the problem plays a crucial role?
- How to find out if there is a phase where supersymmetry is spontaneously broken and which are the physical effects of such a breaking.

It is quite likely the response to these questions would be very important.

## Acknowledgements

I have the pleasure to thank all the people that have worked with me in the subject of random matrices and related problems, i.e. Alessia Annibale, Andrea Cavagna, Andrea Crisanti, Stefano Ciliberti, Barbara Coluzzi, Irene Giardina, Tomas Grigera, Luca Leuzzi, Víctor Martín-Mayor, Marc Mézard, Tommaso Rizzo, Elisa Trevigne, Paolo Verrocchio and Tony Zee.

# References

[1] T. M. Wun and R. F. Loring, J. Chem. Phys 97, 8368 (1992); Y. Wan and R. Stratt, J. Chem. Phys 100, 5123 (1994); A. Cavagna et al., Phys. Rev. Lett. 83, 108 (1999); G. Biroli, R. Monasson, J. Phys. A: Math. Gen. 32, L255 (1999).

[2] M. Mézard, G. Parisi, and A. Zee, Nucl. Phys. B559, 689 (1999),

[3] M. Mézard and G.Parisi, Phys. Rev. Lett. 82 (1999)747.

[4] B.Coluzzi, G.Parisi and P.Verrocchio, J. Chem. Phys. 112, 2933 (2000), B.Coluzzi, G.Parisi and P. Verrocchio, Phys. Rev. Lett. 84,306 (2000)

[5] A. Zee and I. Affleck, J. Phys: Condens. Matter 12, 8863 (2000).

[6] V. Martín-Mayor, M. Mézard, G. Parisi, and P. Verrocchio, "The dynamical structure factor in topologically disordered systems" cond-mat 0008472v1 (unpublished).

[7] V. Martín-Mayor, M. Mézard, G. Parisi, and P. Verrocchio, J. Chem. Phys. 114, 8068 (2001).

[8] T. S. Grigera et al., cond-mat/0104433 (to be published in Philos. Mag. B).

[9] T. S. Grigera et al., to be published.

[10] T.S. Grigera, V. Martín-Mayor, G. Parisi and P. Verrocchio, preprint cond-mat/0110129.

[11] T.S. Grigera, V. Martín-Mayor, G. Parisi and P. Verrocchio, Phys. Rev. Lett. 87 085502 (2001)

[12] T.S. Grigera, V. Martín-Mayor, G. Parisi and P. Verrocchio, Phys. Rev. Lett. 87 085502 (2001), T.S. Grigera, V. Martín-Mayor, G. Parisi and P. Verrocchio, cond-mat/0104433 and cond-mat/0301103 (to be published on Nature).

[13] A. Cavagna, I. Giardina, G. Parisi, J. Phys. A 30 (1997) 7021.

[14] Cavagna A, Giardina I and Parisi G 1998 Phys. Rev. B 57 11251. i

[15] S. Ciliberti, T.S. Grigera, V. Martín-Mayor, G. Parisi and P. Verrocchio,

[16] See Efetof's lectures at thi

[17] W. Götze and L. Sjogren, , Rep. Prog. Phys. 55 241 (1992); W. Kob and H. C. Andersen, Phys. Rev. Lett. 73, 1376 (1994).

[18] G. Parisi J. Phys A 114 (1981) 735, An Introduction to the Statistical Mechanics of Amorphous Systems, in Recent Advances in Field Theory and Statistical Mechanics, edited by J.-B. Zuber and R. Stora (North-Holland, Amsterdam, Netherlands, 1984).

[19] See S.R. Elliot Physics of amorphous materials, Longman (England 1983).

[20] See e.g. J. M. Ziman, Models of disorder, Cambridge University Press, Cambdrige (1979).

[21] T. Keyes, J. Phys. Chem. A 101, 2921 (1997).

[22] A. Cavagna, Europhys. Lett. 53, 490 (2001).

[23] L. Angelani, R. Di Leonardo, G. Ruocco, A. Scala, and F. Sciortino, Phys. Rev. Lett. 85, 5356 (2000), K. Broderix, K. K. Bhattacharya, A. Cavagna, A. Zippelius, and I. Giardina, Phys. Rev. Lett. 85, 5360 (2000), T.S. Grigera, A. Cavagna, I.Giardina and G. Parisi, Phys. Rev. Lett. 88, 055502 (2002).

[24] G. Parisi On the origine of the Boson peak, cond-mat/0301284, Proceeding of the Pisa conference September 2002, to be published on Journal of Physics; cond-mat/030128 Euclidean random matrices, the glass transition and the Boson peak, proceeding of the Messina conference in honour of Gene Stanley, Physica A in press.

[25] A.Cavagna, I.Giardina, G.Parisi, Phys. Rev. Lett. 83, 108 (1999)

[26] G. Parisi and N. Sourlas, Phys. Rev. Lett. 43 (1979) 744; Nucl. Phys. B 206 (1982) 321. J. Cardy, Phys. Lett. 125B (1983) 470. A. Klein and J. Fernando-Perez, Phys.Lett. 125B (1983) 473.

[27] Parisi G and Sourlas N 1982 Nucl. Phys. B 206, 321.

[28] A Poliakiov, Soviet Phys. JEPT 50, 353 (1970).

[29] A.Migdal, (private comunication).

[30] R. Abou-Chacra, P.W. Anderson, and D.J. Thouless, J. Phys. C 5, 1734 (1973). H. Kunz, J. Physique 44, L411 (1883). E.N. Economou and M.H. Cohen, Phys. Rev. B 5, 2931 (1972). A.D. Mirlin and Y.V. Fyodorov, J. Phys. A: Math. Gen. 24, 2273 (1991). A.D. Mirlin and Y.V. Fyodorov, Nucl. Phys. B 366, 507 (1991).

[31] P. Cizeau and J.P. Bouchaud, Phys. Rev. E 50, 1810 (1994). G.J. Rodgers and A.J. Bray, Phys. Rev. B 37, 3557 (1988). A.J. Bray and G.J. Rodgers, Phys. Rev. B 38, 11461 (1988). G. Biroli and R. Monasson, e-print cond-mat/9902032 (1999).

[32] A. Cavagna, I. Giardina and G. Parisi, Phys. Rev. Lett. 83 108 (1999).

[33] A. Zee, I. Affleck, J. Phys.: Cond. Matter 12, 8863 (2000);

[34] Pilla, O. et al., Nature of the Short Wavelength Excitations in Vitreous Silica: An X-Ray Brillouin Scattering Study. Phys. Rev. Lett. 85, 2136–2139 (2000).

[35] G. Ruocco et al., Phys. Rev. Lett. 84, 5788 (2000).

[36] J. Horbach et al., J. Phys. Chem. B 103, 4104 (1999).

[37] S.B. Bembenek and S.D. Laird, J. Chem. Phys 104. 5199 (1996)

[38] W. Schirmacher, G. Diezemann and C. Ganter, Phys. Rev. Lett. 81, 136 (1998)

[39] M. Montagna et al. Phys. Rev. lett. 83, 3450 (1999).

[40] V. Martín-Mayor et al., Phys. Rev. E 62, 2373 (2000).

[41] C.Benoit, E.Royer and G.Poussigue, J. Phys Condens. Matter 4. 3125 (1992), and references therein; C.Benoit, J. Phys Condens. Matter 1. 335 (1989), G.Viliani et al. Phys. Rev. B 52, 3346 (1995). Phys. Rev. E.

[42] P. Turchi, F. Ducastelle and G. Treglia, J. Phys. C 15, 2891 (1982).

[43] W. Götze and M. R. Mayr, Phys. Rev. E 61, 587 (2000).

[44] Mézard M, Parisi G and Virasoro M.A., Spin glass theory and beyond", World Scientific (1987).

[45] G. Parisi A pedagogical introduction to the replica method for fragile glasses, cond-mat/9905318, Phil. Mag. in press.

[46] Cavagna A, Garrahan J P and Giardina I 1998 J. Phys. A: Math. Gen. 32 711.

[47] Zinn-Justin J, 1989 Quantum Field Theory and Critical Phenomena, (Clarendon Press, Oxford).

[48] Kurchan J 1991 J. Phys. A: Math. Gen. 24 4969.

[49] Kurchan J 2002 Preprint cond-mat/0209399.

[50] For an illuminating discussion of the problem of removing the determinant in the supersymmetric formalism see [48, 49] and [27].

[51] A. Bray and M. Moore, e-print cond-mat/0305620.

[52] C. Becchi, A. Stora and R. Rouet, Commun. Math. Phys. 42 (1975) 127. I.V. Tyutin Lebdev preprint FIAN 39 (1975) unpublished.

[53] A. Cavagna, I. Giardina, G. Parisi and M. Mezard, J. Phys. A 36 (2003) 1175.

[54] D. J. Thouless, P. W. Anderson and R. G. Palmer, Phil. Mag. 35 (1977) 593.

[55] A. Crisanti, L. Leuzzi, G. Parisi, T. Rizzo, Phys. Rev. B, 68 (2003) 174401; Phys. Rev. Lett. 92, 127203 (2004), Phys. Rev. B 70, 064423 (2004).

[56] A. Crisanti, L. Leuzzi, T. Rizzo, Eur. Phys. J. B, 36 (2003) 129-136; Complexity in Mean-Field Spin-Glass Models: Ising $p$-spin, cond-mat/0406649.

[57] A. Crisanti, L. Leuzzi, Phys. Rev. B 70, 014409 (2004), The spherical $2 + p$ spin glass model: an exactly solvable model for glass to spin-glass transition cond-mat/0407129.

[58] A, Annibale, A. Cavagna, I. Giardina, G. Parisi, Phys. Rev. E 68, 061103 (2003); A, Annibale, A. Cavagna, I. Giardina, G. Parisi, Elisa Trevigne J. Phys. A 36, 10937 (2003); A. Cavagna, I. Giardina, G. Parisi,Phys. Rev. Lett. 92, 120603 (2004); A. Annibale, G. Gualdi, A. Cavagna J. Phys. A: Math. Gen. 37 (2004) 11311, A. Cavagna, I. Giardina, G. Parisi, Phys. Rev. B 71 (2005) 024422.

[59] G. Parisi, T. Rizzo On Supersymmetry Breaking in the Computation of the Complexity cond-mat/0401509; Zero-Temperature Limit of the SUSY-breaking Complexity in Diluted Spin-Glass Models cond-mat/0411732

[60] T. Rizzo Tap Complexity, the Cavity Method and Supersymmetry cond-mat/0403261, T. Rizzo On the Complexity of the Bethe Lattice Spin Glass cond-mat/0404729;

# MATRIX MODELS AND GROWTH PROCESSES: FROM VISCOUS FLOWS TO THE QUANTUM HALL EFFECT

A. Zabrodin
*Institute of Biochemical Physics, 4 Kosygina st., 119991, Moscow, Russia and ITEP, 25 B. Cheremushkinskaya, 117259, Moscow, Russia*

**Abstract** We review the recent developments in the theory of normal, normal self-dual and general complex random matrices. The distribution and correlations of the eigenvalues at large scales are investigated in the large $N$ limit. The $1/N$ expansion of the free energy is also discussed. Our basic tool is a specific Ward identity for correlation functions (the loop equation), which follows from invariance of the partition function under reparametrizations of the complex eigenvalues plane. The method for handling the loop equation requires the technique of boundary value problems in two dimensions and elements of the potential theory. As far as the physical significance of these models is concerned, we discuss, in some detail, the recently revealed applications to diffusion-controlled growth processes (e.g., to the Saffman-Taylor problem) and to the semiclassical behaviour of electronic blobs in the quantum Hall regime.

## 1. Introduction

The subject matter of random matrix theory is a matrix whose entries are randomly distributed with some probability density. To put it another way, the theory deals with statistical ensembles of matrices. Given such an ensemble, one is typically interested in the distribution of eigenvalues and correlations between them as size of the matrices, $N$, tends to infinity. The distribution and correlation laws obtained in this way turn out to be common to objects and systems of very diverse nature.

The area of applications of the random matrix theory in physics (and mathematics) is enormously vast. It ranges from energy levels statistics in nuclei to number theory, from quantum chaos to string theory. Most extensively employed and best-understood are ensembles of hermitian or unitary matrices, with eigenvalues being confined either to the real line or to the unit circle. Their applications to the level statistics in nuclei go back to Wigner's works of



*E. Brezin et al. (eds.), Applications of Random Matrices in Physics,* 261–318.

early 50-s. For different aspects of random matrix theory, its applications and related topics see e.g. [1]-[4].

In these lectures we consider more general classes of random matrices, with no a priori restrictions to their eigenvalues being imposed. The eigenvalues can be arbitrary complex numbers. Such models are as yet less well understood but they are equally interesting and meaningful. As we shall see, they may exhibit even richer mathematical structures than their Hermitian counterparts. Their physical applications are also many and varied. (A list of the relevant physical problems and corresponding references can be found in, e.g., [5]). The present lectures are based on our recent works [6]-[10] where new applications to diffusion limited growth processes, complex analysis and quantum Hall effect were found.

The progenitor of ensembles of matrices with general complex eigenvalues is the statistical model of complex matrices with the Gaussian weight. It was introduced by Ginibre [11] in 1965. The partition function of this model is

$$Z_N = \int [D\Phi] \exp\left( -\frac{N}{t} \mathrm{tr}\, \Phi^\dagger \Phi \right)$$

Here $[D\Phi] = \prod_{ij} d(\mathcal{R}e\, \Phi_{ij}) d(\mathcal{I}m\, \Phi_{ij})$ is the standard volume element in the space of $N \times N$ matrices with complex entries $\Phi_{ij}$ and $t$ is a (real positive) parameter. Along with the Ginibre ensemble and its generalizations we also consider ensembles of normal matrices, i.e., such that $\Phi$ commutes with its hermitian conjugate $\Phi^\dagger$, and normal self-dual matrices (the definition follows below in Section 2).

Since one is primarily interested in statistics of eigenvalues, it is natural to express the probability density in terms of complex eigenvalues $z_j = x_i + iy_j$ of the matrix $\Phi$. It appears that the volume element can be represented as

$$[D\Phi] \propto \prod_{i<j} |z_i - z_j|^{2\beta} \prod_i d^2 z_i$$

where $\beta = 1$ for complex and normal matrices and $\beta = 2$ for normal self-dual matrices. If the statistical weight depends on the eigenvalues only, as it is usually assumed, the other parameters of the matrix (often referred to as "angular variables") are irrelevant and can be integrated out giving an overall normalization factor. In this case the original matrix problem reduces to statistical mechanics of $N$ particles with complex coordinates $z_j$ in the plane. We thus see that even if the matrix entries $\Phi_{ij}$ are statistically independent, like in the Ginibre ensemble, the eigenvalues are correlated in a nontrivial way. Specifically, the factor $\prod_{i<j} |z_i - z_j|^{2\beta}$, being equal to the exponentiated Coulomb energy in two dimensions, means an effective "repelling" of eigenvalues. This remark leads to the Dyson logarithmic gas interpretation [12], which treats the

matrix ensemble as a two-dimensional "plasma" of eigenvalues in an external electric field.

At $\beta = 1$, there is another important interpretation. Namely, the factor $\prod_{i<j}(z_i - z_j)$ can be thought of as coming from the Slater determinant of one-particle fermionic states. The averaging over matrices then turns into averaging over the ground state of a system of $N$ non-interacting fermions. In the case of the Ginibre ensemble it is the system of $N$ electrons in a uniform magnetic field at the lowest Landau level. In the case of a spin-$\frac{1}{2}$ electron in a non-uniform magnetic field all energy levels split, with the only exception of the lowest one, which remains highly degenerate. We shall see that the normal and complex matrix ensembles with a non-Gaussian statistical weight are equivalent to $N$ polarized electrons in a non-uniform magnetic field confined to the lowest energy level. If the degeneracy of the level equals $N$, i.e., if the level is completely filled, the system of $N$ electrons behaves as an incompressible quantum Hall droplet [13].

When $N$ becomes large some new features emerge, which require a different language for their adequate description, in much the same way as classical thermodynamics results from statistical mechanics. As $N \to \infty$, the eigenvalues densely fill a domain in the complex plane with the mean density outside it being exponentially small in $N$. Around the edge of this domain the density steeply drops down. The width of the transition region tends to zero as $N \to \infty$, so that the density profile in the direction normal to the edge looks like a step function. This fact allows one to introduce the *support of eigenvalues* to be the region where the mean density of eigenvalues does not vanish as $N \to \infty$. Typically, it is a bounded domain (or several disconnected domains) in the complex plane. Its shape is determined by the probability density.

For the Ginibre ensemble, the support of eigenvalues is the disk of radius $\sqrt{t}$ with uniform density. It is the counterpart of the celebrated Wigner "semicircular law". For matrix ensembles with non-Gaussian weights the supports are in general not circular and not connected. Throughout these lectures our attention is mostly restricted to the case when the support of eigenvalues is a connected domain. Even in this relatively simpler case, the shape of this domain depends on parameters of the statistical weight in a rather complicated way. As we shall see in Section 4, the problem to find the support of eigenvalues from a given statistical weight is equivalent to the inverse problem of potential theory in two dimensions. In most cases, solutions of the latter are not available in an explicit form.

Nevertheless, the local dynamical law that governs the evolution of the support of eigenvalues under changes of parameters of the statistical weight (like $t$ in the Ginibre ensemble) can be expressed in terms of the exterior Dirichlet boundary value problem. Namely, the edge of the support moves along gradient of a scalar harmonic field in its exterior, with the velocity being proportional to the absolute value of the gradient. Remarkably, this growth law is known to be common to a wide class of diffusion-limited growth processes of which the most popular example is viscous flow in the Hele-Shaw cell (see [14] for a review). The mentioned above equivalence between the normal matrix ensemble and the quantum Hall droplet suggests that the semiclassical behaviour of electronic droplets in a non-uniform magnetic field follows the same laws as the Hele-Shaw flows do.

This fact allows one to treat the model of normal or complex random matrices as a growth problem. The advantage of this viewpoint is two-fold. First, the hydrodynamic interpretation makes some of the large $N$ matrix model results more illuminating and intuitively accessible. Second and most important, the matrix model perspective may help to suggest new approaches to the long-standing growth problems. In this respect, of special interest is the identification of finite time singularities in some exact solutions to the Hele-Shaw flows with critical points of the normal and complex matrix models.

At last, a few words about the organization of the lectures. The material that follows can be divided into three parts. The first one (Section 2) can be regarded as a continuation of the introduction. We define the main matrix ensembles to be considered and give their physical interpretations. The second part (Section 3) contains exact results valid at any finite $N$. We outline the integrable structure of the normal and complex models at $\beta = 1$ (the Hirota relations for the partition function, the orthogonal polynomials technique and the Lax representation). In addition, we derive the exact relation between correlation functions of the eigenvalue densities (referred to as the loop equation) which holds for arbitrary values of $\beta$. In the third part (Sections 4 and 5) we examine the large $N$ limit of the models of random matrices with complex eigenvalues and discuss the applications to the growth processes and to the semiclassical electronic droplets in magnetic field. The Appendices contain technical details of some proofs and calculations.

## 2. Some ensembles of random matrices with complex eigenvalues

We consider square random matrices $\Phi$ of size $N$ with complex entries $\Phi_{ij}$ subject to certain constraints depending on the particular ensemble. Some ensembles of random matrices are listed in the following table:

| Ensemble | Notation | Condition | Dimension |
|---|---|---|---|
| Hermitian | $\mathscr{H}$ | $\Phi^\dagger = \Phi$ | $N^2$ |
| Unitary | $\mathscr{U}$ | $\Phi^\dagger \Phi = 1$ | $N^2$ |
| Normal | $\mathscr{N}$ | $[\Phi^\dagger, \Phi] = 0$ | $N^2+N$ |
| Normal self-dual | $\mathscr{N}^0$ | $[\Phi^\dagger, \Phi] = 0$ <br> $\Phi$ self-dual | $\frac{1}{2}N^2 + N$ |
| Complex | $\mathscr{C}$ | none | $2N^2$ |

The first two matrix ensembles, $\mathscr{H}$ and $\mathscr{U}$, are the most popular ones. They are given here just for comparison. Eigenvalues of matrices from $\mathscr{H}$ and $\mathscr{U}$ are confined to the real axis and to the unit circle respectively. The last three ensembles (which are the main subject of these lectures) do not imply any a priori restrictions on eigenvalues of the matrices. Normal matrices are defined by the constraint that they commute with their adjoint. The ensemble $\mathscr{N}^0$ is defined for $N$ even only. The meaning of the condition "$\Phi$ is self-dual" is explained below in this section. By dimension of the ensemble we mean the *real* dimension of the matrix variety.

Throughout this paper we consider the probability densities of the form $P(\Phi) \propto e^{\mathrm{tr} W(\Phi)}$, where the function $W(\Phi)$ (often called the potential of the matrix model) is a matrix-valued function of $\Phi$ and $\Phi^\dagger$ such that $(W(\Phi))^\dagger = W(\Phi)$. This form is similar to the one usually employed in Hermitian and unitary ensembles. The partition function is defined as the integral over matrices from one or another ensemble:

$$Z_N = \int [D\Phi] e^{\mathrm{tr}\, W(\Phi)} \tag{1}$$

Summing over $N$ with a suitable weight, one may also define the grand canonical ensembles corresponding to (1) but we do not pursue this possibility here.

We need to specify the integration measure $[D\Phi]$ and the potential $W(\Phi)$. Given the measure and the potential, one is usually interested in the distribution and correlations of the eigenvalues. In general, they can be distributed on the real line for $\mathscr{H}$, on the unit circle for $\mathscr{U}$ and on the whole complex plane for $\mathscr{N}$, $\mathscr{N}^0$ and $\mathscr{C}$.

## 2.1 Integration measures

The integration measure has the most simple form for the ensemble of general complex matrices:

$$[D\Phi] = \prod_{i,j=1}^{N} d(\mathcal{R}e\,\Phi_{ij})\, d(\mathcal{I}m\,\Phi_{ij})$$

This measure is additively invariant and multiplicatively covariant, i.e. for any fixed (nondegenerate) matrix $A \in \mathscr{C}$ we have the properties $[D(\Phi + A)] = [D\Phi]$ and $[D(\Phi A)] = [D(A\Phi)] = |\det A|^{2N}[D\Phi]$. The first one is obvious, to prove the second one is an easy exercise. It is clear that the measure is invariant under transformations of the form $\Phi \to U^\dagger \Phi U$ with a unitary matrix $U$ ("rotations" in the matrix space).

The measure for $\mathscr{N}$ is induced by the standard flat metric in $\mathscr{C}$,

$$||\delta\Phi||^2 = \text{tr}\,(\delta\Phi\delta\Phi^\dagger) = \sum_{ij} |\delta\Phi_{ij}|^2$$

via the embedding $\mathscr{N} \subset \mathscr{C}$. Here $\mathscr{N}$ is regarded as a hypersurface in $\mathscr{C}$ defined by the quadratic relations $\Phi\Phi^\dagger = \Phi^\dagger\Phi$. The measure for the ensemble $\mathscr{N}^0$ is defined in a similar way.

As usual in matrix models, we would like to integrate out the "angular" variables and to express the integration measure through eigenvalues of the matrices.

**The measure for $\mathscr{N}$ through eigenvalues [1, 15].** We derive the explicit representation of the measure in terms of eigenvalues in three steps:

1. Introduce coordinates in $\mathscr{N} \subset \mathscr{C}$.

2. Compute the inherited metric on $\mathscr{N}$ in these coordinates: $||\delta\Phi||^2 = g_{\alpha\beta}d\xi^\alpha d\xi^\beta$.

3. Compute the volume element $[D\Phi] = \sqrt{|\det g_{\alpha\beta}|}\prod_\alpha d\xi^\alpha$.

*Step 1: Coordinates in $\mathscr{N}$.* For any matrix $\Phi$, the matrices $H_1 = \frac{1}{2}(\Phi + \Phi^\dagger)$, $H_2 = \frac{1}{2i}(\Phi - \Phi^\dagger)$ are Hermitian. The condition $[\Phi, \Phi^\dagger] = 0$ is equivalent to $[H_1, H_2] = 0$. Thus $H_{1,2}$ can be simultaneously diagonalized by a unitary matrix $U$:

$$\begin{aligned} H_1 &= UXU^\dagger\,, \quad & X &= \text{diag}\,\{x_1, \ldots, x_N\} \\ H_2 &= UYU^\dagger\,, \quad & Y &= \text{diag}\,\{y_1, \ldots, y_N\} \end{aligned}$$

Introduce the diagonal matrices $Z = X + iY$, $\bar{Z} = X - iY$ with diagonal elements $z_j = x_j + iy_j$ and $\bar{z}_j = x_j - iy_j$ respectively. Note that $z_j$ are eigenvalues of $\Phi$. Therefore, any $\Phi \in \mathcal{N}$ can be represented as

$$\Phi = UZU^\dagger$$

where $U$ is a unitary matrix and $Z$ is the diagonal matrix with eigenvalues of $\Phi$ on the diagonal. In fact normal matrices can be equivalently defined by the property of being the most general matrices that can be diagonalized by a unitary transformation. The matrix $U$ is defined up to multiplication by a diagonal unitary matrix from the right: $U \to U\, U_{\rm diag}$. The dimension of $\mathcal{N}$ is thus

$$\dim(\mathcal{N}) = \dim(\mathcal{U}) - \dim(\mathcal{U}_{\rm diag}) + \dim(\mathcal{C}_{\rm diag}) = N^2 - N + 2N \; = \; N^2 + N$$

Let us make a remark that the naive counting of the number of constraints in the condition $[\Phi, \Phi^\dagger] = 0$ leads to a wrong result for the dimension of $\mathcal{N}$. On the first glance, this condition gives $N^2 - 1$ independent constraints. Indeed, set $H = [\Phi, \Phi^\dagger]$. Then the conditions $H_{lk} = 0$ for $l < k$ give $N(N-1)$ real constraints and the conditions $H_{kk} = 0$ give $N - 1$ real constraints (because $\operatorname{tr} H = 0$ identically), in total $N^2 - 1$ constraints. We thus observe that $\dim(\mathcal{N}) \neq \dim(\mathcal{C}) - (N^2 - 1)$. Therefore, there are only $N^2 - N$ independent constraints among the $N^2 - 1$ equations $[\Phi, \Phi^\dagger] = 0$. This fact can be easily illustrated by the example of $2 \times 2$ matrices.
*Step 2: The induced metric.* Since $\Phi = UZU^\dagger$, the variation is $\delta\Phi = U(\delta u \cdot Z + \delta Z + Z \cdot \delta u^\dagger)U^\dagger$, where $\delta u^\dagger = U \delta U^\dagger = -\delta u^\dagger$. Therefore,

$$\begin{aligned}||\delta\Phi||^2 &= \operatorname{tr}(\delta\Phi\delta\Phi^\dagger) = \operatorname{tr}(\delta Z \delta \bar{Z}) + 2\operatorname{tr}(\delta u Z \delta u \bar{Z} - (\delta u)^2 Z\bar{Z}) \\ &= \sum_{j=1}^{N} |\delta z_j|^2 + 2\sum_{j<k}^{N} |z_j - z_k|^2\, |\delta u_{jk}|^2\end{aligned}$$

(Note that $\delta u_{jj}$ do not enter.) This is the square of the line element $||\delta\Phi||^2 = g_{\alpha\beta}\delta\xi^\alpha\delta\xi^\beta$.
*Step 3. The volume element.* We see that the metric $g_{\alpha\beta}$ is diagonal in the coordinates $\mathcal{R}e(\delta z_j)$, $\mathcal{I}m(\delta z_j)$, $\mathcal{R}e(\delta u_{jk})$, $\mathcal{I}m(\delta u_{jk})$ with $1 \le j < k \le N$, so the determinant of the diagonal matrix $g_{\alpha\beta}$ is easily calculated to be $|\det g_{\alpha\beta}| = 2^{N^2-N} \prod_{j<k}^{N} |z_i - z_k|^4$. Therefore,

$$[D\Phi] \propto [DU]'\, |\Delta_N(z_1, \ldots, z_N)|^2 \prod_{j=1}^{N} d^2 z_j \tag{2}$$

where $d^2 z \equiv dxdy$ is the flat measure in the complex plane, $[DU]' = [DU]/[DU_{\rm diag}]$ is the invariant measure on $\mathcal{U}/\mathcal{U}_{\rm diag}$, and

$$\Delta_N(z_1, \ldots, z_N) = \prod_{j>k}^{N} (z_j - z_k) = \det_{N\times N}(z_j^{k-1}) \tag{3}$$

is the Vandermonde determinant.

Similarly to the ensemble $\mathscr{H}$ of Hermitian matrices, the measure (2) contains the squared modulus of the Vandermonde determinant. The difference is that the eigenvalues are complex numbers. The statistical model of normal random matrices was studied in [15, 16].

**Normal self-dual matrices.** Let $\Gamma$ be the matrix

$$\Gamma = \begin{pmatrix} 0 & 1 & & & & \\ -1 & 0 & & & & \\ & & 0 & 1 & & \\ & & -1 & 0 & & \\ & & & & \cdots & \\ & & & & & \cdots \end{pmatrix}, \quad \Gamma^2 = -1$$

(all other entries are zero). A complex matrix $\Phi$ is called *self-dual* if $\Gamma\Phi^{\mathrm{T}}\Gamma = -\Phi$ (the superscript T means transposition). The size of a self-dual matrix is thus an even number. It can be shown that eigenvalues of self-dual matrices always come in pairs: the diagonal form of $\Phi$ is

$$Z = \mathrm{diag}\,\{z_1, z_1, z_2, z_2, \ldots, z_N, z_N\}$$

As is easy to verify, the condition that $\Phi$ is self-dual is equivalent to the condition that the matrix $\Gamma\Phi$ is anti-symmetric.

*Normal self-dual* matrices are parameterized as $\Phi = UZU^{\dagger}$ with $Z$ as above, where $U$ is unitary and symplectic: $U^{\dagger}U = 1$, $U^{\mathrm{T}}\Gamma U = \Gamma$. In other words, $U$ belongs to the maximal compact subgroup in the complex group of symplectic matrices $\mathrm{Sp}(N)$. The latter is known to have real dimension $4N^2 + 2N$, with the dimension of the maximal compact subgroup being twice less. Therefore, similarly to the calculation for normal matrices, $\dim(\mathscr{N}^0) = 2N^2 + N - N + 2N = 2N^2 + 2N$ (the value given in the table above corresponds to $N$ replaced by $N/2$). The integration measure appears to be

$$[D\Phi] \propto [DU]'\, |\Delta_N(z_1, \ldots, z_N)|^4 \prod_{j=1}^{N} d^2 z_j \tag{4}$$

(for $2N \times 2N$ matrices). Note that the module of the Vandermonde determinant enters in the fourth degree. The statistical model of normal self-dual matrices was discussed in [17].

**The measure for $\mathscr{C}$ through eigenvalues.** A complex matrix $\Phi$ with eigenvalues $z_1, \ldots, z_N$ can be decomposed as

$$\Phi = U(Z + R)U^{\dagger}$$

where $Z = \text{diag}\{z_1, \ldots, z_N\}$ is diagonal, $U$ is unitary, and $R$ is strictly upper triangular, i.e., $R_{ij} = 0$ if $i \geq j$. These matrices are defined up to a "gauge transformation": $U \to U U_{\text{diag}}$, $R \to U^{\dagger}_{\text{diag}} R U_{\text{diag}}$. It is not so easy to see that the measure factorizes. This requires some work, of which the key step is a specific ordering of the independent variables. The final result is:

$$[D\Phi] \propto [DU]' \left( \prod_{k<l} d^2 R_{kl} \right) |\Delta_N(z_i)|^2 \prod_{j=1}^{N} d^2 z_j \tag{5}$$

The details can be found in the Mehta book [1].

## 2.2 Potentials

For the ensembles $\mathcal{N}$, $\mathcal{N}^0$ the "angular variables" (parameters of the unitary matrix $U$) always decouple after taking the trace $\text{tr}\, W(\Phi) = \sum_j W(z_j)$, so the potential $W$ can be a function of $\Phi$, $\Phi^{\dagger}$ of a general form $W(\Phi) = \sum a_{nm} \Phi^n (\Phi^{\dagger})^m$. Two important particular cases arise if the potential is:

- Axially symmetric, $W(\Phi) = W_0(\Phi\Phi^{\dagger})$. In this case the $N$-fold integral essentially reduces to ordinary ones, and so some basic results become available in a quite explicit form.

- Harmonic on the background of $\Phi\Phi^{\dagger}$, i.e., $W(\Phi) = -\Phi\Phi^{\dagger} + V(\Phi) + \bar{V}(\Phi^{\dagger})$. In what follows, we call it *quasiharmonic*. Here $V(z)$ is an *analytic* function of $z$ in some domain containing the origin and $\bar{V}(z) = \overline{V(\bar{z})}$. In terms of the eigenvalues, the quasiharmonic potential is

$$W(z) = -|z|^2 + V(z) + \overline{V(z)} \tag{6}$$

  This case is particularly important for applications. The normal matrix model with quasiharmonic potentials bears some formal similarities with the model of two coupled Hermitian matrices [18] and the matrix quantum mechanics in the singlet sector [19].

The partition function reduces to

$$Z_N = \int |\Delta_N(z_i)|^{2\beta} \prod_{j=1}^{N} e^{W(z_j)} d^2 z_j \tag{7}$$

where $\beta = 1$ for $\mathcal{N}$ and $\beta = 2$ for $\mathcal{N}^0$. From now on this formula is taken as the definition of the partition function. Comparing to (1), we redefine $W \to W/\beta$ and ignore a possible $N$-dependent normalization factor. One may also consider this integral for arbitrary values of $\beta$.

The choice of the potential for the ensemble $\mathscr{C}$ is more restricted. For a general potential, the matrix $U$ in $\Phi = U(Z+R)U^\dagger$ still decouples but $R$ does not. The problem becomes too complicated. An important particular case, when $R$ nevertheless decouples is the quasiharmonic potential (6). Indeed, $\mathrm{tr}\,(\Phi\Phi^\dagger) = \mathrm{tr}\,(Z\bar{Z}) + \mathrm{tr}\,(RR^\dagger)$, $\mathrm{tr}\,(\Phi^n) = \mathrm{tr}\,(Z+R)^n = \mathrm{tr}\,Z^n$, and so

$$\int_{\mathscr{C}} [D\Phi] e^{\mathrm{tr}\,W(\Phi)} = C_N \int |\Delta(z_i)|^2 \prod_k e^{W(z_k)} d^2 z_k \tag{8}$$

where $C_N$ is an $N$-dependent normalization factor proportional to the gaussian integral $\int [DR] e^{-\mathrm{tr}\,(RR^\dagger)}$.

As an example, let us consider the quadratic potential:

$$W(z) = -\sigma|z|^2 + 2\mathcal{R}e\,(t_1 z + t_2 z^2)\,, \quad \sigma > 0$$

The ensemble $\mathscr{C}$ ($\beta = 1$) with this potential is known as *the Ginibre-Girko ensemble* [11, 20]. In this case the partition function (7) can be calculated exactly [21]:

$$Z_N = Z_N^{(0)} (\sigma^2 - 4|t_2|^2)^{-N^2/2} \exp\left( N \frac{t_1^2 \bar{t}_2 + \bar{t}_1^2 t_2 + \sigma |t_1|^2}{\sigma^2 - 4|t_2|^2} \right)$$

where

$$Z_N^{(0)} = \sigma^{(N^2-N)/2} \pi^N \prod_{k=1}^{N} k!$$

To the best of our knowledge, there are no exact results for 2D integrals of this type with $|\Delta(z_i)|^{2\beta}$ for other values of $\beta$, even for the pure Gaussian weight.

Coming back to the general case, we note that some integrals considered above may diverge. In the most important case of quasiharmonic potential, the integral

$$\int e^{-|z|^2 + V(z) + \overline{V(z)}} d^2 z$$

converges for potentials $V(z) = \alpha z^2 + \beta z + \sum_i \mu_i \log(z - a_i)$ with $|\alpha| < \frac{1}{2}$ and $\mu_j > -1$ but it always diverges if $V(z)$ is a polynomial of degree $\geq 3$. As usual in matrix models, really interesting science begins when integrals diverge! Let us say a few words about how one should understand divergent integrals. The conventional viewpoint is to treat all cubic and higher degree terms in the potential as a small perturbation. The integral is then regarded as a perturbative series for a theory which is believed to be well-defined on the nonperturbative level. The nonperturbative definition can be achieved either by introducing an ad hoc cutoff or via more sophisticated methods in the spirit of the Marinari-Parisi approach [22] (the "stochastic stabilization"). As far as the large $N$ limit is concerned, the integral for the partition function can be *defined* through the expansion around the saddle point. This gives the $1/N$-expansion

$$\log Z_N \sim \sum_h N^{-h} F^{(h)} \qquad \text{as } N \to \infty$$

Even if the integral for $Z_N$ diverges, each term of the $1/N$ expansion is often well-defined.

## 2.3 Physical interpretations

The ensembles of random matrices appear to be mathematically equivalent to some important model systems of statistical and quantum mechanics. They are:

- The 2D Coulomb plasma (any $\beta$)
- Non-interacting fermions ($\beta = 1$)
- Electrons in magnetic field ($\beta = 1$)

The equivalence holds for any finite $N$. The first two interpretations are standard and well known. The third one can be regarded as a specification of the second one for models with eigenvalues distributed over the whole complex plane. Remarkably, in this very case the noninteracting fermions picture (which is rather formal for $\mathscr{H}$ and $\mathscr{U}$) acquires a very interesting physical content related to the quantum Hall effect.

**The Dyson gas picture.** This interpretation, first suggested by Dyson [12] for the unitary, symplectic and orthogonal matrix ensembles, relies on rewriting $|\Delta_N(z_i)|^{2\beta}$ as $\exp\left(\beta \sum_{i\neq j} \log|z_i - z_j|\right)$. Clearly, the integral (7) looks then exactly as the partition function of the 2D Coulomb plasma (often called the Dyson gas) at "temperature" $1/\beta$, in the external electric field:

$$Z_N = \int e^{-\beta E(z_1,\ldots,z_N)} \prod d^2 z_j \tag{9}$$

The eigenvalues play the role of the 2D Coulomb charges. The energy is

$$E = -\sum_{i<j} \log|z_i - z_j|^2 - \beta^{-1} \sum_j W(z_j) \tag{10}$$

The first sum is the Coulomb interaction energy, the second one is the energy due to the external field. For the ensembles $\mathscr{H}$ and $\mathscr{U}$ the charges are confined to dimension 1 (the real line or the unit circle) but interact as 2D Coulomb charges. So, the Dyson gas picture for the ensembles $\mathscr{N}$, $\mathscr{N}^0$ and $\mathscr{C}$ looks even more natural. The Dyson gas interpretation becomes especially helpful in the large $N$ limit, where it allows one to apply thermodynamical arguments.

**Non-interacting fermions ($\beta = 1$).** Given any system of polynomials of the form $P_n(z) = z^n +$ lower degrees, the Vandermonde determinant can be written as $\Delta_N(z_i) = \det(z_j^{k-1}) = \det(P_{k-1}(z_j))$. Let us rewrite the statistical weight as

$$|\Delta_N(z_i)|^2 \prod_{j=1}^{N} e^{W(z_j)} = |\Psi(z_1, \ldots, z_N)|^2$$

where $\Psi(z_1, \ldots, z_N) = \det_{N\times N}\left(P_{k-1}(z_j)e^{W(z_j)}\right)$ is the (unnormalized) wave function of $N$ non-interacting fermions (the Slater determinant), with one-particle wave functions being $\psi_k(z) = P_{k-1}(z)e^{W(z)}$. The partition function is the normalization integral:

$$Z_N = \int |\Psi|^2 \prod_i d^2 z_i$$

For the ensembles $\mathscr{N}$ and $\mathscr{C}$ (with quasiharmonic potential) the wave function $\Psi$ has a direct physical meaning as a wave function of 2D electrons in magnetic field.

**Electrons in the plane in magnetic field.** Consider a charged particle with spin $\frac{1}{2}$ (electron) moving in the plane in a strong (not necessarily uniform) magnetic field $B = B(x, y)$ orthogonal to the plane. The Pauli Hamiltonian reads

$$\hat{H} = \frac{1}{2m}\left((i\hbar\nabla + \vec{A})^2 - \hbar\sigma_3 B\right)$$

Here $m$ is mass of the particle, $\sigma_3 = \text{diag}\,(1, -1)$ is the Pauli matrix, $B = \partial_x A_y - \partial_y A_x$. In 2D, the complex notation is convenient: $A = A_x - iA_y$, $\bar{A} = A_x + iA_y$, so that $B = i(\bar{\partial}A - \partial\bar{A})$.

For a uniform magnetic field, $B = B_0$, one can choose the gauge $A = \frac{B_0}{2i}\,\bar{z}$. Solving the Schrodinger equation $\hat{H}\psi = E\psi$, one gets Landau levels:

$$E_n = \frac{\hbar B_0}{m}\left(n + \frac{1}{2} - s\right), \qquad n = 0, 1, 2, \ldots\,, \quad s = \pm\frac{1}{2}$$

The gap between the levels is proportional to $B_0/m$. Each level is highly degenerate. The wave functions at the level $E = 0$ are

$$\psi_n = z^n \exp\left(-\frac{B_0}{4\hbar}|z|^2\right)$$

We note in passing that the system of fermions in the magnetic field at the lowest energy level admits a collective field theory description which was discussed in the literature in different contexts (see e.g. [13, 23, 24]).

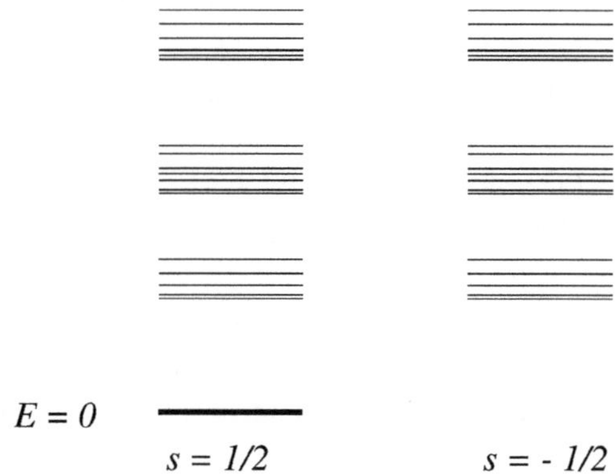


*Figure 1.* *The energy levels of a spin-$\frac{1}{2}$ electron in a non-uniform magnetic field (schematically). The lowest level $E = 0$ remains highly degenerate.*

Let us turn to the problem with a nonuniform magnetic field. Choose the gauge $A = i\,\partial W$ with a real-valued function $W$, then $B = -2\,\partial\bar{\partial}W$ and $\operatorname{div}\vec{A} = \bar{\partial}A + \partial\bar{A} = 0$. Therefore,

$$(i\hbar\nabla + \vec{A})^2 = -4\hbar^2\partial\bar{\partial} + 2i\hbar((\bar{\partial}W)\partial + (\partial W)\bar{\partial}) + |\partial W|^2$$

$$= (2\hbar\partial + \partial W)(-2\hbar\bar{\partial} + \bar{\partial}W) - 2\hbar\,\partial\bar{\partial}W$$

The Hamiltonian can be represented as the 2×2 matrix

$$\hat{H} = \begin{pmatrix} H_+ & 0 \\ 0 & H_- \end{pmatrix}$$

where $2mH_\pm = (i\hbar\nabla + \vec{A})^2 \pm 2\hbar\,\partial\bar{\partial}W$. In general, the spectral problem for this Hamiltonian does not admit an explicit solution. However, the level $E = 0$ is very special. Note that $H_+$ factorizes: $H_+ = (2\hbar\partial + \partial W)(-2\hbar\bar{\partial} + \bar{\partial}W)$, so exact wave functions at the level $E = 0$ can be found by solving the *first order* equation

$$H_+\psi = (2\hbar\bar{\partial} - \bar{\partial}W)\psi = 0$$

The general solution is

$$\psi(z) = P(z)\exp\left(\frac{1}{2\hbar}W(z)\right) \tag{11}$$

where $P(z)$ is an arbitrary *holomorphic* polynomial. The zero energy level remains to be highly degenerate even in the nonuniform magnetic field (Fig. 1). This fact was first observed in [25].

To find degeneracy of the level, we solve the Poisson equation $\Delta W = -2B$,

$$W(z) = -\frac{1}{\pi} \int \log|z-\zeta| B(\zeta) d^2\zeta \tag{12}$$

and observe that $W(z)$ tends to $-\frac{\phi}{\pi} \log|z|$ as $|z| \to \infty$, where $\phi = \int B\, d^2 z$ is the total magnetic flux. If $P(z)$ is of degree $n$, the asymptotics of $\psi$ for $|z|$ large is

$$\psi = P(z)\, e^{\frac{1}{2\hbar}W(z)} \to z^n\, |z|^{-\phi/\phi_0}$$

where $\phi_0 = 2\pi\hbar$ is the flux quantum. We require the wave functions to be normalizable, i.e, $\int |\psi|^2 d^2 z < \infty$ that means $n < \phi/\phi_0 - 1$. Therefore,

$$n_{\max} = [\phi/\phi_0] - 1$$

($[\ldots]$ is the integer part), and the degeneracy is equal to the number of flux quanta in the total flux:

$$N = [\phi/\phi_0]$$

If the Coulomb forces can be ignored, the wave function of $N$ electrons in the plane in the magnetic field at the lowest energy level is constructed as the $N \times N$ Slater determinant of the functions of the type (11) with polynomials of different degrees.

The situation when the lowest energy level $E = 0$ is completely filled, i.e., $N = n_{\max}$, is the (integer) quantum Hall (QH) regime. The notion of the QH droplet [13] implies that the electronic liquid is incompressible, i.e., all states at the lowest energy level are occupied. We come to the following conclusion: the QH droplet consisting of $N$ electrons (in general, in a non-uniform magnetic field) is equivalent to the ensemble of normal $N \times N$ matrices.

# 3. Exact results at finite $N$

## 3.1 Correlation functions: general relations

The main objects to be determined in random matrix models are correlation functions. In general, they are mean values of scalar-valued functions of matrices. The mean value of such a function $A(\Phi)$ of the matrix $\Phi$ is defined, with the help of the statistical weight, in the usual way:

$$\langle A \rangle = \frac{\displaystyle\int [D\Phi] A(\Phi) e^{\operatorname{tr} W(\Phi)}}{\displaystyle\int [D\Phi] e^{\operatorname{tr} W(\Phi)}}$$

We shall consider functions that depend on eigenvalues only – for example, traces of matrices. Correspondingly, typical correlators which we are going to study are mean values of products of traces: $\langle \mathrm{tr}\, f(\Phi)\rangle$, $\langle \mathrm{tr}\, f_1(\Phi)\, \mathrm{tr}\, f_2(\Phi)\rangle$ and so on. Clearly, they are represented as integrals over eigenvalues. For example,

$$\langle \mathrm{tr}\, f(\Phi)\rangle = \frac{N \int |\Delta_N(z_i)|^{2\beta} f(z_1) \prod_{j=1}^{N} e^{W(z_j)} d^2 z_j}{\int |\Delta_N(z_i)|^{2\beta} \prod_{j=1}^{N} e^{W(z_j)} d^2 z_j}$$

Here, $f(\Phi) = f(\Phi, \Phi^\dagger)$ is any function of $\Phi$, $\Phi^\dagger$ which is regarded as the function $f(z_i) = f(z_i, \bar{z}_i)$ of the complex argument $z_i$ (and $\bar{z}_i$) in the r.h.s. A particularly important example is the density function defined as

$$\rho(z) = \sum_j \delta(z - z_j) = \mathrm{tr}\, \delta(z - \Phi) \tag{13}$$

where $\delta(z)$ is the two dimensional $\delta$-function. As it immediately follows from the definition, any correlator of traces is expressed through correlators of $\rho$:

$$\langle \mathrm{tr}\, f_1(\Phi) \ldots \mathrm{tr}\, f_n(\Phi)\rangle = \int \langle \rho(z_1) \ldots \rho(z_n)\rangle\, f_1(z_1) \ldots f_n(z_n) \prod_{j=1}^{n} d^2 z_j \tag{14}$$

Instead of correlations of density it is often convenient to consider correlations of the field

$$\varphi(z) = -\beta \sum_j \log |z - z_j|^2 = -\beta \log |\det(z - \Phi)|^2 \tag{15}$$

from which the correlations of density can be found by means of the relation

$$4\pi\beta\rho(z) = -\Delta\varphi(z) \tag{16}$$

Clearly, $\varphi$ is the 2D Coulomb potential created by the eigenvalues (charges). As it directly follows from the definitions,

$$\langle \rho(z)\rangle_N = N \frac{Z_{N-1}}{Z_N} \left\langle e^{W(z) - \varphi(z)} \right\rangle_{N-1}$$

where $\langle \ldots \rangle_N$ means the expectation value in the ensemble of $N \times N$ matrices.

Handling with multipoint correlation functions, it is customary to pass to their *connected parts*. For example, in the case of 2-point functions, the connected correlation function is defined as

$$\langle \rho(z_1)\rho(z_2)\rangle_c \equiv \langle \rho(z_1)\rho(z_2)\rangle - \langle \rho(z_1)\rangle \langle \rho(z_2)\rangle$$

The connected multi-trace correlators are expressed through the connected density correlators by the same formula (14) with $\langle \rho(z_1) \dots \rho(z_n) \rangle_c$ in the r.h.s. The connected part of the $(n+1)$-point density correlation function is given by the linear response of the $n$-point one to a small variation of the potential. More precisely, the following variational formulas hold true:

$$\langle \rho(z) \rangle = \frac{\delta \log Z_N}{\delta W(z)}, \quad \langle \rho(z_1)\rho(z_2) \rangle_c = \frac{\delta \langle \rho(z_1) \rangle}{\delta W(z_2)} = \frac{\delta^2 \log Z_N}{\delta W(z_1) \delta W(z_2)} \tag{17}$$

Connected multi-point correlators are higher variational derivatives of $\log Z_N$. These formulas follow from the fact that variation of the partition function over a general potential $W$ inserts $\sum_i \delta(z - z_i)$ into the integral. Let us stress that these formulas are exact for any finite $N$.

For a later use, we mention the formula

$$\left\langle e^{\mathrm{tr}\, f(\Phi)} \right\rangle = \exp\left( \sum_{k=1}^{\infty} \frac{1}{k!} \left\langle (\mathrm{tr}\, f(\Phi))^k \right\rangle_c \right) \tag{18}$$

which immediately follows from the expansion

$$\log \left\langle e^{\mathrm{tr}\, f} \right\rangle = \log Z_N(W + f) - \log Z_N(W)$$

$$= \int \frac{\delta \log Z_N}{\delta W(\zeta)} f(\zeta) d^2\zeta + \frac{1}{2!} \int \frac{\delta^2 \log Z_N}{\delta W(\zeta) \delta W(\zeta')} f(\zeta) f(\zeta') d^2\zeta d^2\zeta' + \dots$$

## 3.2 Integrable structure of the $\mathscr{N}$ and $\mathscr{C}$ ensembles ($\beta = 1$)

The partition function (7) for $\beta = 1$,

$$Z_N = \int |\Delta_N(z_i)|^2 \prod_{j=1}^{N} e^{W(z_j)} d^2 z_j$$

regarded as a function of $N$ and Taylor coefficients of the potential $W$, has remarkable properties, which we briefly review below.

**Determinant representation.** The following simple but important determinant representation holds true:

$$Z_N = N! \det_{N \times N} (C_{ij}), \quad 1 \le i, j \le N \tag{19}$$

where

$$C_{ij} = \int z^{i-1} \bar{z}^{j-1} e^{W(z)} d^2 z \tag{20}$$

is the "matrix of moments". The proof is almost a repetition of the corresponding proof for the hermitian model. Complexity of eigenvalues does not cause any difficulties. For completeness, the detailed proof is given in the Appendix.

**Hirota relations.** Let us apply the determinant formula to $\langle \det(\lambda - \Phi)\rangle$:

$$\begin{aligned}\langle \det(\lambda - \Phi)\rangle &= \frac{1}{Z_N}\int |\Delta_N|^2 \prod_j (\lambda - z_j) e^{W(z_j)} d^2 z_j \\ &= \frac{1}{Z_N} \det \left[ z^{i-1}\bar{z}^{j-1}(\lambda - z) e^W d^2 z \right]\end{aligned}$$

Comparing this with the determinant representation of $Z_N$, we see that $\langle \det(\lambda - \Phi)\rangle = \lambda^N Z_N^{-1} \det\left[C_{ij} - \lambda^{-1} C_{i+1,j}\right]$. Taking appropriate linear combinations of the lines, one can reduce this determinant to the determinant of a matrix which differs from $C_{ij}$ only in the last line. Similarly, in the determinant representation of $\langle \det(\lambda_1 - \Phi)\det(\lambda_2 - \Phi)\rangle$ only two last lines change. Now, some standard identities for determinants lead to the following relations:

$$(\lambda_1 - \lambda_2)\,\langle \det(\lambda_1 - \Phi)\det(\lambda_2 - \Phi)\rangle\,\langle \det(\lambda_3 - \Phi)\rangle + \text{cyclic perm-s of } (123) = 0 \qquad (21)$$

$$\left\langle |\det(\lambda - \Phi)|^2 \right\rangle_N - \left|\langle \det(\lambda - \Phi)\rangle_N\right|^2 = \frac{N}{N+1}\,\frac{Z_{N+1}Z_{N-1}}{Z_N^2}\left\langle |\det(\lambda - \Phi)|^2\right\rangle_{N-1} \qquad (22)$$

Write

$$\langle \det(\lambda - \Phi)\rangle_N = \lambda^N \frac{Z_N(W + [\lambda])}{Z_N(W)}$$

where

$$W + [\lambda] \equiv W(z) + \log\left(1 - \frac{z}{\lambda}\right)$$

is the potential modified by the (complex and multi-valued) term $\log\left(1 - \frac{z}{\lambda}\right)$. (Since this term is always under the exp-function, there is no ambiguity in the choice of its branch.) In this notation, the above identities for determinants acquire the form of the *Hirota bilinear equations* [26]:

$$(\lambda_1 - \lambda_2)\, Z_N(W + [\lambda_1] + [\lambda_2])\, Z_N(W + [\lambda_3]) + \text{cyclic perm-s of } 1,2,3 = 0 \qquad (23)$$

$$\begin{aligned}&Z_N(W) Z_N(W + [\lambda] + [\bar\lambda]) - Z_N(W + [\lambda]) Z_N(W + [\bar\lambda]) \\ &= \frac{N}{N+1} |\lambda|^{-2} Z_{N+1}(W)\, Z_{N-1}(W + [\lambda] + [\bar\lambda])\end{aligned} \qquad (24)$$

Let us parameterize the potential as

$$W(z) = W^{(0)}(z) + \sum_k (t_k z^k + \bar t_k \bar z^k)$$

then these equations state that $Z_N/N!$, as a function of $t_k$, $\bar{t}_k$, is the *tau-function* of the 2D Toda lattice hierarchy. The transformation $W \to W + [\lambda]$ is equivalent to the change of variables $t_k \to t_k - \frac{1}{k}\lambda^{-k}$ which is known in the literature as the Miwa transformation.

**Orthogonal polynomials and the kernel function.** The orthogonal polynomials technique is useful not only for hermitian and unitary matrix ensembles but for the normal and complex ensembles as well. The orthogonal polynomials are introduced as mean values of the characteristic polynomials of the random matrices $\Phi$:

$$P_n(\lambda) = \langle \det(\lambda - \Phi) \rangle_n \tag{25}$$

Clearly, $P_n$ are polynomials in $\lambda$ of the form $P_n(\lambda) = \lambda^n +$ lower degrees.

The main property of the polynomials introduced is their *orthogonality* in the complex plane:

$$\int P_n(z)\overline{P_m(z)}e^{W(z)}d^2z = h_n\,\delta_{mn} \tag{26}$$

The square of the norm $h_n = ||P_n||^2$ is connected with the partition function as

$$h_n = \frac{1}{n+1}\frac{Z_{n+1}}{Z_n}, \qquad Z_N = N!\prod_{n=0}^{N-1} h_n$$

Again, the proof is completely parallel to the corresponding proof in the hermitian models. See Appendix for details.

The functions

$$\psi_n(z) = \frac{1}{\sqrt{h_{n-1}}} P_{n-1}(z)e^{W(z)/2}$$

are *orthonormal*:

$$\int \psi_n(z)\overline{\psi_m(z)}d^2z = \delta_{mn} \tag{27}$$

These $\psi_n$'s are "one-particle wave functions" of electrons in the magnetic field. The $N$-particle wave function is $\Psi_N(z_1,\ldots,z_N) \sim \det[\psi_j(z_k)]$. The joint probability to find "particles" at $z_1,\ldots,z_N$ is

$$|\Psi_N(z_1,\ldots,z_N)|^2 = \frac{1}{N!}|\det[\psi_j(z_k)]|^2$$

Since $|\det M|^2 = \det(MM^\dagger)$, we can write: $|\det[\psi_j(z_k)]|^2 = \det\left(\sum_{n=1}^{N}\psi_n(z_j)\overline{\psi_n(z_k)}\right)$. The expression under the determinant is called the *kernel function*:

$$K_N(z,\bar{w}) = \sum_{n=1}^{N}\psi_n(z)\overline{\psi_n(w)} \tag{28}$$

The main properties of the kernel function are:

- Hermiticity: $K_N(z, \bar{w}) = \overline{K_N(w, \bar{z})}$;
- Normalization: $\int K_N(z, \bar{z}) d^2 z = N$;
- Projection property: $\int K_N(z_1, \bar{z}) K_N(z, \bar{z}_2) d^2 z = K_N(z_1, \bar{z}_2)$.

All density correlation functions can be expressed through the kernel function. For example:

$$\langle \rho(z) \rangle_N = K_N(z, \bar{z})$$

$$\langle \rho(z_1)\rho(z_2) \rangle_N = \begin{vmatrix} K_N(z_1, \bar{z}_1) & K_N(z_1, \bar{z}_2) \\ K_N(z_2, \bar{z}_1) & K_N(z_2, \bar{z}_2) \end{vmatrix} + K_N(z_1, \bar{z}_1)\delta(z_1 - z_2)$$

The last term is a *contact term*. It does not contribute if $z_1 \neq z_2$. In general, one has:

$$\langle \rho(z_1) \dots \rho(z_n) \rangle_N = \det(K_N(z_i, \bar{z}_j))_{1 \le i,j \le n} + \text{contact terms}$$

where the contact terms vanish if all the points $z_i$ are different.

Note that for the ensemble $\mathscr{N}$ with an axially-symmetric potential $W(\Phi) = W_0(\Phi\Phi^\dagger)$ the orthogonal polynomials are simply $P_n(z) = z^n$ and $h_n$ is given explicitly:

$$h_n = \int |z|^{2n} e^{W_0(|z|^2)} d^2 z = 2\pi \int_0^\infty r^{2n+1} e^{W_0(r^2)} dr$$

The kernel function is

$$K_N(z, \bar{w}) = e^{\frac{1}{2}(W(z)+W(w))} \sum_{n=0}^{N-1} \frac{(z\bar{w})^n}{h_n}$$

For example, for the Gaussian model with $W(z) = -|z|^2$ the squared norms are $h_n = \pi n!$ and the mean value of density is given by

$$\langle \rho(z) \rangle_N = \frac{1}{\pi} e^{-|z|^2} \sum_{n=0}^{N-1} \frac{|z|^{2n}}{n!}$$

**The Lax representation.** The orthogonal polynomials obey the recurrence relation of the form $zP_n(z) = \sum_{k \le n} c_{nk} P_k(z)$. In terms of the $\psi$-function it reads

$$z\psi_n(z) = r_n \psi_{n+1}(z) + \sum_{k \ge 0} u_k(n) \psi_{n-k}(z)$$

One can represent it as a "spectral problem" $L\psi = z\psi$ for the difference operator

$$L = r_n e^{\partial/\partial n} + \sum_{k\geq 0} u_k(n) e^{-k\partial/\partial n} \tag{29}$$

(the Lax operator). Here $e^{\partial/\partial n}$ is the shift operator $n \to n+1$ with the characteristic property $e^{\partial/\partial n} f(n) = f(n+1)e^{\partial/\partial n}$.

Let

$$W(z) = W^{(0)}(z) + \sum_k (t_k z^k + \bar{t}_k \bar{z}^k)$$

It can be shown that the dependence on the parameters $t_k$, $\bar{t}_k$ is given by the *2D Toda hierarchy*

$$\frac{\partial}{\partial t_k} L = [A_k, L]\,, \qquad \frac{\partial}{\partial \bar{t}_k} L = [L, \bar{A}_k]$$

where $A_k = (L^k)_+ + \frac{1}{2}(L^k)_0$, $\bar{A}_k = (L^{\dagger k})_- + \frac{1}{2}(L^{\dagger k})_0$ and $L^\dagger = e^{-\partial/\partial n} r_n + \sum_{k\geq 0} e^{k\partial/\partial n} \bar{u}_k(n)$ is the conjugate Lax operator. Given an operator of the form $\hat{O} = \sum_k b_k e^{k\partial/\partial n}$, we use the standard definition $(\hat{O})_+ = \sum_{k>0} b_k e^{k\partial/\partial n}$, $(\hat{O})_- = \sum_{k<0} b_k e^{k\partial/\partial n}$ and $(\hat{O})_0 = b_0$. The structure of the Toda hierarchy in models of random matrices was first revealed in [27], see also review [4].

In the case of quasiharmonic potential, it is convenient to modify the $\psi$-functions:

$$\psi_n \to \chi_n = \frac{1}{\sqrt{h_{n-1}}} e^{V(z)} P_{n-1}(z)$$

The functions $\chi_n$ obey the orthogonality condition: $\int \chi_n(z)\overline{\chi_m(z)} e^{-|z|^2} d^2 z = \delta_{nm}$. Then we have two compatible linear problems:

$$(L\chi)_n = z\chi_n\,, \qquad (L^\dagger \chi)_n = \partial_z \chi_n$$

The second equation can be proven by comparing the matrix elements of the both sides using integration by parts.

## 3.3 The loop equation

In matrix models, loop equations are exact relations which follow from the fact that the matrix integral defining the model does not depend on changes of integration variables. In our case, we may start directly from the integral over eigenvalues (7), thereby extending the result to any value of $\beta$.

Clearly, the integral (7) remains the same if we change the integration variables $z_i \to \tilde{z}_i$. In other words, it is invariant under reparametrizations of the $z$-coordinate, which we write in the infinitesimal form as $z_i \to z_i + \epsilon(z_i)$,

$\bar{z}_i \to \bar{z}_i + \bar{\epsilon}(z_i)$. For the integral $Z_N = \int e^{-\beta E(z_1,\ldots,z_N)} \prod_j d^2 z_j$ with $E$ given in (10), the reparametrization yields, in the first order:

$$\prod_j d^2 z_j \longrightarrow \left[1 + \sum_l (\partial\epsilon(z_l) + \bar{\partial}\bar{\epsilon}(z_l))\right] \prod_j d^2 z_j$$

$$E \longrightarrow E + \sum_l \left(\frac{\partial E}{\partial z_l}\,\epsilon(z_i) + \frac{\partial E}{\partial \bar{z}_l}\,\bar{\epsilon}(z_i)\right)$$

The invariance of the integral is then expressed by the identity

$$\sum_i \int \frac{\partial}{\partial z_i}\left(\epsilon(z_i)e^{-\beta E}\right) \prod_j d^2 z_j = 0$$

valid for any $\epsilon$. Introducing a suitable cutoff at infinity, if necessary, one sees that the 2D integral over $z_i$ can be transformed, by virtue of the Green theorem, into a contour integral around infinity and so it does vanish.

Let us take $\epsilon(z_i) = \frac{1}{z - z_i}$, where $z$ is a complex parameter. The singularity at the point $z$ does not destroy the above identity since its contribution is proportional to the vanishing integral $\oint d\bar{z}_i/(z_i - z)$ over a small contour encircling $z$. Therefore, we have the equality

$$\sum_i \int \left[-\frac{\beta \partial_{z_i} E}{z - z_i} + \frac{1}{(z - z_i)^2}\right] e^{-\beta E} \prod_j d^2 z_j = 0$$

where $\partial_{z_i} E = -\sum_{l \neq i} \dfrac{1}{z_i - z_l} - \beta^{-1}\,\partial W(z_i)$ (see (10)). Using the identity

$$\sum_{i,j} \frac{1}{(z - z_i)(z - z_j)} = \sum_{i \neq j} \frac{2}{(z - z_i)(z_i - z_j)} + \sum_i \frac{1}{(z - z_i)^2}$$

we rewrite it in the form $\langle T(z_1, \ldots, z_N)\rangle = 0$, where

$$T = 2\sum_i \frac{\partial W(z_i)}{z - z_i} + \beta\left(\sum_i \frac{1}{z - z_i}\right)^2 + (2 - \beta)\sum_i \frac{1}{(z - z_i)^2}$$

This identity gives an exact relation between one- and two-point correlation functions. To see this, we rewrite it in terms of $\varphi(z) = -\beta \sum_i \log|z - z_i|^2$ using the rule $\sum_i f(z_i) = \int f(z)\rho(z) d^2 z$. The result is *the loop equation*

$$\frac{1}{2\pi}\int \frac{\partial W(\zeta)\,\langle\Delta\varphi(\zeta)\rangle}{z - \zeta}\, d^2\zeta = \langle T(z)\rangle \tag{30}$$

where

$$T(z) = (\partial\varphi(z))^2 + (2-\beta)\partial^2\varphi(z) \tag{31}$$

The correlator at coinciding points is understood as $\langle(\partial\varphi(z))^2\rangle = \lim_{z'\to z}\langle\partial\varphi(z)\,\partial\varphi(z')\rangle$.

We have got an *exact* relation between one- and two-point correlation functions, valid for any finite $N$. For historical reasons, it is called the loop equation. One may read it as a Ward identity obeyed by correlation functions of the model. Being written in the form (30), (31), it resembles conformal Ward identities. Since correlation functions are variational derivatives of the free energy, the loop equation is an implicit functional relation for the free energy. However, it is not a closed relation. It can be made closed by some additional assumptions or approximations. A combination with $1/N$ expansion is particularly meaningful.

## 4. Large $N$ limit

Starting from this section, we study the large $N$ limit of the random matrix models introduced in Section 2. Our main tool is the loop equation. We shall see that in the large $N$ limit meaningful geometric and algebro-geometric structures emerge, as well as important applications in physics.

### 4.1 Preliminaries

In order to be prepared for taking the large $N$ ("quasiclassical") limit, it is convenient to introduce the "Planck constant" $\hbar$ by the rescaling $W(z) \to \frac{1}{\hbar}W(z)$, so the integral for the partition function acquires the form

$$Z_N = \int [D\Phi] e^{\frac{1}{\hbar}\mathrm{tr}\, W(\Phi)} \propto \int |\Delta_N|^{2\beta} \prod_j e^{\frac{1}{\hbar}W(z_j)} d^2 z_j \tag{32}$$

which is ready for an $\hbar$-expansion.

Now we can specify what we mean by the large $N$ limit. Namely, we are going to consider the integral (32) in the limit

$$N \to \infty, \quad \hbar \to 0, \quad \hbar N = t \quad \text{finite},$$

where $t$ is a (positive) parameter having the dimension of area, or, equivalently, the integral

$$Z_N = \int |\Delta_N(z_i)|^{2\beta} \prod_j \exp\left(\frac{N}{t}W(z_j)\right) d^2 z_j$$

as $N \to \infty$. With this convention, the $N \to \infty$ and $\hbar \to 0$ limits mean the same. It is natural to expect that $Z_N \overset{\hbar\to 0}{\longrightarrow} e^{\hbar^{-2}F_0(t)}$, where the rescaled free energy $F_0$ is a smooth function of $t$ (and parameters of the potential) finite as $N \to \infty$.

From now on we *change the normalization* of the functions $\rho$ and $\varphi$ (see (13), (15)) multiplying them by the $\hbar$:

$$\rho(z) = \hbar \sum_i \delta(z - z_i)\,, \qquad \varphi(z) = -\hbar\beta \sum_i \log |z - z_i|^2 \tag{33}$$

and use these definitions hereafter. (The idea is to make their mean values finite as $N \to \infty$.) The relation (16) between these functions remains unchanged. The density is now normalized as follows:

$$\int \rho(z) d^2 z = t$$

Note that in our units the $\hbar$ has dimension of [length]$^2$, $\rho(z)$ is dimensionless and the partition function defined by (32) has dimension [length]$^{N(\beta N + 2 - \beta)}$, i.e., the combination

$$\hbar^{-\frac{1}{2}\beta N^2 + \frac{1}{2}(\beta - 2)N} Z_N = \int \left| \Delta_N \left( z_i / \sqrt{\hbar} \right) \right|^{2\beta} \prod_{j=1}^{N} \left( e^{\frac{1}{\hbar} W(z_j)} \frac{d^2 z_j}{\hbar} \right) \tag{34}$$

is dimensionless.

In terms of the renormalized $\varphi$, the loop equation (30) has the form

$$\frac{1}{2\pi} \int \frac{\partial W(\zeta) \langle \Delta \varphi(\zeta) \rangle}{\zeta - z} d^2 \zeta + \langle (\partial \varphi(z))^2 \rangle + \varepsilon \langle \partial^2 \varphi(z) \rangle = 0 \tag{35}$$

where

$$\varepsilon = (2 - \beta)\hbar \tag{36}$$

is small as $\hbar \to 0$. Note that $\varepsilon$ is exactly zero for the ensemble $\mathcal{N}^0$ ($\beta = 2$), and so the last term does not enter the loop equation in this case. It is convenient to treat $\varepsilon$ and $\hbar$ as independent small parameters.

## 4.2 Solution to the loop equation in the leading order

It is instructive to think about the large $N$ limit under consideration in terms of the Dyson gas picture. Then the limit we are interested in corresponds to a very low temperature of the gas, when fluctuations around equilibrium positions of the charges are negligible. The main contribution to the partition function then comes from a configuration, where the charges are “frozen” at their equilibrium positions. It is also important that the temperature tends to zero simultaneously with increasing the number of charges, so the plasma can be regarded as a continuous fluid at static equilibrium. In the noninteracting fermions picture, this limit has some features of the quasiclassical approximation. Mathematically, all this means that the integral is evaluated by the saddle point method, with only the leading contribution being taken into account. As $\hbar \to 0$, correlation functions take their "classical" values $\langle\varphi(z)\rangle = \varphi_{cl}(z)$, and multipoint correlators factorize in the leading order: $\langle\partial\varphi(z)\,\partial\varphi(z')\rangle = \partial\varphi_{cl}(z)\partial\varphi_{cl}(z')$. Then the loop equation (35) becomes a *closed* relation for $\varphi_{cl}$:

$$\frac{1}{2\pi}\int \frac{\partial W(\zeta)\Delta\varphi_{cl}(\zeta)}{\zeta - z}\, d^2\zeta \;\; + \left(\partial\varphi_{cl}(z)\right)^2 + \varepsilon\,\partial^2\varphi_{cl}(z) = 0 \qquad (37)$$

Note that we hold the last term which is apparently of the next order in $\hbar$. The role of this term will be discussed below.

**The case $\beta = 2$ (the ensemble $\mathcal{N}^0$).** We begin with the case $\beta = 2$, when the last term in the r.h.s. of (37) vanishes exactly. Let us apply $\bar\partial$ to both sides of the equation. This yields:

$$-\partial W(z)\Delta\varphi_{cl}(z) + \partial\varphi_{cl}(z)\Delta\varphi_{cl}(z) = 0$$

Since $\Delta\varphi_{cl}(z) \propto \rho_{cl}(z)$ (see (33)), we obtain

$$\rho_{cl}(z)\;[\partial\varphi_{cl}(z) - \partial W(z)] = 0 \qquad (38)$$

This equation should be solved with the additional constraints $\int \rho_{cl}(z) d^2 z = t$ (normalization) and $\rho_{cl}(z) \geq 0$ (positivity). The equation tells us that either $\partial\varphi_{cl}(z) = \partial W(z)$ or $\rho_{cl}(z) = 0$. Applying $\bar\partial$, we get $\Delta\varphi_{cl}(z) = \Delta W(z)$. Since $\Delta\varphi = -4\pi\beta\rho$, this gives the solution for $\rho_{cl}$:

$$\rho_{cl}(z) = -\frac{\Delta W(z)}{4\pi\beta} \qquad \text{“in the bulk”} \qquad (39)$$

Here, "in the bulk" just means "in the region where $\rho_{cl} > 0$". As we shall see below, this result holds true, up to some details, for other values of $\beta$ as well, so $\beta$ is kept in this formula and in some formulas below.

The physical meaning of the equation $\partial\varphi_{cl}(z) = \partial W(z)$ is clear. It is just the condition that the charges are in equilibrium (the saddle point for the integral). Indeed, the equation states that the total force experienced by a charge at any point $z$, where $\rho_{cl} \neq 0$, is zero. The interaction with the other charges, $\partial\varphi_{cl}(z)$, is compensated by the force $\partial W(z)$ due to the external field.

**Support of eigenvalues.** Let us assume that

$$\sigma(z) := -\frac{1}{4\pi}\Delta W(z) > 0 \qquad (40)$$

For quasiharmonic potentials, $\sigma(z) = 1/\pi$. If, according to (39), $\rho_{cl} = \sigma/\beta$ everywhere, the normalization condition for $\rho_{cl}$ can not be satisfied! So we

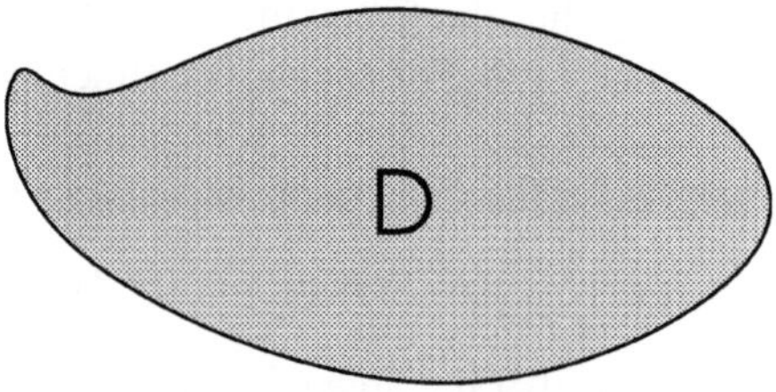


*Figure 2.* *The support of eigenvalues.*

conclude that $\rho_{cl} = \sigma/\beta$ in a compact bounded domain (or domains) only, and outside this domain one should switch to the other solution of (38), $\rho_{cl} = 0$. The domain D where $\rho_{cl} > 0$ is called *support of eigenvalues* (Fig. 2). In general, it may consist of several disconnected components. The *complement* to the support of eigenvalues, $\mathsf{D}^c = \mathbf{C} \setminus \mathsf{D}$, is an unbounded domain in the complex plane. For quasiharmonic potentials, the result is especially simple: $\rho_{cl}$ is constant in D and 0 in $\mathsf{D}^c$.

In terms of the mean value of the function $\varphi(z)$ in the leading order, the above result reads

$$\varphi_{cl}(z) = -\int_{\mathsf{D}} \log|z-\zeta|^2 \sigma(\zeta)\, d^2\zeta$$

As it follows from the theory of potential in two dimensions, this function is continuous across the boundary of D together with its first derivatives. However, the second order derivatives of this function have a jump across the boundary.

To find the shape of D is a much more challenging problem. It appears to be equivalent to the inverse potential problem in two dimensions. The shape of D is determined by the condition $\partial\varphi_{cl}(z) = \partial W(z)$ (imposed for all points $z$ inside D) and by the normalization condition. Since $\partial\varphi_{cl}(z) = -\beta \int \frac{\rho_{cl}(\zeta)d^2\zeta}{z-\zeta}$ (see (33)), we write them in the form

$$\begin{cases} \dfrac{1}{4\pi}\displaystyle\int_{\mathsf{D}} \frac{\Delta W(\zeta) d^2\zeta}{z-\zeta} = \partial W(z) & \text{for all } z \in \mathsf{D} \\[2ex] \displaystyle\int_{\mathsf{D}} \sigma(\zeta) d^2\zeta = \beta t \end{cases}$$

The integral over D in the first equation can be transformed to a contour integral by means of the Cauchy formula (see Appendix B). As a result, one obtains:

$$\oint_{\partial\mathsf{D}} \frac{\partial W(\zeta) d\zeta}{z-\zeta} = 0 \qquad \text{for all } z \in \mathsf{D}. \tag{41}$$

This means that the domain D has the following property: the function $\partial W(z)$ on its boundary is the boundary value of an analytic function in its complement $\mathsf{D}^c$.

We continue our analysis for the quasiharmonic case, where

$$\partial W(z) = -\bar{z} + V'(z), \quad \Delta W(z) = 4\partial\bar{\partial}W(z) = -4$$

The normalization than means that the area of D is equal to $\beta\pi t$. Assume that:

- $V(z) = \sum t_k z^k$ is regular in D (say a polynomial)
- $0 \in \mathsf{D}$ (it is always the case if $-W$ has a local minimum at 0)
- D is connected

Then the first equation in (41) acquires the form

$$\frac{1}{2\pi i}\oint_{\partial \mathsf{D}} \frac{\bar{\zeta} d\zeta}{\zeta - z} = V'(z) \qquad \text{for } z \in \mathsf{D}$$

Expanding it near $z = 0$, we get:

$$t_k = \frac{1}{2\pi i k}\oint_{\partial \mathsf{D}} \bar{\zeta}\zeta^{-k} d\zeta = -\frac{1}{\pi k}\int_{\mathsf{D}} \zeta^{-k} d^2\zeta \tag{42}$$

We see that the "coupling constants" $t_k$ are *harmonic moments* of $\mathsf{D}^c = \mathbf{C} \setminus \mathsf{D}$ and the area of D is $\pi\beta t$.

It is the subject of the inverse potential problem to reconstruct the domain from its area and harmonic moments. (In the case when the support of eigenvalues has several disconnected components, some additional conditions are required.) In general, the problem has many solutions. But it is known that *locally*, i.e., for a small enough change $t \to t + \delta t$, $t_k \to t_k + \delta t_k$ the solution is unique.

**The support of eigenvalues: a fine structure ($\beta \neq 2$).** Let us take a closer look at the support of eigenvalues, taking into account the so far ignored term in (37). Applying $\bar{\partial}$ to the both sides of (37) yields, instead of (38), the equation

$$\partial\varphi_{cl}(z) - \partial W(z) + \frac{\varepsilon}{2}\partial \log \Delta\varphi_{cl}(z) = 0$$

with the extra term proportional to $\varepsilon$ (36). Acting by $\bar{\partial}$ once again, we obtain the equation for $\rho_{cl}$

$$-\frac{\varepsilon}{8\pi}\Delta \log \rho_{cl}(z) + \beta\rho_{cl}(z) = \sigma(z) \tag{43}$$

which looks like the Liouville equation in the "background" $\sigma(z)$. The first term seems to be negligible as $\hbar \to 0$. However, one should be careful since the small parameter stands in front of the term with the highest derivative. As a matter of fact, this term is negligible only if $\rho_{cl}$ is not small! Indeed, in a region where $\rho_{cl} \sim e^{-N}$, the term $\varepsilon\Delta \log \rho_{cl}$ is of order 1 and thus plays the dominant role.

In fact the equation states that $\rho_{cl}$ never vanishes exactly but can be exponentially small as $N \to \infty$. In the bulk, where the charges are distributed with nonzero density at $N = \infty$, equation (43) systematically generates corrections to the value $\sigma(z)/\beta$. If $\sigma \neq$ const, there are power-like corrections in $\varepsilon$, as well as exponentially small ones. We conclude that the effect of the extra term is negligible everywhere except the very vicinity of the edge of the support of eigenvalues. Therefore, the result for $\rho_{cl}$ can still be written in the form

$$\rho_{cl}(z) = \beta^{-1}\sigma(z)\,\Theta(z; \mathsf{D}) \tag{44}$$

where $\Theta(z; \mathsf{D})$ is the characteristic function of the domain $\mathsf{D}$ (which is 1 in $\mathsf{D}$ and 0 in $\mathsf{D}^{c}$). The role of the term $\varepsilon\Delta \log \rho_{cl}$ is to make the edge smooth. Around the edge, the density rapidly (but smoothly) drops down to zero over distances of order $\sqrt{\varepsilon}$. So, the boundary has got a "fine structure".

All this is in agreement with the form of the first nonvanishing correction to the mean density found from the loop equation (35). Let us write $\langle\varphi(z)\rangle = \varphi_{cl}(z) + \varphi_{\hbar}(z)$, where $\varphi_{\hbar}$ is of order $\hbar$. The result for the $\varphi_{\hbar}$ can be compactly written in terms of the function

$$\chi(z) = \log \sqrt{\pi\sigma(z)} \tag{45}$$

and its harmonic continuation $\chi^{H}(z)$ from the boundary of $\mathsf{D}$ to its exterior. Specifically, $\chi^{H}(z)$ is a harmonic function in $\mathsf{D}^{c}$ (regular at $\infty$) such that $\chi^{H}(z) = \chi(z)$ on the boundary. In other words, it is the solution of the (exterior) Dirichlet boundary value problem (see below). The loop equation yields

$$\varphi_{\hbar}(z) = \begin{cases} -\varepsilon\left[\chi(z) - \chi^{H}(\infty) + \frac{1}{2}\right], & z \in \mathsf{D} \\[2ex] -\varepsilon\left[\chi^{H}(z) - \chi^{H}(\infty)\right], & z \in \mathsf{D}^{c} \end{cases} \tag{46}$$

(Note the discontinuity of this function across the boundary.) The corresponding correction to the mean density $\rho_{\hbar} = \langle\rho\rangle - \rho_{cl}$ is

$$\rho_{\hbar}(z) = \frac{\varepsilon}{4\pi\beta}\left(\Theta(z; \mathsf{D})\Delta\chi(z) - \delta(z; \partial\mathsf{D})\partial_n\left(\chi(z) - \chi^{H}(z)\right) - \frac{1}{2}\delta'(z; \partial\mathsf{D})\right) \tag{47}$$

where $\delta(z; \partial\mathsf{D})$ is the delta function with the support on the boundary and $\delta'(z; \partial\mathsf{D})$ is its normal derivative (see Appendix B). Here and below, $\partial_n$ is the normal derivative at the boundary, with the normal vector being directed to the exterior of the domain $\mathsf{D}$. The correction is so singular because the

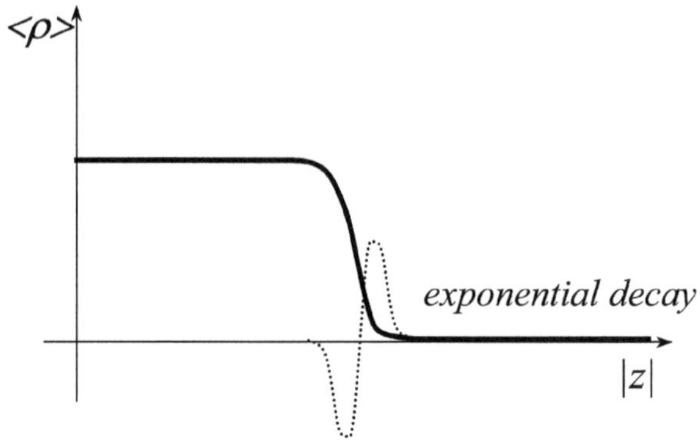


*Figure 3.* *The mean density profile for models with quasiharmonic potential in the large N limit. The correction to the pure step function is shown by the dotted line. The effect of the correction is to form a double layer of charges near the edge.*

zeroth approximation (44) is singular by itself. The singular function $\rho_\hbar$ is to be understood as being integrated with any smooth test function. The first term in (47) is a correction to the bulk density. The second one is a correction to the shape of the support of eigenvalues (it describes a small displacement of the edge). The third term signifies the presence of a double layer of charges around the boundary. This just means that the boundary is smoothed out.

For the normal self-dual matrices ($\beta = 2$) the correction $\rho_\hbar$ vanishes. Certainly, this does not mean that the boundary is sharp. It becomes smooth if higher corrections in $\hbar$ (caused by fluctuations of the particles) are taken into account.

For the normal matrix model ($\beta = 1$) with quasiharmonic potential equation (43) reads

$$-\frac{\hbar}{8}\Delta \log \rho_{cl}(z) + \pi\rho_{cl}(z) = 1$$

The obvious solution is $\rho_{cl} = 1/\pi$. But it is not normalizable! One must look for another solution. The right solution differs from the constant by exponentially small terms in the bulk but exhibits an abrupt drop across the boundary of the domain determined by the harmonic moments (42). So, up to exponentially small corrections, the solution is given by (44) with $\sigma = 1/\pi$ and $\beta = 1$. It is instructive to note that the first two terms in (47) vanish at $\sigma =$ const but the third term does not, so the leading correction in the quasiharmonic case merely makes the edge smooth that results in a double layer of "charges" near the edge (see Fig. 3).

**From the support of eigenvalues to an algebraic curve.** There is an interesting algebraic geometry behind the large $N$ limit of matrix models. For simplicity, here we consider models with quasiharmonic potentials.

In general, the boundary of the support of eigenvalues is a closed curve in the plane without self-intersections. The following important fact holds true. If $V'(z)$ is a rational function, then this curve is a real section of a complex algebraic curve of finite genus. In fact, this curve encodes the $1/N$ expansion of the model. In the context of Hermitian 2-matrix model such a curve was introduced and studied in [28–30].

To explain how the curve comes into play, we start from the equation $\partial\varphi_{cl} = \partial W$, which can be written in the form $\bar{z} - V'(z) = G(z)$ for $z \in \mathsf{D}$, where

$$G(z) = \frac{1}{\pi}\int_{\mathsf{D}} \frac{d^2\zeta}{z-\zeta}$$

Clearly, this function is analytic in $\mathsf{D}^{\mathsf{c}}$. At the same time, $V'(z)$ is analytic in $\mathsf{D}$ and all its singularities in $\mathsf{D}^{\mathsf{c}}$ are poles. Set

$$S(z) = V'(z) + G(z)$$

Then $S(z) = \bar{z}$ on the boundary of the support of eigenvalues. So, $S(z)$ is the analytic continuation of $\bar{z}$ away from the boundary. Assuming that poles of $V'$ are not too close to $\partial\mathsf{D}$, $S(z)$ is well-defined at least in a piece of $\mathsf{D}^{\mathsf{c}}$ adjacent to the boundary. The complex conjugation yields $\overline{S(z)} = z$, so the function $\bar{S}(z) = \overline{S(\bar{z})}$ must be inverse to the $S(z)$:

$$\bar{S}(S(z)) = z$$

("unitarity condition"). The function $S(z)$ is called the *Schwarz function* [31].

Under our assumptions, the $S(z)$ is an algebraic function, i.e., it obeys a polynomial equation $R(z, S(z)) = 0$ of the form

$$R(z, S(z)) = \sum_{n,l=1}^{d+1} a_{nl} z^n (S(z))^l = 0$$

where $\overline{a_{ln}} = a_{nl}$ and $d$ is the number of poles of $V'(z)$ (counted with their multiplicities). Here is the sketch of proof. Consider the Riemann surface $\Sigma = \mathsf{D}^{\mathsf{c}} \cup \partial\mathsf{D} \cup (\mathsf{D}^{\mathsf{c}})^*$ (the *Schottky double* of $\mathsf{D}^{\mathsf{c}}$). Here, $(\mathsf{D}^{\mathsf{c}})^*$ is another copy of $\mathsf{D}^{\mathsf{c}}$, with the local coordinate $\bar{z}$, attached to it along the boundary. On $\Sigma$, there exists an anti-holomorphic involution that interchanges the two copies of $\mathsf{D}^{\mathsf{c}}$ leaving the points of $\partial\mathsf{D}$ fixed. The functions $z$ and $S(z)$ are analytically extendable to $(\mathsf{D}^{\mathsf{c}})^*$ as $\overline{S(z)}$ and $\bar{z}$ respectively. We have two meromorphic functions, each with $d + 1$ poles, on a closed Riemann surface. Therefore, they are connected by a polynomial equation of degree $d+1$ in each variable. Hermiticity of the coefficients follows from the unitarity condition.

The polynomial equation $R(z, \tilde{z}) = 0$ defines a complex curve $\Gamma$ with anti-holomorphic involution $(z, \tilde{z}) \mapsto (\overline{\tilde{z}}, \bar{z})$. The real section is the set of points such that $\tilde{z} = \bar{z}$. It is the boundary of the support of eigenvalues.

It is important to note that for models with non-Gaussian weights (in particular, with polynomial potentials of degree greater than two) the curve has a number of singular points, although the Riemann surface $\Sigma$ (the Schottky double) is smooth. Generically, these are *double points*, i.e., the points where the curve crosses itself. In our case, a double point is a point $z^{(d)} \in \mathsf{D}^{\mathsf{c}}$ such that $S(z^{(d)}) = \overline{z^{(d)}}$ but $z^{(d)}$ does not belong to the boundary of $\mathsf{D}$. Indeed, this condition means that two different points on $\Sigma$, connected by the anti-holomorphic involution, are stuck together on the curve $\Gamma$, which means the self-intersection. The double points play the key role in deriving the nonperturbative (instanton) corrections to the large $N$ matrix models results (see [32] for details).

Finally, let us point out that a complex curve $\Gamma^{(n)}$ can be associated to ensembles of finite matrices as well (at least for $\beta = 1$). For the model of two Hermitian matrices this was done in [33]. If the linear spectral problem for the $L$-operator is of finite order (see the end of Section 3.2), the curve can be defined as the "spectral curve" of the difference spectral problems $(L\chi)_n = z\chi_n$, $(L^\dagger\chi)_n = \tilde{z}\chi_n$. Since the operators $L$ and $L^\dagger$ do not commute, the curve depends on $n$. (See [8] for details.) In contrast to the curve $\Gamma$ the curve $\Gamma^{(n)}$ is in general a smooth curve. A properly performed $n \to \infty$ limit of this curve coincides with the complex curve $\Gamma$ constructed from the support of eigenvalues.

## 4.3 The free energy

In this subsection we assume that the support of eigenvalues is connected. As is known, the free energy admits a $1/N$ expansion. We prefer to work with the equivalent $\hbar$-expansion, thus emphasizing its semiclassical nature. The first few terms of the $\hbar$-expansion for the ensembles of normal matrices with a general potential are

$$\log Z_N = c(N) + \frac{F_0}{\hbar^2} + \frac{F_{1/2}}{\hbar} + F_1 + O(\hbar) \tag{48}$$

The explicit form of the $c(N)$ is given below. In fact this term can be absorbed into a normalization. For example, one may normalize $Z_N$ dividing it by the partition function of the Gaussian model. Although in general the $\hbar$-expansion does not look like a topological one, it appears to be topological (i.e., only even powers of $\hbar$ enter) for the ensemble $\mathscr{N}^0$ with arbitrary potential and for the ensemble $\mathscr{N}$ with quasiharmonic potential.

**The leading order.** The partition function is given by (9), $Z_N = \int e^{-\beta E} \prod d^2 z_j$, where

$$-\beta E(z_1, \ldots, z_N) = \beta \sum_{i \neq j} \log |z_i - z_j| + \hbar^{-1} \sum_j W(z_j) \tag{49}$$

Writing the energy in terms of the density function, we have, in the leading order:

$$-\beta \hbar^2 E[\rho] = \beta \int\!\!\int \rho(z)\rho(\zeta) \log |z - \zeta| d^2 z d^2 \zeta + \int W(z)\rho(z) d^2 z \tag{50}$$

We need to find the minimum of $E[\rho]$ with the constraint $\int \rho \, d^2 z = t$. This is achieved by variation of the functional $E[\rho] + \lambda(\int \rho \, d^2 z - t)$ with the Lagrange multiplier $\lambda$. The resulting equation is

$$2\beta \int \log |z - \zeta| \rho(\zeta) d^2 \zeta + W(z) + \lambda = 0$$

Upon taking the $z$-derivative, we see that the extremal $\rho(z)$ is equal to the $\rho_{cl}(z)$, as expected, and the equation coincides with the previously derived one, $\partial \varphi_{cl}(z) = \partial W(z)$, with

$$\varphi_{cl}(z) = -\beta \int \log |z - \zeta|^2 \rho_{cl}(\zeta) d^2 \zeta = - \int_{\mathsf{D}} \log |z - \zeta|^2 \sigma(\zeta) d^2 \zeta$$

Assuming that $W(0) = 0$ and D is connected, the Lagrange multiplier is fixed to be $\lambda = \varphi_{cl}(0)$, and so $W(z) = \varphi_{cl}(z) - \varphi_{cl}(0)$. Plugging this into (50), we find the leading contribution to the free energy $F_0/\hbar^2 = \max_\rho(-\beta E[\rho]) = -\beta E[\rho_{cl}]$:

$$F_0 = -\frac{1}{\beta} \int_{\mathsf{D}}\!\int_{\mathsf{D}} \sigma(z) \log \left| \frac{1}{z} - \frac{1}{\zeta} \right| \sigma(\zeta) d^2 z d^2 \zeta \tag{51}$$

which is basically the electrostatic energy of the domain D charged with the density $\sigma(z)$ with a point-like compensating charge at the origin.

Since the $t$-derivative of the extremal value of the functional is equal to the Lagrange multiplier (with the sign minus), $\partial_t F_0 = -\lambda$, we incidentally obtain the useful formula

$$\partial_t F_0 = 2 \int_{\mathsf{D}} \log |z| \, \sigma(z) \, d^2 z \tag{52}$$

which will be rederived below by a more direct method.

**Corrections to the leading term.** Taking into account the discrete "atomic" structure of the Dyson gas, one is able to find the subleading corrections to the free energy.

The first correction comes from a more accurate integral representation of the sum $\sum_{i\neq j}\log|z_i - z_j|$, when passing to the continuous theory. Namely, one should exclude the terms with $i = j$, writing

$$\sum_{i\neq j}\log|z_i - z_j| = \sum_{i,j}\log|z_i - z_j| - \sum_{j}\log|\ell(z_j)|$$

where $\ell$ is a short-distance cutoff (which may depend on the point $z_j$). It is natural to take the cutoff to be

$$\ell(z) \sim \sqrt{\frac{\hbar}{\rho_{cl}(z)}} \qquad (53)$$

which is the mean distance between the charges around the point $z$. (In the context of the quantum Hall effect, $\ell \sim \sqrt{\hbar/B}$ is called the magnetic length.) This gives the improved estimate for $E[\rho_{cl}]$:

$$-\beta\hbar^2 E[\rho_{cl}] = F_0 + \beta\hbar\int \rho_{cl}(z)\sqrt{\rho_{cl}(z)}\, d^2z - \frac{1}{2}\beta N\log\hbar + \alpha_1 N \qquad (54)$$

where $\alpha_1$ is a numerical constant which can not be determined by this argument.

Another correction comes from the integration measure when one passes from the integration over $z_j$ to the integration over macroscopic densities[1]. We can write

$$\prod_j d^2z_j = N!\, J[\rho]\, [D\rho]$$

where $[D\rho]$ is an integration measure in the space of densities, $J[\rho]$ is the Jacobian of this change of variables and the factor $N!$ takes into account the symmetry under permutations (all the states that differ by a permutation of the charges are identical). To estimate the Jacobian, we divide the plane into $N$ microscopic "cells" such that $j$-th particle occupies a cell of size $\ell(z_j)$, where $\ell(z_j)$ is the mean distance (53) between the particles around the point $z_j$. All the microscopic states in which the particles remain in their cells are macroscopically indistinguishable. Given a macroscopic density $\rho$, $J[\rho]$ is then approximately equal to the integral $\int_{\text{cells}}\prod_j d^2z_j$, with each particle being confined to its own cell. Therefore, $J[\rho] \sim \prod_j \ell^2(z_j)$, and thus $\log J[\rho]$ (sometimes referred to as entropy of the state with the macroscopic density $\rho$) is given by

$$\log J[\rho] = -\frac{1}{\hbar}\int \rho_{cl}(z)\log\rho_{cl}(z)\, d^2z \,+\, N\log\hbar + \alpha_2 N \qquad (55)$$

where $\alpha_2$ is a numerical constant. This result agrees with the corresponding Jacobian obtained within the collective field theory approach [34].

Combining (54), (55) with $\rho = \rho_{cl}$, and taking into account the factor $N!$ in the measure, we obtain:

$$c(N) = \log N! + \frac{N}{2}(2-\beta)\log\hbar + \alpha N \tag{56}$$

where $\alpha$ is a numerical constant, and

$$F_{1/2} = -\frac{2-\beta}{2\beta}\int_{\mathsf{D}} \sigma(z)\log(\pi\sigma(z))\, d^2z \tag{57}$$

The term $F_{1/2}$ is thus the sum of the contribution due to the short-distance cutoff and the entropy contribution, which cancel each other in the ensemble of normal self-dual matrices (at $\beta = 2$). Another remarkable case when $F_{1/2}$ vanishes exactly[2] is the case of quasiharmonic potentials.

The result for $F_{1/2}$ can be derived in a more rigorous way from the loop equation (see Appendix C). Here we simply note that the variation of (57) over the potential $W$ does yield the correction to the mean density given by (47). One can verify this using the variational technique presented below in Section 4.4.

The result for $c(N)$ is in agreement with the dimensionality argument. It is easy to see that $e^{F_0/\hbar^2}$ carries the dimension of $[\text{length}]^{\beta N^2}$, and higher terms are dimensionless. The dimension of $Z_N$ is given by (34). Therefore, $e^{c(N)}$ must carry the residual dimension $[\text{length}]^{N(2-\beta)}$, which agrees with (56). For quasiharmonic potentials, the constant $\alpha$ can be found explicitly: $\alpha = \log\sqrt{2\pi^3}$, as is readily seen from the Gaussian case.

To summarize, the asymptotic expansion of the partition function as $\hbar \to 0$ has the form

$$Z_N = N!\hbar^{\frac{1}{2}(2-\beta)N} e^{\alpha N} \exp\left(\frac{F_0}{\hbar^2} + \frac{F_{1/2}}{\hbar} + F_1 + \sum_{k\geq 3}\hbar^{k-2}F_{k/2}\right) \tag{58}$$

where $F_0$ and $F_{1/2}$ are given by (51) and (57) respectively. The higher corrections are due to fluctuations of the eigenvalues around the equilibrium configuration. No simple method to find their explicit form is known. In principle, these corrections can be found by expanding the loop equation in powers of $\hbar$ (see Appendix C), similarly to how it goes for the model of one Hermitian matrix [35, 36]. However, the calculations are rather tedious, even in the first two orders. At present only fragmentary results are available. Some of them look quite suggestive.

For example, the result for the $F_1$-correction obtained in [37] for the case of the normal matrix model with quasiharmonic potential and a connected support of eigenvalues has a clean interpretation as the free energy of the theory of free

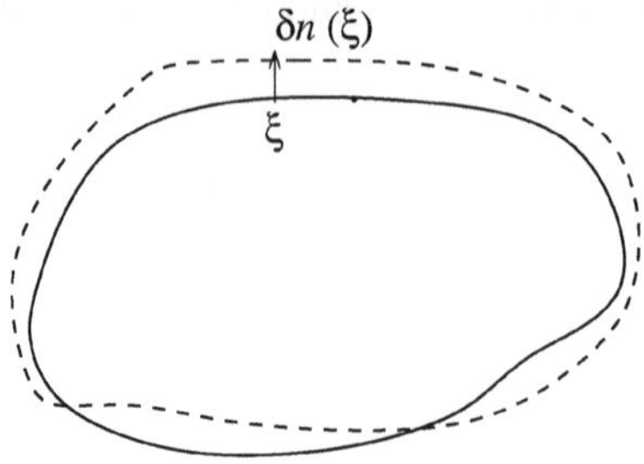


*Figure 4.* *The normal displacement of the boundary.*

bosons in the domain $\mathsf{D}^c$ with the Dirichlet boundary conditions. Namely, the calculations yield the result

$$F_1 = -\frac{1}{24\pi}\oint_{|w|=1}\Big(\log|z'(w)|\partial_n \log|z'(w)| + 2\log|z'(w)|\Big)|dw| \quad (59)$$

where $z(w)$ is the univalent conformal map from $\mathsf{D}^c$ onto the exterior of the unit circle. Comparison with the Polyakov-Alvarez formula [38, 39] allows one to identify this quantity with $-\frac{1}{2}\log\det(-\Delta_{\mathsf{D}^c})$, where $\det(-\Delta_{\mathsf{D}^c})$ is the regularized determinant of the Laplace operator in $\mathsf{D}^c$ with the Dirichlet boundary conditions. This suggests the interpretation through free bosons[3]. Presumably, the higher corrections to the free energy are connected with spectral geometry of the Laplace operator, too. Recently, some progress in computation of $F_1$ and, more generally, in understanding the structure of the whole series (including the case of disconnected supports) in models of Hermitian matrices was achieved [41]. Conjecturally, the answer is to be expressed in terms of a (conformal?) field theory on the complex curve $\Gamma$ introduced at the end of Section 4.2.

Finally, we note that the structure of the loop equation suggests to rearrange the $\hbar$-expansion of the free energy and to write it in the "topological" form $F = \sum_{g\geq 0}\hbar^{2g}F_g$, where each term has its own expansion in $\varepsilon = (2-\beta)\hbar$: $F_g = F_g^0 + \sum_{n\geq 1}\varepsilon^n F_g^{(n)}$.

## 4.4 Correlation functions in the large $N$ limit

**Variational technique and the Dirichlet boundary value problem.** Correlation functions in the leading order in $\hbar$ can be obtained from the free energy by variation w.r.t. $W(z)$ according to the formulas from Section 3.1. For a variation of the potential, $W \to W + \delta W$ with $N\hbar$ fixed we ask how $\mathsf{D}$ changes. It is convenient to describe small deformations $\mathsf{D} \to \tilde{\mathsf{D}}$, by the normal displacement $\delta n(\xi)$ of the boundary at a boundary point $\xi$ (Fig. 4).

Consider a small variation of the potential $W$ in the condition (41), which determines the shape of D at a fixed $t$. To take into account the deformation of the domain, $\delta \mathsf{D} = \tilde{\mathsf{D}} \setminus \mathsf{D}$, we write, for any fixed function $f$,

$$\delta \left( \oint_{\partial \mathsf{D}} f(\zeta) d\zeta \right) = \oint_{\partial(\delta \mathsf{D})} f(\zeta) d\zeta = 2i \int_{\delta \mathsf{D}} \bar{\partial} f(\zeta) d^2\zeta \approx 2i \oint_{\partial \mathsf{D}} \bar{\partial} f(\zeta) \delta n(\zeta) |d\zeta|$$

and thus obtain from (41):

$$\oint_{\partial \mathsf{D}} \frac{\partial \, \delta W(\zeta) d\zeta}{z - \zeta} + \frac{i}{2} \oint_{\partial \mathsf{D}} \frac{\Delta W(\zeta) \delta n(\zeta)}{z - \zeta} |d\zeta| = 0 \tag{60}$$

Here the first term comes from the variation of $W$ and the second one comes from the change of D.

This is an integral equation for the $\delta n(\zeta)$. It can be solved in terms of the exterior Dirichlet boundary value problem. Given any smooth function $f(z)$, let $f^H(z)$ be its *harmonic continuation* (already introduced in Section 4.2) from the boundary of D to its exterior, i.e., the function such that it is harmonic in $\mathsf{D}^c$, $\Delta f^H = 0$, and regular at $\infty$, and $f^H(z) = f(z)$ for all $z \in \partial \mathsf{D}$. The harmonic continuation is known to be unique. Explicitly, a harmonic function can be reconstructed from its boundary value by means of the *Dirichlet formula*

$$f^H(z) = -\frac{1}{2\pi} \oint_{\partial \mathsf{D}} f(\xi) \partial_n G(z, \xi) |d\xi| \tag{61}$$

The main ingredient of this formula is $G(z, \xi)$, which is the Green function of the domain $\mathsf{D}^c$:

$$\Delta_z G(z, \zeta) = 2\pi \delta(z - \zeta) \quad \text{in } \mathsf{D}^c, \qquad G(z, \zeta) = 0 \quad \text{if } z \in \partial \mathsf{D}$$

As $\zeta \to z$, it has the logarithmic singularity $G(z, \zeta) \to \log |z - \zeta|$. The well known properties of harmonic functions imply that $\partial_n G(z, \xi) \leq 0$ for all $\xi \in \partial \mathsf{D}$ (the operator of the normal derivative acts here to the second argument).

Consider the integral

$$\oint_{\partial \mathsf{D}} \frac{\partial (\delta W^H) d\zeta}{z - \zeta}$$

which is obviously equal to 0 for all $z$ inside D, subtract it from the first term in (60) and rewrite the latter in the form

$$\oint_{\partial \mathsf{D}} \frac{\partial (\delta W - \delta W^H) d\zeta}{z - \zeta} = \frac{i}{2} \oint_{\partial \mathsf{D}} \frac{\partial_n (\delta W - \delta W^H)}{z - \zeta} |d\zeta|$$

Here the integral over $d\zeta$ is transformed to the integral over the line element $|d\zeta|$. The normal derivative is taken in the exterior of the boundary. After this simple transformation our condition acquires the form

$$\oint_{\partial \mathsf{D}} \frac{\delta n_1(\zeta) + \hat{R}(\zeta)}{z - \zeta} |d\zeta| = 0 \qquad \text{for all } z \in \mathsf{D} \tag{62}$$

where $\delta n_1(z) := \Delta W(z)\delta n(z)$ (just for brevity) and $\hat{R}$ is the Neumann jump operator. Acting on a smooth function $f$, this operator gives the difference between the normal derivative of this function and the normal derivative of its harmonic extension:

$$\hat{R}f(z) = \partial_n^-(f(z) - f^H(z)) \tag{63}$$

The superscript indicates that the derivative is taken in the exterior of the boundary.

By properties of Cauchy integrals, it follows from (62) that $\left[\delta n_1(z) + \hat{R}\delta W(z)\right]\frac{|dz|}{dz}$ is the boundary value of an analytic function $h(z)$ in $\mathsf{D}^c$ such that $h(\infty) = 0$. For $z \in \mathsf{D}^c$, this function is given by

$$h(z) = \frac{1}{2\pi i}\oint_{\partial\mathsf{D}} \frac{\delta n_1(\zeta) + \hat{R}(\zeta)}{z - \zeta}\,|d\zeta|$$

If $h$ is not identically zero, the number of zeros of $h$ in $\mathsf{D}^c$, counted with multiplicities, is given by the contour integral $\frac{1}{2\pi}\oint_{\partial\mathsf{D}} d(\arg h)$. Since $\arg h(z) = \arg(|dz|/dz) = -\theta(z)$, where $\theta$ is the angle between the tangent vector and the real axis, this integral is equal to 1. We conclude that the function $h$ has exactly one simple zero outside the domain D. It is just the zero at $\infty$.

On the other hand, the variation of the normalization condition yields, in a similar manner:

$$\oint_{\partial\mathsf{D}} (\delta n_1 + \partial_n \delta W)|d\zeta| = 0 \tag{64}$$

This relation implies that the zero at $\infty$ is at least of the 2-nd order. Indeed, expanding the Cauchy integral around $\infty$,

$$2\pi i h(z) \to \frac{1}{z}\oint_{\partial\mathsf{D}} (\delta n_1 + \partial_n \delta W - \partial_n \delta W^H)|d\zeta| + O(z^{-2})$$

one concludes, using the Gauss law $\oint_{\partial\mathsf{D}} \partial_n \delta W^H |d\zeta| = -\int_{\mathsf{D}^c} \Delta \delta W^H d^2\zeta = 0$, that the coefficient in front of $1/z$ vanishes. We have got a contradiction.

Therefore, $h(z) \equiv 0$, and so $\delta n_1(z) + \hat{R}\delta W(z) = 0$. This gives the following result for the normal displacement of the boundary caused by small changes of the potential $W \to W + \delta W$:

$$\delta n(z) = \frac{\partial_n^-(\delta W^H(z) - \delta W(z))}{\Delta W(z)} \tag{65}$$

**Some results for the correlation functions.** In order to find the correlation functions of traces, we use the general variational formulas (17), where the exact free energy is replaced by the leading contribution (51):

$$\lim_{\hbar\to 0}\langle\rho(z)\rangle = \frac{\delta F_0}{\delta W(z)} = \rho_{cl}(z)\,, \qquad \lim_{\hbar\to 0}\langle\rho(z_1)\rho(z_2)\rangle_c = \hbar^2\,\frac{\delta\rho_{cl}(z_1)}{\delta W(z_2)}$$

Basically, these are linear response relations used in the Coulomb gas theory [42]. In this approximation, the eigenvalue plasma is represented as a continuous charged fluid, so the information about its discrete microscopic structure is lost. So, these formulas give "smoothed" correlation functions in the first non-vanishing order in $\hbar$. They are correct at distances much larger than the mean distance between the charges.

Here are the main results for the correlation functions obtained by the variational technique. For details of the derivation see [7] and Appendix C.

The leading contribution to the one-trace function was already found in Section 4.2. Here we present the general result including the first subleading correction which can be found by variation of (57):

$$\beta\,\langle \mathrm{tr} f(\Phi)\rangle \;=\; \frac{1}{\hbar}\int_{\mathsf{D}}\sigma(z)f(z)\,d^2z$$

$$+\frac{2-\beta}{8\pi}\left[\int_{\mathsf{D}}(1+\log\sigma(z))\Delta f(z)\,d^2z - \oint_{\partial\mathsf{D}}\log\sigma(z)\,\hat{R}f(z)|dz|\right] + O(\hbar) \tag{66}$$

where $\hat{R}$ is the Neumann jump operator (63). Applying this formula to the function $\varphi(z)$, we get the familiar result $\langle\varphi(z)\rangle = \varphi_{cl}(z) + \varphi_{\hbar}(z) + O(\hbar^2)$, where $\varphi_{\hbar}$ is given by (46). The connected two-trace function is:

$$\beta\,\langle \mathrm{tr} f\;\mathrm{tr} g\rangle_c = \frac{1}{4\pi}\int_{\mathsf{D}}\nabla f\nabla g d^2z - \frac{1}{4\pi}\oint_{\partial\mathsf{D}} f\partial_n g^H|dz| + O(\hbar) \tag{67}$$

In particular, for the connected correlation functions of the fields $\varphi(z_1)$, $\varphi(z_2)$ (see (33)) this formula gives (if $z_{1,2}\in\mathsf{D}^c$):

$$\frac{1}{2\beta\hbar^2}\langle\varphi(z_1)\varphi(z_2)\rangle_c = G(z_1,z_2) - G(z_1,\infty) - G(\infty,z_2) - \log\frac{|z_1-z_2|}{r} + O(\hbar) \tag{68}$$

where $G$ is the Green function of the Dirichlet boundary value problem and

$$r = \exp\left[\lim_{\xi\to\infty}(\log|\xi| + G(\xi,\infty))\right] \tag{69}$$

is the (external) conformal radius of the domain D. The 2-trace functions are *universal*, i.e., they depend on the shape of the support of eigenvalues only and do not depend on the potential $W$ explicitly. They resemble the two-point functions of the Hermitian 2-matrix model found in [18]; they were also obtained in [43] in the study of thermal fluctuations of a confined 2D Coulomb gas. The structure of the formulas indicates that there are local correlations in the bulk as well as strong long range correlations at the edge of the support of eigenvalues. (See [44] for a similar result in the context of classical Coulomb systems).

From the mathematical point of view, the significance of formula (68) is to provide a link between such seemingly unrelated disciplines as classical analysis in two dimensions and the random matrix theory. Namely, different limits or certain specifications of the arguments in this formula allow one to represent some important objects of classical analysis associated with the domains D and $\mathsf{D}^c$ (e.g., the conformal map onto the unit circle and its Schwarzian derivative, the Bergman kernel) in terms of correlations between eigenvalues of random matrices.

Further variation of the pair density correlation function suggests that, starting from $n = 3$, the connected $n$-point density correlations vanish in the bulk in all orders of $\hbar$ (in fact they are exponential in $1/\hbar$). The entire leading contribution comes from the boundary. The result for the connected three-trace function is:

$$\beta \left\langle \prod_{i=1}^{3} \operatorname{tr} f_i \right\rangle_c = \frac{\hbar}{16\pi^2} \oint_{\partial \mathsf{D}} \frac{|dz|}{\sigma(z)} \prod_{j=1}^{3} \hat{R} f_j(z) + O(\hbar^2) \tag{70}$$

## 5. The matrix model as a growth problem

### 5.1 Growth of the support of eigenvalues

When $N$ increases at a fixed potential $W$, one may say that the support of eigenvalues grows. More precisely, we are going to find how the shape of the support of eigenvalues changes under $t \to t + \delta t$, where $t = N\hbar$, if $W$ stays fixed.

The starting point is the same as for the variations of the potential, and the calculations are very similar as well. Variation of the condition (41) and of the normalization condition yields

$$\oint_{\partial \mathsf{D}} \frac{\Delta W(\zeta)\delta n(\zeta)}{z - \zeta} |d\zeta| = 0 \text{ (for all } z \in \mathsf{D}), \oint_{\partial \mathsf{D}} \Delta W(\zeta)\delta n(\zeta)\, |d\zeta| = -4\pi\beta\delta t$$

The first equation means that $\Delta W(z)\delta n(z)\frac{|dz|}{dz}$ is the boundary value of an analytic function $h(z)$ such that $h(z) = -4\pi\beta\delta t/z + O(z^{-2})$ as $z \to \infty$. The solution for the $\delta n(z)$ is:

$$\delta n(z) = -\frac{\beta \delta t}{2\pi\sigma(z)} \partial_n G(\infty, z) \tag{71}$$

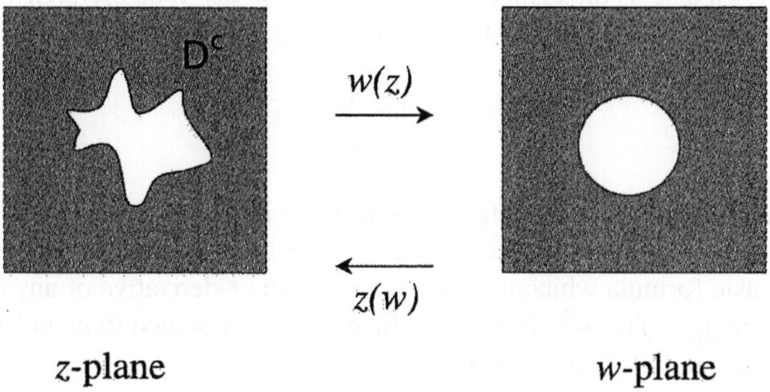


*Figure 5. The mutually inverse conformal maps $w(z)$ and $z(w)$.*

where $G$ is the Green function of the Dirichlet boundary problem in $\mathsf{D}^{\mathsf{c}}$. For quasiharmonic potentials (with $\sigma = 1/\pi$), the formula simplifies:

$$\delta n(z) = -\frac{\beta}{2}\,\delta t\,\partial_n G(\infty, z) \tag{72}$$

Identifying $t$ with time, one can say that the normal velocity of the boundary, $V_n = \delta n/\delta t$, is proportional to gradient of the Green function: $V_n \propto -\partial_n G(\infty, z)$. This result is quite general. It holds for any (not necessarily connected) domains of eigenvalues with a smooth boundary.

If the domain is connected, the Green function can be expressed through the conformal map $w(z)$ from $\mathsf{D}^{\mathsf{c}}$ onto the exterior of the unit circle:

$$G(z_1, z_2) = \log\left|\frac{w(z_1) - w(z_2)}{1 - w(z_1)\overline{w(z_2)}}\right| \tag{73}$$

In particular, $G(\infty, z) = -\log|w(z)|$. As $|z| \to \infty$, $w(z) = z/r + O(1)$, where $r$ is the external conformal radius of the domain $\mathsf{D}$ which enters eq. (68). It is easy to see that $\partial_n \log|w(z)| = |w'(z)|$ on $\partial\mathsf{D}$, so one can rewrite the growth law (72) as follows:

$$\delta n(z) = \frac{\beta}{2}\delta t\,|w'(z)| \tag{74}$$

It is worth noting that the inverse map, $z(w)$, is the classical ($\hbar \to 0$) limit of the Lax operator (29) of the 2D Toda hierarchy. Indeed, making the rescaling $n\hbar = t$, $\partial/\partial n = \hbar\partial_t$, we see that the shift operator $e^{\hbar\partial_t}$ can be replaced by a commuting variable $w$ with the Poisson bracket $\{\log w, t\} = 1$. In this limit, known also as the dispersionless limit of the 2D Toda hierarchy, the Lax operator $L(e^{\hbar\partial_t})$ converts into the function $z(w)$ given by the series of the form

$$z(w) = r(t)w + \sum_{k\geq 0} u_k(t)w^{-k} \tag{75}$$

It defines the one-to-one conformal map from the exterior of the unit circle onto $\mathsf{D}^c$, which is inverse to the map $w(z)$ (Fig. 5). For more details see [8].

The basic formula which allows one to find the $t$-derivative of any quantity of the form $\int_{\mathsf{D}(t)} f(z)\, d^2z$ (where the function $f$ is assumed to be independent of $t$) immediately follows from (71):

$$\delta\left(\int_{\mathsf{D}} f(z)\, d^2z\right) = \oint_{\mathsf{D}} f(z)\delta n(z)\, |dz| = -\frac{\beta\delta t}{2\pi}\oint_{\partial\mathsf{D}} \frac{f(z)}{\sigma(z)}\partial_n G(\infty, z)\, |dz|$$

In the r.h.s. we recognize the value at infinity of the harmonic continuation of the function $f(z)/\sigma(z)$, so the result is

$$\frac{\partial}{\partial t}\left(\int_{\mathsf{D}} f(z)\, d^2z\right) = \beta\, (f/\sigma)^H(\infty) \tag{76}$$

Using this formula and the integral representation of $F_0$, we find $t$-derivatives of the free energy. The first derivative,

$$\partial_t F_0 = 2\int_{\mathsf{D}} \sigma(z)\log|z|\, d^2z \tag{77}$$

was already found in Section 4.3 by other means (see (52)). The second derivative is proportional to the logarithm of the conformal radius (69):

$$\partial_t^2 F_0 = 2\beta\, (\log|z|)^H\Big|_{z=\infty} = 2\beta \lim_{z\to\infty}(\log|z| + G(z,\infty)) = 2\beta\log r \tag{78}$$

In this connection let us also mention the nice formula for the conformal map $w(z)$,

$$w(z) = \lim_{N\to\infty}\frac{\psi_{N+1}(z)}{\psi_N(z)} \tag{79}$$

which follows from the relation $\log\left(P_{N+1}(z)/P_N(z)\right) = \hbar\partial_t\, \langle \mathrm{tr}\log(z-\Phi)\rangle + O(\hbar)$ after calculating the $t$-derivative according to the above rule. This connection between conformal maps and orthogonal polynomials goes back to the classical theory of analytic functions (see e.g. [45]). It is the context of the random matrix theory where it looks natural and easily understandable.

## 5.2 Laplacian growth

The growth law (72) is common to many important problems in physics. The class of growth processes, in which dynamics of a moving front (an interface) between two distinct phases is driven by a harmonic scalar field is known under the name *Laplacian growth.* The most known examples are viscous flows in the Hele-Shaw cell (the Saffman-Taylor problem), filtration processes in porous media, electrodeposition and solidification of undercooled liquids. A comprehensive list of relevant papers published prior to 1998 can be found in [46]. Recently, the Laplacian growth mechanism was recognized [10] in a purely quantum evolution of semiclassical electronic blobs in the QH regime.

**The Saffman-Taylor problem.** Let us describe the main features of the Laplacian growth on the example of viscous flows. To be specific, we shall speak about an interface between two incompressible fluids with very different viscosities on the plane (say, oil and water). In practice, the 2D geometry is realized in the Hele-Shaw cell – a narrow gap between two parallel glass plates. In this version, the problem is also known as the Saffman-Taylor problem or viscous fingering. For a review, see [14]. The velocity field in a viscous fluid in the Hele-Shaw cell is proportional to the gradient of pressure $p$ (Darcy's law):

$$\vec{V} = -K\nabla p\,, \quad K = \frac{b^2}{12\mu}$$

Here the constant $K$ is called the filtration coefficient, $\mu$ is viscosity and $b$ is the size of the gap between the two plates. Note that if $\mu \to 0$, then $\nabla p \to 0$, i.e., pressure in a fluid with negligibly small viscosity is uniform. Incompressibility of the fluids ($\nabla \vec{V} = 0$) implies that the pressure field is harmonic: $\Delta p = 0$. By continuity, the velocity of the interface between the two fluids is proportional to the normal derivative of the pressure field on the boundary: $V_n = -K\partial_n p$.

To be definite, we assume that the Hele-Shaw cell contains a bounded droplet of water surrounded by an infinite "sea" of oil (another possible experimental set-up is an air bubble surrounded by water). Water is injected into the droplet while oil is withdrawn at infinity at a constant rate, as is shown schematically in Fig. 6. The latter means that the pressure field behaves as $p \propto -\log|z|$ at large distances. We also assume that the interface between oil and water is a smooth closed curve $\gamma$ which depends on time. As it was mentioned above, if viscosity of water is negligible, then one may set $p = 0$ inside the water droplet. However, pressure usually has a jump across the interface, so $p$ in general does not tend to zero if one approaches the boundary from outside. This effect is due to *surface tension.* It is hard to give realistic estimates of the surface tension effect from first principles, so one often employs certain ad hoc assumptions. The most popular one is to say that the pressure jump is proportional to the local curvature of the interface.

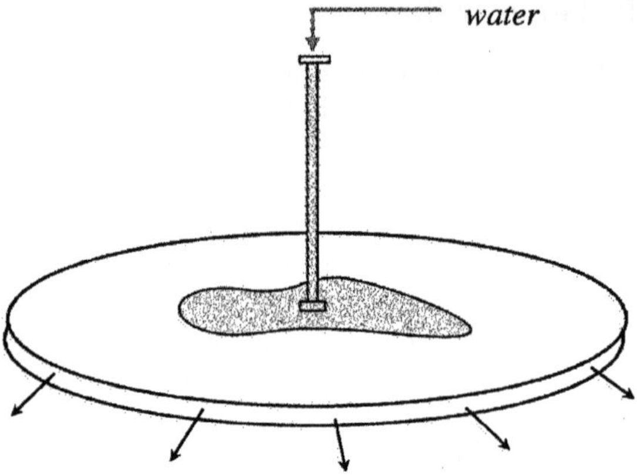


*Figure 6.* *The Hele-Shaw cell.*

To summarize, the mathematical setting of the Saffman-Taylor problem is as follows:

$$\begin{cases} V_n = -\partial_n p & \text{on } \gamma \\ \Delta p = 0 & \text{in oil} \\ p \to -\log|z| & \text{in oil as } z \to \infty \\ p = 0 & \text{in water} \\ p^{(+)} - p^{(-)} = -\nu\kappa & \text{across } \gamma \end{cases} \tag{80}$$

Here $\nu$ is the surface tension coefficient and $\kappa$ is the local curvature of the interface. (The filtration coefficient is set to be 1.) The experimental evidence suggests that when surface tension is small enough, the dynamics becomes unstable. Any initial domain develops an unstable fingering pattern. The fingers split into new ones, and after a long lapse of time the water droplet attains a fractal-like structure. This phenomenon is similar to the formation of fractal patterns in the diffusion-limited aggregation.

Comparing (72) and (80), we identify the D and $\mathsf{D}^{\mathsf{c}}$ with the domains occupied by water and oil respectively, and conclude that the growth laws are identical, with the pressure field being given by the Green function: $p(z) = G(\infty, z)$, and $p = 0$ on the interface. The latter means that supports of eigenvalues grow according to (80) with *zero surface tension*, i.e., with $\nu = 0$ in (80).

Neglecting the surface tension effects, one obtains a good approximation unless the curvature of the interface becomes large. We see that the idealized Laplacian growth problem, i.e., the one with zero surface tension, is mathematically equivalent to the growth of the support of eigenvalues in ensembles of random matrices $\mathscr{N}$, $\mathscr{N}^0$ and $\mathscr{C}$. This fact clarifies the origin of the integrable structure of the Laplacian growth with zero surface tension discovered in [47]. The link to the normal matrix model has been established in [6], see also [48].

**The finite-time singularities.** As a matter of fact, the Laplacian growth problem with zero surface tension is ill-posed since an initially smooth interface often becomes singular in the process of evolution, and the solution blows up. The role of surface tension is to inhibit a limitless increase of the interface curvature. In the absense of such a cutoff, the tip of the most rapidly growing finger typically grows to a singularity (a cusp). In particular, a singularity necessarily occurs for any initial interface that is the image of the unit circle under a rational conformal map, with the only exception of an ellipse.

An important fact is that the cusp-like singularity occurs at a finite time $t = t_c$, i.e., at a finite area of the droplet. It can be shown that the conformal radius of the droplet $r$ (as well as some other geometric parameters), as $t \to t_c$, exhibits a singular behaviour

$$r - r_c \propto (t_c - t)^{-\gamma}$$

characterized by a critical exponent $\gamma$. The generic singularity is the cusp $(2, 3)$, which in suitable local coordinates looks like $y^2 = x^3$. In this case $\gamma = -\frac{1}{2}$. The evolution can not be extended beyond $t_c$.

A similar phenomenon was well-known in the theory of random matrices for quite a long time, and in fact it was the key to their applications to 2D quantum gravity and string theory. In the large $N$ limit, the random matrix models have *critical points* – the points where the free energy is not analytic as a function of a coupling constant. As we have seen, the Laplacian growth time $t$ should be identified with a coupling constant of the normal or complex matrix model. In a vicinity of a critical point,

$$F_0 \sim F_0^{\rm reg} + \alpha(t_c - t)^{2-\gamma}$$

where the critical index $\gamma$ (often denoted by $\gamma_{\rm str}$ in applications to string theory) depends on the type of the critical point. Accordingly, the singularities show up in correlation functions. Using the equivalence established above, we can say that the finite-time blow-up (a cusp-like singularity) of the Laplacian growth with zero surface tension is a *critical point* of the normal and complex matrix models.

## 5.3 The semiclassical limit for electrons in magnetic field

As we have seen in Section 2.3, the system of $N$ electrons in the plane in non-uniform magnetic field, which fully occupy the lowest energy level, is equivalent to the ensemble of normal $N \times N$ matrices, where $N$ is degeneracy of the level. In this section we study the QH droplet in the semiclassical regime.

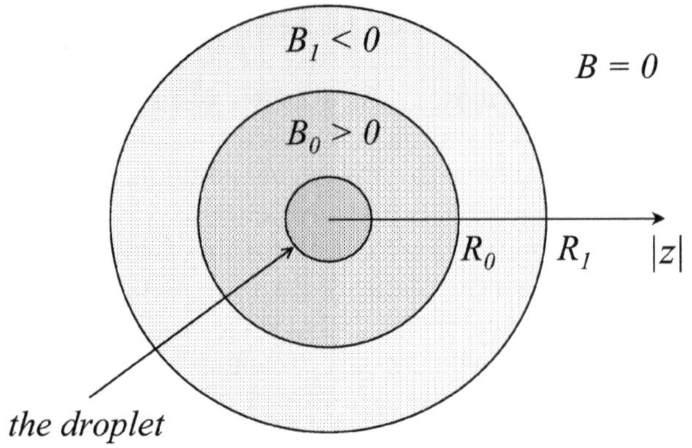


*Figure 7. The configuration of magnetic fields.*

The equivalence with the large $N$ limit of the matrix model suggests to identify the semiclassical QH droplet with the support of eigenvalues. Remarkably, it is this limit where one makes contact with the purely classical Saffman-Taylor problem.

A remark is in order. The limit $\hbar \to 0$ we are talking about is really a *semi*classical limit, or better to say "partially classical". Although the Planck constant $\hbar$ tends to zero, all the particles remain at the lowest (most quantum) energy level, assuming their mass is small or the magnetic field is large. The true quasiclassical limit would imply that the particles occupy higher energy levels.

We recall that the joint probability to find electrons at the points $z_i$ is $|\Psi_N (z_1, \ldots, z_N)|^2$, with $\Psi_N \propto \det_{N\times N}[\psi_n(z_k)]$, where

$$\psi_n(z) = \frac{1}{\sqrt{h_{n-1}}} P_{n-1}(z) e^{W(z)/(2\hbar)} \tag{81}$$

are orthogonal one-particle wave functions for electrons in the magnetic field $B = -\frac{1}{2}\Delta W$ at the lowest level $E = 0$. The level is assumed to be completely filled, i.e., $n = 0, 1 \ldots, N = [\phi/\phi_0]$, where $\phi_0$ is the flux quantum. Then the mean density of the electrons coincides with the expectation value of the density of eigenvalues in the normal or complex matrix model.

The degeneracy of the level cam be controlled in different ways. One of them is to assume the following arrangement (see Fig. 7). Let a strong uniform magnetic field $B_0 > 0$ be applied in a large disk of radius $R_0$. The disk is surrounded by a large annulus $R_0 < |z| < R_1$ with a magnetic field $B_1 < 0$ such that the total magnetic flux through the system is $N\phi_0$: $\pi B_0 R_0^2 - \pi B_1(R_1^2 - R_0^2) = N\phi_0$. The magnetic field outside the largest disk $|z| < R_1$ vanishes. The disk is connected through a tunnel barrier to a large capacitor that maintains a small positive chemical potential slightly above the zero energy. If the field $B_0$ is strong enough, the gap between the energy levels is large, and the higher levels can be neglected. In the case of the uniform fields $B_0$ and $B_1$ the QH droplet is a disk of radius $r_0 = \sqrt{N\hbar} \ll R_0$ trapped at the origin.

In this set-up, let us apply a non-uniform magnetic field, $\delta B$, somewhere inside the disk $|z| < R_0$ but well away from the droplet. This leads to the following two effects.

**The Aharonov-Bohm effect in the QH regime [10, 49].** Suppose that the nonuniform magnetic field $\delta B$ does not change the total flux: $\int \delta B d^2 z = 0$. As is argued above, the shape of the droplet is the same as that of the support of eigenvalues in the ensemble $\mathcal{N}$ with the potential

$$W(z) = -\frac{B_0}{2}|z|^2 - \frac{1}{\pi}\int \log|z-\zeta|\,\delta B(\zeta)\,d^2\zeta$$

The second term is harmonic inside and around the droplet. One may have in mind thin solenoids carrying magnetic flux ("magnetic impurities"). In the case of point-like magnetic fluxes $q_i$ at points $a_i$ the change of the potential is $\delta W(z) = \sum_i q_i \log|z - a_i|$.

Let us stress that in the presence of the fluxes, the shape of the droplet is no longer circular although the magnetic field inside the droplet and not far from it remains uniform and is not changed at all (Fig. 8). In this respect this phenomenon is similar to the Aharonov-Bohm effect. Due to the quantum interference the electronic fluid is attracted to positive fluxes and is repelled by negative ones. The response of the droplet to an infinitesimal change of the magnetic field $\delta B$ is described by eq. (65) in which

$$\delta W^H(z) - \delta W(z) = \frac{1}{\pi}\int_{\mathrm{D^c}} G(z,\zeta)\delta B(\zeta) d^2\zeta$$

In fact this formula holds for arbitrary $\delta B$, not necessarily vanishing inside the droplet. In particular, for small point-like fluxes $\delta q_i$ at some points $a_i$ we have $\delta W = \sum_i \delta q_i \log|z - a_i|$, $\delta B = -\pi \sum_i \delta q_i \delta^{(2)}(z - a_i)$, and $\delta W^H(z) - \delta W(z) = -\sum_i G(z, a_i)\delta q_i$. If $a_i$ is inside, $G(z, a_i)$ is set to be zero. The sum, therefore, is over outside fluxes only. The fluxes inside the droplet, if any, appear to be completely screened and do not have any influence on its shape.

**Growth of the electronic droplet.** If the total magnetic flux increases, with magnetic impurities kept fixed, the electronic droplet grows. For example, one may adiabatically increase $B_1$, with $B_0$ and $\delta B$ fixed. Then the droplet grows

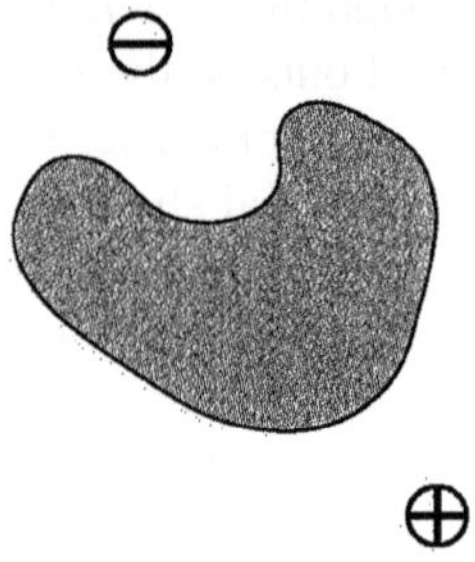

*Figure 8.* *The electronic droplet in the presence of magnetic impurities.*

because the degeneracy of the lowest level is enlarged and new electrons enter the system. The growth is described by eq. (71) with $\Delta W(z) = -2B_0$ which is equivalent to the Darcy law. This phenomenon is purely quantum. Like the Aharonov-Bohm effect, it is caused by quantum interference. Its characteristic scale is less than that of the Saffman-Taylor fingering by a factor of $10^9$. The correspondence established above suggests that the edge of the QH droplet may develop unstable features similar to the fingers in the Hele-Shaw cell.

## 5.4 Semiclassical $\psi$-function

In this Subsection we derive the semiclassical asymptotics of the $\psi$-function (81), i.e.,

$$\psi_{N+1}(z) \stackrel{\hbar\to 0}{\longrightarrow} \psi(z)$$

To avoid cumbersome technical details, we mainly consider models with quasiharmonic potential.

First of all, it is necessary to know the large $N$ limit of the orthogonal polynomials $P_N$. Truncating the general formula (18) at the second term in the exponent, we can write:

$$\begin{aligned} P_N(z) &= \langle \det(z-\Phi) \rangle = \left\langle e^{\mathrm{tr}\,\log(z-\Phi)} \right\rangle \\ &\stackrel{\hbar\to 0}{=} \exp\left( \langle \mathrm{tr}\,\log(z-\Phi) \rangle + \frac{1}{2} \left\langle (\mathrm{tr}\,\log(z-\Phi))^2 \right\rangle_c + \ldots \right) \end{aligned} \qquad (82)$$

The first term in the r.h.s. is $O(\hbar^{-1})$, the second one is $O(\hbar^0)$ and the ignored terms vanish as $\hbar \to 0$. However, this formula should be applied with some care since the logarithm is not a single-valued function. This formula is correct if one can fix a single-valued branch of the logarithm. It is possible if $z$ is outside D. For $z$ inside D an analytic continuation should be used. To take care of the normalization, we also need

$$h_N = \frac{Z_{N+1}}{(N+1)Z_N} = (2\pi^3\hbar)^{1/2} \exp\left(\hbar^{-1}\partial_t F_0 + \frac{1}{2}\partial_t^2 F_0 + \dots\right) \quad (83)$$

which is written here with the same precision as (82). (We have used (58) with $\alpha = \log\sqrt{2\pi^3\hbar}$ and have taken into account that $F_{1/2} = 0$ for quasiharmonic potentials.)

Let us first keep the dominant terms in the r.h.s. of (82), (83) and ignore the $O(\hbar^0)$ terms for a while. We have, for $z \in \mathsf{D}^c$: $|\psi(z)|^2 \sim e^{-\frac{1}{\hbar}\partial_t F_0}$ $e^{\frac{1}{\hbar}(W(z)-\langle\varphi(z)\rangle)}$, or, using the results of the previous section,

$$|\psi(z)|^2 \sim e^{-2\mathcal{A}(z)/\hbar}, \qquad 2\mathcal{A}(z) = \varphi_{cl}(z) - \varphi_{cl}(0) - W(z)$$

As we know, $\mathcal{A}(z)$ defined by this formula is zero on the boundary. The analytic continuation of the $\mathcal{A}(z)$ inside the contour $\gamma = \partial\mathsf{D}$ can be done using the Schwarz function. Namely, since $2\partial\mathcal{A}(z) = \bar{z} - S(z)$, the desired analytic continuation can be defined (up to a constant) by the formula

$$2\mathcal{A}(z) = |z|^2 - 2\mathcal{R}e \int^z S(\zeta)\, d\zeta$$

Clearly, $\mathcal{A}(z)$ defined in this way is constant on the contour $\gamma$. Indeed, if $z_{1,2} \in \gamma$, then

$$2(\mathcal{A}(z_2) - \mathcal{A}(z_1)) = |z_2|^2 - |z_1|^2 - 2\mathcal{R}e \int_{z_1}^{z_2} S(z)dz$$

$$= z_2\bar{z}_2 - z_1\bar{z}_1 - \int_{z_1}^{z_2} \bar{z}dz - \int_{z_1}^{z_2} zd\bar{z} = z_2\bar{z}_2 - z_1\bar{z}_1 - \int_{z_1}^{z_2} d(z\bar{z}) = 0$$

Since $\mathcal{A}(z)$ should be zero on $\gamma$, we finally define

$$\mathcal{A}(z) = \frac{1}{2}|z|^2 - \frac{1}{2}|\xi_0|^2 - \mathcal{R}e \int_{\xi_0}^{z} S(\zeta)d\zeta \quad (84)$$

where $\xi_0$ is an arbitrary point on the contour. We call the function defined by (84) *the effective action*. From the above it follows that its first derivatives vanish for all $z \in \gamma$: $\partial\mathcal{A}(z) = \bar{\partial}\mathcal{A}(z) = 0$. This means that $|\psi|^2 \sim e^{-2\mathcal{A}/\hbar}$ has a sharp maximum on the contour $\gamma$. We may say that purely quantum particles are in general delocalized in the plane, purely classical particles are localized at some points in the plane while partially classical particles, like our electrons, are localized on closed curves in the plane.

Let us turn to the $O(\hbar^0)$-corrections, which give the subexponential factor in the asymptotics of the $\psi$-function[4]. We assume, for simplicity, that the domain $\mathsf{D}$ is connected. Extracting analytic and anti-analytic parts of eq. (68), we get

$$\left\langle (\text{tr}\log(z-\Phi))^2 \right\rangle_c = \log(rw'(z)) + O(\hbar)$$

where $r$ is the external conformal radius of D and $w'(z)$ is the derivative of the conformal map from $\mathsf{D}^c$ onto the exterior of the unit circle. Plugging this into (82) and taking into account eq. (78), we finally obtain:

$$|\psi(z)|^2 = \frac{|w'(z)|}{\sqrt{2\pi^3\hbar}}\, e^{-2\mathcal{A}(z)/\hbar} \tag{85}$$

This formula does resemble the WKB asymptotics in quantum mechanics. If a singularity of the conformal map is sufficiently close to the boundary from inside, the asymptotics becomes invalid in this region.

The effective action can be expanded near the contour, where it takes the minimal value:

$$\mathcal{A}(z+\delta_n z) = |\delta_n z|^2 \mp \frac{1}{3}\kappa(z)|\delta_n z|^3 + \frac{1}{4}\kappa^2(z)|\delta_n z|^4 + \ldots \tag{86}$$

Here $\kappa(z)$ is the local curvature of the contour at the point $z$ and $\delta_n z$ is a small deviation from the point $z \in \gamma$ in the normal direction. (The upper and lower signs correspond to the outward and inward deviations respectively.) A similar expansion of $\log|w'(z)|$ reads

$$\log|w'(z+\delta_n z)| = \log|w'(z)| \pm (|w'(z)| - \kappa(z))\,|\delta_n z| + \ldots \tag{87}$$

Some details of the derivation are given in Appendix E. Therefore, if $\kappa(z) \ll \hbar^{-1/2}$, the squared modulus of the $\psi$-function is well approximated by the sharp Gaussian distribution in the normal direction with the amplitude slowly modulated along the curve:

$$|\psi(z+\delta_n z)|^2 \simeq \frac{|w'(z)|}{\sqrt{2\pi^3\hbar}}\, e^{-2|\delta_n z|^2/\hbar} \tag{88}$$

In the case of general potentials the calculations lead to a similar result:

$$|\psi(z+\delta_n z)|^2 \simeq \sqrt{\frac{\sigma(z)}{2\pi^2\hbar}}\,|w'(z)|\, e^{-2\pi\sigma(z)|\delta_n z|^2/\hbar} \tag{89}$$

We see that the width of the Gaussian distribution depends on the point of the curve through the function $\sigma$ (which is proportional to the magnetic field in the QH interpretation). Note that this asymptotics is consistent with the normalization $\int |\psi(z)|^2 d^2 z = 1$. The easiest way to see this is to notice that the Gaussian function in (88) (as well as the one in (89)) tends to the delta function $\delta(z;\gamma)$ with the support on the curve (see Appendix B). Hence one can formally write the limiting $\psi$-function in the form

$$|\psi(z)|^2 = \frac{|w'(z)|}{2\pi}\,\delta(z;\gamma) \tag{90}$$

and the right normalization is transparent.

Finally, we note that the growth law (74) (the Darcy law) can be written in the suggestive form

$$V_n(z) \propto |\psi(z)|^2 \tag{91}$$

The normal velocity $V_n$ is defined by the relation

$$\partial_t \Theta(z; \mathsf{D}(t)) = V_n \delta(z; \partial \mathsf{D}(t))$$

where $\Theta(z; \mathsf{D})$ is the characteristic function of the domain D. Since the classical value of $\langle \rho(z) \rangle$ is $\Theta(z; \mathsf{D})/\pi$, we can represent the Darcy law in yet another form:

$$\partial_t \langle \rho(z) \rangle = |\psi(z)|^2$$

In fact it is the $\hbar \to 0$ limit of the exact relation

$$\langle \rho(z) \rangle_N - \langle \rho(z) \rangle_{N-1} = \hbar |\psi_N(z)|^2 \tag{92}$$

which immediately follows from the fact that $\langle \rho(z) \rangle_N = \hbar K_N(z, \bar{z})$ and from the definition of the kernel function (28). In fact it is equivalent to the Hirota equation (24). We see that (92) can be regarded as a "quantization" of the Darcy law.

## Acknowledgements

These lectures were presented at the Les Houches School "Applications of Random Matrices in Physics" in June 2004. It is a pleasure to acknowledge the nice and productive atmosphere of the School. I am grateful to O.Agam, E.Bettelheim, I.Kostov, I.Krichever, A.Marshakov, M.Mineev-Weinstein, R.Teodorescu and P.Wiegmann for collaboration and to A.Abanov, A.Boyarsky, L.Chekhov, A.Gorsky, V.Kazakov, O.Ruchaisky for discussions. This work was supported in part by RFBR grant 04-01-00642, by grant for support of scientific schools NSh-1999.2003.2 and by grant INTAS 03-51-6346.

## Appendices

## Appendix A

**The proof of the determinant representation (19).** It is simpler to start with $\det C_{ij}$, where $C_{ij}$ is given in (20), and to show that $N! \det C_{ij}$ coincides with $Z_N$. The proof is a chain of obvious equalities. By definition,

$$N! \det C_{ij} = \sum_{Q,P} (-)^Q (-)^P \, C_{Q(1),P(1)} \, \cdots \, C_{Q(N),P(N)}$$

Plugging the explicit form of the $C_{ij}$, we write:

$$N!\det C_{ij} = \sum_{Q,P}(-)^Q(-)^P \int \prod_{i=1}^{N} \left[ z_i^{Q(i)-1} \bar{z}_i^{P(i)-1} e^{W(z_i)} d^2 z_i \right]$$

and, interchanging the order of summation and integration, obtain the result:

$$N!\det C_{ij} = \int \underbrace{\left[ \sum_Q (-)^Q \prod_{i=1}^{N} z_i^{Q(i)-1} \right]}_{\Delta_N(z_i)} \underbrace{\left[ \sum_P (-)^P \prod_{i=1}^{N} \bar{z}_i^{P(i)-1} \right]}_{\overline{\Delta_N(z_i)}} \prod_{k=1}^{N} e^{W(z_k)} d^2 z_k \;=\; Z_N$$

**Orthogonality of the polynomials (25).** The definition (25) implies

$$P_n(\lambda) = \frac{1}{Z_n} \int |\Delta_n(z_i)|^2 \prod_{j=1}^{n} (\lambda - z_j) e^{W(z_j)} d^2 z_j$$

We want to show that these polynomials are orthogonal in the complex plane.

Set $e^{W(z)} d^2 z = d\mu$ for brevity. It is enough to show that $\int P_n(z) \bar{z}^m d\mu = 0$ for all $m < n$, i.e.,

$$\int d\mu(z) \bar{z}^m \int |\Delta_n(z_i)|^2 \prod_{j=1}^{n} (z - z_j) d\mu(z_j) = 0$$

Note that $\Delta_n(z_1, \ldots, z_n) \prod_{j=1}^{n} (z - z_j) = \Delta_{n+1}(z_1, \ldots, z_n, z)$. Setting $z \equiv z_{n+1}$, we have:

$$\text{LHS} = \int \Delta_{n+1}(z_i) \overline{\Delta_n(z_i)}\, \bar{z}_{n+1}^m \prod_{j=1}^{n+1} d\mu(z_j)$$

$$= \frac{1}{n+1} \sum_{l=1}^{n+1} (-1)^{l+n+1} \int \Delta_{n+1}(z_1, \ldots, z_{n+1}) \overline{\Delta_n(z_1, \ldots, \not{z}_l, \ldots, z_{n+1}) z_l^m} \prod_{j=1}^{n+1} d\mu(z_j)$$

One can notice that the summation gives the expansion of the determinant

$$\begin{vmatrix} 1 & z_1 & \ldots & z_1^{n-1} & z_1^m \\ 1 & z_2 & \ldots & z_2^{n-1} & z_2^m \\ \ldots & \ldots & \ldots & \ldots & \ldots \\ 1 & z_{n+1} & \ldots & z_{n+1}^{n-1} & z_{n+1}^m \end{vmatrix}$$

under the bar. If $m < n$, it vanishes. If $m = n$, it equals $\Delta_{n+1}(z_i)$, and we get LHS $= \frac{1}{n+1} Z_{n+1}$.

## Appendix B

Here we list some standard formulas often used in Section 4.

**The complex notation.**

- Complex coordinates: $z = x + iy$, $\bar{z} = x - iy$, $\partial_z = \frac{1}{2}(\partial_x - i\partial_y)$, $\partial_{\bar{z}} = \frac{1}{2}(\partial_x + i\partial_y)$.
- The Laplace operator: $\Delta = \partial_x^2 + \partial_y^2 = 4\partial_z\partial_{\bar{z}}$
- Contour integrals: let $f$ and $g$ be any smooth functions defined in some neighborhood of the contour $\gamma$, then

$$\oint_\gamma g\partial_n f|dz| = -2i \oint_\gamma g\partial_z f dz - i \oint f dg$$

**Singular functions.**

- Two-dimensional $\delta$-function: $\delta(z) = \frac{1}{2\pi}\Delta \log|z| = \frac{1}{\pi}\partial_{\bar{z}}(1/z)$. The characteristic property of the delta-function is $f(z)\delta(z-a)\, d^2z = f(a)$ for any (smooth) function $f$.
- The $\delta$-function with the support on a curve (a closed contour) $\gamma$: a function $\delta(z;\gamma)$ such that

$$\int f(z)\delta(z;\gamma)d^2z = \oint_\gamma f(z)|dz|$$

  for any smooth function $f$.
- The "normal derivative" of the $\delta$-function of the contour $\gamma$: a function $\delta'(z;\gamma)$ such that

$$\int f(z)\delta'(z;\gamma)d^2z = -\oint_\gamma \partial_n f(z)|dz|$$

  for any smooth function $f$, with the normal vector being directed to the exterior of the contour.
- The characteristic function of the domain D: $\Theta(z;\mathsf{D}) = 1$ if $z \in \mathsf{D}$ and 0 otherwise; $\nabla\Theta(z;\mathsf{D}) = -\vec{n}\,\delta(z;\partial\mathsf{D})$.

**Integral formulas.**

- Cauchy's integral formula ($f$ is any smooth function):

$$\frac{1}{2\pi i}\oint_{\partial\mathsf{D}} \frac{f(\zeta)d\zeta}{z-\zeta} - \frac{1}{\pi}\int_{\mathsf{D}} \frac{\partial_{\bar{\zeta}} f(\zeta)d^2\zeta}{z-\zeta} = \begin{cases} -f(z), & z \in \mathsf{D} \\ 0, & z \in \mathbf{C}\setminus\mathsf{D} \end{cases}$$

In particular, $\oint_{\partial \mathsf{D}} f(\zeta)d\zeta = 2i\int_{\mathsf{D}} \bar{\partial} f(\zeta)\, d^2\zeta$.

- The Green formula:

$$\int_{\mathsf{D}} f\Delta g d^2 z = -\int_{\mathsf{D}} \nabla f\, \nabla g\, d^2 z + \oint_{\partial \mathsf{D}} f\partial_n g |dz|$$

where the normal vector looks outward D.

- The Dirichlet formula:

$$u(z) = -\frac{1}{2\pi}\oint_{\gamma} u(\zeta)\partial_n G(z,\zeta)|d\zeta|$$

for any function $u$ harmonic in $\mathsf{D}^{\mathsf{c}} = \mathbf{C} \setminus \mathsf{D}$. Here $G(z,\zeta)$ is the Green function of the Dirichlet boundary value problem in $\mathsf{D}^{\mathsf{c}}$.

## Appendix C

Let us present some details of the $\hbar$-expansion of the loop equation (35). First of all we rewrite it in the form

$$\frac{1}{2\pi}\int L(z,\zeta)\,\langle \Delta\varphi(\zeta)\rangle\, d^2\zeta = (\partial\varphi_{cl}(z))^2 - \langle (\partial(\varphi(z)-\varphi_{cl}(z))^2\rangle - (2-\beta)\hbar\,\langle \partial^2\varphi(z)\rangle$$

which is ready for the $\hbar$-expansion. Here

$$L(z,\zeta) = \frac{\partial W(\zeta) - \partial\varphi_{cl}(z)}{\zeta - z}$$

is the kernel of the integral operator in the l.h.s. (the "loop operator"). The zeroth order in $\hbar$ gives equation (38) which implies the familiar result $\varphi_{cl}(z) = -\int_{\mathsf{D}} \log|z-\zeta|^2\sigma(\zeta)d^2\zeta$ for the $\varphi_{cl}$. To proceed, one should insert the series

$$\langle\varphi(z)\rangle = \varphi_{cl}(z) + \hbar\varphi_{1/2}(z) + \hbar^2\varphi_1(z) + O(\hbar^3)$$

(which corresponds to the $\hbar$-expansion (48) of the free energy) into the loop equation and separate terms of order $\hbar$, $\hbar^2$ etc. (In the notation adopted in the main body of the paper $\hbar\varphi_{1/2} = \varphi_{\hbar} + O(\hbar^2)$.) The terms of order $\hbar$ and $\hbar^2$ give:

$$\frac{1}{2\pi}\int L(z,\zeta)\,\langle \Delta\varphi_{1/2}(\zeta)\rangle\, d^2\zeta = -(2-\beta)\partial^2\varphi_{cl}(z)$$

$$\frac{1}{2\pi}\int L(z,\zeta)\,\langle \Delta\varphi_1(\zeta)\rangle\, d^2\zeta = -\left[(\partial\varphi_{1/2}(z))^2 + (2-\beta)\partial^2\varphi_{1/2}(z)\right] - \omega(z)$$

where

$$\omega(z) = \lim_{\hbar\to 0}\left[\hbar^{-2}\lim_{z'\to z}\langle\partial\varphi(z)\,\partial\varphi(z')\rangle_c\right]$$

is the connected part of the pair correlator at merging points. If the point $z$ is in $\mathsf{D}^{\mathsf{c}}$, then eq. (68) yields

$$\left\langle \partial\varphi(z)\,\partial\varphi(z')\right\rangle_c = 2\beta\hbar^2\partial_z\partial_{z'}\Big(G(z,z') - \log|z-z'|\Big) + O(\hbar^3))$$

Since the r.h.s. is regular for all $z, z' \in \mathsf{D}^{\mathsf{c}}$, the points can be merged without any regularization and the result does not depend on the particular limit $z' \to z$. We thus obtain that the function $\omega(z)$ is proportional to the Schwarzian derivative of the conformal map $w(z)$:

$$\omega(z) = \frac{\beta}{6}\left(\frac{w'''(z)}{w'(z)} - \frac{3}{2}\left(\frac{w''(z)}{w'(z)}\right)^2\right)$$

The expansion of the loop equation can be continued order by order. In principle, this gives a recurrence procedure to determine the coefficients $\varphi_k(z)$. However, the equations of the chain are integral equations in the plane, and it is not easy to solve them explicitly. Another difficulty is that in general one can not extend these equations to the interior of the support of eigenvalues because the $\hbar$-expansion may break down or change its form there. Indeed, in the domain filled by the gas of eigenvalues the microscopic structure of the gas becomes essential, and one needs to know correlation functions at small scales. Nevertheless, at least in the first two orders in $\hbar^2$ the equations above can be solved assuming that $z \in \mathsf{D}^{\mathsf{c}}$. Note that in this region all the functions $\varphi_k(z)$ are harmonic. If these functions are known, the corresponding expansion coefficients of the free energy in (48) can be obtained by "integration" of the variational formulas (17).

The procedure of solving the loop equation in the first two orders in $\hbar$ is too technical to be presented here. In the order $\hbar$ one is able to find a complete solution which gives formulas (46) and (57) mentioned in the main text. The solution in the next order, $\hbar^2$, is much more difficult to obtain. The results for $\varphi_1$ and $F_1$ are still not available in full generality (i.e., for general $\beta$ and $W$). Nevertheless, for normal matrices with a general potential ($\beta = 1$) and with a connected support of eigenvalues the $F_1$-correction to the free energy can be found explicitly by the method outlined above. Here we present the result (mostly for an illustrative purpose), using the notation introduced in the main text (see (45), (75)):

$$\begin{aligned} F_1 &= -\frac{1}{24\pi}\oint_{|w|=1}\Big(\log|z'(w)|\partial_n \log|z'(w)| + 2\log|z'(w)|\Big)|dw| \\ &\quad - \frac{1}{24\pi}\left[\int_{\mathsf{D}} |\nabla\chi|^2\, d^2 z + 2\oint_{\partial\mathsf{D}} \kappa\chi\,|dz|\right] \\ &\quad + \frac{1}{8\pi}\left[\int_{\mathsf{D}} |\nabla\chi|^2\, d^2 z - \oint_{\partial\mathsf{D}} \chi\partial_n\chi^H\,|dz|\right] - \frac{1}{16\pi}\int_{\mathsf{D}}\Delta\chi d^2 z + c_0 \end{aligned}$$

where $c_0$ is a numerical constant and $\kappa(z) = \partial_n \log\left|\frac{w(z)}{w'(z)}\right|$ is the local curvature of the boundary. For quasiharmonic potentials only the first integral survives.

## Appendix D

Here we demonstrate how the variational technique works with the 2-trace correlator (67). We shall use the variational formulas (17) in the following equivalent version. Set $\delta W(z) = \epsilon g(z)$, where $g$ is an arbitrary smooth function and $\epsilon \to 0$. Then, in the first order in $\epsilon$,

$$\hbar\,\delta\,\langle \mathrm{tr}\, f\rangle = \epsilon\,\langle \mathrm{tr}\, f\ \mathrm{tr}\, g\rangle_c$$

This relation allows one to find the connected part of the two-trace correlation function by variation of the known one-trace function. Similar formulas hold for variations of multi-trace functions.

We have, in the leading order in $\hbar$:

$$\hbar\beta\delta\,\langle \mathrm{tr}\, f\rangle = \delta\left(\int_{\mathsf{D}} \sigma f\, d^2 z\right) = \int_{\mathsf{D}} \delta\sigma f\, d^2 z + \int_{\delta\mathsf{D}} \sigma f\, d^2 z \equiv I_1 + I_2$$

The first integral, $I_1$, can be transformed using the definition of $\sigma$ (40) and the Green formula:

$$I_1 = -\frac{1}{4\pi}\int_{\mathsf{D}} \Delta(\delta W) f\, d^2 z = \frac{\epsilon}{4\pi}\int_{\mathsf{D}} \nabla g\,\nabla f\, d^2 z - \frac{\epsilon}{4\pi}\oint_{\partial\mathsf{D}} f\partial_n g\,|dz|$$

The second integral is

$$I_2 = -\frac{1}{4\pi}\int_{\mathsf{D}} \Delta W(z) f(z)\delta n(z)|dz|$$

where $\delta n(z)$ is to be taken from eq. (65): $\delta n(z) = \epsilon\partial_n(g^H(z) - g(z))/\Delta W(z)$. Summing the two contributions, we get (67).

## Appendix E

In this Appendix we obtain the expansions of the effective action $\mathcal{A}(z)$

$$\mathcal{A}(z) = \frac{1}{2}|z|^2 - \frac{1}{2}|\xi_0|^2 - \mathcal{R}e \int_{\xi_0}^{z} S(\zeta)\, d\zeta$$

near the contour $\gamma = \partial \mathrm{D}$. We know that the first variation of $\mathcal{A}(z)$ vanishes on $\gamma$. To find the second variation, we write

$$2\delta\mathcal{A}(z) = |\delta z|^2 - \mathcal{R}e\,(S'(z)(\delta z)^2)\,, \quad z \in \gamma$$

Let us represent $\delta z$ as a sum of normal and tangential deviations w.r.t. the curve: $\delta z = \delta_n z + \delta_t z$, then

$$2\mathcal{R}e\,(S'(z)(\delta z)^2) = \mathcal{R}e\,(S'(z)(\delta_n z)^2) + \mathcal{R}e\,(S'(z)(\delta_t z)^2) + \mathcal{R}e\,(S'(z)\delta_n z \delta_t z)$$

Using the obvious relations

$$\frac{\delta_t z}{|\delta_t z|} = \sqrt{\frac{\delta_t z}{\delta_t \bar{z}}} = \frac{1}{\sqrt{S'(z)}}\,, \qquad \delta_n z = \mp i \left|\frac{\delta_n z}{\delta_t z}\right| \delta_t z$$

where the upper (lower) sign should be taken for the outward (inward) deviation, and the formula for the scalar product of 2-vectors $\vec{x}, \vec{y}$ represented as complex numbers $x, y$, $(\vec{x}, \vec{y}) = \mathcal{R}e\,(x\bar{y})$, we see that the first and second terms in the r.h.s. are equal to $-|\delta_n z|^2$ and $|\delta_t z|^2$ respectively while the third one vanishes since the vectors $\delta_n z$ and $\delta_t z$ are orthogonal. Since $|\delta z|^2 = |\delta_n z|^2 + |\delta_t z|^2$, we obtain the desired result $\delta\mathcal{A}(z) = |\delta_n z|^2$.

The next terms of the expansion of the $\mathcal{A}(z)$ around the contour can be found in a similar way. They are expressed through the curvature $\kappa$ and its derivatives w.r.t. the arc length $s$ along the curve. To perform the calculations in next two orders we need the following formulas for the $\kappa$ and $\kappa' = d\kappa/ds$ through the Schwarz function [31]:

$$\kappa(z) = \frac{i}{2}\frac{S''(z)}{(S'(z))^{3/2}}\,, \qquad \kappa'(z) = \frac{i}{2}\,\frac{S'''(z)S'(z) - \frac{3}{2}(S''(z))^2}{(S'(z))^3}\,, \qquad z \in \gamma$$

For $z$ on the contour we have

$$\mathcal{A}(z + \delta_n z) = |\delta_n z|^2 - \frac{1}{6}\mathcal{R}e\,(S''(z)(\delta_n z)^3) - \frac{1}{24}\mathcal{R}e\,(S'''(z)(\delta_n z)^4) + \ldots$$

Now, with the help of the formulas for the curvature, it is easy to find that

$$\begin{aligned}
\mathcal{R}e\,(S''(z)(\delta_n z)^3) &= 2\kappa(z)\mathcal{R}e\left(\frac{1}{i}\left(\sqrt{S'(z)}\delta_n z\right)^3\right) \\
&= 2\kappa(z)\mathcal{R}e\left(\frac{1}{i}\left(\frac{\overline{\delta_t z}}{\delta_t z}\right)^{3/2}(\delta_n z)^3\right)
\end{aligned}$$

whence

$$\mathcal{R}e\,(S''(z)(\delta_n z)^3) = \pm 2\kappa(z)|\delta_n z|^3$$

A similar computation gives $\mathcal{R}e\,(S'''(z)(\delta_n z)^4) = -6\kappa^2(z)|\delta_n z|^4$ and we obtain the expansion (86). The expansion of $|w'(z)|$ around the contour can be easily performed with the help of the relations

$$\kappa(z) = \partial_n \log\left|\frac{w(z)}{w'(z)}\right|, \qquad \partial_n \log|w(z)| = |w'(z)|$$

valid for $z \in \gamma$.

## References


[1] M.L.Mehta, Random matrices and the statistical theory of energy levels, 2-nd edition, Academic Press, NY, 1991

[2] T.Guhr, A.Müller-Groeling and H.Weidenmüller, Phys. Rep. 299 (1998) 189-428, e-print archive: cond-mat/9707301

[3] P.Di Francesco, P.Ginsparg and J.Zinn-Justin, Phys. Rep. 254 (1995) 1-133

[4] A.Morozov, Phys. Usp. 37 (1994) 1-55, e-print archive: hep-th/9303139

[5] Y.Fyodorov, B.Khoruzhenko and H.-J.Sommers, Phys. Rev. Lett. 79 (1997) 557, e-print archive: cond-mat/9703152; J.Feinberg and A.Zee, Nucl. Phys. B504 (1997) 579-608, e-print archive: cond-mat/9703087; G.Akemann, J. Phys. A 36 (2003) 3363, e-print archive: hep-th/0204246; G.Akemann, Phys. Rev. Lett. 89 (2002) 072002, e-print archive: hep-th/0204068 A.M.Garcia-Garcia, S.M.Nishigaki and J.J.M.Verbaarschot, Phys. Rev. E66 (2002) 016132 e-print archive: cond-mat/0202151

[6] I.Kostov, I.Krichever, M.Mineev-Weinstein, P.Wiegmann and A.Zabrodin, $\tau$-function for analytic curves, Random matrices and their applications, MSRI publications, eds. P.Bleher and A.Its, vol.40, p. 285-299, Cambridge Academic Press, 2001, e-print archive: hep-th/0005259

[7] P.Wiegmann and A.Zabrodin, J. Phys. A 36 (2003) 3411, e-print archive: hep-th/0210159

[8] R.Teodorescu, E.Bettelheim, O.Agam, A.Zabrodin and P.Wiegmann, Nucl. Phys. B704/3 (2005) 407-444, e-print archive: hep-th/0401165; Nucl. Phys. B700 (2004) 521-532, e-print archive: hep-th/0407017

[9] A.Marshakov, P.Wiegmann and A.Zabrodin, Commun. Math. Phys. 227 (2002) 131-153, e-print archive: hep-th/0109048

[10] O.Agam, E.Bettelheim, P.Wiegmann and A.Zabrodin, Phys. Rev. Lett. 88 (2002) 236801. e-print archive: cond-mat/0111333

[11] J.Ginibre, J. Math. Phys. 6 (1965) 440

[12] F.Dyson, J. Math. Phys. 3 (1962) 140, 157, 166

[13] R.B.Laughlin, in: The Quantum Hall Effect, R.E.Prange and S.M.Girvin (eds.), Springer, New York, 1987; S.Iso, D.Carabali and B.Sakita, Phys. Lett. B296 (1992) 143; A.Cappelli, C.Trugenberger and G.Zemba, Nucl. Phys. B396 (1993) 465

[14] D.Bensimon, L.P.Kadanoff, S.Liang, B.I.Shraiman, and C.Tang, Rev. Mod. Phys. 58 (1986) 977

[15] L.Chau and Y.Yu, Phys. Lett. A167 (1992) 452-458; L.Chau and O.Zaboronsky, Commun. Math. Phys. 196 (1998) 203, e-print arhive: hep-th/9711091

[16] G.Oas, Universal cubic eigenvalue repulsion for random normal matrices, e-print archive: cond-mat/9610073

[17] M.Hastings, J. Stat. Phys 103 (2001) 903, e-print archive: cond-mat/9909234

[18] J.-M.Daul, V.Kazakov and I.Kostov, Nucl. Phys. B409 (1993) 311

[19] S.Alexandrov, V.Kazakov and I.Kostov, Nucl. Phys. B640 (2002) 119-144, e-print archive: hep-th/0205079; Nucl. Phys. B667 (2003) 90-110, e-print archive: hep-th/0302106

[20] V.Girko, Theor. Prob. Appl. 29 (1985) 694

[21] P.Di Francesco, M.Gaudin, C.Itzykson and F.Lesage, Int. J. Mod. Phys. A9 (1994) 4257-4351

[22] E.Marinari and G.Parisi, Phys. Lett. B240 (1990) 375-380; E.Marinari and G.Parisi, Phys. Lett. B247 (1990) 537-541

[23] M.Hastings, Non-Hermitian fermion mapping for one-component plasma, e-print archive: cond-mat/9703006

[24] A.Boyarsky, V.Cheianov and O.Ruchayskiy, Fermions in the harmonic potential and string theory, e-print archive: hep-th/0409129

[25] Y.Aharonov and A.Casher, Phys. Rev. A19 (1979) 2461; J.Avron and R.Seiler, Phys. Rev. Lett. 42 (1979) 931

[26] R.Hirota, J. Phys. Soc. Japan, 50 (1981) 3785-3791

[27] A.Gerasimov, A.Marshakov, A.Mironov, A.Morozov and A.Orlov, Nucl. Phys. B357 (1991) 565

[28] M.Staudacher, Phys. Lett B305 (1993) 332-338, e-print archive: hep-th/9301038

[29] V.Kazakov and A.Marshakov, J. Phys. A 36 (2003) 3107, e-print archive: hep-th/0211236

[30] B.Eynard, JHEP 0301 (2003) 051, e-print archive: hep-th/0210047

[31] P.J.Davis, The Schwarz function and its applications, The Carus Math. Monographs, No. 17, The Math. Assotiation of America, Buffalo, N.Y., 1974

[32] F.David, Phys. Lett. B302 (1993) 403-410, e-print archive: hep-th/9212106; B.Eynard and J.Zinn-Justin, Phys. Lett. B302 (1993) 396-402, e-print archive: hep-th/9301004; V.Kazakov and I.Kostov, e-print archive: hep-th/0403152; S.Alexandrov, JHEP 0405 (2004) 025, e-print archive: hep-th/0403116

[33] M.Bertola, B.Eynard and J.Harnad Commun. Math. Phys. 229 (2002) 73-120, e-print archive: nlin.SI/0108049; Theor. Math. Phys. 134 (2003) 27-38, e-print archive: nlin.SI/0112006

[34] A.Jevicki and B.Sakita, Nucl. Phys. B165 (1980) 511; A.Jevicki and B.Sakita, Nucl. Phys. B185 (1981) 89; A.Jevicki, Collective field theory and Schwinger-Dyson equations in matrix models, preprint Brown-HET-777, Proceedings of the meeting "Symmetries, quarks and strings" held at the City College of New York, Oct. 1-2, 1990

[35] J.Ambjorn, L.Chekhov, C.F.Kristjansen and Yu.Makeenko, Nucl. Phys. B404 (1993) 127-172; Erratum: ibid. B449 (1995) 681, e-print archive: hep-th/9302014

[36] G.Akemann, Nucl. Phys. B482 (1996) 403-430, e-print archive: hep-th/9606004; L.Chekhov, e-print archive: hep-th/0401089

[37] P.Wiegmann and A.Zabrodin, Large $N$ expansion for normal and complex matrix ensembles, e-print archive: hep-th/0309253

[38] A.Polyakov, Phys. Lett. B103 (1981) 207-210

[39] O.Alvarez, Nucl. Phys. B216 (1983) 125-184; P.Di Vecchia, B.Durhuus, P.Olesen and J.Petersen, Nucl.Phys. B207 (1982) 77; J.Ambjorn, B.Durhuus, J.Frölich and P.Orland, Nucl.Phys. B270 (1986) 457

[40] I.Kostov, Conformal Field Theory Techniques in Random Matrix models, e-print archive: hep-th/9907060

[41] B.Eynard, A.Kokotov, D.Korotkin, Nucl. Phys. B694 (2004) 443-472, e-print archive: hep-th/0403072; B.Eynard, JHEP 0411 (2004) 031, e-print archive: hep-th/0407261

[42] P.J.Forrester, Physics Reports, 301 (1998) 235-270; Log gases and random matrices webpage http://www.ms.unimelb.edu.au/ matjpf/matjpf.html

[43] A.Alastuey and B.Jancovici, J. Stat. Phys. 34 (1984) 557; B.Jancovici J. Stat. Phys. 80 (1995) 445

[44] B.Jancovici, J. Stat. Phys. 28 (1982) 43

[45] E.Hille, Analytic function theory, vol. 2, Needham Heights, MA: Ginn (1962)

[46] K.A.Gillow and S.D.Howison, A bibliography of free and moving boundary problems for Hele-Shaw and Stokes flow, http://www.maths.ox.ac.uk/ howison/Hele-Shaw/

[47] M.Mineev-Weinstein, P.B.Wiegmann and A.Zabrodin, Phys. Rev. Lett. 84 (2000) 5106-5109, e-print archive: nlin.SI/0001007

[48] A.Zabrodin, Ann. Henri Poincaré 4 Suppl. 2 (2003) S851-S861, e-print archive: cond-mat/0210331

[49] P.Wiegmann, in: Statistical Field Theories, pp. 337-349, A. Cappelli and G. Mussardo (eds.), Kluwer Academic Publishers, 2002, e-print archive: cond-mat/0204254

# MATRIX MODELS AND TOPOLOGICAL STRINGS

Marcos Mariño
*Department of Physics, CERN*
*Geneva 23, CH-1211 Switzerland**

marcos@mail.cern.ch

**Abstract** In these lecture notes for the Les Houches School on Applications of Random Matrices in Physics we give an introduction to the connections between matrix models and topological strings. We first review some basic results of matrix model technology and then we focus on type B topological strings. We present the main results of Dijkgraaf and Vafa describing the spacetime string dynamics on certain Calabi-Yau backgrounds in terms of matrix models, and we emphasize the connection to geometric transitions and to large $N$ gauge/string duality. We also use matrix model technology to analyze large $N$ Chern-Simons theory and the Gopakumar-Vafa transition.

## 1. Introduction

Topological string theory was introduced by Witten in [70, 72] as a simplified model of string theory which captures topological information of the target space, and it has been intensively studied since then. There are three important lessons that have been learned in the last few years about topological strings:

1) Topological string amplitudes are deeply related to physical amplitudes of type II string theory.

2) The spacetime description of open topological strings in terms of string field theory reduces in some cases to very simple gauge theories.

3) There is an open/closed topological string duality which relates open and closed string backgrounds in a precise way

In these lectures we will analyze a particular class of topological string theories where the gauge theory description in (2) above reduces in fact to a matrix model. This was found by Dijkgraaf and Vafa in a series of beautiful papers [27–29], where they also showed that, thanks to the connection to physical

*Also at Departamento de Matemática, IST, Lisboa, Portugal.



*E. Brezin et al. (eds.), Applications of Random Matrices in Physics,* 319–378.

strings mentioned in (1), the computation of nonperturbative superpotentials in a wide class of $\mathcal{N} = 1$ gauge theories reduces to perturbative computations in a matrix model. This aspect of the work of Dijkgraaf and Vafa was very much explored and exploited, and rederived in the context of supersymmetric gauge theories without using the connection to topological strings. In these lectures we will focus on the contrary on (2) and (3), emphasizing the string field theory construction and the open/closed string duality. The applications of the results of Dijkgraaf and Vafa to supersymmetric gauge theories have been developed in many papers and reviewed for example in [6], and we will not cover them here. Before presenting the relation between matrix models and topological strings, it is worthwhile to give a detailed conceptual discussion of the general ideas behind (2) and (3) and their connections to large $N$ dualities.

In closed string theory we study maps from a Riemann surface $\Sigma_g$ to a target manifold $X$, and the quantities we want to compute are the free energies at genus $g$, denoted by $F_g(t_i)$. Here, the $t_i$ are geometric data of the target space $X$, and the free energies are computed as correlation functions of a two-dimensional conformal field theory coupled to gravity. In topological string theory there are two different models, the A and the B model, the target space is a Calabi-Yau manifold (although this condition can be relaxed in the A model), and the parameters $t_i$ are Kahler and complex parameters, respectively. The free energies are assembled together into a generating functional

$$F(g_s, t_i) = \sum_{g=0}^{\infty} g_s^{2g-2} F_g(t_i), \tag{1.1}$$

where $g_s$ is the string coupling constant.

In open string theory we study maps from an open Riemann surface $\Sigma_{g,h}$ to a target $X$, and we have to provide boundary conditions as well. For example, we can impose Dirichlet conditions by using a submanifold $S$ of $X$ where the open strings have to end. In addition, we can use Chan-Paton factors to introduce a $U(N)$ gauge symmetry. The open string amplitudes are now $F_{g,h}$, and in the cases that will be studied in these lectures the generating functional will have the form

$$F(g_s, N) = \sum_{g=0}^{\infty} \sum_{h=1}^{\infty} F_{g,h} g_s^{2g-2} N^h. \tag{1.2}$$

Physically, the introduction of Chan-Paton factors and boundary conditions through a submanifold $S$ of $X$ means that we are wrapping $N$ (topological) D-branes around $S$. A slightly more general situation arises when there are $n$ submanifolds $S_1, \cdots, S_n$ where the strings can end. In this case, the open string amplitude is of the form $F_{g,h_1,\cdots,h_n}$ and the total free energy is now given by

$$F(g_s, N_i) = \sum_{g=0}^{\infty} \sum_{h_1,\cdots,h_n=1}^{\infty} F_{g,h_1,\cdots,h_n} g_s^{2g-2} N_1^{h_1} \cdots N_n^{h_n}. \tag{1.3}$$

In the case of open strings one can in some situations use string field theory to describe the spacetime dynamics. The open string field theory of Witten [69], which was originally constructed for the open bosonic string theory, can also be applied to topological string theory, and on some particular Calabi-Yau backgrounds the full string field theory of the topological string reduces to a simple $U(N)$ gauge theory, where $g_s$ plays the role of the gauge coupling constant and $N$ is the rank of the gauge group. In particular, the string field reduces in this case to a finite number of gauge fields. As a consequence of this, the open string theory amplitude $F_{g,h}$ can be computed from the gauge theory by doing perturbation theory in the double line notation of 't Hooft [66]. More precisely, $F_{g,h}$ is the contribution of the fatgraphs of genus $g$ and $h$ holes. The idea that fatgraphs of a $U(N)$ gauge theory correspond to open string amplitudes is an old one, and it is very satisfying to find a concrete realization of this idea in the context of a string field theory description of topological strings, albeit for rather simple gauge theories.

The surprising fact that the full string field theory is described by a simple gauge theory is typical of topological string theory, and does not hold for conventional string models. There are two examples where this description has been worked out:

1) The A model on a Calabi-Yau of the form $X = T^*M$, where $M$ is a three-manifold, and there are $N$ topological D-branes wrapping $M$. In this case, the gauge theory is Chern-Simons theory on $M$ [74].

2) The B model on a Calabi-Yau manifold $X$ which is the small resolution of a singularity characterized by the hyperelliptic curve $y^2 = (W'(x))^2$. If $W'(x)$ has degree $n$, the small resolution produces $n$ two-spheres, and one can wrap $N_i$ topological D-branes around each two-sphere, with $i = 1, \cdots, n$. In this case Dijkgraaf and Vafa showed that the gauge theory is a multicut matrix model with potential $W(x)$ [27].

In both examples, the open string amplitudes $F_{g,h}$ are just numbers computed by the fatgraphs of the corresponding gauge theories.

The fatgraph expansion of a $U(N)$ gauge theory can be resummed formally by introducing the so called 't Hooft parameter $t = g_s N$. For example, in the case of the free energy, we can rewrite (1.2) in the form (1.1) by defining

$$F_g(t) = \sum_{h=1}^{\infty} F_{g,h} t^h. \tag{1.4}$$

In other words, starting from an *open* string theory expansion we can obtain a *closed* string theory expansion by resumming the hole expansion as indicated in (1.4). This idea was proposed by 't Hooft [66] and gives a closed string theory interpretation of a gauge theory.

What is the interpretation of the above resummation for the gauge theories that describe the spacetime dynamics of topological open string theories? As was explained in [35] (for the A model example above) and in [15] (for the B model example), there is a *geometric* or *large* $N$ transition that relates the open string Calabi-Yau background $X$ underlying the gauge theory to a closed string Calabi-Yau background $X'$. The geometric transition typically relates two different ways of smoothing out a singular geometry (the "resolved" geometry and the "deformed" geometry). Moreover, the "master field" that describes the large $N$ limit [68] turns out to encode the target space geometry of the closed string background, and the 't Hooft parameter becomes a geometric parameter of the resulting closed geometry. The idea that an open string background

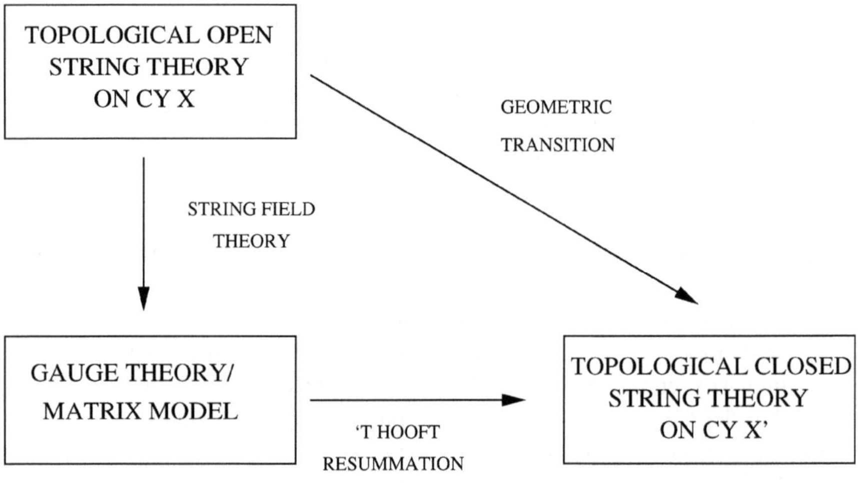


*Figure 1.* This diagram summarizes the different relations between closed topological strings, open topological strings, and gauge theories.

with D-branes is equivalent to a different, geometric closed string background (therefore with no D-branes) appeared originally in the AdS/CFT correspondence [3]. In this correspondence, type IIB theory in flat space in the presence of D-branes is conjectured to be equivalent to type IIB theory in $\mathrm{AdS}_5 \times \mathbf{S}^5$ with no D-branes, and where the radius of the $\mathbf{S}^5$ is related to the 't Hooft parameter. The reason this holds is that, at large $N$, the presence of the D-branes can be traded for a deformation of the background geometry. In other words, we can make the branes disappear if we change the background geometry at the same time. Therefore, as emphasized by Gopakumar and Vafa in [35], large $N$ dualities relating open and closed strings should be associated to transitions in the geometry. The logical structure of all the connections we have sketched is depicted in Fig. 1.

In these lectures we will mostly focus on the B topological string, the Dijkgraaf Vafa scenario, and the geometric transition of [15]. For a detailed review of a similar story for the A string, we refer the reader to [54]. The organization of these lectures is as follows. In section 2 we review some basic ingredients of matrix models, including saddle-point techniques and orthogonal polynomials. In section 3 we explain in detail the connection between matrix models and topological strings due to Dijkgraaf and Vafa. We first review the topological B model and its string field theory description, and we show that in the Calabi-Yau background associated to the resolution of a polynomial singularity, the string field theory reduces to a matrix model. We develop some further matrix model technology to understand all the details of this description, and we make the connection with geometric transitions. In section 4 we briefly consider the geometric transition of Gopakumar and Vafa [35] from the point of view of the matrix model description of Chern-Simons theory. This allows us to use matrix model technology to derive some of the results of [35].

# 2. Matrix models

In this section we develop some aspects and techniques of matrix models which will be needed in the following. There are excellent reviews of this material, such as for example [21, 22].

## 2.1 Basics of matrix models

Matrix models are the simplest examples of quantum gauge theories, namely, they are quantum gauge theories in zero dimensions. The basic field is a Hermitian $N \times N$ matrix $M$. We will consider an action for $M$ of the form:

$$\frac{1}{g_s}W(M) = \frac{1}{2g_s}\mathrm{Tr}\, M^2 + \frac{1}{g_s}\sum_{p\geq 3}\frac{g_p}{p}\mathrm{Tr}\, M^p. \tag{2.1}$$

where $g_s$ and $g_p$ are coupling constants. This action has the obvious gauge symmetry

$$M \to UMU^\dagger, \tag{2.2}$$

where $U$ is a $U(N)$ matrix. The partition function of the theory is given by

$$Z = \frac{1}{\mathrm{vol}(U(N))}\int dM\, e^{-\frac{1}{g_s}W(M)} \tag{2.3}$$

where the factor $\mathrm{vol}(U(N))$ is the usual volume factor of the gauge group that arises after fixing the gauge. In other words, we are considering here a *gauged* matrix model. The measure in the "path integral" is the Haar measure

$$dM = 2^{\frac{N(N-1)}{2}} \prod_{i=1}^{N} dM_{ii} \prod_{1 \leq i < j \leq N} d\mathrm{Re}\, M_{ij} d\mathrm{Im}\, M_{ij}. \tag{2.4}$$

The numerical factor in (2.4) is introduced to obtain a convenient normalization.

A particularly simple example is the *Gaussian matrix model*, defined by the partition function

$$Z_G = \frac{1}{\mathrm{vol}(U(N))} \int dM\, e^{-\frac{1}{2g_s} \mathrm{Tr}\, M^2}. \tag{2.5}$$

We will denote by

$$\langle f(M) \rangle_G = \frac{\int dM\, f(M)\, e^{-\mathrm{Tr}\, M^2/2g_s}}{\int dM\, e^{-\mathrm{Tr}\, M^2/2g_s}} \tag{2.6}$$

the normalized vevs of a gauge-invariant functional $f(M)$ in the Gaussian matrix model. This model is of course exactly solvable, and the vevs (2.6) can be computed systematically as follows. Any gauge-invariant function $f(M)$ can be written as a linear combination of traces of $M$ in arbitrary representations $R$ of $U(N)$. If we represent $R$ by a Young tableau with rows of lengths $\lambda_i$, with $\lambda_1 \geq \lambda_2 \geq \cdots$, and with $\ell(R)$ boxes in total, we define the set of $\ell(R)$ integers $f_i$ as follows

$$f_i = \lambda_i + \ell(R) - i, \quad i = 1, \cdots, \ell(R). \tag{2.7}$$

Following [23], we will say that the Young tableau associated to $R$ is even if the number of odd $f_i$'s is the same as the number of even $f_i$'s. Otherwise, we will say that it is odd. If $R$ is even, one has the following result [42, 23]:

$$\langle \mathrm{Tr}_R M \rangle_G = c(R) \dim R, \tag{2.8}$$

where

$$c(R) = (-1)^{\frac{A(A-1)}{2}} \frac{\prod_{f \text{ odd}} f!! \prod_{f' \text{ even}} f'!!}{\prod_{f \text{ odd}, f' \text{ even}} (f - f')} \tag{2.9}$$

and $A = \ell(R)/2$ (notice that $\ell(R)$ has to be even in order to have a non-vanishing result). Here $\dim R$ is the dimension of the irreducible representation of $SU(N)$ associated to $R$, and can be computed for example by using the hook formula. On the other hand, if $R$ is odd, the above vev vanishes.

The partition function $Z$ of more general matrix models with action (2.1) can be evaluated by doing perturbation theory around the Gaussian point: one expands the exponential of $\sum_{p \geq 3} (g_p/g_s) \mathrm{Tr} M^p/p$ in (2.3), and computes the partition function as a power series in the coupling constants $g_p$. The evaluation of each term of the series involves the computation of vevs like (2.6). Of course, this computation can be interpreted in terms of Feynman diagrams, and as usual the perturbative expansion of the free energy

$$F = \log Z$$

will only involve connected vacuum bubbles.

Since we are dealing with a quantum theory of a field in the adjoint representation we can reexpress the perturbative expansion of $F$ in terms of fatgraphs, by using the double line notation due to 't Hooft [66]. The purpose of the fatgraph expansion is the following: in $U(N)$ gauge theories there is, in addition to the coupling constants appearing in the model (like for example $g_s, g_p$ in (2.1)), a hidden variable, namely $N$, the rank of the gauge group. The $N$ dependence in the perturbative expansion comes from the group factors associated to Feynman diagrams, but in general a single Feynman diagram gives rise to a polynomial in $N$ involving different powers of $N$. Therefore, the standard Feynman diagrams, which are good in order to keep track of powers of the coupling constants, are not good in order to keep track of powers of $N$. If we want to keep track of the $N$ dependence we have to "split" each diagram into different pieces which correspond to a definite power of $N$. To do that, one writes the Feynman diagrams of the theory as "fatgraphs" or double line graphs, as first indicated by 't Hooft [66]. Let us explain this in some detail,

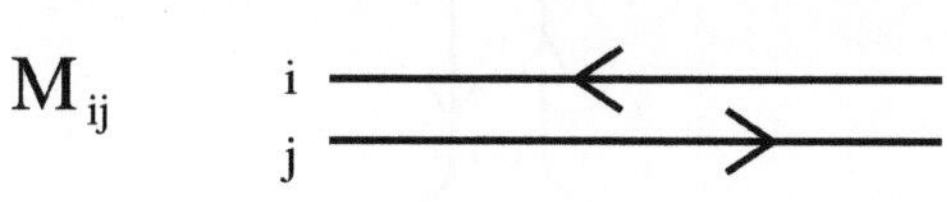


*Figure 2.* The index structure of the field $M_{ij}$ in the adjoint representation of $U(N)$ is represented through a double line.

taking the example of the matrix model with a cubic potential (*i.e.* $g_p = 0$ in (2.1) for $p > 3$). The fundamental field $M_{ij}$ is in the adjoint representation. Since the adjoint representation of $U(N)$ is the tensor product of the fundamental $N$ and the antifundamental $\overline{N}$, we can look at $i$ (resp. $j$) as an index of the fundamental (resp. antifundamental) representation. We will represent this double-index structure by a double line notation as shown in Fig. 2. The only thing we have to do now is to rewrite the Feynman rules of the theory by taking into account this double-line notation. For example, the kinetic term of the theory is of the form

$$\frac{1}{g_s}\mathrm{Tr}\, M^2 = \frac{1}{g_s}\sum_{i,j} M_{ij}M_{ji}. \tag{2.10}$$

This means that the propagator of the theory is

$$\langle M_{ij} M_{kl} \rangle = g_s \delta_{il} \delta_{jk} \tag{2.11}$$

and can be represented in the double line notation as in Fig. 3. Next, we con-

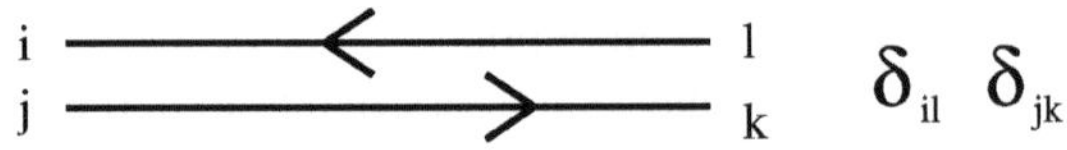


*Figure 3.* The propagator in the double line notation.

sider the vertices of the theory. For example, the trivalent vertex given by

$$\frac{g_3}{g_s} \mathrm{Tr}\, M^3 = \frac{g_3}{g_s} \sum_{i,j,k} M_{ij}\, M_{jk}\, M_{ki} \tag{2.12}$$

can be represented in the double line notation as in Fig. 4. A vertex of order $p$ can be represented in a similar way by drawing $p$ double lines joined together. Once we have rewritten the Feynman rules in the double-line notation, we can

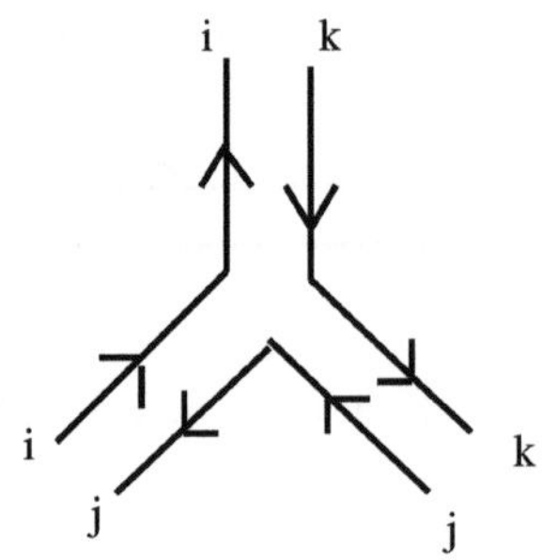


*Figure 4.* The cubic vertex in the double line notation.

construct the corresponding graphs, which look like ribbons and are called ribbon graphs or fatgraphs. It is clear that in general a usual Feynman diagram can give rise to many different fatgraphs. Consider for example the one-loop vacuum diagram 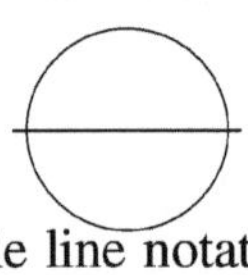, which comes from contracting two cubic vertices. In the double line notation the contraction can be done in two different

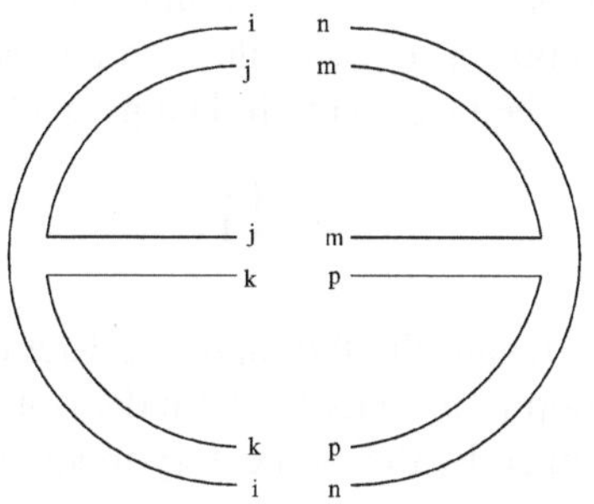


*Figure 5.* Contracting two cubic vertices in the double line notation: the $N^3$ contribution.

ways. The first one is illustrated in Fig. 5 and gives a factor

$$\sum_{ijkmnp} \langle M_{ij} M_{mn} \rangle \langle M_{jk} M_{pm} \rangle \langle M_{ki} M_{np} \rangle = g_s^3 N^3. \tag{2.13}$$

The second one is shown in Fig. 6 and gives a factor

$$\sum_{ijkmnp} \langle M_{ij} M_{mn} \rangle \langle M_{jk} M_{np} \rangle \langle M_{ki} M_{pm} \rangle = g_s^3 N. \tag{2.14}$$

In this way we have split the original diagram into two different fatgraphs with a well-defined power of $N$ associated to them. The number of factors of $N$ is simply equal to the number of closed loops in the graph: there are three closed lines in the fatgraph resulting from the contractions in Fig. 5 (see the first graph in Fig. 7), while there is only one in the diagram resulting from Fig. 6. In general, fatgraphs turn out to be characterized topologically by the

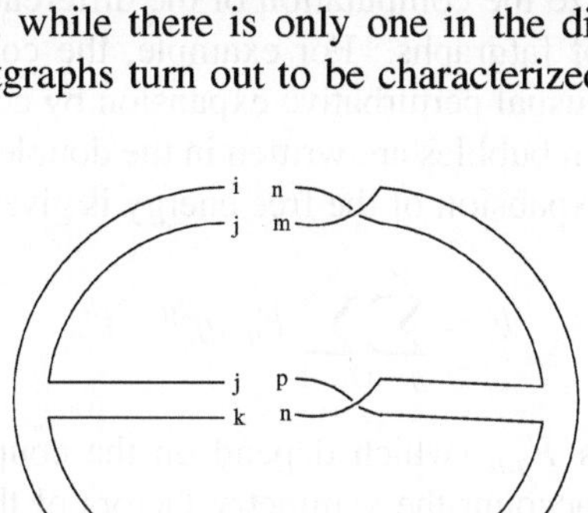


*Figure 6.* Contracting two cubic vertices in the double line notation: the $N$ contribution.

number of propagators or edges $E$, the number of vertices with $p$ legs $V_p$, and the number of closed loops $h$. The total number of vertices is $V = \sum_p V_p$. Each propagator gives a power of $g_s$, while each interaction vertex with $p$ legs gives a power of $g_p/g_s$. The fatgraph will then give a factor

$$g_s^{E-V} N^h \prod_p g_p^{V_p}. \tag{2.15}$$

The key point now is to regard the fatgraph as a Riemann surface with holes, in which each closed loop represents the boundary of a hole. The genus $g$ of such a surface is determined by the elementary topological relation

$$2g - 2 = E - V - h \tag{2.16}$$

therefore we can write (2.15) as

$$g_s^{2g-2+h} N^h \prod_p g_p^{V_p} = g_s^{2g-2} t^h \prod_p g_p^{V_p} \tag{2.17}$$

where we have introduced the 't Hooft parameter

$$t = N g_s \tag{2.18}$$

The fatgraphs with $g = 0$ are called *planar*, while the ones with $g > 0$ are called *nonplanar*. The graph giving the $N^3$ contribution in Fig. 5 is planar: it has $E = 3$, $V_3 = 2$ and $h = 3$, therefore $g = 0$, and it is a sphere with three holes. The graph in Fig. 6 is nonplanar: it has $E = 3$, $V_3 = 2$ and $h = 1$, therefore $g = 1$, and represents a torus with one hole (it is easy to see this by drawing the diagram on the surface of a torus).

We can now organize the computation of the different quantities in the matrix model in terms of fatgraphs. For example, the computation of the free energy is given in the usual perturbative expansion by connected vacuum bubbles. When the vacuum bubbles are written in the double line notation, we find that the perturbative expansion of the free energy is given by

$$F = \sum_{g=0}^{\infty} \sum_{h=1}^{\infty} F_{g,h} g_s^{2g-2} t^h, \tag{2.19}$$

where the coefficients $F_{g,h}$ (which depend on the coupling constants of the model $g_p$) takes into account the symmetry factors of the different fatgraphs. We can now formally define the free energy at genus $g$, $F_g(t)$, by keeping $g$ fixed and summing over all closed loops $h$ as in (1.4), so that the total free energy can be written as

$$F = \sum_{g=0}^{\infty} F_g(t) g_s^{2g-2}. \tag{2.20}$$

This is the *genus expansion* of the free energy of the matrix model. In (2.20) we have written the diagrammatic series as an expansion in $g_s$ around $g_s = 0$, keeping the 't Hooft parameter $t = g_s N$ fixed. Equivalently, we can regard it as an expansion in $1/N$, keeping $t$ fixed, and then the $N$ dependence appears as $N^{2-2g}$. Therefore, for $t$ fixed and $N$ large, the leading contribution comes from planar diagrams with $g = 0$, which go like $\mathcal{O}(N^2)$. The nonplanar diagrams give subleading corrections. Notice that $F_g(t)$, which is the contribution to $F$ to a given order in $g_s$, is given by an infinite series where we sum over all possible numbers of holes $h$, weighted by $t^h$.

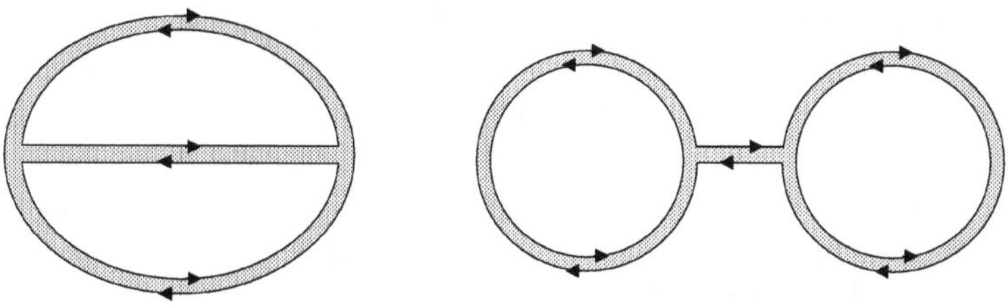

*Figure 7.* Two planar diagrams in the cubic matrix model.

**Example**. One can show that

$$\langle (\mathrm{Tr}\, M^3)^2 \rangle_G = g_s^3 (12N^3 + 3N),$$

where the first term corresponds to the two planar diagrams shown in Fig. 7 (contributing $3N^3$ and $9N^3$, respectively), and the second term corresponds to the nonplanar diagram shown in Fig. 6. Therefore, in the cubic matrix model the expansion of the free energy reads, at leading order,

$$F - F_G = \frac{2}{3} g_s g_3^2 N^3 + \frac{1}{6} g_s g_3^2 N + \cdots \tag{2.21}$$

There is an alternative way of writing the matrix model partition function which is very useful. The original matrix model variable has $N^2$ real parameters, but using the gauge symmetry we can see that, after modding out by gauge transformations, there are only $N$ parameters left. We can for example take advantage of our gauge freedom to diagonalize the matrix $M$

$$M \to U M U^\dagger = D, \tag{2.22}$$

with $D = \mathrm{diag}(\lambda_1, \cdots, \lambda_N)$, impose this as a gauge choice, and use standard Faddeev-Popov techniques in order to compute the gauge-fixed integral (see for example [9]). The gauge fixing (2.22) leads to the delta-function constraint

$$\delta({}^U M) = \prod_{i<j} \delta^{(2)}({}^U M_{ij}) \tag{2.23}$$

where ${}^U M = UMU^\dagger$. We then introduce

$$\Delta^{-2}(M) = \int dU\, \delta({}^U M). \tag{2.24}$$

It then follows that the integral of any gauge-invariant function $f(M)$ can be written as

$$\int dM\, f(M) = \int dM\, f(M) \Delta^2(M) \int dU\, \delta({}^U M) = \Omega_N \int \prod_{i=1}^N d\lambda_i \Delta^2(\lambda) f(\lambda), \tag{2.25}$$

where we have used the gauge invariance of $\Delta(M)$, and

$$\Omega_N = \int dU \tag{2.26}$$

is proportional to the volume of the gauge group $U(N)$, as we will see shortly. We have to evaluate the the factor $\Delta(\lambda)$, which can be obtained from (2.24) by choosing $M$ to be diagonal. If

$$F(M) = 0$$

is the gauge-fixing condition, the standard Faddeev-Popov formula gives

$$\Delta^2(M) = \det \left( \frac{\delta F({}^U M)}{\delta A} \right)_{F=0} \tag{2.27}$$

where we write $U = e^A$, and $A$ is a anti-Hermitian matrix. Since

$$F_{ij}({}^U D) = (UDU^\dagger)_{ij} = A_{ij}(\lambda_i - \lambda_j) + \cdots . \tag{2.28}$$

(2.27) leads immediately to

$$\Delta^2(\lambda) = \prod_{i<j} (\lambda_i - \lambda_j)^2, \tag{2.29}$$

the square of the Vandermonde determinant. Finally, we fix the factor $\Omega_N$ as follows. The Gaussian matrix integral can be computed explicitly by using the Haar measure (2.4), and is simply

$$\int dM\, e^{-\frac{1}{2g_s} \mathrm{Tr}\, M^2} = (2\pi g_s)^{N^2/2}. \tag{2.30}$$

On the other hand, by (2.25) this should equal

$$\Omega_N \int \prod_{i=1}^N d\lambda_i \Delta^2(\lambda) e^{-\frac{1}{2g_s} \sum_{i=1}^N \lambda_i^2}. \tag{2.31}$$

The integral over eigenvalues can be evaluated in various ways, using for example the Selberg function [55] or the technique of orthogonal polynomials that we describe in the next subsection, and its value is

$$g_s^{N^2/2}(2\pi)^{N/2}G_2(N+2) \tag{2.32}$$

where $G_2(z)$ is the Barnes function, defined by

$$G_2(z+1) = \Gamma(z)G_2(z), \quad G_2(1) = 1. \tag{2.33}$$

Comparing these results, we find that

$$\Omega_N = \frac{(2\pi)^{\frac{N(N-1)}{2}}}{G_2(N+2)}. \tag{2.34}$$

Using now (see for example [60]):

$$\text{vol}(U(N)) = \frac{(2\pi)^{\frac{1}{2}N(N+1)}}{G_2(N+1)}. \tag{2.35}$$

we see that

$$\frac{1}{\text{vol}(U(N))}\int dM\, f(M) = \frac{1}{N!}\frac{1}{(2\pi)^N}\int \prod_{i=1}^N d\lambda_i\, \Delta^2(\lambda) f(\lambda). \tag{2.36}$$

The factor $N!$ in the r.h.s. of (2.36) has an obvious interpretation: after fixing the gauge symmetry of the matrix integral by fixing the diagonal gauge, there is still a residual symmetry given by the Weyl symmetry of $U(N)$, which is the symmetric group $S_N$ acting as permutation of the eigenvalues. The "volume" of this discrete gauge group is just its order, $|S_N| = N!$, and since we are considering gauged matrix models we have to divide by it as shown in (2.36). As a particular case of the above formula, it follows that one can write the partition function (2.3) as

$$Z = \frac{1}{N!}\frac{1}{(2\pi)^N}\int \prod_{i=1}^N d\lambda_i\, \Delta^2(\lambda) e^{-\frac{1}{2g_s}\sum_{i=1}^N W(\lambda_i)}. \tag{2.37}$$

The partition function of the gauged Gaussian matrix model (2.5) is given essentially by the inverse of the volume factor. Its free energy to all orders can be computed by using the asymptotic expansion of the Barnes function

$$\begin{aligned}\log G_2(N+1) &= \frac{N^2}{2}\log N - \frac{1}{12}\log N - \frac{3}{4}N^2 + \frac{1}{2}N\log 2\pi + \zeta'(-1)\\ &+ \sum_{g=2}^{\infty}\frac{B_{2g}}{2g(2g-2)}N^{2-2g},\end{aligned} \tag{2.38}$$

where $B_{2g}$ are the Bernoulli numbers. Therefore, we find the following expression for the total free energy:

$$
\begin{aligned}
F_G &= \frac{N^2}{2}\Big(\log(Ng_s)-\frac{3}{2}\Big)-\frac{1}{12}\log N+\zeta'(-1) \\
&+ \sum_{g=2}^{\infty}\frac{B_{2g}}{2g(2g-2)}N^{2-2g}.
\end{aligned}
\tag{2.39}
$$

If we now put $N = t/g_s$, we obtain the following expressions for $F_g(t)$:

$$
\begin{aligned}
F_0(t) &= \frac{1}{2}t^2\Big(\log t-\frac{3}{2}\Big), \\
F_1(t) &= -\frac{1}{12}\log t, \\
F_g(t) &= \frac{B_{2g}}{2g(2g-2)}t^{2-2g}, \quad g>1.
\end{aligned}
$$

## 2.2 Matrix model technology I: saddle-point analysis

The computation of the functions $F_g(t)$ in closed form seems a difficult task, since in perturbation theory they involve summing up an infinite number of fatgraphs (with different numbers of holes $h$). However, in the classic paper [12] it was shown that, remarkably, $F_0(t)$ can be obtained by solving a Riemann-Hilbert problem. In this section we will review this procedure.

Let us consider a general matrix model with action $W(M)$, and let us write the partition function after reduction to eigenvalues (2.37) as follows:

$$
Z = \frac{1}{N!}\int\prod_{i=1}^{N}\frac{d\lambda_i}{2\pi}e^{N^2 S_{\rm eff}(\lambda)}
\tag{2.40}
$$

where the effective action is given by

$$
S_{\rm eff}(\lambda) = -\frac{1}{tN}\sum_{i=1}^{N}W(\lambda_i)+\frac{2}{N^2}\sum_{i<j}\log|\lambda_i-\lambda_j|.
\tag{2.41}
$$

Notice that, since a sum over $N$ eigenvalues is roughly of order $N$, the effective action is of order $\mathcal{O}(1)$. We can now regard $N^2$ as a sort of $\hbar^{-1}$ in such a way that, as $N\to\infty$, the integral (2.40) will be dominated by a saddle-point configuration that extremizes the effective action. Varying $S_{\rm eff}(\lambda)$ w.r.t. the eigenvalue $\lambda_i$, we obtain the equation

$$
\frac{1}{2t}W'(\lambda_i) = \frac{1}{N}\sum_{j\neq i}\frac{1}{\lambda_i-\lambda_j}, \quad i=1,\cdots,N.
\tag{2.42}
$$

The eigenvalue distribution is formally defined for finite $N$ as

$$\rho(\lambda) = \frac{1}{N}\sum_{i=1}^{N}\delta(\lambda - \lambda_i), \tag{2.43}$$

where the $\lambda_i$ solve (2.42). In the large $N$ limit, it is reasonable to expect that this distribution becomes a continuous function with compact support. We will assume that $\rho(\lambda)$ vanishes outside an interval $\mathcal{C}$. This is the so-called *one-cut solution.*

Qualitatively, what is going on is the following. Assume for simplicity that $W(x)$, the potential, has only one minimum $x_*$. We can regard the eigenvalues as coordinates of a system of $N$ classical particles moving on the real line. The equation (2.42) says that these particles are subject to an effective potential

$$W_{\text{eff}}(\lambda_i) = W(\lambda_i) - \frac{2t}{N}\sum_{j\neq i}\log|\lambda_i - \lambda_j| \tag{2.44}$$

which involves a logarithmic Coulomb repulsion between eigenvalues. For small 't Hooft parameter, the potential term dominates over the Coulomb repulsion, and the particles tend to be in the minimum $x_*$ of the potential $W'(x_*) = 0$. This means that, for $t = 0$, the interval $\mathcal{C}$ collapses to the point $x_*$. As $t$ grows, the Coulomb repulsion will force the eigenvalues to be apart from each other and to spread out over an interval $\mathcal{C}$.

We can now write the saddle-point equation in terms of continuum quantities, by using the standard rule

$$\frac{1}{N}\sum_{i=1}^{N} f(\lambda_i) \to \int_{\mathcal{C}} f(\lambda)\rho(\lambda)d\lambda. \tag{2.45}$$

Notice that the distribution of eigenvalues $\rho(\lambda)$ satisfies the normalization condition

$$\int_{\mathcal{C}} \rho(\lambda)d\lambda = 1. \tag{2.46}$$

The equation (2.42) then becomes

$$\frac{1}{2t}W'(\lambda) = \mathrm{P}\int \frac{\rho(\lambda')d\lambda'}{\lambda - \lambda'} \tag{2.47}$$

where P denotes the principal value of the integral. The above equation is an integral equation that allows one in principle to compute $\rho(\lambda)$, given the potential $W(\lambda)$, as a function of the 't Hooft parameter $t$ and the coupling constants. Once $\rho(\lambda)$ is known, one can easily compute $F_0(t)$: in the saddle-point approximation, the free energy is given by

$$\frac{1}{N^2}F = S_{\text{eff}}(\rho) + \mathcal{O}(N^{-2}), \tag{2.48}$$

where the effective action in the continuum limit is a functional of $\rho$:

$$S_{\text{eff}}(\rho) = -\frac{1}{t}\int_{\mathcal{C}} d\lambda \rho(\lambda) W(\lambda) + \int_{\mathcal{C}\times\mathcal{C}} d\lambda d\lambda' \rho(\lambda)\rho(\lambda') \log|\lambda - \lambda'|. \quad (2.49)$$

Therefore, the planar free energy is given by

$$F_0(t) = t^2 S_{\text{eff}}(\rho), \quad (2.50)$$

Since the effective action is evaluated on the distribution of eigenvalues which solves (2.47), one can simplify the expression to

$$F_0(t) = -\frac{t}{2}\int_{\mathcal{C}} d\lambda \rho(\lambda) W(\lambda). \quad (2.51)$$

Similarly, averages in the matrix model can be computed in the planar limit as

$$\frac{1}{N}\langle \text{Tr}\, M^{\ell}\rangle = \int_{\mathcal{C}} d\lambda\, \lambda^{\ell} \rho(\lambda). \quad (2.52)$$

We then see that the planar limit is characterized by a *classical* density of states $\rho(\lambda)$, and the planar piece of quantum averages can be computed as a moment of this density. The fact that the planar approximation to a quantum field theory can be regarded as a classical field configuration was pointed out in [68] (see [20] for a beautiful exposition). This classical configuration is often called the *master field*. In the case of matrix models, the master field configuration is given by the density of eigenvalues $\rho(\lambda)$, and as we will see later it can be encoded in a complex algebraic curve with a deep geometric meaning.

The density of eigenvalues is obtained as a solution to the saddle-point equation (2.47). This equation is a singular integral equation which has been studied in detail in other contexts of physics (see, for example, [57]). The way to solve it is to introduce an auxiliary function called the *resolvent*. The resolvent is defined as a correlator in the matrix model:

$$\omega(p) = \frac{1}{N}\langle \text{Tr} \frac{1}{p-M}\rangle, \quad (2.53)$$

which is in fact a generating functional of the correlation functions (2.52):

$$\omega(p) = \frac{1}{N}\sum_{k=0}^{\infty}\langle \text{Tr} M^k\rangle p^{-k-1} \quad (2.54)$$

Being a generating functional of connected correlators, it admits an expansion of the form [20]:

$$\omega(p) = \sum_{g=0}^{\infty} g_s^{2g}\omega_g(p), \quad (2.55)$$

and the genus zero piece can be written in terms of the eigenvalue density as

$$\omega_0(p) = \int d\lambda \frac{\rho(\lambda)}{p-\lambda} \tag{2.56}$$

The genus zero resolvent (2.56) has three important properties. First of all, as a function of $p$ it is an analytic function on the whole complex plane except on the interval $\mathcal{C}$, since if $\lambda \in \mathcal{C}$ one has a singularity at $\lambda = p$. Second, due to the normalization property of the eigenvalue distribution (2.46), it has the asymptotic behavior

$$\omega_0(p) \sim \frac{1}{p}, \qquad p \to \infty. \tag{2.57}$$

Finally, one can compute the discontinuity of $\omega_0(p)$ as one crosses the interval $\mathcal{C}$. This is just the residue at $\lambda = p$, and one then finds the key equation

$$\rho(\lambda) = -\frac{1}{2\pi i}\big(\omega_0(\lambda + i\epsilon) - \omega_0(\lambda - i\epsilon)\big). \tag{2.58}$$

Therefore, if the resolvent at genus zero is known, the eigenvalue distribution follows from (2.58), and one can compute the planar free energy. On the other hand, by looking again at the resolvent as we approach the discontinuity, we see that the r.h.s. of (2.47) is given by $-(\omega_0(p+i\epsilon) + \omega_0(p-i\epsilon))/2$, and we then find the equation

$$\omega_0(p+i\epsilon) + \omega_0(p-i\epsilon) = -\frac{1}{t}W'(p), \tag{2.59}$$

which determines the resolvent in terms of the potential. In this way we have reduced the original problem of computing $F_0(t)$ to the Riemann-Hilbert problem of computing $\omega_0(\lambda)$. There is in fact a closed expression for the resolvent in terms of a contour integral [56] which is very useful. Let $\mathcal{C}$ be given by the interval $b \le \lambda \le a$. Then, one has

$$\omega_0(p) = \frac{1}{2t}\oint_{\mathcal{C}} \frac{dz}{2\pi i}\frac{W'(z)}{p-z}\left(\frac{(p-a)(p-b)}{(z-a)(z-b)}\right)^{\frac{1}{2}}. \tag{2.60}$$

This equation is easily proved by converting (2.59) into a discontinuity equation:

$$\widehat{\omega}_0(p+i\epsilon) - \widehat{\omega}_0(p-i\epsilon) = -\frac{1}{t}\frac{W'(p)}{\sqrt{(p-a)(p-b)}}, \tag{2.61}$$

where $\widehat{\omega}_0(p) = \omega_0(p)/\sqrt{(p-a)(p-b)}$. This equation determines $\omega_0(p)$ to be given by (2.60) up to regular terms, but because of the asymptotics (2.57), these regular terms are absent. The asymptotics of $\omega_0(p)$ also gives two more conditions. By taking $p \to \infty$, one finds that the r.h.s. of (2.60) behaves like

$c + d/p + \mathcal{O}(1/p^2)$. Requiring the asymptotic behavior (2.57) imposes $c = 0$ and $d = 1$, and this leads to

$$\begin{aligned} \oint_{\mathcal{C}} \frac{dz}{2\pi i} \frac{W'(z)}{\sqrt{(z-a)(z-b)}} &= 0, \\ \oint_{\mathcal{C}} \frac{dz}{2\pi i} \frac{zW'(z)}{\sqrt{(z-a)(z-b)}} &= 2t. \end{aligned} \tag{2.62}$$

These equations are enough to determine the endpoints of the cuts, $a$ and $b$, as functions of the 't Hooft coupling $t$ and the coupling constants of the model.

The above expressions are in fact valid for very general potentials (we will apply them to logarithmic potentials in section 4), but when $W(z)$ is a polynomial, one can find a very convenient expression for the resolvent: if we deform the contour in (2.60) we pick up a pole at $z = p$, and another one at infinity, and we get

$$\omega_0(p) = \frac{1}{2t} W'(p) - \frac{1}{2t} \sqrt{(p-a)(p-b)}\, M(p), \tag{2.63}$$

where

$$M(p) = \oint_0 \frac{dz}{2\pi i} \frac{W'(1/z)}{1-pz} \frac{1}{\sqrt{(1-az)(1-bz)}}. \tag{2.64}$$

Here, the contour is around $z = 0$. These formulae, together with the expressions (2.61) for the endpoints of the cut, completely solve the one-matrix model with one cut in the planar limit, for polynomial potentials.

Another way to find the resolvent is to start with (2.42), multiply it by $1/(\lambda_i - p)$, and sum over $i$. One finds, in the limit of large $N$,

$$(\omega_0(p))^2 - \frac{1}{t} W'(p)\omega_0(p) + \frac{1}{4t^2} R(p) = 0, \tag{2.65}$$

where

$$R(p) = 4t \int d\lambda\, \rho(\lambda) \frac{W'(p) - W'(\lambda)}{p - \lambda}. \tag{2.66}$$

Notice that (2.65) is a quadratic equation for $\omega_0(p)$ and has the solution

$$\omega_0(p) = \frac{1}{2t}\Big( W'(p) - \sqrt{(W'(p))^2 - R(p)} \Big), \tag{2.67}$$

which is of course equivalent to (2.63).

A useful way to encode the solution to the matrix model is to define

$$y(p) = W'(p) - 2t\, \omega_0(p). \tag{2.68}$$

Notice that the force on an eigenvalue is given by

$$f(p) = -W'_{\text{eff}}(p) = -\frac{1}{2}(y(p + i\epsilon) + y(p - i\epsilon)). \tag{2.69}$$

In terms of $y(p)$, the quadratic equation (2.65) determining the resolvent can be written as

$$y^2 = W'(p)^2 - R(p). \tag{2.70}$$

This is nothing but the equation of a hyperelliptic curve given by a certain deformation (measured by $R(p)$) of the equation $y^2 = W'(p)^2$ typical of singularity theory. We will see in the next section that this result has a beautiful interpretation in terms of topological string theory on certain Calabi-Yau manifolds.

**Example**. *The Gaussian matrix model.* Let us now apply this technology to the simplest case, the Gaussian model with $W(M) = M^2/2$. Let us first look for the position of the endpoints from (2.61). Deforming the contour to infinity and changing $z \to 1/z$, we find that the first equation in (2.61) becomes

$$\oint_0 \frac{dz}{2\pi i}\frac{1}{z^2}\frac{1}{\sqrt{(1-az)(1-bz)}} = 0, \tag{2.71}$$

where the contour is now around $z = 0$. Therefore $a + b = 0$, in accord with the symmetry of the potential. Taking this into account, the second equation becomes:

$$\oint_0 \frac{dz}{2\pi i}\frac{1}{z^3}\frac{1}{\sqrt{1-a^2z^2}} = 2t, \tag{2.72}$$

and gives

$$a = 2\sqrt{t}. \tag{2.73}$$

We see that the interval $\mathcal{C} = [-a, a] = [-2\sqrt{t}, 2\sqrt{t}]$ opens as the 't Hooft parameter grows up, and as $t \to 0$ it collapses to the minimum of the potential at the origin, as expected. We immediately find from (2.63)

$$\omega_0(p) = \frac{1}{2t}\left(p - \sqrt{p^2 - 4t}\right), \tag{2.74}$$

and from the discontinuity equation we derive the density of eigenvalues

$$\rho(\lambda) = \frac{1}{2\pi t}\sqrt{4t - \lambda^2}. \tag{2.75}$$

The graph of this function is a semicircle of radius $2\sqrt{t}$, and the above eigenvalue distribution is the famous Wigner-Dyson semicircle law. Notice also that the equation (2.70) is in this case

$$y^2 = p^2 - 4t. \tag{2.76}$$

This is the equation for a curve of genus zero, which resolves the singularity $y^2 = p^2$. We then see that the opening of the cut as we turn on the 't Hooft parameter can be interpreted as a deformation of a geometric singularity. This will be later interpreted in section 3.5 from the point of view of topological string theory on Calabi-Yau manifolds.

**Exercise.** *Resolvent for the cubic matrix model.* Consider the cubic matrix model with potential $W(M) = M^2/2 + g_3 M^3/3$. Derive an expression for the endpoints of the one-cut solution as a function of $t$, $g_3$, and find the resolvent and the planar free energy. The solution is worked out in [12].

Although we will not need it in this review, there are well-developed techniques to obtain the higher genus $F_g(t)$ as systematic corrections to the saddle-point result $F_0(t)$ [5, 32]. Interestingly enough, these corrections can be computed in terms of integrals of differentials defined on the hyperelliptic curve (2.70).

We have so far considered the so-called one cut solution to the one-matrix model. This is not, however, the most general solution, and we now will consider the multicut solution in the saddle-point approximation. Recall from our previous discussion that the cut appearing in the one-matrix model was centered around a minimum of the potential. If the potential has many minima, one can have a solution with various cuts, centered around the different minima. The most general solution has then $s$ cuts (where $s$ is lower or equal than the number of minima $n$), and the support of the eigenvalue distribution is a disjoint union of $s$ intervals

$$\mathcal{C} = \cup_{i=1}^{s} \mathcal{C}_i, \tag{2.77}$$

where

$$\mathcal{C}_i = [x_{2i}, x_{2i-1}] \tag{2.78}$$

and $x_{2s} < \cdots < x_1$. The equation (2.67) still gives the solution for the resolvent, and it is easy to see that the way to have multiple cuts is to require $\omega_0(p)$ to have $2s$ branch points corresponding to the roots of the polynomial $W'(z)^2 - R(z)$. Therefore we have

$$\omega_0(p) = \frac{1}{2t} W'(p) - \frac{1}{2t} \sqrt{\prod_{k=1}^{2s} (p - x_k) M(p)}, \tag{2.79}$$

which can be solved in a compact way by

$$\omega_0(p) = \frac{1}{2t} \oint_{\mathcal{C}} \frac{dz}{2\pi i} \frac{W'(z)}{p - z} \left( \prod_{k=1}^{2s} \frac{p - x_k}{z - x_k} \right)^{\frac{1}{2}}. \tag{2.80}$$

In order to satisfy the asymptotics (2.57) the following conditions must hold:

$$\delta_{\ell s} = \frac{1}{2t} \oint_{\mathcal{C}} \frac{dz}{2\pi i} \frac{z^{\ell} W'(z)}{\prod_{k=1}^{2s} (z - x_k)^{\frac{1}{2}}}, \qquad \ell = 0, 1, \cdots, s. \tag{2.81}$$

In contrast to the one-cut case, these are only $s + 1$ conditions for the $2s$ variables $x_k$ representing the endpoints of the cut. For $s > 1$, there are not enough conditions to determine the solution of the model, and we need extra input to determine the positions of the endpoints $x_k$. Usually, the extra condition which is imposed is that the different cuts are at equipotential lines (see for example [11, 4]). It is easy to see that in general the effective potential is constant on each cut,

$$W_{\text{eff}}(p) = \Gamma_i, \qquad p \in \mathcal{C}_i, \tag{2.82}$$

but the values of $\Gamma_i$ will be in general different for the different cuts. This means that there can be eigenvalue tunneling from one cut to the other. The way to guarantee equilibrium is to choose the endpoints of the cuts in such a way that $\Gamma_i = \Gamma$ for all $i = 1, \cdots, s$. This gives the $s - 1$ conditions:

$$W_{\text{eff}}(x_{2i+1}) = W_{\text{eff}}(x_{2i}), \qquad i = 1, \cdots, s-1, \tag{2.83}$$

which, together with the $s + 1$ conditions (2.81) provide $2s$ constraints which allow one to find the positions of the $2s$ endpoints $x_i$. We can also write the equation (2.83) as

$$\int_{x_{2i+1}}^{x_{2i}} dz\, M(z) \prod_{k=1}^{2s} (z - x_k)^{\frac{1}{2}} = 0. \tag{2.84}$$

In the context of the matrix models describing topological strings, the multicut solution is determined by a different set of conditions and will be described in section 3.4.

## 2.3 Matrix model technology II: orthogonal polynomials

Another useful technique to solve matrix models involves orthogonal polynomials. This technique was developed in [8, 9] (which we follow quite closely), and provides explicit expressions for $F_g(t)$ at least for low genus. This technique turns out to be particularly useful in the study of the so-called double-scaling limit of matrix models [13]. We will use this technique to study Chern-Simons matrix models, in section 4, therefore this subsection can be skipped by the reader who is only interested in the conventional matrix models involved in the Dijkgraaf-Vafa approach.

The starting point of the technique of orthogonal polynomials is the eigenvalue representation of the partition function

$$Z = \frac{1}{N!} \int \prod_{i=1}^{N} \frac{d\lambda_i}{2\pi}\, \Delta^2(\lambda) e^{-\frac{1}{g_s} \sum_{i=1}^{N} W(\lambda_i)}, \tag{2.85}$$

where $W(\lambda)$ is an arbitrary potential. If we regard

$$d\mu = e^{-\frac{1}{g_s} W(\lambda)} \frac{d\lambda}{2\pi} \tag{2.86}$$

as a measure in **R**, one can introduce *orthogonal polynomials* $p_n(\lambda)$ defined by

$$\int d\mu \, p_n(\lambda) p_m(\lambda) = h_n \delta_{nm}, \quad n \geq 0, \tag{2.87}$$

where $p_n(\lambda)$ are normalized by requiring the behavior $p_n(\lambda) = \lambda^n + \cdots$. One can now compute $Z$ by noting that

$$\Delta(\lambda) = \det p_{j-1}(\lambda_i). \tag{2.88}$$

By expanding the determinant as

$$\sum_{\sigma \in S_N} (-1)^{\epsilon(\sigma)} \prod_k p_{\sigma(k)-1}(\lambda_k) \tag{2.89}$$

where the sum is over permutations $\sigma$ of $N$ indices and $\epsilon(\sigma)$ is the signature of the permutation, we find

$$Z = \prod_{i=0}^{N-1} h_i = h_0^N \prod_{i=1}^{N} r_i^{N-i}, \tag{2.90}$$

where we have introduced the coefficients

$$r_k = \frac{h_k}{h_{k-1}}, \qquad k \geq 1. \tag{2.91}$$

One of the most important properties of orthogonal polynomials is that they satisfy recursion relations of the form

$$(\lambda + s_n) p_n(\lambda) = p_{n+1}(\lambda) + r_n p_{n-1}(\lambda). \tag{2.92}$$

It is easy to see that the coefficients $r_n$ involved in this relation are indeed given by (2.91). This follows from the equality

$$h_{n+1} = \int d\mu \, p_{n+1}(\lambda) \lambda p_n(\lambda), \tag{2.93}$$

together with the use of the recursion relation for $\lambda p_{n+1}(\lambda)$. For even potentials, $s_n = 0$.

As an example of this technique, we can consider again the simple case of the Gaussian matrix model. The orthogonal polynomials of the Gaussian model are well-known: they are essentially the Hermite polynomials $H_n(x)$, which are defined by

$$H_n(x) = (-1)^n e^{x^2} \frac{d^n}{dx^n} e^{-x^2}. \tag{2.94}$$

More precisely, one has

$$p_n(x) = \left(\frac{g_s}{2}\right)^{n/2} H_n(x/\sqrt{2g_s}), \tag{2.95}$$

and one can then check that

$$h_n^G = \left(\frac{g_s}{2\pi}\right)^{\frac{1}{2}} n! g_s^n, \quad r_n^G = n\, g_s. \tag{2.96}$$

Using now (2.90) we can confirm the result (2.33) that we stated before.

It is clear that a detailed knowledge of the orthogonal polynomials allows the computation of the partition function of the matrix model. It is also easy to see that the computation of correlation functions also reduces to an evaluation in terms of the coefficients in the recursion relation. To understand this point, it is useful to introduce the *orthonormal* polynomials

$$\mathcal{P}_n(\lambda) = \frac{1}{\sqrt{h_n}} p_n(\lambda), \tag{2.97}$$

which satisfy the recursion relation

$$\lambda \mathcal{P}_n(\lambda) = -s_n \mathcal{P}_n(\lambda) + \sqrt{r_{n+1}} \mathcal{P}_{n+1}(\lambda) + \sqrt{r_n} \mathcal{P}_{n-1}(\lambda). \tag{2.98}$$

Let us now consider the normalized vev $\langle \mathrm{Tr}\, M^\ell \rangle$, which in terms of eigenvalues is given by the integral

$$\langle \mathrm{Tr}\, M^\ell \rangle = \frac{1}{N! Z} \int \prod_{i=1}^N e^{-\frac{1}{g_s} W(\lambda_i)} \frac{d\lambda_i}{2\pi} \Delta^2(\lambda) \Big( \sum_{i=1}^N \lambda_i^\ell \Big). \tag{2.99}$$

By using (2.88) it is easy to see that this equals

$$\sum_{j=0}^{N-1} \int d\mu \lambda^\ell \mathcal{P}_j^2(\lambda). \tag{2.100}$$

This integral can be computed in terms of the coefficients in (2.97). For example, for $\ell = 2$ we find

$$\langle \mathrm{Tr}\, M^2 \rangle = \sum_{j=0}^{N-1} (s_j^2 + r_{j+1} + r_j), \tag{2.101}$$

where we put $r_0 = 0$. A convenient way to encode this result is by introducing the Jacobi matrix

$$\mathcal{J} = \begin{pmatrix} 0 & r_1^{1/2} & 0 & 0 & \cdots \\ r_1^{1/2} & 0 & r_2^{1/2} & 0 & \cdots \\ 0 & r_2^{1/2} & 0 & r_3^{1/2} & \cdots \\ \cdots & \cdots & \cdots & \cdots & \cdots \end{pmatrix} \tag{2.102}$$

as well as the diagonal matrix

$$\mathcal{S} = \begin{pmatrix} s_0 & 0 & 0 & 0 & \cdots \\ 0 & s_1 & 0 & 0 & \cdots \\ 0 & 0 & s_2 & 0 & \cdots \\ \cdots & \cdots & \cdots & \cdots & \cdots \end{pmatrix}. \tag{2.103}$$

It then follows that

$$\langle \mathrm{Tr}\, M^\ell \rangle = \mathrm{Tr}\, (\mathcal{J} - \mathcal{S})^\ell. \tag{2.104}$$

The results we have presented so far give the exact answer for the correlators and the partition function, at all orders in $1/N$. As we have seen, we are particularly interested in computing the functions $F_g(t)$ which are obtained by resumming the perturbative expansion at fixed genus. As shown in [8, 9], one can in fact use the orthogonal polynomials to provide closed expressions for $F_g(t)$ in the one-cut case. We will now explain how to do this in some detail.

The object we want to compute is

$$\mathcal{F} = F - F_G = \log Z - \log Z_G. \tag{2.105}$$

If we write the usual series $\mathcal{F} = \sum_{g\geq 0} \mathcal{F}_g g_s^{2g-2}$, we have

$$g_s^2 \mathcal{F} = \frac{t^2}{N^2}(\log Z - \log Z_G) = \frac{t^2}{N} \log \frac{h_0}{h_0^G} + \frac{t^2}{N}\sum_{k=1}^{N} (1 - \frac{k}{N}) \log \frac{r_k(N)}{k g_s}. \tag{2.106}$$

The planar contribution to the free energy $\mathcal{F}_0(t)$ is obtained from (2.106) by taking $N \to \infty$. In this limit, the variable

$$\xi = \frac{k}{N}$$

becomes a continuous variable, $0 \leq \xi \leq 1$, in such a way that

$$\frac{1}{N}\sum_{k=1}^{N} f(k/N) \to \int_0^1 d\xi f(\xi)$$

as $N$ goes to infinity. Let us assume as well that $r_k(N)$ has the following asymptotic expansion as $N \to \infty$:

$$r_k(N) = \sum_{s=0}^{\infty} N^{-2s} R_{2s}(\xi). \tag{2.107}$$

We then find

$$\mathcal{F}_0(t) = -\frac{1}{2}t^2 \log t + t^2 \int_0^1 d\xi (1-\xi) \log \frac{R_0(\xi)}{\xi}. \tag{2.108}$$

This provides a closed expression for the planar free energy in terms of the large $N$ limit of the recursion coefficients $r_k$.

It is interesting to see how to recover the density of states $\rho(\lambda)$ in the saddle-point approximation from orthogonal polynomials. Let us first try to evaluate (2.104) in the planar approximation, following [9]. A simple argument based on the recursion relations indicates that, at large $N$,

$$(\mathcal{J}^\ell)_{nn} \sim \frac{\ell!}{(\ell/2)!^2} r_n^{\ell/2}. \tag{2.109}$$

Using now the integral representation

$$\frac{\ell!}{(\ell/2)!^2} = \int_{-1}^{1} \frac{dy}{\pi} \frac{(2y)^\ell}{\sqrt{1-y^2}},$$

we find

$$\frac{1}{N}\langle \mathrm{Tr}\, M^\ell \rangle = \int_0^1 d\xi \int_{-1}^{1} \frac{dy}{\pi} \frac{1}{\sqrt{1-y^2}} (2yR_0^{1/2}(\xi) - s(\xi))^\ell,$$

where we have denoted by $s(\xi)$ the limit as $N \to \infty$ of the recursion coefficients $s_k(N)$ which appear in (2.92). Since the above average can be also computed by (2.52), by comparing we find

$$\rho(\lambda) = \int_0^1 d\xi \int_{-1}^{1} \frac{dy}{\pi} \frac{1}{\sqrt{1-y^2}} \delta(\lambda - (2yR_0^{1/2}(\xi) - s(\xi))),$$

or, more explicitly,

$$\rho(\lambda) = \int_0^1 \frac{d\xi}{\pi} \frac{\theta[4R_0(\xi) - (\lambda + s(\xi))^2]}{\sqrt{4R_0(\xi) - (\lambda + s(\xi))^2}}. \tag{2.110}$$

Here, $\theta$ is the step function. It also follows from this equation that $\rho(\lambda)$ is supported on the interval $[b(t), a(t)]$, where

$$b(t) = -2\sqrt{R_0(1)} - s(1), \quad a(t) = 2\sqrt{R_0(1)} - s(1). \tag{2.111}$$

**Example.** In the Gaussian matrix model $R_0(\xi) = t\xi$, and $s(\xi) = 0$. We then find that the density of eigenvalues is supported in the interval $[-2\sqrt{t}, 2\sqrt{t}]$ and it is given by

$$\rho(\lambda) = \frac{1}{\pi} \int_0^1 d\xi \frac{\theta[4\xi t - \lambda^2]}{\sqrt{4\xi t - \lambda^2}} = \frac{1}{2\pi t}\sqrt{4t - \lambda^2}$$

which reproduces of course Wigner's semicircle law.

As shown in [8, 9], orthogonal polynomials can be used as well to obtain the higher genus free energies $\mathcal{F}_g$. The key ingredient to do that is simply the Euler-MacLaurin formula, which reads

$$\frac{1}{N}\sum_{k=1}^{N} f\left(\frac{k}{N}\right) = \int_0^1 f(\xi)d\xi + \frac{1}{2N}[f(1)-f(0)] + \sum_{p=1}^{\infty} \frac{1}{N^{2p}}\frac{B_{2p}}{(2p)!}[f^{(2p-1)}(1)-f^{(2p-1)}(0)], \tag{2.112}$$

and should be regarded as an asymptotic expansion for $N$ large which gives a way to compute systematically $1/N$ corrections. We can then use it to calculate (2.106) at all orders in $1/N$, where

$$f(k/N) = \left(1 - \frac{k}{N}\right)\log\frac{Nr_k(N)}{k}, \tag{2.113}$$

and we use the fact that $r_k$ has an expansion of the form (2.107). In this way, we find for example that

$$\mathcal{F}_1(t) = t^2 \int_0^1 d\xi(1-\xi)\frac{R_2(\xi)}{R_0(\xi)} + \frac{t^2}{12}\frac{d}{d\xi}\left[(1-\xi)\log\frac{R_0(\xi)}{\xi}\right]_0^1,$$

and so on. We will use this formulation in section 4 to compute $\mathcal{F}_g(t)$ in the matrix model that describes Chern-Simons theory on $\mathbf{S}^3$.

It is clear from the above analysis that matrix models can be solved with the method of orthogonal polynomials, in the sense that once we know the precise form of the coefficients in the recursion relation we can compute all quantities in an $1/N$ expansion. Since the recursion relation is only known exactly in a few cases, we need methods to determine its coefficients for general potentials $W(M)$. In the case of *polynomial* potentials, of the form

$$W(M) = \sum_{p\geq 0}\frac{g_p}{p}\mathrm{Tr}\, M^p,$$

there are well-known techniques to obtain explicit results [9], see [21, 22] for reviews. We start by rewriting the recursion relation (2.92) as

$$\lambda p_n(\lambda) = \sum_{m=0}^{n+1} B_{nm}p_m,$$

where $B$ is a matrix. The identities

$$\begin{aligned} r_n \int d\lambda e^{-\frac{1}{g_s}W(\lambda)}W'(\lambda)p_n(\lambda)p_{n-1}(\lambda) &= nh_n g_s, \\ \int d\lambda \frac{d}{d\lambda}(p_n e^{-\frac{1}{g_s}W(\lambda)}p_n) &= 0 \end{aligned} \tag{2.114}$$

lead to the matrix equations

$$\begin{aligned}(W'(B))_{nn-1} &= ng_s,\\ (W'(B))_{nn} &= 0.\end{aligned}\tag{2.115}$$

These equations are enough to determine the recursion coefficients. Consider for example a quartic potential

$$W(\lambda) = \frac{g_2}{2}\lambda^2 + \frac{g_4}{4}\lambda^4.$$

Since this potential is even, it is easy to see that the first equation in (2.115) is automatically satisfied, while the second equation leads to

$$r_n\{g_2 + g_4(r_n + r_{n-1} + r_{n+1})\} = ng_s$$

which at large $N$ reads

$$R_0(g_2 + 3g_4 R_0) = \xi t.$$

In general, for an even potential of the form

$$W(\lambda) = \sum_{p\geq 0} \frac{g_{2p+2}}{2p+2}\lambda^{2p+2}\tag{2.116}$$

one finds

$$\xi t = \sum_{p\geq 0} g_{2p+2}\binom{2p+1}{p} R_0^{p+1}(\xi),\tag{2.117}$$

which determines $R_0$ as a function of $\xi$. The above equation is sometimes called –especially in the context of double-scaled matrix models– the *string equation*, and by setting $\xi = 1$ we find an explicit equation for the endpoints of the cut in the density of eigenvalues as a function of the coupling constants and $t$.

**Exercise**. Verify, using saddle-point techniques, that the string equation correctly determines the endpoints of the cut. Compute $R_0(\xi)$ for the quartic and the cubic matrix model, and use it to obtain $\mathcal{F}_0(t)$ (for the quartic potential, the solution is worked out in detail in [9]).

# 3. Type B topological strings and matrix models

## 3.1 The topological B model

The topological B model was introduced in [49, 73] and can be constructed by twisting the $\mathcal{N} = 2$ superconformal sigma model in two dimensions. There are in fact two different twists, called the A and the B twist in [49, 73], and in these lectures we will focus on the second one. A detailed review of topological sigma models and topological strings can be found in [39].

The topological B model is a theory of maps from a Riemann surface $\Sigma_g$ to a Calabi-Yau manifold $X$ of complex dimension $d$. The Calabi-Yau condition arises in order to cancel an anomaly that appears after twisting (see for example Chapter 3 of [52] for a detailed analysis of this issue). Indices for the real tangent bundle of $X$ will be denoted by $i = 1, \cdots, 2d$, while holomorphic and antiholomorphic indices will be denoted respectively by $I, \overline{I} = 1, \cdots, d$. The holomorphic tangent bundle will be simply denoted by $TX$, while the antiholomorphic tangent bundle will be denoted by $\overline{TX}$. One of the most important properties of Calabi-Yau manifolds (which can actually be taken as their defining feature) is that they have a holomorphic, nonvanishing section $\Omega$ of the canonical bundle $K_X = \Omega^{3,0}(X)$. Since the section is nowhere vanishing, the canonical line bundle is trivial and $c_1(K_X) = 0$. We will always consider examples with complex dimension $d = 3$.

The field content of the topological B model is the following. First, since it is a nonlinear sigma model, we have a map $x : \Sigma_g \to X$, which is a scalar, commuting field. Besides the field $x$, we have two sets of Grassmann fields $\eta^{\overline{I}}, \theta^{\overline{I}} \in x^*(\overline{TX})$, which are scalars on $\Sigma_g$, and a Grassmannian one-form on $\Sigma_g$, $\rho^I_\alpha$, with values in $x^*(TX)$. We also have commuting auxiliary fields $F^I, F^{\overline{I}}$ (we will follow here the off-shell formulation of [49, 50]). The action for the theory is:

$$\begin{aligned}\mathcal{L} &= t\int_{\Sigma_g} d^2z\Big[G_{I\overline{J}}(\partial_z x^I \partial_{\bar z} x^{\overline{J}} + \partial_{\bar z} x^I \partial_z x^{\overline{J}}) - \rho^I_z (G_{I\overline{J}} D_{\bar z}\eta^{\overline{J}} + D_{\bar z}\theta_I) \\ &\quad -\rho^I_{\bar z}(G_{I\overline{J}} D_z \eta^{\overline{J}} - D_z\theta_I) - R^I{}_{\overline{J}\overline{L}K}\eta^{\overline{L}}\rho^J_z\rho^K_{\bar z}\theta_I - G_{I\overline{J}}F^I F^{\overline{J}}\Big], \end{aligned} \tag{3.1}$$

In this action, we have picked local coordinates $z, \bar z$ on $\Sigma_g$, and $d^2z$ is the measure $-idz \wedge d\bar z$. $t$ is a parameter that plays the role of $1/\hbar$, the field $\theta_I$ is given by $\theta_I = G_{I\overline{J}}\theta^J$, and the covariant derivative $D_\alpha$ acts on sections $\psi^i$ of the tangent bundle as

$$D_\alpha \psi^i = \partial_\alpha \psi^i + \partial_\alpha x^j \Gamma^i_{jk}\psi^k. \tag{3.2}$$

The theory also has a BRST, or topological, charge $\mathcal{Q}$ which acts on the fields according to

$$\begin{aligned}
[\mathcal{Q}, x^I] &= 0, & [\mathcal{Q}, x^{\overline{I}}] &= \eta^{\overline{I}}, \\
\{\mathcal{Q}, \eta^{\overline{I}}\} &= 0, & \{\mathcal{Q}, \theta_I\} &= G_{I\overline{J}}F^{\overline{J}}, \\
\{\mathcal{Q}, \rho^I_z\} &= \partial_z x^I, & [\mathcal{Q}, F^I] &= D_z\rho^I_{\bar z} - D_{\bar z}\rho^I_z + R^I{}_{\overline{J}\overline{L}K}\eta^{\overline{L}}\rho^J_z\rho^K_{\bar z}, \\
\{\mathcal{Q}, \rho^I_{\bar z}\} &= \partial_{\bar z} x^I, & [\mathcal{Q}, F^{\overline{I}}] &= -\Gamma^{\overline{I}}_{\overline{J}\overline{K}}\eta^{\overline{J}}F^{\overline{K}},
\end{aligned}$$

The action of $\mathcal{Q}$ explicitly depends on the splitting between holomorphic and antiholomorphic coordinates on $X$, in other words, it depends explicitly on the choice of complex structure on $X$. It is easy to show that $\mathcal{Q}^2 = 0$, and that the action of the model is $\mathcal{Q}$-exact:

$$\mathcal{L} = \{\mathcal{Q}, V\} \tag{3.3}$$

where $V$ (sometimes called the gauge fermion) is given by

$$V = t \int_{\Sigma_g} d^2z \left[ G_{I\bar{J}} (\rho_z^I \partial_{\bar{z}} x^{\bar{J}} + \rho_{\bar{z}}^I \partial_z x^{\bar{J}}) - F^I \theta_I \right]. \tag{3.4}$$

Finally, we also have a $U(1)$ ghost number symmetry, in which $x$, $\eta$, $\theta$ and $\rho$ have ghost numbers 0, 1, 1, and $-1$, respectively. The Grassmannian charge $\mathcal{Q}$ then has ghost number 1. Notice that, if we interpret $\eta^{\bar{I}}$ as a basis for antiholomorphic differential forms on $X$, the action of $\mathcal{Q}$ on $x^I$, $x^{\bar{I}}$ may be interpreted as the Dolbeault antiholomorphic differential $\bar{\partial}$.

It follows from (3.3) that the energy-momentum tensor of this theory is given by

$$T_{\alpha\beta} = \{\mathcal{Q}, b_{\alpha\beta}\}, \tag{3.5}$$

where $b_{\alpha\beta} = \delta V / \delta g^{\alpha\beta}$ and has ghost number $-1$. The fact that the energy-momentum tensor is $\mathcal{Q}$-exact means that the theory is *topological*, in the sense that the partition function does not depend on the background two-dimensional metric. This is easily proved: the partition function is given by

$$Z = \int \mathcal{D}\phi \, e^{-\mathcal{L}}, \tag{3.6}$$

where $\phi$ denotes the set of fields of the theory, and we compute it in the background of a two-dimensional metric $g_{\alpha\beta}$ on the Riemann surface. Since $T_{\alpha\beta} = \delta\mathcal{L}/\delta g^{\alpha\beta}$, we find that

$$\frac{\delta Z}{\delta g^{\alpha\beta}} = -\langle \{\mathcal{Q}, b_{\alpha\beta}\} \rangle, \tag{3.7}$$

where the bracket denotes an unnormalized vacuum expectation value. Since $\mathcal{Q}$ is a symmetry of the theory, the above vacuum expectation value vanishes, and we find that $Z$ is metric-independent, at least formally.

The $\mathcal{Q}$-exactness of the action itself also has an important consequence: the same argument that we used above implies that the partition function of the theory is independent of $t$. Now, since $t$ plays the role of $1/\hbar$, the limit of $t$ large corresponds to the semiclassical approximation. Since the theory does not depend on $t$, the semiclassical approximation is *exact*. The classical configurations for the above action are *constant* maps $x : \Sigma_g \to X$. Therefore, it follows that path integrals of the above theory reduce to integrals over $X$ [73].

What are the operators to consider in this theory? Since the most interesting aspect of this model is the independence w.r.t. to the two-dimensional metric, we want to look for operators whose correlation functions satisfy this condition. It is easy to see that the operators in the cohomology of $\mathcal{Q}$ do the job: topological invariance requires them to be $\mathcal{Q}$-closed, and on the other hand they cannot be $\mathcal{Q}$-exact, since otherwise their correlation functions would vanish. One can also check that the $\mathcal{Q}$-cohomology is given by operators of the form

$$\mathcal{O}_\phi = \phi_{\overline{I}_1 \cdots \overline{I}_p}^{J_1 \cdots J_q} \eta^{\overline{I}_1} \cdots \eta^{\overline{I}_p} \theta_{J_1} \cdots \theta_{J_q}, \tag{3.8}$$

where

$$\phi = \phi_{\overline{I}_1 \cdots \overline{I}_p}^{J_1 \cdots J_q} dx^{\overline{I}_1} \wedge \cdots \wedge dx^{\overline{I}_p} \frac{\partial}{\partial x^{J_1}} \wedge \cdots \wedge \frac{\partial}{\partial x^{J_q}} \tag{3.9}$$

is an element of $H^p_{\overline{\partial}}(X, \wedge^q TX)$. Therefore, the $\mathcal{Q}$-cohomology is in one-to-one correspondence with the twisted Dolbeault cohomology of the target manifold $X$. We can then consider correlation functions of the form

$$\langle \prod_a \mathcal{O}_{\phi_a} \rangle. \tag{3.10}$$

This correlation function vanishes unless the following selection rule is satisfied

$$\sum_a p_a = \sum_a q_a = d(1-g), \tag{3.11}$$

where $g$ is the genus of the Riemann surface. This selection rule comes from a $U(1)_L \times U(1)_R$ anomalous global current. Due to the arguments presented above, this correlation function can be computed in the semiclassical limit, where the path integral reduces to an integration over the target $X$. The product of operators in (3.10) corresponds to a form in $H^d_{\overline{\partial}}(X, \wedge^d TX)$. To integrate such a form over $X$ we crucially need the Calabi-Yau condition. This arises as follows. In a Calabi-Yau manifold we have an invertible map

$$\begin{array}{ccc} \Omega^{0,p}(\wedge^q TX) & \longrightarrow & \Omega^{d-q,p}(X) \\ \phi^{I_1 \cdots I_q}_{\overline{J}_1 \cdots \overline{J}_p} & \mapsto & \Omega_{I_1 \cdots I_q I_{q+1} \cdots I_d} \phi^{I_1 \cdots I_q}_{\overline{J}_1 \cdots \overline{J}_p} \end{array} \tag{3.12}$$

where the $(d, 0)$-form $\Omega$ is used to contract the indices. Since $\Omega$ is holomorphic, this descends to the $\overline{\partial}$-cohomology. It then follows that an element in $H^d_{\overline{\partial}}(X, \wedge^d TX)$ maps to an element in $H^d_{\overline{\partial}}(X)$. After further multiplication by $\Omega$, one can then integrate a $(d, d)$-form over $X$. This is the prescription to compute correlation functions like (3.10). A simple and important example of this procedure is the case of a Calabi-Yau threefold, $d = 3$, and operators associated to forms in $H^1_{\bar\partial}(X, TX)$, or by using (3.12), to forms in $H^{2,1}_{\bar\partial}(X)$. These operators are important since they correspond to infinitesimal deformations of the complex structure of $X$. The selection rule (3.11) says that we have to integrate three of these operators, and the correlation function reads in this case

$$\langle \mathcal{O}_{\phi_1}\mathcal{O}_{\phi_2}\mathcal{O}_{\phi_3}\rangle = \int_X (\phi_1)^{I_1}_{\bar J_1}(\phi_2)^{I_2}_{\bar J_2}(\phi_3)^{I_3}_{\bar J_3}\,\Omega_{I_1 I_2 I_3} dz^{\bar J_1} dz^{\bar J_2} dz^{\bar J_3}\wedge\Omega. \quad (3.13)$$

It turns out that the full information of the correlators (3.13) at genus zero can be encoded in a single function called the *prepotential*. We will quickly review here some of the basic results of special geometry and the theory of the prepotential for the topological B model, and we refer the reader to [17, 39] for more details. The correlation functions in the B model, like for example (3.13), depend on a choice of complex structure, as we have already emphasized. The different complex structures form a moduli space $\mathcal{M}$ of dimension $h^{2,1}$. A convenient parametrization of $\mathcal{M}$ is the following. Choose first a symplectic basis for $H_3(X)$, denoted by $(A_a, B^a)$, with $a = 0, 1, \cdots, h^{2,1}$, and such that $A_a \cap B^b = \delta^b_a$. We then define the *periods* of the Calabi-Yau manifold as

$$z_a = \int_{A_a}\Omega, \qquad \mathcal{F}^a = \int_{B^a}\Omega, \quad a = 0, \cdots, h^{2,1}. \quad (3.14)$$

Of course, the symplectic group $\mathrm{Sp}(2h^{2,1}+2, \mathbf{R})$ acts on the vector $(z^a, \mathcal{F}_a)$. A basic result of the theory of deformation of complex structures says that the $z^a$ are (locally) complex projective coordinates for $\mathcal{M}$. Inhomogeneous coordinates can be introduced in a local patch where one of the projective coordinates, say $z_0$, is different from zero, and taking

$$t_a = \frac{z_a}{z_0}, \qquad a = 1, \cdots, h^{2,1}. \quad (3.15)$$

The coordinates $z_a$ are called *special projective coordinates*, and since they parametrize $\mathcal{M}$ we deduce that the other set of periods must depend on them, *i.e.* $\mathcal{F}^a = \mathcal{F}^a(z)$. Using the periods (3.14) we can define the *prepotential* $\mathcal{F}(z)$ by the equation

$$\mathcal{F} = \frac{1}{2}\sum_{a=0}^{h^{2,1}} z_a \mathcal{F}^a. \quad (3.16)$$

The prepotential satisfies

$$\mathcal{F}^a(z) = \frac{\partial \mathcal{F}}{\partial z_a} \quad (3.17)$$

and turns out to be a homogeneous function of degree two in the $z_a$. Therefore, one can rescale it in order to obtain a function of the inhomogeneous coordinates $t_a$:

$$F_0(t_a) = \frac{1}{z_0^2}\,\mathcal{F}(z_a). \tag{3.18}$$

The fact that the coordinates $z_a$ are projective is related to the freedom in normalizing the three-form $\Omega$. In order to obtain expressions in terms of the inhomogeneous coordinates $t_a$, we simply have to rescale $\Omega \to \frac{1}{z_0}\Omega$, and the periods $(z_a, \mathcal{F}^a)$ become

$$(1, t_a, 2F_0 - \sum_{a=1}^{h^{2,1}} t_a \frac{\partial F_0}{\partial t_a}, \frac{\partial F_0}{\partial t_a}). \tag{3.19}$$

One of the key results in special geometry is that the correlation functions (3.13) can be computed in terms of the prepotential $F_0(t)$. Given a deformation of the complex structure parametrized by $t_a$, the corresponding tangent vector $\partial/\partial t_a$ is associated to a differential form of type $(2,1)$. This form leads to an operator $\mathcal{O}_a$, and the three-point functions involving these operators turn out to be given by

$$\langle \mathcal{O}_a \mathcal{O}_b \mathcal{O}_c \rangle = \frac{\partial^3 F_0}{\partial t_a \partial t_b \partial t_c}. \tag{3.20}$$

The prepotential $F_0(t)$ encodes the relevant information about the B model on the sphere, and it has an important physical meaning, since it gives the four-dimensional supergravity prepotential of type IIB string theory compactified on $X$ (and determines the leading part of the vector multiplet effective action).

In order to obtain interesting quantities at higher genus one has to couple the topological B model to two-dimensional gravity, using the fact that the structure of the twisted theory is very close to that of the bosonic string [30, 74, 7]. In the bosonic string, there is a nilpotent BRST operator, $\mathcal{Q}_{\rm BRST}$, and the energy-momentum tensor turns out to be a $\mathcal{Q}_{\rm BRST}$-commutator: $T(z) = \{\mathcal{Q}_{\rm BRST}, b(z)\}$. In addition, there is a ghost number with anomaly $3\chi(\Sigma_g) = 6 - 6g$, in such a way that $\mathcal{Q}_{\rm BRST}$ and $b(z)$ have ghost number $1$ and $-1$, respectively. This is precisely the same structure that we found in (3.5), and the composite field $b_{\alpha\beta}$ plays the role of an antighost. Therefore, one can just follow the prescription of coupling to gravity for the bosonic string and define a genus $g \geq 1$ free energy as follows:

$$F_g = \int_{\overline{M}_g} \langle \prod_{k=1}^{6g-6} (b, \mu_k) \rangle, \tag{3.21}$$

where

$$(b, \mu_k) = \int_{\Sigma_g} d^2 z (b_{zz} (\mu_k)_{\bar{z}}^{z} + b_{\bar{z}\bar{z}} (\overline{\mu}_k)_{z}^{\bar{z}}), \tag{3.22}$$

and $\mu_k$ are the usual Beltrami differentials. The vacuum expectation value in (3.21) refers to the path integral over the fields of the topological B model, and gives a differential form on the moduli space of Riemann surfaces of genus $g$, $\overline{M}_g$, which is then integrated over. The free energies $F_g$ of the B model coupled to gravity for $g \geq 1$ are also related to variation of complex structures. A target space description of this theory, called Kodaira-Spencer theory of gravity, was found in [7], and can be used to determine recursively the $F_g$ in terms of special geometry data.

## 3.2 The open type B model and its string field theory description

The topological B model can be formulated as well for open strings, *i.e.*, when the worldsheet is an open Riemann surface with boundaries $\Sigma_{g,h}$ [74, 59]. In order to construct the open string version we need boundary conditions (b.c.) for the fields. It turns out that the appropriate b.c. for the B model are Dirichlet along holomorphic cycles of $X$, $S$, and Neumann in the remaining directions. Moreover, one can add Chan-Paton factors to the model, and this is implemented by considering a $U(N)$ holomorphic bundle over the holomorphic cycle $S$. The resulting theory can then be interpreted as a topological B model in the presence of $N$ topological D-branes wrapping $S$. Since we will be interested in finding a spacetime description of the open topological B model, we can consider the case in which the branes fill spacetime (the original case considered in [74]) and deduce the spacetime action for lower dimensional branes by dimensional reduction. In the spacetime filling case, when $S = X$, the boundary conditions for the fields are $\theta = 0$ along $\partial\Sigma_{g,h}$ and that the pullback to $\partial\Sigma_{g,h}$ of $*\rho$ vanishes (where $*$ is the Hodge operator).

The open topological B model can also be coupled to gravity following the same procedure that is used in the closed case, and one obtains in this way the open type B topological string propagating along the Calabi-Yau manifold $X$. We are now interested in providing a description of this model when the $N$ branes are spacetime filling. As shown by Witten in [74], the most efficient way to do that is to use the cubic string field theory introduced in [69].

In bosonic open string field theory we consider the worldsheet of the string to be an infinite strip parameterized by a spatial coordinate $0 \leq \sigma \leq \pi$ and a time coordinate $-\infty < \tau < \infty$, and we pick the flat metric $ds^2 = d\sigma^2 + d\tau^2$. We then consider maps $x : I \rightarrow X$, with $I = [0, \pi]$ and $X$ the target of the string. The string field is a functional of open string configurations $\Psi[x(\sigma)]$, of ghost number one (the string functional depends as well on the ghost fields, but we do not indicate this dependence explicitly). In [69], Witten defines two operations on the space of string functionals. The first one is the *integration*, which is defined formally by folding the string around its midpoint and gluing the two halves:

$$\int \Psi = \int \mathcal{D}x(\sigma) \prod_{0\leq\sigma\leq\pi/2} \delta[x(\sigma) - x(\pi - \sigma)]\Psi[x(\sigma)]. \tag{3.23}$$

The integration has ghost number $-3$, which is the ghost number of the vacuum. This corresponds to the usual fact that in open string theory on the disk one has to soak up three zero modes. One also defines an associative, noncommutative *star product* $\star$ of string functionals through the following equation:

$$\int \Psi_1\star\cdots\star\Psi_N = \int \prod_{i=1}^{N} \mathcal{D}x_i(\sigma) \prod_{i=1}^{N} \prod_{0\leq\sigma\leq\pi/2} \delta[x_i(\sigma)-x_{i+1}(\pi-\sigma)]\Psi_i[x_i(\sigma)], \tag{3.24}$$

where $x_{N+1} \equiv x_1$. The star product simply glues the strings together by folding them around their midpoints, and gluing the first half of one with the second half of the following (see for example the review [65] for more details), and it doesn't change the ghost number. In terms of these geometric operations, the string field action is given by

$$S = \frac{1}{g_s} \int \left( \frac{1}{2} \Psi \star Q_{\rm BRST} \Psi + \frac{1}{3} \Psi \star \Psi \star \Psi \right) \tag{3.25}$$

where $g_s$ is the string coupling constant. Notice that the integrand has ghost number 3, while the integration has ghost number $-3$, so that the action (3.25) has zero ghost number. If we add Chan-Paton factors, the string field is promoted to a $U(N)$ matrix of string fields, and the integration in (3.25) includes a trace ${\rm Tr}$. The action (3.25) has all the information about the spacetime dynamics of open bosonic strings, with or without D-branes. In particular, one can derive the Born-Infeld action describing the dynamics of D-branes from the above action [64].

We will not need all the technology of string field theory in order to understand open topological strings. The only piece of relevant information is the following: the string functional is a function of the zero mode of the string (which corresponds to the position of the string midpoint), and of the higher oscillators. If we decouple all the oscillators, the string functional becomes an ordinary function of spacetime, the $\star$ product becomes the usual product of functions, and the integral is the usual integration of functions. The decoupling of the oscillators is in fact the point-like limit of string theory. As we will see, this is the relevant limit for topological open type B strings on $X$.

We can now exploit again the analogy between open topological strings and the open bosonic string that we used to define the coupling of the topological B model to gravity (*i.e.*, that both have a nilpotent BRST operator and an energy-momentum tensor that is $\mathcal{Q}_{\rm BRST}$-exact). Since both theories have a similar structure, the spacetime dynamics of open topological type B strings is governed as well by (3.25), where $\mathcal{Q}_{\rm BRST}$ is given in this case by the topological charge defined in (3.3), and where the star product and the integration operation are as in the bosonic string. The construction of the cubic string field theory also requires the existence of a ghost number symmetry, which is also present in the topological sigma model in the form of a $U(1)_R$ symmetry, as we discussed in 3.1. It is convenient to consider the $U(1)_R$ charge of the superconformal algebra in the Ramond sector, which is shifted by $-d/2$ with respect to the assignment presented in 3.1 (here, $d$ is the dimension of the target). When $d=3$ this corresponds to the normalization used in [69], in which the ghost vacuum of the $bc$ system is assigned the ghost number $-1/2$.

In order to provide the string field theory description of open topological type B strings on $X$, we have to determine the precise content of the string field, the $\star$ algebra and the integration of string functionals for this particular model. As in the conventional string field theory of the bosonic string, we have to consider the Hamiltonian description of topological open strings. We then take $\Sigma$ to be an infinite strip and consider maps $x: I \rightarrow X$, with $I=[0,\pi]$. The Hilbert space is made up out of functionals $\Psi[x(\sigma),\cdots]$, where $x$ is a map from the interval as we have just described, and the $\cdots$ refer to the Grassmann fields (which play here the rôle of ghost fields). Notice that, since $\rho^I_{z,\bar z}$ are canonically conjugate to $\eta$, $\theta$, we can choose our functional to depend only on $\eta$, $\theta$. It is easy to see that the Hamiltonian has the form

$$H=\int_0^\pi d\sigma\biggl(tG_{ij}\frac{dx^i}{d\sigma}\frac{dx^j}{d\sigma}+\cdots\biggr). \tag{3.26}$$

We then see that string functionals with $dx^i/d\sigma \neq 0$ cannot contribute: as we saw in the previous subsection, the physics is $t$-independent, therefore we can take $t\rightarrow\infty$. In this limit the functional gets infinitely massive and decouples from the spectrum, unless $dx^i/d\sigma=0$. Therefore, the map $x: I\rightarrow X$ has to be constant and in particular it must be a point in $X$. A similar analysis holds for the Grassmann fields as well. Since $\theta=0$ at the boundary, it follows that string functionals are functions of the commuting zero modes $x^i$ and $\eta^{\bar I}$, and can be written as

$$\Psi=A^{(0)}(x)+\sum_{p\geq 1}\eta^{\bar I_1}\cdots\eta^{\bar I_p}A^{(p)}_{\bar I_1\cdots\bar I_p}(x). \tag{3.27}$$

These functionals can be interpreted as a sum of $(0,p)$-forms on $X$. If we have $N$ D-branes wrapping $X$, these forms will be valued in $\mathrm{End}(E)$ (where $E$ is a holomorphic $U(N)$ bundle). The $\mathcal{Q}$ symmetry acts as on these functionals as the Dolbeault operator $\bar\partial$ with values in $\mathrm{End}(E)$. Notice that a differential form of degree $p$ will have ghost number $p$.

We are now ready to write the string field action for topological open type B strings on $X$ with $N$ spacetime filling branes. We have seen that the relevant string functionals are of the form (3.27). Since in string field theory the string field has ghost number one, we must have

$$\Psi = \eta^{\overline{I}} A_{\overline{I}}(x), \tag{3.28}$$

where $A_{\overline{I}}(x)$ is a $(0,1)$-form taking values in the endomorphisms of some holomorphic vector bundle $E$. In other words, the string field is just the $(0,1)$ piece of a gauge connection on $E$. Since the string field only depends on commuting and anticommuting zero modes, the star product becomes the wedge products of forms in $\Omega^{(0,p)}(\mathrm{End}(E))$, and the integration of string functionals becomes ordinary integration of forms on $X$ wedged by $\Omega$. We then have the following dictionary:

$$\begin{aligned} \Psi &\to A, \qquad && \mathcal{Q}_{\mathrm{BRST}} \to \overline{\partial} \\ \star &\to \wedge, \qquad && \int \to \int_X \Omega\wedge . \end{aligned} \tag{3.29}$$

The string field action (3.25) is then given by

$$S = \frac{1}{2g_s}\int_X \Omega\wedge \mathrm{Tr}\Big(A\wedge\overline{\partial}A + \frac{2}{3}A\wedge A\wedge A\Big). \tag{3.30}$$

This is the so-called *holomorphic Chern-Simons action*. It is a rather peculiar quantum field theory in six dimensions, but as we will see, when we consider D-branes of lower dimension, we will be able to obtain from (3.30) more conventional theories by dimensional reduction.

## 3.3 Topological strings and matrix models

We have seen that the spacetime description of the open B model with spacetime filling branes reduces to a six-dimensional theory (3.30). We will see now that, in some circumstances, this theory simplifies drastically and reduces to a matrix model.

In order to simplify the spacetime description one should study simple Calabi-Yau manifolds. The simplest example of a local Calabi-Yau threefold is a Riemann surface together with an appropriate bundle over it. The motivation for considering this kind of models is the following. Consider a Riemann surface $\Sigma_g$ holomorphically embedded inside a Calabi-Yau threefold $X$, and let us consider the holomorphic tangent bundle of $X$ restricted to $\Sigma_g$. We then have

$$TX|_{\Sigma_g} = T\Sigma_g \oplus N_{\Sigma_g} \tag{3.31}$$

where $N_{\Sigma_g}$ is a holomorphic rank two complex vector bundle over $\Sigma_g$, called the normal bundle of $\Sigma_g$, and the CY condition $c_1(X)=0$ gives

$$c_1(N_{\Sigma_g}) = 2g-2. \tag{3.32}$$

The Calabi-Yau $X$ "near $\Sigma_g$" looks precisely like the total space of the bundle

$$N \to \Sigma_g \tag{3.33}$$

where $N$ is regarded here as a bundle over $\Sigma_g$ satisfying (3.32). The space (3.33) is an example of a *local* Calabi-Yau threefold, and it is noncompact.

When $g = 0$ and $\Sigma_g = \mathbb{P}^1$ it is possible to be more precise about the bundle $N$. A theorem due to Grothendieck says that any holomorphic bundle over $\mathbb{P}^1$ splits into a direct sum of line bundles (for a proof, see for example [36], pp. 516-7). Line bundles over $\mathbb{P}^1$ are all of the form $\mathcal{O}(n)$, where $n \in \mathbf{Z}$. The bundle $\mathcal{O}(n)$ can be easily described in terms of two charts on $\mathbb{P}^1$: the north pole chart, with coordinates $z, \Phi$ for the base and the fiber, respectively, and the south pole chart, with coordinates $z', \Phi'$. The change of coordinates is given by

$$z' = 1/z, \quad \Phi' = z^{-n}\Phi. \tag{3.34}$$

We also have that $c_1(\mathcal{O}(n)) = n$. We then find that local Calabi-Yau manifolds that are made out of a two-sphere together with a bundle over it are all of the form

$$\mathcal{O}(-a) \oplus \mathcal{O}(a-2) \to \mathbb{P}^1, \tag{3.35}$$

since the degree of the bundles have to sum up to $-2$ due to (3.32).

Let us now consider the string field theory of type B open topological strings on the Calabi-Yau manifold (3.35). We will consider a situation where we have Dirichlet boundary conditions associated to $\mathbb{P}^1$, in other words, there are $N$ topological D-branes wrapping $\mathbb{P}^1$. Since the normal directions to the D-brane worldvolume are noncompact, the spacetime description can be obtained by considering the dimensional reduction of the original string field theory action (3.30). As usual in D-brane physics, the gauge potential $A$ splits into a gauge potential on the worldvolume of the brane and Higgs fields describing the motion along the noncompact, transverse directions. In a nontrivial geometric situation like the one here, the Higgs fields are sections of the normal bundle. We then get three different fields:

$$A, \quad \Phi_0, \quad \Phi_1, \tag{3.36}$$

where $A$ is a $U(N)$ $(0,1)$ gauge potential on $\mathbb{P}^1$, $\Phi_0$ is a section of $\mathcal{O}(-a)$, and $\Phi_1$ is a section of $\mathcal{O}(a-2)$. Both fields, $\Phi_0$ and $\Phi_1$, take values in the adjoint representation of $U(N)$. It is easy to see that the action (3.30) becomes

$$S = \frac{1}{g_s}\int_{\mathbb{P}^1} \mathrm{Tr}\,(\Phi_0 \overline{D}_A \Phi_1), \tag{3.37}$$

where $\overline{D}_A = \overline{\partial} + [A, \cdot]$ is the antiholomorphic covariant derivate. Notice that this theory is essentially a gauged $\beta\gamma$ system, since $\Phi_0$, $\Phi_1$ are quasiprimary conformal fields of dimensions $a/2$, $1 - a/2$, respectively.

We will now consider a more complicated geometry. We start with the Calabi-Yau manifold (3.35) with $a = 0$, *i.e.*

$$\mathcal{O}(0) \oplus \mathcal{O}(-2) \to \mathbb{P}^1. \tag{3.38}$$

In this case, $\Phi_0$ is a scalar field on $\mathbb{P}^1$, while $\Phi_1$ is a $(1,0)$ form (since $K_{\mathbb{P}^1} = \mathcal{O}(-2)$). If we cover $\mathbb{P}^1$ with two patches with local coordinates $z, z'$ related by $z' = 1/z$, the fields in the two different patches, $\Phi_0, \Phi_1$, and $\Phi_0', \Phi_1'$ will be related by

$$\Phi_0' = \Phi_0, \quad \Phi_1' = z^2\Phi_1. \tag{3.39}$$

We can regard this geometry as a *family* of $\mathbb{P}^1$s located at $\Phi_1' = 0$ (the zero section of the nontrivial line bundle $\mathcal{O}(-2)$) parametrized by $\Phi_0 = \Phi_0' = x \in \mathbf{C}$. The idea is to obtain a geometry where we get $n$ isolated $\mathbb{P}^1$s at fixed positions of $x$. To do that, we introduce an arbitrary polynomial of degree $n+1$ on $\Phi_0$, $W(\Phi_0)$, and we modify the gluing rules above as follows [15]:

$$z' = 1/z, \quad \Phi_0' = \Phi_0, \quad \Phi_1' = z^2\Phi_1 + W'(\Phi_0)z. \tag{3.40}$$

Before, the $\mathbb{P}^1$ was in a family parameterized by $\Phi_0 \in \mathbf{C}$. Now, we see that there are $n$ isolated $\mathbb{P}^1$s located at fixed positions of $\Phi_0$ given by $W'(\Phi_0) = 0$, since this is the only way to have $\Phi_1 = \Phi_1' = 0$.

The geometry obtained by imposing the gluing rules (3.40) can be interpreted in yet another way. Call $\Phi_0 = x$ and define the coordinates

$$u = 2\Phi_1', \quad v = 2\Phi_1, \quad y = i(2z'\Phi_1' - W'(x)). \tag{3.41}$$

The last equation in (3.40) can now be written as

$$uv + y^2 + W'(x)^2 = 0. \tag{3.42}$$

This is a singular geometry, since there are singularities along the line $u = v = y = 0$ for every $x_*$ such that $W'(x_*) = 0$. For example, if $W'(x) = x$, (3.42) becomes, after writing $u, v \to u - iv, u + iv$

$$u^2 + v^2 + x^2 + y^2 = 0. \tag{3.43}$$

This Calabi-Yau manifold is called the *conifold*, and it is singular at the origin. For arbitrary polynomials $W(x)$, the equation (3.42) describes more general, singular Calabi-Yau manifolds. Notice that locally, around the singular points $u = v = y = 0$, $x = x_*$, the geometry described by (3.42) looks like a conifold (whenever $W''(x_*) = 0$). The manifold described by (3.40) is obtained after blowing up the singularities in (3.42), *i.e.* we modify the geometry by "inflating" a two-sphere $\mathbb{P}^1$ at each singularity. This process is called resolution of singularities in algebraic geometry, and for this reason we will call the manifold specified by (3.40) the *resolved manifold* $X_{\text{res}}$.

We can now consider the dynamics of open type B topological strings on $X_{\text{res}}$. We will consider a situation in which we have in total $N$ D-branes in such a way that $N_i$ D-branes are wrapped around the $i$-th $\mathbb{P}^1$, with $i = 1, \cdots, n$. As before, we have three fields in the adjoint representation of $U(N)$, $\Phi_0$, $\Phi_1$ and the gauge connection $A$. The action describing the dynamics of the D-branes turns out to be given by

$$S = \frac{1}{g_s} \int_{\mathbb{P}^1} \text{Tr}\left( \Phi_1 \overline{D}_A \Phi_0 + \omega W(\Phi_0) \right) \tag{3.44}$$

where $\omega$ is a Kahler form on $\mathbb{P}^1$ with unit volume. This action was derived in [43, 27]. A quick way to see that the modification of the gluing rules due to adding the polynomial $W'(\Phi_0)$ leads to the extra term in (3.44) is to use standard techniques in CFT [27]. The fields $\Phi_0, \Phi_1$ are canonically conjugate and on the conformal plane they satisfy the OPE

$$\Phi_0(z)\Phi_1(w) \sim \frac{g_s}{z - w}. \tag{3.45}$$

Let us now regard the geometry described in (3.40) as two disks (or conformal planes) glued through a cylinder. Since we are in the cylinder, we can absorb the factors of $z$ in the last equation of (3.40). The operator that implements the transformation of $\Phi$ is

$$U = \exp \frac{1}{g_s} \oint \text{Tr}\, W(\Phi_0(z))\, dz, \tag{3.46}$$

since from (3.45) it is easy to obtain

$$\Phi_1' = U \Phi_1 U^{-1}. \tag{3.47}$$

We can also write

$$U = \exp \frac{1}{g_s} \int_{\mathbb{P}^1} \text{Tr}\, W(\Phi_0(z)) \omega \tag{3.48}$$

where $\omega$ is localized to a band around the equator of $\mathbb{P}^1$ (as we will see immediately, the details of $\omega$ are unimportant, as long as it integrates to 1 on the two-sphere).

One easy check of the above action is that the equations of motion lead to the geometric picture of D-branes wrapping $n$ holomorphic $\mathbb{P}^1$s in the geometry. The gauge connection is just a Lagrange multiplier enforcing the condition

$$[\Phi_0, \Phi_1] = 0, \tag{3.49}$$

therefore we can diagonalize $\Phi_0$ and $\Phi_1$ simultaneously. The equation of motion for $\Phi_0$ is simply

$$\overline{\partial} \Phi_0 = 0, \tag{3.50}$$

and since we are on $\mathbb{P}^1$, we have that $\Phi_0$ is a constant, diagonal matrix. Finally, the equation of motion for $\Phi_1$ is

$$\overline{\partial}\Phi_1 = W'(\Phi_0)\omega, \tag{3.51}$$

and for nonsingular $\Phi_1$ configurations both sides of the equation must vanish simultaneously, as we can see by integrating both sides of the equation over $\mathbb{P}^1$. Therefore, $\Phi_1 = 0$ and the constant eigenvalues of $\Phi_0$ satisfy

$$W'(\Phi_0) = 0 \tag{3.52}$$

*i.e.* they must be located at the critical points of $W(x)$. In general, we will have $N_i$ eigenvalues of $\Phi_0$ at the $i$-th critical point, $i = 1, \cdots, n$, and this is precisely the D-brane configuration we are considering.

What happens in the quantum theory? In order to analyze it, we will use the approach developed in [10] for the analysis of two-dimensional gauge theories[1]. First of all, we choose the maximally Abelian gauge for $\Phi_0$, *i.e.* we write

$$\Phi_0 = \Phi_0^{\mathbf{k}} + \Phi_0^{\mathbf{t}}, \tag{3.53}$$

where $\Phi_0^{\mathbf{t}}$ is the projection on the Cartan subalgebra $\mathbf{t}$, and $\Phi_0^{\mathbf{k}}$ is the projection on the complementary part $\mathbf{k}$. The maximally Abelian gauge is defined by the condition

$$\Phi_0^{\mathbf{k}} = 0 \tag{3.54}$$

which means that the nondiagonal entries of $\Phi_0$ are gauge-fixed to be zero. This is in fact the same gauge that we used before to write the matrix model in the eigenvalue basis. After fixing the gauge the usual Faddeev-Popov techniques lead to a ghost functional determinant given by

$$\frac{1}{N!}\mathrm{Det}_{\mathbf{k}}(\mathrm{ad}(\Phi_0^{\mathbf{t}}))_{\Omega^0(\mathbb{P}^1)} \tag{3.55}$$

where the subscript $\mathbf{k}$ means that the operator $\Phi_0^{\mathbf{t}}$ acts on the space $\mathbf{k}$, and the normalization factor $1/N!$ is the inverse of the order of the residual symmetry group, namely the Weyl group which permutes the $N$ entries of $\Phi_0^t$. The integrand of (3.44) reads, after gauge fixing,

$$\mathrm{Tr}\left(\Phi_1^{\mathbf{t}}\overline{\partial}\Phi_0^{\mathbf{t}} + W(\Phi_0^{\mathbf{t}})\right) + 2\sum_{\alpha} A^{\alpha}\Phi_1^{-\alpha}\alpha(\Phi_0^{\mathbf{t}}), \tag{3.56}$$

where $\alpha$ are roots, $E_\alpha$ is a basis of $\mathbf{k}$, and we have expanded $\Phi_1^{\mathbf{k}} = \sum_\alpha \Phi_1^\alpha E_\alpha$ as well as $A^{\mathbf{k}}$. We can now integrate out the $A^\alpha$ to obtain

$$\frac{1}{\mathrm{Det}_{\mathbf{k}}(\mathrm{ad}(\Phi_0^{\mathbf{t}}))_{\Omega^{1,0}(\mathbb{P}^1)}}\prod_{\alpha>0}\delta(\Phi_1^\alpha). \tag{3.57}$$

Here we have used the functional generalization of the standard formula $\delta(ax) = |a|^{-1}\delta(x)$. We can now trivially integrate over $\Phi_1^{\mathbf{k}}$. The inverse determinant in (3.57) combines with (3.55) to produce

$$\frac{\mathrm{Det}_{\mathbf{k}}(\mathrm{ad}(\Phi_0^{\mathbf{t}}))_{H^0(\mathbb{P}^1)}}{\mathrm{Det}_{\mathbf{k}}(\mathrm{ad}(\Phi_0^{\mathbf{t}}))_{H^{1,0}(\mathbb{P}^1)}} \tag{3.58}$$

where (as usual) nonzero modes cancel (since they are paired by $\partial$) and one ends with the determinants evaluated at the cohomologies. Similarly, integrating out $\Phi_1^{\mathbf{t}}$ in (3.56) leads to $\overline{\partial}\Phi_0^{\mathbf{t}} = 0$, therefore $\Phi_0^{\mathbf{t}}$ must be constant. The quotient of determinants is easy to evaluate in this case, and one finds

$$\left[\prod_{i<j}(\lambda_i - \lambda_j)^2\right]^{h^0(\mathbb{P}^1)-h^{1,0}(\mathbb{P}^1)}, \tag{3.59}$$

where $\lambda_i$ are the constant eigenvalues of $\Phi_0^{\mathbf{t}}$. Since $h^0(\mathbb{P}^1) = 1$, $h^{1,0}(\mathbb{P}^1) = 0$, we just get the square of the Vandermonde determinant and the partition function reads:

$$Z = \frac{1}{N!}\int \prod_{i=1}^{N} d\lambda_i \prod_{i<j}(\lambda_i - \lambda_j)^2\, e^{-\frac{1}{g_s}\sum_{i=1}^{N} W(\lambda_i)}. \tag{3.60}$$

In principle, as explained in [10], one has to include a sum over nontrivial topological sectors of the Abelian gauge field $A^{\mathbf{t}}$ in order to implement the gauge fixing (3.54) correctly. Fortunately, in this case the gauge-fixed action does not depend on $A^{\mathbf{t}}$, and the inclusion of topological sectors is irrelevant. The expression (3.60) is (up to a factor $(2\pi)^N$) the gauge-fixed version of the matrix model

$$Z = \frac{1}{\mathrm{vol}(U(N))}\int \mathcal{D}\Phi\, e^{-\frac{1}{g_s}\mathrm{Tr}\, W(\Phi)} \tag{3.61}$$

We have then derived a surprising result due to Dijkgraaf and Vafa [27]: the string field theory action for open topological B strings on the Calabi-Yau manifold described by (3.40) is a matrix model with potential $W(\Phi)$.

## 3.4 Open string amplitudes and multicut solutions

The total free energy $F(N_i, g_s)$ of topological B strings on the Calabi-Yau (3.40) in the background of $N = \sum_i N_i$ branes wrapped around $n$ $\mathbb{P}^1$'s is of the form (1.3), and as we have just seen it is given by the free energy of the matrix model (3.61). In particular, the coefficients $F_{g,h_1,\cdots,h_n}$ can be computed perturbatively in the matrix model. We have to be careful however to specify the classical vacua around which we are doing perturbation theory. Remember from the analysis of the matrix model that the classical solution which describes the brane configuration is characterized by having $N_i$ eigenvalues of the matrix located at the $i$-th critical point of the potential $W(x)$. In the saddle-point approximation, this means that we have to consider a *multicut* solution, with eigenvalues "condensed" around *all* the extrema of the potential. Therefore, in contrast to the multicut solution discussed in 2.2, we have that (1) all critical points of $W(x)$ have to be considered, and not only the minima, and (2) the number of eigenvalues in each cut is not determined dynamically as in (2.83), but it is rather fixed to be $N_i$ in the $i$-th cut. In other words, the integral of the density of eigenvalues $\rho(\lambda)$ along each cut equals a *fixed* filling fraction $\nu_i = N_i/N$:

$$\int_{x_{2i}}^{x_{2i-1}} d\lambda\, \rho(\lambda) = \nu_i, \tag{3.62}$$

where $N = \sum_{i=1}^{n} N_i$ is the total number of eigenvalues. Let us introduce the partial 't Hooft couplings

$$t_i = g_s N_i = t\nu_i. \tag{3.63}$$

Taking into account (2.58) and (2.68), we can write (3.62) as

$$t_i = \frac{1}{4\pi i}\oint_{A_i} y(\lambda) d\lambda, \quad i = 1, \cdots, n, \tag{3.64}$$

where $A_i$ is the closed cycle of the hyperelliptic curve (2.70) which surrounds the cut $C_i$. Assuming for simplicity that all the $t_i$ are different from zero, and taking into account that $\sum_i t_i = t$, we see that (3.64) gives $n-1$ independent conditions, where $n$ is the number of critical points of $W(x)$. These conditions, together with (2.81), determine the positions of the endpoints $x_i$ as functions of the $t_i$ and the coupling constants in $W(x)$. It is clear that the solution obtained in this way is not an equilibrium solution of the matrix model, since cuts can be centered around local maxima and different cuts will have different values of the effective potential. This is not surprising, since we are not considering the matrix model as a quantum mechanical system *per se*, but as an effective description of the original brane system. The different choices of filling fractions correspond to different choices of classical vacua for the brane system.

A subtle issue concerning the above matrix model is the following. The matrix field $\Phi$ in (3.61) comes from the B model field $\Phi_0$, which is a holomorphic field. Therefore, the matrix integral (3.60) should be understood as a contour integral, and in order to define the theory a choice of contour should be made. This can be done in perturbation theory, by choosing for example a contour that leads to the usual results for Gaussian integration, and therefore at this level the matrix model is not different from the usual Hermitian matrix model [27, 75]. In some cases, however, regarding (3.61) as a holomorphic matrix model can be clarifying, see [51] for an exhaustive discussion.

The above description of the multicut solution refers to the saddle-point approximation. What is the meaning of the multicut solutions from the point of view of perturbation theory? To address this issue, let us consider for simplicity the case of the cubic potential:

$$\frac{1}{g_s}W(\Phi) = \frac{1}{2g_s}\mathrm{Tr}\,\Phi^2 + \frac{1}{3}\frac{\beta}{g_s}\mathrm{Tr}\,\Phi^3. \tag{3.65}$$

This potential has two critical points, $a_1 = 0$ and $a_2 = -1/\beta$. The most general multicut solution will have two cuts. There will $N_1$ eigenvalues sitting at $\Phi = 0$, and $N_2$ eigenvalues sitting at $\Phi = -1/\beta$. The partition function $Z$ of the matrix model is:

$$Z = \frac{1}{N!}\int \prod_{i=1}^{N} \frac{d\lambda_i}{2\pi} \Delta^2(\lambda) e^{-\frac{1}{2g_s}\sum_i \lambda_i^2 - \frac{\beta}{3g_s}\sum_i \lambda_i^3}, \tag{3.66}$$

where $\Delta(\lambda) = \prod_{i<j}(\lambda_i - \lambda_j)$ is the Vandermonde determinant. We can now expand the integrand around the vacuum with $\lambda_i = 0$ for $i = 1, \ldots, N_1$ and $\lambda_i = -\frac{1}{\beta}$ for $i = N_1 + 1, \ldots, N$. Denoting the fluctuations by $\mu_i$ and $\nu_j$, the Vandermonde determinant becomes

$$\Delta^2(\lambda) = \prod_{1\le i_1<i_2\le N_1} (\mu_{i_1} - \mu_{i_2})^2 \prod_{1\le j_1<j_2\le N_2} (\nu_{j_1} - \nu_{j_2})^2 \prod_{\substack{1\le i\le N_1 \\ 1\le j\le N_2}} \left(\mu_i - \nu_j + \frac{1}{\beta}\right)^2. \tag{3.67}$$

We also expand the potential around this vacuum and get

$$W = \sum_{i=1}^{N_1}\left(\frac{1}{2g_s}\mu_i^2 + \frac{\beta}{3g_s}\mu_i^3\right) - \sum_{i=1}^{N_2}\left(\frac{m}{2g_s}\nu_i^2 - \frac{\beta}{3g_s}\nu_i^3\right) + \frac{1}{6\beta^2 g_s}N_2. \tag{3.68}$$

Notice that the propagator of the fluctuations around $-1/\beta$ has the 'wrong' sign, since we are expanding around a local maximum. The interaction between the two sets of eigenvalues, which is given by the last factor in (3.67), can be exponentiated and included in the action. This generates an interaction term between the two eigenvalue bands

$$W_{\mathrm{int}} = 2N_1N_2\log\beta + 2\sum_{k=1}^{\infty}\frac{1}{k}\beta^k \sum_{i,j}\sum_{p=0}^{k}(-1)^p\binom{k}{p}\mu_i^p\nu_j^{k-p}. \tag{3.69}$$

By rewriting the partition function in terms of matrices instead of their eigenvalues, we can represent this model as an effective two-matrix model, involving an $N_1 \times N_1$ matrix $\Phi_1$, and an $N_2 \times N_2$ matrix $\Phi_2$:

$$Z = \frac{1}{\mathrm{Vol}(U(N_1)) \times \mathrm{Vol}(U(N_2))}\int D\Phi_1 D\Phi_2 e^{-W_1(\Phi_1) - W_2(\Phi_2) - W(\Phi_1,\Phi_2)}, \tag{3.70}$$

where

$$
\begin{aligned}
W_1(\Phi_1) &= +\mathrm{Tr}\Big(\frac{1}{2g_s}\Phi_1^2 + \frac{\beta}{3g_s}\Phi_1^3\Big), \\
W_2(\Phi_2) &= -\mathrm{Tr}\Big(\frac{1}{2g_s}\Phi_2^2 - \frac{\beta}{3g_s}\Phi_2^3\Big), \\
W_{\mathrm{int}}(\Phi_1, \Phi_2) &= 2\sum_{k=1}^{\infty}\frac{\beta^k}{k}\sum_{p=0}^{k}(-1)^p\binom{k}{p}\mathrm{Tr}\,\Phi_1^p\,\mathrm{Tr}\,\Phi_2^{k-p} \\
&+ N_2 W(a_2) + N_1 W(a_1) - 2N_1N_2\ln\beta. \qquad (3.71)
\end{aligned}
$$

Here, $\mathrm{Tr}\,\Phi_1^0 = N_1$, $\mathrm{Tr}\,\Phi_2^0 = N_2$, $W(a_1) = 0$ and $W(a_2) = 1/(6g_s\beta^2)$. Although the kinetic term for $\Phi_2$ has the 'wrong' sign, we can still make sense of the model in perturbation theory by using formal Gaussian integration, and this can in fact be justified in the framework of holomorphic matrix models [51]. Therefore, the two-cut solution of the cubic matrix model can be formally represented in terms of an effective two-matrix model. It is now straightforward to compute the free energy $F_{\mathrm{pert}} = \log\big(Z(\beta)/Z(\beta=0)\big)$ in perturbation theory. It can be expanded as

$$
F_{\mathrm{pert}} = -N_1W(a_1) - N_2W(a_2) - 2N_1N_2\ln\beta + \sum_{h=1}^{\infty}\sum_{g\geq 0}(g_s\beta^2)^{2g-2+h}F_{g,h}(N_1,N_2) \qquad (3.72)
$$

where $F_{g,h}$ is a homogeneous polynomial in $N_1$ and $N_2$ of degree $h$. One finds, up to fourth order in the coupling constant $\beta$, the following result [45]:

$$
\begin{aligned}
F_{\mathrm{pert}} &= -N_1W(a_1) - N_2W(a_2) - 2N_1N_2\ln\beta \\
&+ g_s\beta^2\left[\Big(\frac{2}{3}N_1^3 - 5N_1^2N_2 + 5N_1N_2^2 - \frac{2}{3}N_2^3\Big) + \frac{1}{6}(N_1 - N_2)\right] \\
&+ g_s^2\beta^4\left[\Big(\frac{8}{3}N_1^4 - \frac{91}{3}N_1^3N_2 + 59N_1^2N_2^2 - \frac{91}{3}N_1N_2^3 + \frac{8}{3}N_2^4\Big)\right. \\
&\left. + \Big(\frac{7}{3}N_1^2 - \frac{31}{3}N_1N_2 + \frac{7}{3}N_2^2\Big)\right] + \cdots \qquad (3.73)
\end{aligned}
$$

From this explicit perturbative computation one can read off the first few coefficients $F_{g,h_1,h_2}$. Of course, this procedure can be generalized, and the $n$-cut solution can be represented by an effective $n$ matrix model with interactions among the different matrices that come from the expansion of the Vandermonde determinant. These interactions can be also incorporated in terms of ghost fields, as explained in [24]. This makes possible to compute corrections to the saddle-point approximation in perturbation theory. One can also use the multicut solution to the loop equations [4, 47] with minor modifications to compute the genus one correction in closed form [45, 26, 18].

## 3.5 Master field and geometric transition

We have seen that the open topological string amplitudes on the Calabi-Yau manifold $X_{\rm res}$ are computed by a multicut matrix model whose planar solution (or, equivalently, its master field configuration) is given by a hyperelliptic curve

$$y^2 = W'(x)^2 - R(x). \qquad (3.74)$$

Moreover, we also saw in (3.64) that the partial 't Hooft couplings can be understood as integrals around the $A_i$ cycles of this curve, with $i = 1, \cdots, n$. Let us now compute the variation of the free energy $F_0(t_i)$ when we vary $t_i$. The variation w.r.t. $t_i$ (keeping the $t_j$, $j \neq i$, fixed) can be obtained by computing the variation in the free energy as we move one eigenvalue from the cut $C_i$ to infinity [27]. This variation is given by (minus) the integral of the force exerted on an eigenvalue, as we move it from the endpoint of the cut to infinity. The path from the endpoint of $C_i$ to infinity, which does not intersect the other cuts $C_j$, will be denoted by $B_i$. Taking into account (2.69), and the fact that $y(p)$ has no discontinuities outside the cuts $C_j$, we find

$$\frac{\partial F_0}{\partial t_i} = \int_{B_i} y(x)dx. \qquad (3.75)$$

Usually this integral is divergent, but can be easily regularized by taking $B_i$ to run up to a cutoff point $x = \Lambda$, and subtracting the divergent pieces as the cutoff $\Lambda$ goes to infinity. For example, for the Gaussian matrix model one has

$$\frac{\partial F_0}{\partial t} = \int_{2\sqrt{t}}^{\Lambda} dx\, \sqrt{x^2 - 4t} = t(\log t - 1) - 2t\log\Lambda + \frac{1}{2}\Lambda^2 + \mathcal{O}(1/\Lambda^2). \qquad (3.76)$$

Therefore, the regularized integral gives $t(\log t - 1)$, which is indeed the right result. It is now clear that (3.64) and (3.75) look very much like the relations (3.14) that define the periods (therefore the prepotential) in special geometry. What is the interpretation of the appearance of special geometry?

Recall that our starting point was a Calabi-Yau geometry obtained as a blowup of the singularity given in (3.42). However, there is another way of smoothing out singularities in algebraic geometry, which is by *deforming* them rather than by resolving then. For example, the conifold singularity given in (3.43) can be smoothed out by deforming the geometry to

$$x^2 + y^2 + u^2 + v^2 = \mu. \qquad (3.77)$$

This is the so called *deformed conifold.* Geometrically, turning on $\mu$ corresponds to inflating a three-sphere in the geometry, since the real section of the conifold is indeed an $\mathbf{S}^3$. As $\mu \to 0$, the three-sphere collapses to zero size, so we can interpret the singularity as arising from a collapsing three-cycle in the geometry. In the more general singularity (3.42), the generic deformation requires turning on a generic polynomial of degree $n-1$ $R(x)$, and we get the Calabi-Yau manifold

$$u^2 + v^2 + y^2 + W'(x)^2 = R(x). \tag{3.78}$$

We will call this geometry the *deformed manifold* $X_{\rm def}$. The deformation by $R(x)$ introduces in fact $n$ three-spheres in the geometry, one for each singularity (recall that each of the singular points in (3.42) is locally like the conifold). The noncompact Calabi-Yau manifold (3.78) has a holomorphic three-form:

$$\Omega = \frac{1}{2\pi}\frac{dxdydu}{v} \tag{3.79}$$

The three-spheres created by the deformation can be regarded as two-spheres fibered over an interval in the complex $x$-plane. To see this, let us consider for simplicity the case of the deformed conifold (3.77), with $\mu$ real. This geometry contains a three-sphere which is given by the restriction of (3.77) to real values of the variables. If we now consider a fixed, real value of $x$ in the interval $-\sqrt{\mu} < x < \sqrt{\mu}$, we get of course a two-sphere of radius $\sqrt{\mu - x^2}$. The sphere collapses at the endpoints of the interval, $x = \pm\sqrt{\mu}$, and the total geometry of the two-sphere together with the interval $[-\sqrt{\mu}, \sqrt{\mu}]$ is a three-sphere. In the more general case, the curve $W'(x)^2 - R(x)$ has $n$ cuts with endpoints $x_{2i}, x_{2i-1}$, $i = 1, \cdots, n$, and the $n$ three-spheres are $\mathbf{S}^2$ fibrations over these cuts.

Let us now consider *closed* type B topological strings propagating on $X_{\rm def}$. As we saw in 3.1, the genus zero theory is determined by the periods of the three-form $\Omega$ given in (3.79). We then choose a symplectic basis of three-cycles $\widehat{A}_i$, $\widehat{B}^j$, with $\widehat{A}_i \cap \widehat{B}^j = \delta_i^j$. Here, the $\widehat{A}_i$ cycles are the $n$ three-spheres, and they project to cycles $A_i$ surrounding the cut $C_i = [x_{2i}, x_{2i-1}]$ in the $x$-plane. The $\widehat{B}_i$ cycles are dual cycles which project in the $x$ plane to the $B_i$ paths [15]. The periods of $\Omega$ are then given by

$$t_i = \frac{1}{4\pi}\oint_{\widehat{A}_i}\Omega, \quad \frac{\partial F_0}{\partial t_i} = \int_{\widehat{B}^i}\Omega. \tag{3.80}$$

It is easy to see that these periods reduce to the periods (3.63) and (3.75) on the hyperelliptic curve (3.74), respectively. Let us consider again the case of the deformed conifold (3.77), which is simpler since there is only one three-sphere. Let us compute the $A$-period over this three-sphere, which is an $\mathbf{S}^2$ fibration over the cut $[-\sqrt{\mu}, \sqrt{\mu}]$, by first doing the integral over $\mathbf{S}^2$, and then doing the integral over the cut. Since $v = \sqrt{\mu - x^2 - \rho^2}$, where $\rho^2 = y^2 + u^2$, the integral of $\Omega$ over $\mathbf{S}^2$ is simply

$$\frac{1}{2\pi}\int_{\mathbf{S}^2}\frac{dydz}{\sqrt{\mu - x^2 - \rho^2}} = \sqrt{\mu - x^2}. \tag{3.81}$$

Therefore, the $A$-period becomes

$$t = \frac{1}{2\pi} \int_{-\sqrt{\mu}}^{\sqrt{\mu}} y(x) dx, \tag{3.82}$$

where $y$ is now given by $y^2 + x^2 = \mu$. This is nothing but the A-period (3.63) (up to a redefinition $y \to -iy$). The general case is very similar, and one finally obtains that the special geometry (3.80) of the deformed Calabi-Yau geometry (3.78) is equivalent to the planar solution of the matrix model, given by the hyperelliptic curve (3.74) and the equations for the partial 't Hooft couplings (3.64) and the planar free energy (3.75).

The physical interpretation of this result is that there is an equivalence between an *open* topological string theory on the manifold $X_{\rm res}$, with $N$ D-branes wrapping the $n$ spheres obtained by blowup, and a *closed* topological string theory on the manifold $X_{\rm def}$, where the $N$ D-branes have disappeared. Moreover, the 't Hooft couplings $t_i$ in the open string theory become geometric periods in the closed string theory. Since the open topological strings on $X_{\rm res}$ are described by a matrix model, the fact that the planar solution reproduces very precisely the deformed geometry is important evidence for this interpretation. This duality relating an open and a closed string theory is an example of a *geometric*, or *large N*, transition. Notice that, as a consequence of this duality, the 't Hooft resummation of the matrix model corresponds to a closed string theory propagating on $X_{\rm def}$. The master field controlling the planar limit (which is encoded in the planar resolvent, or equivalently in the quantity $y(\lambda)$) leads to an algebraic equation that describes very precisely the *target* of the closed string theory dual. The large $N$ transition between these two geometries was proposed in [15]. The fact that the open string side can be described by a matrix model was discovered in [27].

## 3.6 Extensions and applications

The results derived above can be extended to more complicated Calabi-Yau backgrounds with branes [28, 29]. For example, one can consider ADE type geometries with branes wrapping two-spheres [16, 14], and the string field theory description reduces to the ADE matrix models considered in [46]. In the one-matrix model described before, the master field is given by a hyperelliptic curve $F(x, y) = 0$ which is then regarded as the Calabi-Yau manifold

$$uv + F(x, y) = 0 \tag{3.83}$$

in disguise. In some of the examples considered in [28, 29], however, the master field is no longer described by a hyperelliptic curve, but involves a more complicated geometry. This geometry is the Calabi-Yau closed string background that is obtained by geometric transition from the open string background with branes. A detailed study of the more complicated master field geometries that arise in multimatrix models can be found in [33].

Another consequence of the result of Dijkgraaf and Vafa, together with the geometric transition of [15], is that the Kodaira-Spencer theory of gravity [7] on the noncompact Calabi-Yau manifold (3.78) is equivalent to the 't Hooft resummation of the matrix model with potential $W(x)$. For the simple example of the cubic potential, this was explicitly checked at genus one in [45]. The formalism developed in [32] seems to be very appropriate to establish this equivalence in detail.

As we mentioned in the introduction, the main application of the results of Dijkgraaf and Vafa has been the computation of effective superpotentials in supersymmetric gauge theories by using matrix model techniques. This is based on the fact [7, 29] that the resummation $F_0(t)$ of the open string amplitudes is deeply related to the superpotential of the gauge theory which can be obtained from string backgrounds with branes. We refer the reader to [6, 62] for an exposition of these results.

# 4. Type A topological strings, Chern-Simons theory and matrix models

The conceptual structure of what we have seen in the B model is the following: first one shows, by using string field theory, that the target space description of open topological B strings reduces to a matrix model in certain backgrounds. Then one solves the model in the planar limit, and a geometry emerges which is interpreted as a closed string dual to the original open string theory. Both geometries are related by a large $N$ transition. The first transition of this type was discovered in the context of topological A strings by Gopakumar and Vafa [35]. What we will do here is to rederive their result by using the language and technology of matrix models. The key ingredient is the fact pointed out in [53] that the partition function of Chern-Simons theory can be written in terms of a somewhat exotic matrix model. We will only focus on the matrix model aspects of this correspondence. A detailed review of Chern-Simons theory and the geometric transition for the A model can be found in [54].

## 4.1 Solving the Chern-Simons matrix model

The Chern-Simons action with gauge group $G$ on a generic three-manifold $M$ is defined by

$$S = \frac{k}{4\pi}\int_M \mathrm{Tr}\Big(A\wedge dA + \frac{2}{3}A\wedge A\wedge A\Big) \tag{4.1}$$

Here, $k$ is the coupling constant, and $A$ is a $G$-gauge connection on the trivial bundle over $M$. We will consider Chern-Simons theory with gauge group $G = U(N)$. As noticed in [71], since the action (4.1) does not involve the metric, the resulting quantum theory is topological, at least formally. In particular, the partition function

$$Z(M) = \int [\mathcal{D}A] e^{iS} \tag{4.2}$$

should define a topological invariant of $M$. A detailed analysis shows that this is in fact the case, with an extra subtlety related to a choice of framing of the three-manifold.

The partition function of Chern-Simons theory can be computed in a variety of ways. In [71] it was shown that in fact the theory is exactly solvable by using nonperturbative methods and the relation to the Wess-Zumino-Witten (WZW) model. In particular, the partition function of the $U(N)$ theory on the three-sphere $\mathbf{S}^3$ is given by

$$Z(\mathbf{S}^3) = \frac{1}{(k+N)^{N/2}} \sum_{w \in \mathcal{W}} \epsilon(w) \exp\Big( -\frac{2\pi i}{k+N} \rho \cdot w(\rho) \Big), \tag{4.3}$$

where the sum over $w$ is a sum over the elements of the Weyl group $\mathcal{W}$ of $U(N)$, $\epsilon(w)$ is the signature of $w$, and $\rho$ is the Weyl vector of $SU(N)$. By using Weyl's denominator formula,

$$\sum_{w \in \mathcal{W}} \epsilon(w) e^{w(\rho) \cdot u} = \prod_{\alpha > 0} 2 \sinh \frac{\alpha \cdot u}{2}, \tag{4.4}$$

where $\alpha$ are positive roots, one finds

$$Z(\mathbf{S}^3) = \frac{1}{(k+N)^{N/2}} \prod_{\alpha > 0} 2 \sinh\Big( \frac{(\alpha \cdot \rho)}{2} g_s \Big) \tag{4.5}$$

where

$$g_s = \frac{2\pi i}{k+N}. \tag{4.6}$$

It was found by Witten that open topological type A strings on $T^*\mathbf{S}^3$ (which is nothing but the deformed conifold geometry (3.77)) in the presence of $N$ D-branes wrapping $\mathbf{S}^3$ are in fact described by $U(N)$ Chern-Simons theory on $\mathbf{S}^3$ [74]. This is the type A model analog to the fact that open type B strings on the geometry described by (3.40) are captured by a matrix model, and in both cases this is shown by using open string field theory. The free energy of Chern-Simons theory on $\mathbf{S}^3$ has an expansion of the form (1.2), with $g_s$ given in (4.6), and the coefficients $F_{g,h}$, which can be computed by standard perturbation theory, have the interpretation of open string amplitudes on $T^*\mathbf{S}^3$.

The analogy between the A story and the B story can be taken even further, since it turns out that the partition function of Chern-Simons on $\mathbf{S}^3$, as well as on many other three-manifolds, can be represented as a matrix integral [53]. In the case of $\mathbf{S}^3$ most of the physical information in $Z(\mathbf{S}^3)$ can be obtained by other means, but for other three-manifolds like lens spaces and Seifert spaces, the matrix model representation is crucial in order to extract the coefficients $F_{g,h}$ [53]. The Chern-Simons matrix model on $\mathbf{S}^3$ gives however a particularly clean way to derive the resummed free energies $F_g(t)$ and the geometry of the master field, and we will devote the rest of these lectures to presenting this analysis.

In the case of $\mathbf{S}^3$ the easiest way to derive the matrix model representation of the Chern-Simons partition function is through direct computation. Consider the following integral:

$$Z_{CS} = \frac{e^{-\frac{g_s}{12}N(N^2-1)}}{N!} \int \prod_{i=1}^{N} \frac{d\beta_i}{2\pi}\, e^{-\sum_i \beta_i^2/2g_s} \prod_{i<j} \left(2\sinh\frac{\beta_i - \beta_j}{2}\right)^2. \quad (4.7)$$

It can easily be seen that this reproduces the partition function of $U(N)$ Chern-Simons theory on $\mathbf{S}^3$, given in (4.5), and the derivation is left as an exercise.

**Exercise**. Use the Weyl formula (4.4) to write (4.7) as a Gaussian integral, and show that it reproduces (4.3).

The measure factor in (4.7)

$$\prod_{i<j} \left(2\sinh\frac{\beta_i - \beta_j}{2}\right)^2 \quad (4.8)$$

is not the standard Vandermonde determinant, although it reduces to it for small separations among the eigenvalues. In fact, for very small $g_s$, the Gaussian potential in (4.7) will be very narrow, forcing the eigenvalues to be close to each other, and one can expand the sinh in (4.8) in power series. At leading order we find the usual Gaussian matrix model, while the corrections to it can be evaluated systematically by computing correlators in the Gaussian theory. In this way one obtains the perturbative expansion of Chern-Simons theory, see [53] for details.

Here we will take a slightly different route in order to analyze the model. First of all, we want to write the above integral as a standard matrix integral with the usual Vandermonde discriminant. This can be achieved with the change of variables [67]

$$\exp(\beta_i + t) = \lambda_i, \quad (4.9)$$

where $t = Ng_s$, as usual. It is easy to see that the above integral becomes, up to a factor $\exp(-N^3 g_s/2)$,

$$Z_{SW} = \frac{1}{N!}\int \prod_{i=1}^{N} \frac{d\lambda_i}{2\pi}\, \Delta^2(\lambda)\, \exp\Big(-\sum_{i=1}^{N}(\log\lambda_i)^2/2g_s\Big), \qquad (4.10)$$

therefore we are considering the matrix model

$$Z_{SW} = \frac{1}{\mathrm{vol}(U(N))}\int dM\, e^{-\frac{1}{2g_s}\mathrm{Tr}\,(\log M)^2}. \qquad (4.11)$$

We will call this model the *Stieltjes-Wigert matrix model*, hence the subscript in (4.10) and (4.11). This is because it can be exactly solved with the so-called Stieltjes-Wigert polynomials, as we will explain in a moment.

Matrix integrals with logarithmic potentials are somewhat exotic, but have appeared before in connection with the Penner model [61], with the $c = 1$ string at the self-dual radius [25, 41], and with the $\mathbb{P}^1$ model [31]. We want to analyze now the saddle-point approximation to the matrix integral (4.7), or equivalently to (4.10). Since the model in (4.10) has the standard Vandermonde, we can use the techniques of section 2.2. Although the formulae there were obtained for a polynomial potential, some of them generalize to arbitrary polynomials. In particular, to obtain the resolvent $\omega_0(p)$ we can use the formula (2.60) with

$$W'(z) = \frac{\log z}{z}. \qquad (4.12)$$

Notice that this potential has a minimum at $z = 1$. We then expect a one-

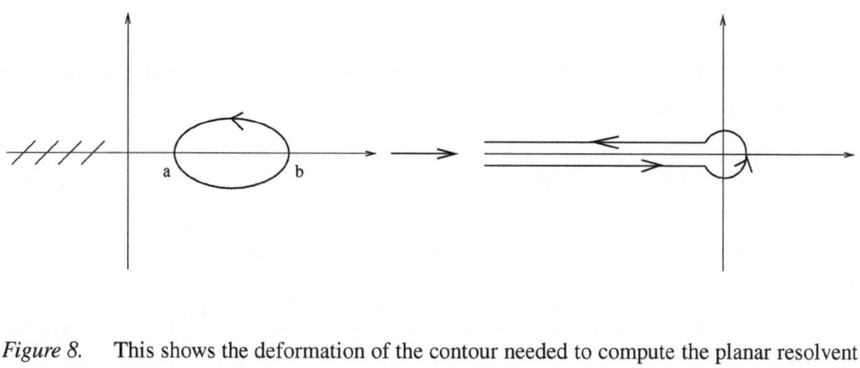


*Figure 8.* This shows the deformation of the contour needed to compute the planar resolvent of the Chern-Simons matrix integral. We pick a residue at $z = p$, and we have to encircle the singularity at the origin as well as the branch cut of the logarithm, which on the left hand side is represented by the dashed lines.

cut solution where the endpoints of the interval $a(t)$, $b(t)$ will satisfy $a(0) = b(0) = 1$. In order to compute the integral (2.60) we deform the integration contour. In the case of polynomial potentials, we picked a residue at $z = p$ and at infinity. Here, since the logarithm has a branch cut, we cannot push the contour to infinity. Instead, we deform the contour as indicated in Fig. 8: we pick the pole at $z = p$, and then we surround the cut of the logarithm along the negative real axis and the singularity at $z = 0$ with a small circle $C_\epsilon$ of radius $\epsilon$. This kind of situation is typical of the solution of matrix models with the character expansion [44]. The resulting integrals are:

$$\frac{1}{2t}\left\{-\int_{-\infty}^{-\epsilon}\frac{dz}{z(z-p)\sqrt{(z-a)(z-b)}}+\oint_{C_\epsilon}\frac{dz\,\log z}{z(z-p)\sqrt{(z-a)(z-b)}}\right\}. \tag{4.13}$$

Both are singular as $\epsilon \to 0$, but singularities cancel, and after some computations one finds for the resolvent:

$$\omega_0(p) = -\frac{1}{2tp}\log\left[\frac{(\sqrt{a}\sqrt{p-b}-\sqrt{b}\sqrt{p-a})^2}{(\sqrt{p-a}-\sqrt{p-b})^2p^2}\right]+\frac{\sqrt{(p-a)(p-b)}}{2tp\sqrt{ab}}\log\left[\frac{4ab}{2\sqrt{ab}+a+b}\right]. \tag{4.14}$$

In order to satisfy the asymptotics (2.57) the second term must vanish, and the first one must go like $1/p$. This implies

$$\begin{aligned} 4ab &= 2\sqrt{ab}+a+b, \\ \sqrt{a}+\sqrt{b} &= 2e^t, \end{aligned} \tag{4.15}$$

and from here we obtain the positions of the endpoints of the cut $a, b$ as a function of the 't Hooft parameter:

$$\begin{aligned} a(t) &= 2e^{2t} - e^t + 2e^{\frac{3t}{2}}\sqrt{e^t-1}, \\ b(t) &= 2e^{2t} - e^t - 2e^{\frac{3t}{2}}\sqrt{e^t-1}. \end{aligned} \tag{4.16}$$

Notice that, for $t = 0$, $a(0) = b(0) = 1$, as expected. The final expression for the resolvent is then:

$$\omega_0(p) = -\frac{1}{tp}\log\left[\frac{1+e^{-t}p+\sqrt{(1+e^{-t}p)^2-4p}}{2p}\right], \tag{4.17}$$

and from here we can easily find the density of eigenvalues

$$\rho(\lambda) = \frac{1}{\pi t\lambda}\tan^{-1}\left[\frac{1+e^{-t}\lambda+\sqrt{(1+e^{-t}\lambda)^2-4\lambda}}{2\lambda}\right]. \tag{4.18}$$

If we now define

$$u(p) = t(1-p\omega_0(p)) + \pi i \tag{4.19}$$

we see that it solves the equation

$$e^u + e^v + e^{v-u+t} + 1 = 0 \tag{4.20}$$

where we put $p = e^{t-v}$. This was found in [2] by a similar analysis. The equation (4.20) is the analog of (3.74) in the case of polynomial matrix models, and can be regarded as an algebraic equation describing a noncompact Riemann surface. In fact, (4.20) is nothing but the *mirror* of the resolved conifold geometry (see for example [40, 1]), and $t$ is the Kahler parameter of the geometry. This is of course in agreement with the result of [35], who argued that the 't Hooft resummation of Chern-Simons theory leads to a closed string theory propagating on the resolved conifold. As in the B model that we analyzed before, the master field of the matrix model encodes the information about the target geometry of the closed string description, and provides evidence for the geometric transition relating $T^*\mathbf{S}^3$ and the resolved conifold geometry.

As we mentioned before, the matrix model (4.11) can be solved exactly with a set of orthogonal polynomials called the Stieltjes-Wigert polynomials. The fact that the Chern-Simons matrix model is essentially equivalent to the Stieltjes-Wigert matrix model was pointed out by Tierz in [67]. The Stieltjes-Wigert polynomials are defined as follows [63]:

$$p_n(x) = (-1)^n q^{n^2+\frac{n}{2}} \sum_{\nu=0}^{n} \begin{bmatrix} n \\ \nu \end{bmatrix} q^{\frac{\nu(\nu-n)}{2}-\nu^2} (-q^{-\frac{1}{2}}x)^\nu \tag{4.21}$$

and satisfy the orthogonality condition (2.87) with

$$d\mu(x) = e^{-\frac{1}{2g_s}(\log x)^2} \frac{dx}{2\pi} \tag{4.22}$$

and

$$h_n = q^{\frac{3}{4}n(n+1)+\frac{1}{2}} [n]! \left(\frac{g_s}{2\pi}\right)^{\frac{1}{2}},$$

where

$$q = e^{g_s}. \tag{4.23}$$

In the above equations,

$$[n] = q^{\frac{n}{2}} - q^{-\frac{n}{2}}, \quad \begin{bmatrix} n \\ m \end{bmatrix} = \frac{[n]!}{[m]![n-m]!}. \tag{4.24}$$

The recursion coefficients appearing in (2.92) are in this case

$$r_n = q^{3n}(q^n - 1), \qquad s_n = -q^{\frac{1}{2}+n}(q^{n+1} + q^n - 1).$$

The Stieltjes-Wigert ensemble can be regarded as a $q$-deformation (in the sense of quantum group theory) of the usual Gaussian ensemble. For example, as $g_s \to 0$ one has that $[n] \to n g_s$, therefore

$$h_n \to h_n^G, \tag{4.25}$$

where $h_n^G$ is given in (2.96). Also, one can easily check that the normalized vev of $\mathrm{Tr}_R\, M$ in this ensemble is given by

$$\langle \mathrm{Tr}_R\, M \rangle_{SW} = e^{\frac{3t\ell(R)}{2}} q^{\frac{\kappa_R}{2}} \dim_q R, \tag{4.26}$$

where $\ell(R)$ is the number of boxes of $R$, $\kappa_R$ is a quantity defined by

$$\kappa_R = \ell(R) + \sum_i \lambda_i(\lambda_i - 2i) \tag{4.27}$$

in terms of lengths of rows $\lambda_i$ in $R$, and $\dim_q R$ is the quantum dimension of the representation $R$

$$\dim_q R = \prod_{\alpha>0} \frac{[\alpha \cdot (\Lambda + \rho)]}{[\alpha \cdot \rho]} \tag{4.28}$$

where $\Lambda$ is the highest weight associated to $R$. As $g_s \to 0$, the vev (4.26) becomes just $\dim R$, the classical dimension of $R$, which is essentially the vev in the Gaussian ensemble (2.8).

Notice that, for this set of orthogonal polynomials, the expansion (2.107) is very simple since

$$\begin{aligned} R_0(\xi) &= e^{4t\xi}(1 - e^{-t\xi}), \quad R_{2s}(\xi) = 0, \quad s > 0, \\ s(\xi) &= e^{t\xi}(1 - 2e^{t\xi}). \end{aligned} \tag{4.29}$$

As we pointed out in section 2.3, $R_0(\xi)$ and $s(\xi)$ can be used to determine the endpoints of the cut in the resolvent through (2.111). It is easy to see that (4.29) indeed lead to (4.15), and that by using (2.110) one obtains (4.18). In fact, it is well-known that the expression (4.18) is the density of zeroes of the Stieltjes-Wigert polynomials [48, 19].

We can now use the technology developed in section 2.3 to compute $\mathcal{F}_g(t)$. Since

$$F_{CS} = F_{SW} - \frac{7}{12}t^3 + \frac{1}{12}t, \tag{4.30}$$

the formula (2.108) gives

$$F_0^{CS}(t) = \frac{t^3}{12} - \frac{\pi^2 t}{6} - \mathrm{Li}_3(e^{-t}) + \zeta(3), \tag{4.31}$$

where the polylogarithm of index $j$ is defined by:

$$\mathrm{Li}_j(x) = \sum_{n=1}^{\infty} \frac{x^n}{n^j}. \tag{4.32}$$

The above result is in precise agreement with the result in [35] obtained by resumming the perturbative series. With some extra work we can also compute

$F_g^{CS}(t)$, for all $g > 0$, starting from (2.112). We just have to compute $f^{(p)}(1) - f^{(p)}(0)$, for $p$ odd, where

$$f(\xi) = (1-\xi)\phi(\xi,t), \quad \phi(\xi,t) = \log \frac{1-e^{-t\xi}}{\xi} + 4t\xi.$$

It is easy to see that

$$\phi^{(p)}(\xi,t) = (-1)^{p+1}\Big\{ \mathrm{Li}_{1-p}(e^{-t\xi})t^p - \frac{(p-1)!}{\xi^p} \Big\},$$

and by using the expansion

$$\frac{1}{1-e^{-t}} = \frac{1}{t} + \sum_{k=0}^{\infty} (-1)^{k+1} B_{k+1} \frac{t^k}{(k+1)!}$$

one gets

$$\phi^{(p)}(0,t) = \frac{(-1)^p B_p}{p} t^p.$$

Putting everything together, we find for $g > 1$

$$\mathcal{F}_g(t) = \frac{B_{2g}B_{2g-2}}{2g(2g-2)(2g-2)!} + \frac{B_{2g}}{2g(2g-2)!}\mathrm{Li}_{3-2g}(e^{-t}) - \frac{B_{2g}}{2g(2g-2)}t^{2-2g}.$$

Since the last piece is the free energy at genus $g$ of the Gaussian model, we conclude that the Chern-Simons free energy at genus $g$ is given by

$$F_g^{CS}(t) = \frac{B_{2g}B_{2g-2}}{2g(2g-2)(2g-2)!} + \frac{B_{2g}}{2g(2g-2)!}\mathrm{Li}_{3-2g}(e^{-t}) \tag{4.33}$$

which agrees with the resummation of [35] and also with the genus $g$ closed string amplitude of type A topological strings on the resolved conifold (see [54] for more details).

## 4.2 Extensions

We have seen that the matrix model reformulation of Chern-Simons theory provides an efficient way to obtain the master field geometry and to resum the perturbative expansion. The result (4.33) can be derived as well from the perturbation series [35, 34], but the existence of a matrix model description of Chern-Simons theory turns out to be useful in other situations as well. For example, one can easily write a matrix integral for Chern-Simons theory for other gauge groups [53], and the corresponding models have been analyzed in [37]. Moreover, the matrix representation of Chern-Simons partition functions can be extended to lens spaces and Seifert spaces, and provides a useful way to study perturbative expansions around nontrivial flat connections. The matrix models that describe these expansions have been studied in perturbation theory in [53, 2] and the saddle-point approximation to lens space matrix models has been studied in [38]. There are as well multimatrix models describing A topological strings on some noncompact Calabi-Yau geometries [2] that can be studied by using saddle-point techniques [76], and it is possible as well to formulate the Chern-Simons partition function on $\mathbf{S}^3$ in terms of a unitary model [58]. However, all these matrix models are usually much harder to analyze than conventional ones, and more work is needed to understand their large $N$ properties.

## Acknowledgements

I would like to thank the organizers of the Les Houches School for inviting me to present these lectures in an extraordinary environment. I want to thank in particular Volodya Kazakov and Paul Wiegmann for the opportunity to lecture in the evenings on Dovjenko, Godard and *The Matrix* besides my regular lectures on matrix models during the day. I'm also grateful to the participants for their enthusiasm, their questions and comments, and the fun. Thanks too to Arthur Greenspoon and Niclas Wyllard for a detailed reading of the manuscript. Finally, I would like to thank Mina Aganagic, Robbert Dijkgraaf, Sergei Gukov, Volodya Kazakov, Albrecht Klemm, Ivan Kostov, Stefan Theisen, George Thompson, Miguel Tierz and Cumrun Vafa for educating me about matrix models over the last two years.

## References

[1] M. Aganagic, A. Klemm and C. Vafa, "Disk instantons, mirror symmetry and the duality web," Z. Naturforsch. A **57**, 1 (2002) [arXiv:hep-th/0105045].

[2] M. Aganagic, A. Klemm, M. Mariño and C. Vafa, "Matrix model as a mirror of Chern-Simons theory," JHEP **0402**, 010 (2004) [arXiv:hep-th/0211098].

[3] O. Aharony, S. S. Gubser, J. M. Maldacena, H. Ooguri and Y. Oz, "Large N field theories, string theory and gravity," Phys. Rept. **323**, 183 (2000) [arXiv:hep-th/9905111].

[4] G. Akemann, "Higher genus correlators for the Hermitian matrix model with multiple cuts," Nucl. Phys. B **482**, 403 (1996) [arXiv:hep-th/9606004].

[5] J. Ambjorn, L. Chekhov, C. F. Kristjansen and Y. Makeenko, "Matrix model calculations beyond the spherical limit," Nucl. Phys. B **404**, 127 (1993) [Erratum-ibid. B **449**, 681 (1995)] [arXiv:hep-th/9302014].

[6] R. Argurio, G. Ferretti and R. Heise, "An introduction to supersymmetric gauge theories and matrix models," Int. J. Mod. Phys. A **19**, 2015 (2004) [arXiv:hep-th/0311066].

[7] M. Bershadsky, S. Cecotti, H. Ooguri and C. Vafa, "Kodaira-Spencer theory of gravity and exact results for quantum string amplitudes," Commun. Math. Phys. **165**, 311 (1994) [arXiv:hep-th/9309140].

[8] D. Bessis, "A New Method In The Combinatorics Of The Topological Expansion," Commun. Math. Phys. **69**, 147 (1979).

[9] D. Bessis, C. Itzykson and J. B. Zuber, "Quantum Field Theory Techniques In Graphical Enumeration," Adv. Appl. Math. **1**, 109 (1980).

[10] M. Blau and G. Thompson, "Lectures on 2-d gauge theories: Topological aspects and path integral techniques," arXiv:hep-th/9310144.

[11] G. Bonnet, F. David and B. Eynard, "Breakdown of universality in multi-cut matrix models," J. Phys. A **33**, 6739 (2000) [arXiv:cond-mat/0003324].

[12] E. Brezin, C. Itzykson, G. Parisi and J. B. Zuber, "Planar Diagrams," Commun. Math. Phys. **59**, 35 (1978).

[13] E. Brezin and V. A. Kazakov, "Exactly Solvable Field Theories Of Closed Strings," Phys. Lett. B **236**, 144 (1990) . M. R. Douglas and S. H. Shenker, "Strings In Less Than One-Dimension," Nucl. Phys. B **335**, 635 (1990). D. J. Gross and A. A. Migdal, "Nonperturbative Two-Dimensional Quantum Gravity," Phys. Rev. Lett. **64**, 127 (1990).

[14] F. Cachazo, B. Fiol, K. A. Intriligator, S. Katz and C. Vafa, "A geometric unification of dualities," Nucl. Phys. B **628**, 3 (2002) [arXiv:hep-th/0110028].

[15] F. Cachazo, K. A. Intriligator and C. Vafa, "A large N duality via a geometric transition," Nucl. Phys. B **603**, 3 (2001) [arXiv:hep-th/0103067].

[16] F. Cachazo, S. Katz and C. Vafa, "Geometric transitions and N = 1 quiver theories," arXiv:hep-th/0108120.

[17] P. Candelas and X. C. De La Ossa, "Moduli Space Of Calabi-Yau Manifolds," Nucl. Phys. B **355**, 455 (1991).

[18] L. Chekhov, "Genus one correlation to multi-cut matrix model solutions," arXiv:hep-th/0401089.

[19] Y. Chen and N. Lawrence, "Density of zeros of some orthogonal polynomials," Methods Appl. Anal. **5**, No.4, 367 (1998).

[20] S. Coleman, "$1/N$," in *Aspects of symmetry*, Cambridge University Press, 1988.

[21] P. Di Francesco, "Matrix model combinatorics: Applications to folding and coloring," arXiv:math-ph/9911002; "2D quantum gravity, matrix models and graph combinatorics," arXiv:math-ph/0406013.

[22] P. Di Francesco, P. Ginsparg and J. Zinn-Justin, "2-D Gravity and random matrices," Phys. Rept. **254**, 1 (1995) [arXiv:hep-th/9306153].

[23] P. Di Francesco and C. Itzykson, "A Generating function for fatgraphs," Annales Poincare Phys. Theor. **59**, 117 (1993) [arXiv:hep-th/9212108].

[24] R. Dijkgraaf, S. Gukov, V. A. Kazakov and C. Vafa, "Perturbative analysis of gauged matrix models," Phys. Rev. D **68**, 045007 (2003) [arXiv:hep-th/0210238].

[25] R. Dijkgraaf, G. W. Moore and R. Plesser, "The Partition function of 2-D string theory," Nucl. Phys. B **394**, 356 (1993) [arXiv:hep-th/9208031].

[26] R. Dijkgraaf, A. Sinkovics and M. Temurhan, "Matrix models and gravitational corrections," Adv. Theor. Math. Phys. **7**, 1155 (2004) [arXiv:hep-th/0211241].

[27] R. Dijkgraaf and C. Vafa, "Matrix models, topological strings, and supersymmetric gauge theories," Nucl. Phys. B **644**, 3 (2002) [arXiv:hep-th/0206255].

[28] R. Dijkgraaf and C. Vafa, "On geometry and matrix models," Nucl. Phys. B **644**, 21 (2002) [arXiv:hep-th/0207106].

[29] R. Dijkgraaf and C. Vafa, "A perturbative window into non-perturbative physics," arXiv:hep-th/0208048.

[30] R. Dijkgraaf, E. Verlinde and H. Verlinde, "Notes on topological string theory and two-dimensional topological gravity," in Trieste Spring School 1990.

[31] T. Eguchi and S. K. Yang, "The Topological $\mathbb{P}^1$ model and the large $N$ matrix integral," Mod. Phys. Lett. A **9**, 2893 (1994) [arXiv:hep-th/9407134]. T. Eguchi, K. Hori and S. K. Yang, "Topological sigma models and large $N$ matrix integral," Int. J. Mod. Phys. A **10**, 4203 (1995) [arXiv:hep-th/9503017].

[32] B. Eynard, "Topological expansion for the 1-hermitian matrix model correlation functions," arXiv:hep-th/0407261.

[33] F. Ferrari, "Planar diagrams and Calabi-Yau spaces," Adv. Theor. Math. Phys. **7**, 619 (2004) [arXiv:hep-th/0309151].

[34] R. Gopakumar and C. Vafa, "M-theory and topological strings. I," arXiv:hep-th/9809187.

[35] R. Gopakumar and C. Vafa, "On the gauge theory/geometry correspondence," Adv. Theor. Math. Phys. **3**, 1415 (1999) [arXiv:hep-th/9811131].

[36] P. Griffiths and J. Harris, *Principles of algebraic geometry*, Wiley, 1977.

[37] N. Halmagyi and V. Yasnov, "Chern-Simons matrix models and unoriented strings," JHEP **0402**, 002 (2004) [arXiv:hep-th/0305134].

[38] N. Halmagyi and V. Yasnov, "The spectral curve of the lens space matrix model," arXiv:hep-th/0311117. N. Halmagyi, T. Okuda and V. Yasnov, "Large N duality, lens spaces and the Chern-Simons matrix model," JHEP **0404**, 014 (2004) [arXiv:hep-th/0312145].

[39] K. Hori, S. Katz, A. Klemm, R. Pandharipande, R. Thomas, C. Vafa, R. Vakil, and E. Zaslow, *Mirror symmetry*, American Mathematical Society, 2003.

[40] K. Hori and C. Vafa, "Mirror symmetry," arXiv:hep-th/0002222.

[41] C. Imbimbo and S. Mukhi, "The Topological matrix model of c = 1 string," Nucl. Phys. B **449**, 553 (1995) [arXiv:hep-th/9505127].

[42] C. Itzykson and J. B. Zuber, "Matrix Integration And Combinatorics Of Modular Groups," Commun. Math. Phys. **134**, 197 (1990).

[43] S. Kachru, S. Katz, A. E. Lawrence and J. McGreevy, "Open string instantons and superpotentials," Phys. Rev. D **62**, 026001 (2000) [arXiv:hep-th/9912151].

[44] V. A. Kazakov, M. Staudacher and T. Wynter, "Character expansion methods for matrix models of dually weighted graphs," Commun. Math. Phys. **177**, 451 (1996) [arXiv:hep-th/9502132].

[45] A. Klemm, M. Mariño and S. Theisen, "Gravitational corrections in supersymmetric gauge theory and matrix models," JHEP **0303**, 051 (2003) [arXiv:hep-th/0211216].

[46] I. K. Kostov, "Gauge invariant matrix model for the A-D-E closed strings," Phys. Lett. B **297**, 74 (1992) [arXiv:hep-th/9208053]. S. Kharchev, A. Marshakov, A. Mironov, A. Morozov and S. Pakuliak, "Conformal matrix models as an alternative to conventional multimatrix models," Nucl. Phys. B **404**, 717 (1993) [arXiv:hep-th/9208044].

[47] I. K. Kostov, "Conformal field theory techniques in random matrix models," arXiv:hep-th/9907060.

[48] A.B.J. Kuijlaars and W. Van Assche, "The asymptotic zero distribution of orthogonal polynomials with varying recurrence coefficients," J. Approximation Theory **99**, No.1, 167-197 (1999).

[49] J. M. F. Labastida and P. M. Llatas, "Topological matter in two-dimensions," Nucl. Phys. B **379**, 220 (1992) [arXiv:hep-th/9112051].

[50] J. M. F. Labastida and M. Mariño, "Type B topological matter, Kodaira-Spencer theory, and mirror symmetry," Phys. Lett. B **333**, 386 (1994) [arXiv:hep-th/9405151].

[51] C. I. Lazaroiu, "Holomorphic matrix models," JHEP **0305**, 044 (2003) [arXiv:hep-th/0303008].

[52] M. Mariño, "The geometry of supersymmetric gauge theories in four dimensions," arXiv:hep-th/9701128.

[53] M. Mariño, "Chern-Simons theory, matrix integrals, and perturbative three-manifold invariants," arXiv:hep-th/0207096.

[54] M. Mariño, "Chern-Simons theory and topological strings," arXiv:hep-th/0406005.

[55] M.L. Mehta, *Random matrices*, 2nd edition, Academic Press, 1991.

[56] A. Migdal, "Loop equations and $1/N$ expansion," Phys. Rept. **102**, 199 (1983).

[57] N.I. Muskhelishvili, *Singular integral equations*, Noordhoff, 1953.

[58] T. Okuda, "Derivation of Calabi-Yau Crystals from Chern-Simons Gauge Theory," arXiv:hep-th/0409270.

[59] H. Ooguri, Y. Oz and Z. Yin, "D-branes on Calabi-Yau spaces and their mirrors," Nucl. Phys. B **477**, 407 (1996) [arXiv:hep-th/9606112].

[60] H. Ooguri and C. Vafa, "Worldsheet derivation of a large N duality," Nucl. Phys. B **641**, 3 (2002) [arXiv:hep-th/0205297].

[61] R.C. Penner, "Perturbative series and the moduli space of Riemann surfaces," J. Differ. Geom. **27**, No.1, 35-53 (1988).

[62] P. B. Rfinne, "On the Dijkgraaf-Vafa conjecture," arXiv:hep-th/0408103.

[63] G. Szegө, *Orthogonal polynomials*, American Mathematical Society, 1959.

[64] W. Taylor, "D-brane effective field theory from string field theory," Nucl. Phys. B **585**, 171 (2000) [arXiv:hep-th/0001201].

[65] W. Taylor and B. Zwiebach, "D-branes, tachyons, and string field theory," arXiv:hep-th/0311017.

[66] G. 't Hooft, "A planar diagram theory for strong interactions," Nucl. Phys. B **72** (1974) 461.

[67] M. Tierz, "Soft matrix models and Chern-Simons partition functions," Mod. Phys. Lett. A **19**, 1365 (2004) [arXiv:hep-th/0212128].

[68] E. Witten, "The $1/N$ Expansion In Atomic And Particle Physics," in G. 't Hooft *et al.* (eds.), *Recent developments in gauge theories*, Plenum Press, 1980.

[69] E. Witten, "Noncommutative Geometry And String Field Theory," Nucl. Phys. B **268**, 253 (1986).

[70] E. Witten, "Topological Sigma Models," Commun. Math. Phys. **118**, 411 (1988).

[71] E. Witten, "Quantum Field Theory And The Jones Polynomial," Commun. Math. Phys. **121**, 351 (1989).

[72] E. Witten, "On The Structure Of The Topological Phase Of Two-Dimensional Gravity," Nucl. Phys. B **340**, 281 (1990).

[73] E. Witten, "Mirror manifolds and topological field theory," arXiv:hep-th/9112056.

[74] E. Witten, "Chern-Simons gauge theory as a string theory," Prog. Math. **133**, 637 (1995) [arXiv:hep-th/9207094].

[75] E. Witten, "Perturbative gauge theory as a string theory in twistor space," arXiv:hep-th/0312171.

[76] V. Yasnov, "Chern-Simons matrix model for local conifold," arXiv:hep-th/0409136.

# MATRIX MODELS OF MODULI SPACE

Sunil Mukhi
*Tata Institute of Fundamental Research,*
*Mumbai, India*

**Abstract** We review matrix models corresponding to triangulations of the moduli space of Riemann surfaces: primarily the Kontsevich model that computes intersection numbers on moduli space, and the Penner model that computes the virtual Euler characteristic of moduli space. Generalisations of the former model describe noncritical strings with $c < 1$ matter, while the latter can be generalised to describe amplitudes of $c = 1$ strings at selfdual radius.

## 1. Introduction

Random matrices have played an important role in string theory. The primary physical intuition underlying their role is that certain random-matrix integrals correspond to summing over triangulated Riemann surfaces. In a suitable continuum limit, these simulate the integrals over inequivalent Riemann surfaces that are at the heart of perturbative string theory[1].

However, if properly treated, the matrix integral can do more (as we learn from Emil Martinec's lectures in this volume), and actually performs a sum over the contributions of all Riemann surfaces of arbitrary genus[2–4]. This makes it more powerful than conventional continuum formulations of string theory where at best one can compute amplitudes for low genus. The ultimate hope, in the matrix approach, has always been to understand nonperturbative string theory. Earlier attempts to do this ran into stability problems, but with the new understanding of type 0 backgrounds of superstring theory[5, 6] hopes have been revived of extracting genuine nonperturbative results. Eventually one would of course hope that some insights will extend to the more physically interesting case of critical string theory.

A rather independent line of development arose from studies by the mathematicians Kontsevich[7] and Penner[8] of topological properties of the moduli spaces of Riemann surfaces. In these studies, random matrix integrals were constructed starting from triangulations of the moduli spaces, rather than of the Riemann surfaces themselves. This involves beautiful mathematics and also makes use of time-honoured techniques in random matrix theory such as



*E. Brezin et al. (eds.), Applications of Random Matrices in Physics,* 379–401.

orthogonal polynomials and the like. The principal idea is to build up a matrix integral whose expansion computes topological invariants associated to moduli space. The Penner and Kontsevich models will be the main focus of the present lectures.

Soon after these matrix models of moduli space were written down, it was found[9, 10] that they could be related to the partition functions and amplitudes of noncritical string theory. In particular, there appeared to be such a "topological" matrix model for every noncritical string background. Given that there was already a "conventional" matrix model for each noncritical string, this seemed like an unnecessary duplication. But about a year ago, a reason was proposed for the existence of both types of matrix models of noncritical strings [11–13], related to the existence of two types of static D-branes in noncritical string theory[14–16]. But the entire open-string explanation of these models is relatively recent and is perhaps of greater interest to string theorists than to students of random matrices. So, while it will remain a motivation in what follows, the bulk of these notes will review the elegant mathematical ideas that go into the construction of the Kontsevich and Penner models and their generalisations. In the last section there will be some discussion of the string-theoretic issues involved.

## 2. Moduli space of Riemann surfaces and its topology

We start by discussing some preliminaries regarding Riemann surfaces, which play a primary role in the discussion. From a physicist's point of view, a Riemann surface is basically a 1-complex-dimensional manifold. We will deal with surfaces that are compact and without boundary.

Topologically, these are classified by their genus or number of handles: These manifolds admit a many-parameter family of complex structures: dif-

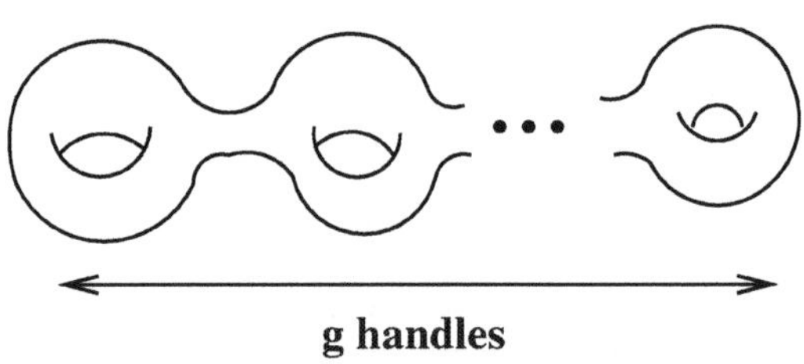


*Figure 1.* A generic compact Riemann surface.

ferent ways to define complex coordinates that are analytically inequivalent to each other. The **moduli space** of a compact Riemann surface of genus $g$ and $n$

punctures, $\mathcal{M}_{g,n}$, is the space of inequivalent complex structures that one can put on the surface.

$\mathcal{M}_{g,n}$ is known to be a (singular) complex manifold of complex dimension $3g-3+n$ (whenever this number is $\geq 0$). It arises as the quotient of a covering space, the *Teichmuller space* $\mathcal{T}_{g,n}$, by a discrete group, the *mapping class group* $MC_{g,n}$:

$$\mathcal{M}_{g,n} = \frac{\mathcal{T}_{g,n}}{MC_{g,n}}$$

This action typically has fixed points, hence the moduli space $\mathcal{M}_{g,n}$ has "orbifold" singularities.

Let us look at some simple examples of such moduli spaces. $\mathcal{M}_{0,n}$ is the moduli space of the sphere ($g = 0$) with $n$ punctures. This has complex dimension $n - 3$. For the simplest case of $n = 3$ one can fix all the punctures at arbitrary locations using the $SL(2, C)$ invariance of the sphere, so the moduli space is a single point. That point in turn is fixed under the action of the mapping class group $S_3$ that permutes the punctures. Locally, $\mathcal{M}_{0,n}$ has the structure of $n-3$ copies of the complex plane, but with a singularity whenever a pair of punctures coalesces on the original sphere.

Another example is $g = 1, n = 0$, the moduli space of a torus. We have:

$$\dim(\mathcal{M}_{1,0}) = 1$$

In this case, the Teichmuller space $\mathcal{T}_{1,0}$ is the upper half plane. The mapping

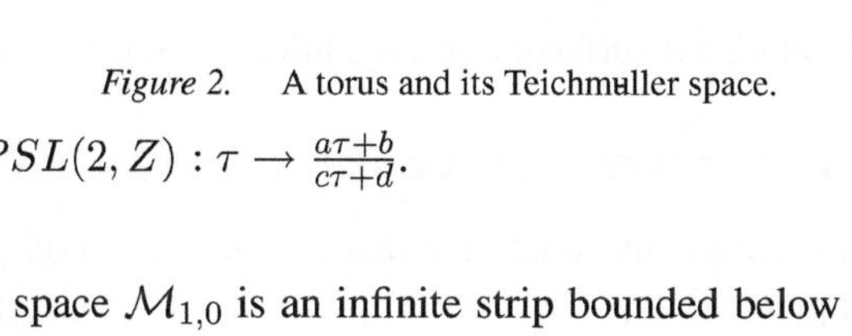


*Figure 2.* A torus and its Teichmuller space.

class group is $PSL(2, Z) : \tau \to \frac{a\tau+b}{c\tau+d}$.

The quotient space $\mathcal{M}_{1,0}$ is an infinite strip bounded below by an arc of a semicircle (see Fig.2).

The remaining $\mathcal{M}_{g,n}$ are much more complicated. In the absence of simple pictures for those spaces, mathematicians are interested in characterising them by their topological invariants.

What is the simplest topological invariant of $\mathcal{M}_{g,n}$? For a smooth manifold, we can define the the Euler characteristic $\chi$. For this, we make a simplicial

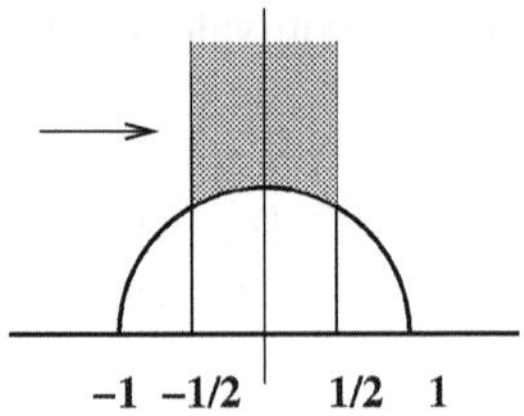


*Figure 3.* Moduli space of a torus, $\mathcal{M}_{1,0}$.

decomposition, or triangulation, $\mathcal{S}$ of the manifold, and evaluate:

$$\chi = \sum_{I \in \mathcal{S}} (-1)^{d_I}$$

where $d_I$ is the dimension of the $I$ th simplex, and the sum is over all the simplices in the complex $\mathcal{S}$. As is well known, this is a topological invariant, independent of how we triangulate the manifold.

A triangulation of a two-dimensional sphere is illustrated in Fig.4.

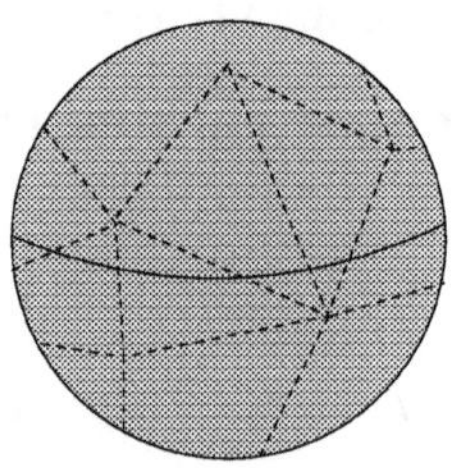

*Figure 4.* Triangulation of a two-dimensional sphere.

For a two-dimensional surface, a triangulation is literally made of triangles, and

$$\chi = \#(\text{vertices}) - \#(\text{edges}) + \#(\text{faces}) = 2 - 2g$$

Topological invariance is the statement that the answer is completely independent of the triangulation.

Now, we are not interested in triangulations of the Riemann surface itself, but of its moduli space, which is a manifold of arbitrarily high dimension. So we need to consider higher-dimensional simplicial complexes. In general dimensions, a simplicial complex involves "solid triangles", or simplices, of all dimensions upto the dimension of the manifold. Each simplex has a boundary whose components are simplices of one lower dimension.

Moreover, because the mapping class group can have fixed points, we need to consider spaces with orbifold singularities. In this case, the natural topological invariant to define is the virtual Euler characteristic $\chi_V$. This differs from the usual Euler characteristic in that each term in the sum over simplices is divided by the order of the discrete group $\Gamma_I$ that fixes the $I$ th simplex.

Thus,

$$\chi_V = \sum_{I \in \mathcal{S}} \frac{(-1)^{d_I}}{\#(\Gamma_I)}$$

Using combinatoric methods, it was found by Harer and Zagier[17] that the virtual Euler characteristic of $\mathcal{M}_{g,n}$ is:

$$\chi_V(\mathcal{M}_{g,n}) = (-1)^n \frac{(n+2g-3)!(2g-1)}{n!(2g)!} B_{2g}$$

where $B_{2g}$ are the Bernoulli numbers.

## 3. Quadratic differentials and fatgraphs

The above results were obtained by triangulating the moduli space of punctured Riemann surfaces using the so-called quadratic differentials. We will now survey in some detail how the triangulation was carried out. That will set the stage to construct a matrix integral whose expansion reproduces the Harer-Zagier formula above, the Penner model.

Triangulation of $\mathcal{M}_{g,n}$ was carried out by Harer[18], using a theorem due to Strebel[19], as follows[1]. On a Riemann surface with a finite number of marked points, one can define a meromorphic quadratic differential

$$\eta = \eta_{z,z}(z)dz^2$$

with poles at the marked points.

Under a change of coordinates $z \to z'(z)$, a quadratic differential transforms as:

$$\eta'_{z',z'}(z') = \left(\frac{\partial z}{\partial z'}\right)^2 \eta_{z,z}(z)$$

For a fixed complex structure on the surface, such a differential (with certain extra properties) is unique upto multiplication by a positive real number.

This differential can be used to invariantly define the length of a curve $\gamma$ on the Riemann surface:

$$|\gamma|_\eta = \int_\gamma \sqrt{|\eta(z)|}|dz|$$

Indeed, defining a new coordinate via

$$dw = \sqrt{\eta(z)}dz$$

we see that this length is the ordinary length of the curve in the Euclidean sense, in the $w$ coordinate.

Now consider a geodesic curve under the metric defined above. At any point, such a curve will be called *horizontal* if $\eta$ is real and positive along it, and *vertical* if $\eta$ is real and negative. The horizontal curves define *flows* along the Riemann surface.

The flow pattern is regular except at zeroes and poles of $\eta$. Here the flows exhibit interesting properties. At an $n$ th-order zero of the quadratic differential, precisely $n+2$ horizontal curves meet at a point. To see this, consider the differential near this zero and along the radial direction:

$$\eta \sim z^n (dz)^2 \sim e^{i(n+2)\theta} dr^2$$

As we encircle the zero, there are precisely $n+2$ values of the angle $\theta$ at which this differential is positive.

On the other hand, at a double pole of the quadratic differential, if the coefficient is real and negative, the flows form concentric circles around the point. We see that near such a pole, and along the angular direction, the differential looks like:

$$\eta \sim -c\frac{dz^2}{z^2} \sim c\, d\theta^2$$

Thus, in the $\theta$ direction, the differential is positive, or horizontal, at all points surrounding the double pole.

Other behaviours are possible at poles other than double poles, or if the coefficient of $\eta$ at a double pole is complex. But we will restrict our attention to quadratic differentials with a double pole at a point $P$, with the coefficient $c$ being real and negative.

We also require that all smooth horizontal trajectories (i.e., those that do not pass through zeroes of $\eta$) form closed curves. Quadratic differentials satisfying all these conditions exist, and are called *horocyclic*. An example of the flow pattern of a horocyclic quadratic differential is illustrated in Fig.5. In the figure, the vertex has five lines meeting at a point, indicating a third-order zero.

A key result about such differentials is Strebel's theorem, which states that on every Riemann surface of genus $g$ with 1 puncture, for fixed complex structure, there exists a *unique* horocyclic quadratic differential with a double pole at the puncture. The uniqueness is upto multiplication by a real positive number.

Thus, by studying how these quadratic differentials vary as we vary the moduli, we get information about the moduli space $\mathcal{M}_{g,1}$ of a once-punctured Riemann surface. Similar considerations apply for $\mathcal{M}_{g,n}$.

We can now see the emergence of "fatgraphs" and hence random matrices. Most of the flows are closed and smooth, but there are singular ones that branch into $n+2$-point vertices at $n$ th order zeroes of $\eta$. We can think of these

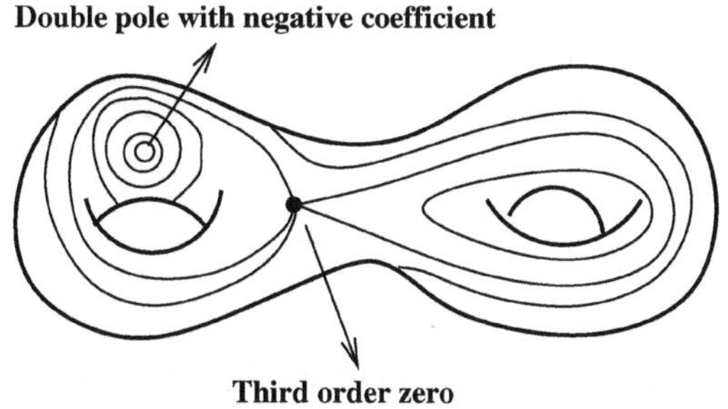


*Figure 5.* Riemann surface with the flow pattern of a horocyclic quadratic differential.

singular flows as defining a Feynman diagram, whose vertices are the branch points, and whose edges are the singular flow lines. Each double pole of $\eta$ is a point around which the flows form a loop. Hence the number of loops of the diagram is the number of double poles, which is the number of punctures of the original Riemann surface. Finally, because the flows that do not pass through a zero are closed and smooth, each singular flow can be "thickened" into a smooth ribbon in a unique way, and we arrive at a fatgraph.

The fatgraphs with a single loop triangulate the moduli space $\mathcal{M}_{g,1}$ in the following way. Consider the lengths of each edge of a fatgraph, as computed in the metric defined earlier. Scaling the whole Riemann surface clearly does not change the complex structure. So to vary the complex structure, we must change the lengths of the different edges keeping the total length fixed. This sweeps out a region of the moduli space of the Riemann surface. The (real) dimensionality of this region will be $E - 1$ where $E$ is the number of edges of the graph. This region is a simplex of the moduli space.

In a simplicial decomposition, at the boundary of a simplex we find a lower-dimensional simplex. In terms of fat graphs, a boundary occurs whenever a length goes to zero and two vertices meet. An example is given in Fig.6.

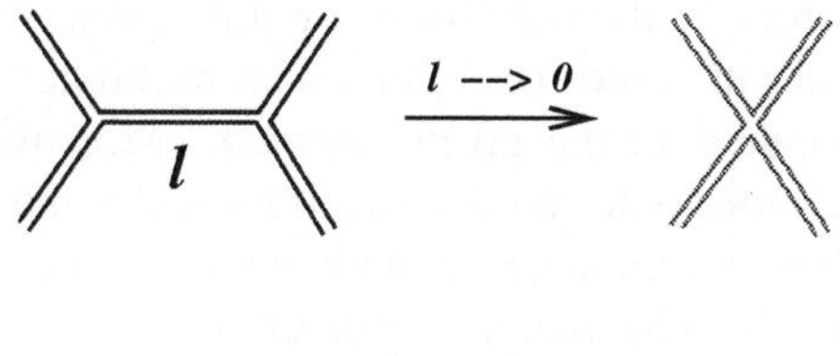


*Figure 6.* Collapse of a line takes us to a boundary of moduli space.

Now the virtual Euler characteristic of $\mathcal{M}_{g,n}$ can be defined directly in terms of fatgraphs. We consider the set of all fatgraphs of a given genus $g$ and a single puncture. Call the set $S$, and label each distinct graph by an integer $I \in S$. Let

$\Gamma_I$ be the automorphism group of a fatgraph. We will define it more precisely later.

Then, defining $d_I = (E-1)_I$, we claim that:

$$\chi_V(\mathcal{M}_{g,1}) = \sum_{I \in \mathcal{S}} \frac{(-1)^{d_I}}{\#(\Gamma_I)}$$

This is analogous to the original definition of $\chi_V$, except that now the sum is over fatgraphs rather than over simplices. In particular, the automorphism group of the fatgraph is the same as the group that fixes the corresponding simplex.

Let us check how this correspondence between fatgraphs and quadratic differentials works out in practice. The fatgraphs we have been considering have $V$ vertices, $E$ edges and 1 face. These integers satisfy:

$$V - E + 1 = 2 - 2g$$

where $g$ is the genus of the Riemann surface on which the graph is drawn.

We also have the relations:

$$V = \sum_k v_k, \quad E = \frac{1}{2}\sum_k k v_k$$

where $v_k$ is the number of $k$-point vertices. From these relations, we get:

$$\sum_k (k-2)v_k = 4g - 2$$

All integer solutions of this equation, i.e. all choices of the set $\{v_k\}$ for fixed $g$, are valid graphs that correspond to simplices in the triangulation of $\mathcal{M}_{g,1}$.

Let us recast the above equation as

$$\sum_k (k-2)v_k - 2 = 4g - 4$$

Since $k-2$ is the order of the zero for a $k$-point vertex, the first term on the left is the total number of zeroes (weighted with multiplicity) of the quadratic differential corresponding to the given fatgraph. Moreover, the differential has precisely one double pole, so the second term is minus the (weighted) number of poles. Thus this result agrees with the theorem that for meromorphic quadratic differentials on a Riemann surface of genus $g$,

$$\#(\text{zeroes}) - \#(\text{poles}) = 4g - 4$$

A particular solution that is always available is

$$v_3 = V, \quad v_k = 0, k \geq 4$$

This gives the maximum possible number of vertices, and therefore also of edges. In this case,

$$V = 4g - 2, \qquad E = \frac{3}{2}V = 6g - 3$$

Thus the dimension of the space spanned by varying the lengths of the graph keeping the overall length fixed, is:

$$E - 1 = 6g - 4 = 2(3g - 3 + 1)$$

which is the real dimension of $\mathcal{M}_{g,1}$. Thus, graphs with only cubic vertices span a top-dimensional simplex in moduli space.

All other graphs arise by collapse of one or more lines, merging two or more 3-point vertices to create higher $n$-point vertices. These correspond to simplices of lower dimension in the moduli space.

To conclude this part, let us see as an example how $\chi_V(\mathcal{M}_{1,1})$ is obtained from fatgraphs. From the Harer-Zagier formula, we expect to find:

$$\chi_V(\mathcal{M}_{1,1}) = -\frac{1}{2}B_2 = -\frac{1}{12}$$

In genus 1, there are two possible ways to satisfy

$$\sum_k (k-2)v_k = 4g - 2 = 2$$

namely $v_3 = 2$ or $v_4 = 1$. In the first case we find $V = 2, E = 3$ and in the second, $V = 1, E = 2$.

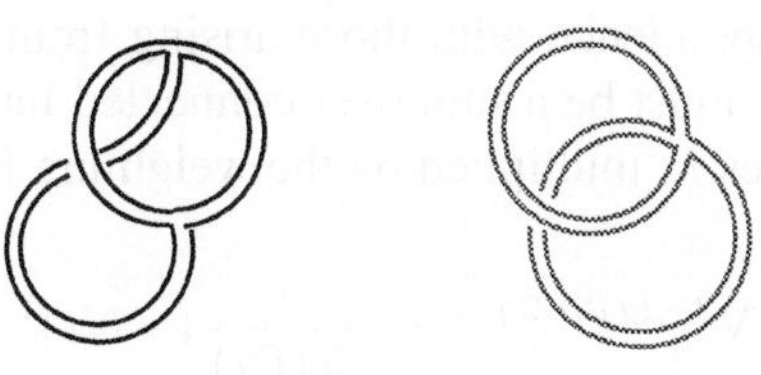

*igure 7.* Two graphs that together triangulate $\mathcal{M}_1,$

The graphs are shown in Fig.7. We see explicitly that they have genus 1. It is left as an exercise to the reader to show that the automorphism groups of these graphs are of order 6 and 4 respectively. Then, $\chi_V(\mathcal{M}_{1,1}) = \frac{(-1)^2}{6} + \frac{(-1)^1}{4} = -\frac{1}{12}$.

In the above we have restricted our attention to moduli spaces of once-punctured Riemann surfaces, but the above considerations can be extended to all $\mathcal{M}_{g,n}$.

## 4. The Penner model

In 1986, Penner[8] constructed a model of random matrices that provides a generating functional for $\chi_V(\mathcal{M}_{g,s})$. The Penner model is defined in terms of $N \times N$ matrices whose fatgraphs are precisely the ones described in the previous subsection. The free energy $\mathcal{F} = \log \mathcal{Z}$ of this model therefore must have the expansion:

$$\mathcal{F} = \sum_g \mathcal{F}_g = \sum_{g,n} \chi_{g,n}\, N^{2-2g}\, t^{2-2g-n}$$

where $t$ is a parameter of the model. The term $n = 0$ is not present in the sum.

The model is given by an integral over Hermitian random matrices:

$$\begin{aligned} \mathcal{Z}_{\text{Penner}} &= \mathcal{N}_P \int [dQ]\, e^{-Nt\, \mathbf{tr} \sum_{k=2}^{\infty} \frac{1}{k} Q^k} \\ &= \mathcal{N}_P \int [dQ]\, e^{Nt\, \mathbf{tr}\, (\log(1-Q)+Q)} \end{aligned}$$

where $\mathcal{N}_P$ is a normalisation factor given by:

$$\mathcal{N}_P^{-1} = \int [dQ]\, e^{-Nt\, \mathbf{tr}\, \frac{1}{2} Q^2}$$

and the matrix measure $[dQ] \equiv \prod_i dQ_{ii} \prod_{i<j} dQ_{ij}\, dQ_{ij}^*$ as usual.

This action has all powers of the random matrix appearing in it! The model is to be considered as a perturbation series around $Q \sim 0$.

To show that this model is correct, we must show that its fatgraphs are in one-to-one correspondence with those arising from quadratic differentials. Thus the free energy must be a sum over connected fatgraphs of a fixed genus $g$ and number of faces $n$, multiplied by the weighting factor

$$\frac{(-1)^{E-n}}{\#(\Gamma_I)}\, N^{2-2g} t^{2-2g-n} = \frac{1}{\#(\Gamma_I)}\, (-Nt)^V (Nt)^{-E} (N)^n$$

Here $\Gamma_I$, the automorphism group, is the collection of maps of a given fatgraph to itself such that:

(i) the set of vertices is mapped onto itself,

(ii) the set of edges is mapped to itself,

(iii) the cyclic ordering of each vertex is preserved.

A key result due to Penner is that the order of $\Gamma_I$ is given by:

$$\frac{1}{\#(\Gamma_I)} = C \times \prod_k \left(\frac{1}{k}\right)^{v_k} \frac{1}{v_k!}$$

where $C$ is the combinatoric factor labelling how many distinct contractions lead to the same graph. Now this is exactly the factor that arises if we obtain our fatgraphs by expanding the Penner matrix integral:

$$\begin{array}{cl} \frac{1}{v_k!} & : \text{order of expansion of the } k \text{ th term in the exponent} \\ \frac{1}{k} & : \text{weight per vertex appearing in the action} \\ C & : \text{combinatoric factor from contractions} \\ (-Nt)^V & : \text{from weight of each vertex} \\ (Nt)^{-E} & : \text{from each propagator} \\ N^n & : \text{from the index sum on each face} \end{array}$$

This proves that the Penner model computes the desired quantity, $\chi_V(\mathcal{M}_{g,n})$.

In his paper, Penner constructed the orthogonal polynomials for this model. They turn out to be Laguerre polynomials. Using the above facts, Penner was able to deduce, directly from his matrix model, that

$$\chi_V(\mathcal{M}_{g,n}) = (-1)^n \frac{(n+2g-3)!(2g-1)}{n!(2g)!} B_{2g}$$

where $B_{2g}$ are the Bernoulli numbers. This is precisely the Harer-Zagier formula.

## 5. Penner model and matrix gamma function

Recall the definition of the Penner matrix integral:

$$\mathcal{Z}_{\text{Penner}} = \mathcal{N}_P \int [dQ]\, e^{Nt\, \mathbf{tr}\, (\log(1-Q)+Q)}$$

Let us make the following change of variables:

$$Q = 1 - \frac{t+1}{t} M, \qquad t = -1 + \frac{\nu}{N}$$

This replaces the original matrix $Q$ and parameter $t$ by a new matrix $M$ and parameter $\nu$. The change in measure is a trivial factor since the change of variables is linear. Hence the Penner action becomes:

$$Nt\, \mathbf{tr}\, (\log(1-Q)+Q) = \mathbf{tr}\, ((\nu - N) \log M - \nu M) + \text{constant}$$

The additive constant depends on $\nu, N$.

Thus we can write:

$$\mathcal{Z}_{\text{Penner}} = \mathcal{N}'_P \int [dM]\, e^{\,\mathbf{tr}\, ((\nu-N)\log M - \nu M)}$$

where the new normalisation $\mathcal{N}'_P$ has absorbed the constant factors in the exponential and also the simple Jacobian.

For a $1 \times 1$ matrix $M = m$, the integral is just the Euler $\Gamma$-function:

$$\int dm\, m^{\nu-1}\, e^{-\nu m} = \Gamma(\nu)$$

as long as we choose the correct limits $m \in (0, \infty)$. Hence we make the same restriction on the matrix $M$ in $\mathcal{Z}_{\text{Penner}}$ above, namely its eigenvalues must be positive. Then $\mathcal{Z}_{\text{Penner}}$ can be called the Matrix $\Gamma$-Function.

We can remove the positivity restriction on $M$ by defining:

$$M = e^{\Phi}$$

where $\Phi$ is a generic Hermitian matrix. In this case there is a nontrivial Jacobian:

$$[dM] = (\det e^{\Phi})^N\, [d\Phi]$$

Writing this equivalently as:

$$[d\Phi] = [dM](\det M)^{-N} = [dM]e^{-N\, \mathbf{tr}\, \log M}$$

we see that the Penner integral takes its simplest form:

$$\mathcal{Z}_{\text{Penner}} = \mathcal{N}'_P \int [d\Phi]\, e^{\nu\, \mathbf{tr}\, (\Phi - e^{\Phi})}$$

which we call the Liouville Matrix Model[20].

This matrix model has some intriguing properties that are familiar from string theory. The integral is like a matrix version of the Liouville path integral occurring in string theory, when restricted to the constant mode of the Liouville field.

It converges at $\Phi \to +\infty$ because of the exponential term, and at $\Phi \to -\infty$ because of the linear term. It has an $N$-independent coefficient $\nu$, suggestive of D-brane actions in string theory, if $\nu$ is interpreted as the inverse string coupling. We will see later that this interpretation of $\nu$ does hold in a string theory setting of this model, though the D-brane interpretation has not been developed beyond the tentative identification suggested above, and in some more detail in Ref.[20].

## 6. The Kontsevich model

Another interesting topological problem associated to the moduli space $\mathcal{M}_{g,n}$ is the following. It is known that $\mathcal{M}_{g,n}$ can be compactified, and the resulting space is called $\overline{\mathcal{M}}_{g,n}$. Topological invariants can then be defined as integrals of cohomology classes on $\overline{\mathcal{M}}_{g,n}$.

The problem of intersection theory on moduli space[21–23] is defined as follows[2]. Let $\mathcal{L}_i, i = 1, 2, \ldots, n$ be line bundles on $\overline{\mathcal{M}}_{g,n}$. The fibre for each $i$ is the cotangent space to the Riemann surface at the puncture. Each such bundle has its associated top Chern class $c_1(\mathcal{L}_i)$. This is a two-form (intuitively, the field strength associated to the $U(1)$ connection on this bundle).

Now construct the integral

$$\int_{\overline{\mathcal{M}}_{g,n}} c_1(\mathcal{L}_1)^{d_1} \wedge \cdots \wedge c_1(\mathcal{L}_n)^{d_n}$$

where $d_i \geq 0$ are a set of integers satisfying:

$$\sum_{i=1}^{n} d_i = 3g - 3 + n$$

This means that the integrand is a $6g - 6 + 2n$ form, equal in degree to the real dimension of $\overline{\mathcal{M}}_{g,n}$. So the integral is well-defined and is a topological invariant of the moduli space.

Next we give this invariant a suggestive name:

$$\int_{\overline{\mathcal{M}}_{g,n}} c_1(\mathcal{L}_1)^{d_1} \wedge \cdots \wedge c_1(\mathcal{L}_n)^{d_n} = \langle \tau_{d_1} \cdots \tau_{d_n} \rangle$$

as if it is a correlation function of some observables $\tau_i$ in a quantum field theory. (We define the RHS to be 0 if $\sum d_i \neq 3g - 3 + n$ for any integer $g$.) There is actually such a quantum field theory, the so-called "topological 2d gravity"[9], but we will not go into its definition here.

Let us now define a generating functional for these invariants by summing them up.

$$\begin{aligned} F(t_0, t_1, \cdots) &\equiv \left\langle \exp\Big(\sum_{i=0}^{\infty} t_i \tau_i\Big) \right\rangle = \sum_{k_0, k_1, \cdots} \langle \tau_0^{k_0} \tau_1^{k_1} \cdots \rangle \prod_{i=0}^{\infty} \frac{t_i^{k_i}}{k_i!} \\ &= \sum_{n=1}^{\infty} \sum_{\{d_i\}} \frac{1}{n!} \langle \tau_{d_1} \tau_{d_2} \cdots \tau_{d_n} \rangle \, t_{d_1} \ldots t_{d_n} \end{aligned}$$

It is known that:

$$U(t_0, t_1, \cdots) \equiv \frac{\partial^2 F}{\partial t_0^2}(t_0, t_1, \cdots)$$

satisfies the KdV equation:

$$\frac{\partial U}{\partial t_1} = U \frac{\partial U}{\partial t_0} + \frac{1}{12} \frac{\partial^3 U}{\partial t_0^3}$$

Also, the series $\exp(F)$ in terms of the variables

$$T_{2i+1} \equiv \frac{1}{(2i+1)!!}\, t_i$$

is a $\tau$-function of the KdV hierarchy. This constitutes the "solution" of the problem.

Kontsevich[7] proposed a matrix model whose connected fatgraphs generate the function $F(t_0, t_1, \cdots)$. Clearly the model must depend on infinitely many parameters $t_i$. However, these are encoded in a nontrivial way. Introduce an $N \times N$ positive-definite Hermitian matrix $\Lambda$ and let:

$$t_i = -(2i-1)!!\ \mathbf{tr}\,\Lambda^{-(2i+1)}$$

Clearly the $t_i$ obtained in this way are not all independent of each other if the rank of $\Lambda$ is finite. Only as $N \to \infty$ can they be chosen independently. This is a new role for the large-$N$ limit!

The Kontsevich matrix model, depending on the fixed matrix $\Lambda$, is:

$$\mathcal{Z}_{\text{Kontsevich}}(\Lambda) = \mathcal{N}_K(\Lambda) \int [dX]\, e^{\,\mathbf{tr}\left(-\frac{1}{2}X^2\Lambda + \frac{i}{6}X^3\right)}$$

where $X$ is an $N \times N$ Hermitian random matrix, and:

$$\mathcal{N}_K(\Lambda) = \left\{ \int [dX]\, e^{\,\mathbf{tr}\left(-\frac{1}{2}X^2\Lambda\right)} \right\}^{-1}$$

By a change of variables, the above model can also be written:

$$\mathcal{Z}_{\text{Kontsevich}}(\tilde{\Lambda}) = \mathcal{N}'_K(\tilde{\Lambda}) \int [d\tilde{X}]\, e^{i\,\mathbf{tr}\left(\frac{1}{3}\tilde{X}^3 - \tilde{X}\tilde{\Lambda}\right)}$$

Comparing this with the Airy Function:

$$\mathcal{A}(\lambda) = \int_{-\infty}^{\infty} dx\, e^{i\left(\frac{1}{3}x^3 - x\lambda\right)}$$

we see that the Kontsevich model is a Matrix Airy Function.

Without loss of generality, the fixed matrix $\Lambda$ can be taken to be diagonal:

$$\Lambda = \text{diag}(\Lambda_1, \Lambda_2, \cdots, \Lambda_N)$$

Then, using

$$\mathbf{tr}\, X^2\Lambda = \frac{1}{2}\sum_{i,j}(\Lambda_i + \Lambda_j)X_{ij}X_{ji}$$

we see that the matrix propagator in this model is:

$$\langle X_{ij}X_{kl}\rangle = \delta_{jk}\delta_{li}\,\frac{2}{\Lambda_i + \Lambda_j}$$

The vertices, unlike in the Penner model, are all cubic.

In his paper, Kontsevich showed that:

$$F(t_0, t_1, \cdots) \equiv \left\langle \exp(\sum_{i=0}^{\infty} t_i \tau_i) \right\rangle = \log \mathcal{Z}_{\text{Kontsevich}}(\Lambda)$$

He also showed that $\mathcal{Z}_{\text{Kontsevich}}(\Lambda)$ is a $\tau$-function of the KdV hierarchy. Let us sketch the derivation. From the definition of $F(t_0, t_1, \cdots)$ and the change of variables

$$t_{d_i} = -(2d_i - 1)!! \sum_{j=1}^{N} \frac{1}{\Lambda_j^{2d_i+1}}$$

we see that:

$$\begin{aligned} F(t_0, t_1, \cdots) &= \sum_{n=1}^{\infty} \sum_{\{d_i\}} \frac{1}{n!} \langle \tau_{d_1} \tau_{d_2} \cdots \tau_{d_n} \rangle \, t_{d_1} \ldots t_{d_n} \\ &= \sum_{n=1}^{\infty} \sum_{\{d_i\}} \frac{(-1)^n}{n!} \langle \tau_{d_1} \tau_{d_2} \cdots \tau_{d_n} \rangle \sum_{\{j_i\}} \prod_{i=1}^{n} \frac{(2d_i - 1)!!}{\Lambda_{j_i}^{2d_i+1}} \end{aligned}$$

Now given a 3-valent graph, we first "unravel" it into polygons: On the

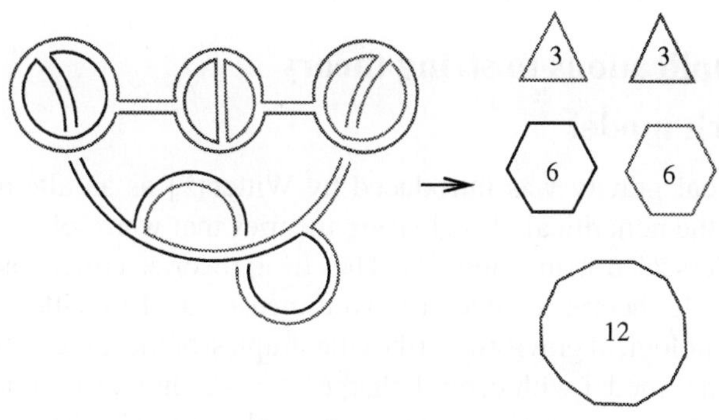


*Figure 8.* Unravelling a 3-valent graph.

polygons, we associate lengths $l_a$ via the metric induced from the horocyclic quadratic differentials. The unravelling defines a map:

$$\mathcal{M}_{g,n} \times R_+^n \to (\text{space of polygons with marked lengths})^n$$

Next, for each polygon we define a 2-form:

$$\omega_i \sim \sum_{a<b} dl_a \wedge dl_b$$

and pull the form back to $\mathcal{M}_{g,n} \times R_+^n$. Kontsevich then proves that the resulting 2-form projects to a 2-form on $\mathcal{M}_{g,n}$, and is in fact just equal to $c_1(\mathcal{L}_i)$.

This sets up a correspondence between the desired Chern classes and properties of fatgraphs. From this he then shows that:

$$\sum_{n=1}^{\infty}\sum_{\{d_i\}}\frac{(-1)^n}{n!}\langle\tau_{d_1}\tau_{d_2}\cdots\tau_{d_n}\rangle\sum_{\{j_i\}}\prod_{i=1}^{n}\frac{(2d_i-1)!!}{\Lambda_{j_i}^{2d_i+1}}=\sum_{\substack{3-\text{valent}\\ \text{graphs } I}}\frac{\left(\frac{i}{2}\right)^V}{\#(\Gamma_I)}\prod_{\substack{\text{edges}\\ \langle i,j\rangle}}\frac{2}{\Lambda_i+\Lambda_j}$$

The RHS is the graphical expansion of $F(t_0, t_1, \cdots) = \log \mathcal{Z}_{\text{Kontsevich}}(\Lambda)$.

Finally, Kontsevich provides an asymptotic expansion of the Matrix Airy Function using the famous Harish-Chandra formula:

$$\int [dX]\ \mathbf{tr}\, p(X)\, e^{-i\ \mathrm{tr}\, X\Lambda} = \mathrm{C}\int\prod_i dx_i \prod_{i<j}\frac{(x_i-x_j)}{(\Lambda_i-\Lambda_j)}\sum_i p(x_i)\, e^{-i\sum_i x_i\Lambda_i}$$

He then identifies this with the asymptotic expansion of the $\tau$-function of the KdV hierarchy. This proves that $\mathcal{Z}_{\text{Kontsevich}}$ is a KdV $\tau$-function.

In a subsequent paper, Witten[25] showed directly that the partition function of the Kontsevich model solves the Virasoro identities, a sequence of *linear* differential equations which imply both the KdV equation and an additional equation called the "string equation" that is known to characterise string theory in a class of minimal-model backgrounds.

## 7. Applications to string theory

### Kontsevich model

Topological gravity was introduced by Witten[9] as an alternative way to understand the noncritical closed-string theories that were solved around 1990 using double-scaled matrix models. The string theories corresponded to $c < 1$ conformal field theories coupled to two-dimensional (Liouville) gravity.

"Pure" topological gravity describes the simplest of these theories, the $(p, q) = (2, 1)$ minimal model with central charge $c = -2$. In matrix model language, one gets this theory by not going to any critical point. The theory is non-trivial (though its critical exponents are trivial), and its operators are the $\tau_i$ mentioned before. By construction, the Kontsevich model gives us all its correlators in every genus.

However, the entire chain of $(2, q)$ minimal models coupled to gravity, for all odd $q$, can be studied using the same model. As Witten argued, to go to higher $q$, one only has to give an expectation value to some of the $t_i$. Thus, the Kontsevich model expanded around different "vacua" i.e. choices of expectation values $\langle t_i\rangle$ generates all $(2, q)$ minimal models coupled to gravity.

For $(p, 1)$ noncritical strings with $p > 2$, one needs a model proposed by Adler and van Moerbeke[26] and independently by Kharchev-Marshakov-

Mironov-Morozov-Zabrodin[27]:

$$\mathcal{Z}_{\text{AvM-KMMMZ}}(\tilde{\Lambda}) = \mathcal{N}'_{\text{AvM-KMMMZ}}(\tilde{\Lambda}) \int [d\tilde{X}]\, e^{i\,\mathbf{tr}\left(\frac{1}{p+1}\tilde{X}^{p+1} - \tilde{X}\tilde{\Lambda}\right)}$$

and again one recovers the $(p, q)$ case by going to suitable critical points.

## Penner model

In 1990, it was noticed by Distler and Vafa[10] that starting with the Penner free energy:

$$\begin{aligned} \mathcal{F} &= \sum_g \mathcal{F}_g = \sum_{g,n} \chi_{g,n}\, N^{2-2g}\, t^{2-2g-n} \\ \chi_{g,n} &= \frac{(-1)^n(2g-3+n)!(2g-1)}{(2g)!\, n!} B_{2g} \end{aligned}$$

one can perform the sum over $n$ explicitly, to get:

$$\mathcal{F}_g = \frac{B_{2g}}{2g(2g-2)}(Nt)^{2-2g}\left((1+\frac{1}{t})^{2-2g} - 1\right)$$

For $g > 1$ , they took the limit $N \to \infty$ and $t \to t_c = -1$, keeping fixed the product $N(1+t) = \nu$. This led to the simpler result:

$$\mathcal{F}_g = \frac{B_{2g}}{2g(2g-2)}\,\nu^{2-2g}$$

But this, for $g > 1$, is precisely the virtual Euler characteristic of unpunctured Riemann surfaces!

Thus the Penner model, originally designed to study the moduli space of punctured Riemann surfaces, describes unpunctured ones too. This happens in the special double-scaling limit above. More remarkably, we see that its free energy in the double scaling limit:

$$\mathcal{F} = \sum_g \frac{B_{2g}}{2g(2g-2)}\,\nu^{2-2g}$$

is almost identical to a well-known quantity in string theory: the free energy of the $c = 1$ noncritical string compactified at self-dual radius[3]:

$$\mathcal{F} = \sum_g \frac{|B_{2g}|}{2g(2g-2)}\,\mu^{2-2g}$$

However, there is an issue of alternating signs. We have:

$$|B_{2g}| = (-1)^{g-1} B_{2g}$$

Therefore if we define $\mu = i\nu$, we can write:

$$\begin{aligned}\mathcal{F}(\nu)_{c=1} &= \sum_{g=0}^{\infty} \frac{B_{2g}}{2g(2g-2)}\, \nu^{2-2g} \\ &= \sum_{g=0}^{\infty} \chi_g\, \nu^{2-2g}\end{aligned}$$

Thus the genus $g$ contribution to the free energy of the $c = 1, R = 1$ string at imaginary cosmological constant is the (virtual) Euler characteristic of genus-$g$ moduli space, which in turn is the Penner free energy after double-scaling.

The $c = 1$ string is a background of bosonic string theory with two spacetime dimensions, one of which (the spatial direction) is the Liouville coordinate along which translation invariance is broken by a potential. The other direction is translation-invariant and timelike. Therefore there is a conserved energy corresponding to it. Here we are dealing with a Euclideanised time direction that is moreover compact. So the Euclidean momentum should be conserved and integral. Now if the Penner model is associated to the $c = 1$ string, it should describe correlators of its observables: the so-called "discrete tachyons" $\mathcal{T}_k$, where $k$ labels the conserved Euclidean momentum. But as formulated, it does not depend on the necessary (infinitely many) parameters.

There is a deformation of the model that does precisely this job. This was constructed in Ref.[29] starting with the generating functional for all tachyon correlators to all genus obtained in Ref.[30]. Such a functional $\mathcal{F}(t, \bar{t})$ depends on couplings $t_k, \bar{t}_k$ such that:

$$\langle \mathcal{T}_{k_1} \ldots \mathcal{T}_{k_n} \mathcal{T}_{-l_1} \ldots \mathcal{T}_{-l_m} \rangle = \frac{\partial}{\partial t_{k_1}} \cdots \frac{\partial}{\partial t_{k_n}} \frac{\partial}{\partial \bar{t}_{l_1}} \cdots \frac{\partial}{\partial \bar{t}_{l_m}} \mathcal{F}(t, \bar{t}) \bigg|_{t=\bar{t}=0}$$

where on the LHS we have connected amplitudes.

In the same spirit as Kontsevich, start by defining a constant $N \times N$ matrix $A$ that satisfies:

$$t_k = \frac{1}{\nu k}\, \mathbf{tr}\, A^{-k}$$

This matrix can encode infinitely many parameters $t_k$ in the limit $N \to \infty$. However, we do not perform a similar transformation on $\bar{t}_k$, rather we allow the model to depend directly on these parameters.

Using the conventional matrix description of $c = 1$ string theory, Matrix Quantum Mechanics, at $R = 1$, it was shown by Dijkgraaf, Moore and Plesser [30] that $\mathcal{Z}(t, \bar{t}) = e^{\mathcal{F}(t, \bar{t})}$ satisfies the $W_\infty$ equation:

$$\frac{1}{(-\nu)} \frac{\partial \mathcal{Z}}{\partial \bar{t}_n} = \frac{1}{(-\nu)^n} (\det A)^{\nu}\, \mathbf{tr} \left( \frac{\partial}{\partial A} \right)^n (\det A)^{-\nu}\, \mathcal{Z}(t, \bar{t})$$

where $\nu = -i\mu$ and $\mu$ is the cosmological constant.

Let us postulate that $\mathcal{Z}(t, \bar{t})$ is an integral over Hermitian matrices $M$ of the form:

$$\mathcal{Z}(t,\bar{t}) = (\det A)^{\nu} \int [dM]\, e^{\,\mathbf{tr}\, V(M,A,\bar{t})}$$

for some $V(M, A, \bar{t})$. The function $V$ is determined by imposing the above differential equation:

$$\left[\frac{1}{(-\nu)}\frac{\partial}{\partial \bar{t}_n} - \frac{1}{(-\nu)^n}\,\mathbf{tr}\left(\frac{\partial}{\partial A}\right)^n\right]\int [dM]\, e^{\,\mathbf{tr}\, V(M,A,\bar{t})} = 0$$

This can be solved to give:

$$V(M,A,\bar{t}) = -\nu\Big(MA + \sum_{k=1}^{\infty} \bar{t}_k M^k\Big) + f(M)$$

where $f(M)$ is a function independent of $A, \bar{t}$ that we determine using a boundary condition.

From conservation of the tachyon momentum, we know that $Z(t, 0)$ must be independent of $t_k$. Using:

$$Z(t,0) = (\det A)^{\nu} \int [dM]\, e^{-\nu\, \mathbf{tr}\, MA + \,\mathbf{tr}\, f(M)}$$

and changing variables $M \to MA^{-1}$, we have

$$[dM] \to (\det A)^{-N} [dM]$$

Then:

$$\begin{aligned} Z(t,0) &= (\det A)^{\nu - N} \int [dM]\, e^{-\nu\, \mathbf{tr}\, M + \,\mathbf{tr}\, f(MA^{-1})} \\ &= \int [dM]\, e^{-\nu\, \mathbf{tr}\, M + \,\mathbf{tr}\, f(MA^{-1}) + (\nu - N)\,\mathbf{tr}\,\log A} \end{aligned}$$

This uniquely determines the remaining unknown function:

$$f(M) = (\nu - N)\log M$$

In summary, we have found that the generating function of all tachyon amplitudes in the $c = 1, R = 1$ string theory is:

$$\begin{aligned} \mathcal{Z}(t,\bar{t}) &= (\det A)^{\nu} \int [dM]\, e^{\,\mathbf{tr}\,(-\nu MA + (\nu - N)\log M - \nu \sum_{k=1}^{\infty} \bar{t}_k M^k)} \\ &= \int [dM]\, e^{\,\mathbf{tr}\,(-\nu M + (\nu - N)\log M - \nu \sum_{k=1}^{\infty} \bar{t}_k (MA^{-1})^k)} \end{aligned}$$

We see that the first two terms of the matrix potential precisely correspond to the Penner model! So it has reappeared from a completely independent starting point, and now with infinitely many parameters $A, \bar{t}_k$. This deformed Penner model can be called the $W_\infty$ model. As we saw before, the matrix $M$ (more precisely, its eigenvalues) must be positive semidefinite.

At the selfdual radius, $c = 1$ string theory has additional states besides the momentum-carrying tachyons $T_k$. There are winding tachyons quantised in the same way as the momentum states. and there are also the famous "discrete states" that are like two-dimensional remnants of gravitons and other tensor states in critical string theory. Unfortunately and extension of the $W_\infty$ model to incorporate these states is not known at present.

There is a different (2-matrix) model that also describes the $c =$ string at selfdual radius (and other radii) - due to Alexandrov, Kazakov, Kostov[31]. This is the "normal matrix model", for a complex matrix $Z$ satisfying:

$$[Z, Z^\dagger] = 0$$

For selfdual radius, the partition function of this model is:

$$\mathcal{Z}(t,\bar{t})_{NMM} = \int [dZ\, dZ^\dagger]\, e^{\,\mathbf{tr}\left(-\nu ZZ^\dagger + (\nu - N)\log ZZ^\dagger - \nu \sum_{k=1}^{\infty}(t_k Z^k + \bar{t}_k Z^{\dagger k})\right)}$$

This is different from the $W_\infty$ model and yet describes the same correlation functions. Also, it has no Kontsevich-type constant matrix in it. At present its relation to the $W_\infty$ model is not known, despite obvious similarities in the matrix action.

## 8. Conclusions

We conclude with a few general comments and open problems.

*Universality of $W_\infty$ model.* The $W_\infty$ or deformed Penner model is universal, in that it contains many other models including the Kontsevich model as special cases. Setting $A, \bar{t}_k = 0$ we recover the original Penner model. Setting $\nu = N, \bar{t}_3 = const, \bar{t}_i = 0\ (i \neq 3)$ we recover the matrix Airy function, or Kontsevich model. And setting $\nu = N, \bar{t}_{p+1} = const, \bar{t}_i = 0\ (i \neq p+1)$ we recover the $p$-th AvM-KMMMZ model. Setting $\nu = N, A = 0, \bar{t}_k = 0, k > m$ we recover the polynomial 1-matrix model of any degree $m$. This might be a pointer to a unification of all noncritical strings via this model[29]. A related point of view has been advocated in Ref.[27]. However, it is fair to say that a complete picture of this unification, and a good understanding of how it relates to background-independence and string field theory, has still to emerge.

*Open string origin of matrix models.* In these notes we have not emphasized the recent discovery that matrix models describing noncritical string theory can be understood in terms of D-branes. During the first incarnation of matrix models for noncritical strings, in the early 1990's, the notion of D-branes did not exist, therefore nor did the subsequent understanding of holography as open-closed string duality. But in recent times, starting with the work of Refs. [11, 12], it was shown that the matrix of the "conventional" matrix models is just the tachyonic mode living on $N$ unstable D-branes in noncritical string backgrounds. These branes are localised in the Liouville direction and can be thought of as D0-branes[16]. The fact that this model describes closed strings was then a consequence of open-closed duality.

The above discussion says nothing about the matrix models of moduli space that are the subject of the present notes. One may ask if open strings are responsible for these matrix models too. Progress was made in Ref.[13], where it was argued (for the $(2,p)$ models) that the Kontsevich model is just the open-string field theory evaluated on the so-called FZZT[14, 15] branes. Schematically this amounts to the statement that:

$$\int \left(\frac{1}{2}\Psi Q\Psi + \frac{1}{3}\Psi * \Psi * \Psi\right) \to \mathbf{tr}\left(\frac{1}{2}\Lambda X^2 + \frac{1}{3}X^3\right)$$

From this point of view, the existence of two types of matrix models for noncritical strings is related to the existence of two types of static D-branes in noncritical string theory, a satisfying conclusion.

The considerations above were extended in Ref.[32] to $c = 1$ string theory. One may expect to reproduce the $W_\infty$ model in this way, but in fact something different seems to emerge: a matrix model where the $SU(2)$ symmetry of the $c = 1$ string at selfdual radius is manifest. This is not surprising since the open-string field theory action evaluated on physical states will always lead to a matrix model with all the symmetries of the continuum theory. On the other hand, the $W_\infty$ model does not exhibit this $SU(2)$ symmetry (that would mix tachyons with discrete states, while the $W_\infty$ model has only tachyons and no discrete states). It is not totally clear how to reconcile the two models at present.

*Topological D-branes.* An interpretation of the matrix models of moduli space, specifically the Kontsevich and $W_\infty$ models, has been advanced in Ref. [34] in terms of topological D-branes. In particular, it is argued that topological strings on Calabi-Yau geometries can unify the "ordinary" matrix models and the Kontsevich-like (moduli space) matrix models. This approach, and its relationship to other approaches that do not make use of topological string theory, needs to be clarified.

*Nonperturbative effects.* The matrix models of moduli space have in all cases been invented to describe the topology of moduli space genus by genus. This is natural for mathematicians, for whom each genus Riemann surface can be treated as a quite distinct entity. Physicists, however, like to think of summing over Riemann surfaces and even going beyond such a sum to explore nonperturbative phenomena. This is clearly motivated by string theory. So one might ask if the matrix models discussed in these notes include nonperturbative effects, or exhibit nonperturbative ambiguities in the sense of Borel resummation. Recently a formula has been advanced for the nonperturbative partition function of $c = 1$ noncritical string theories (at self-dual radius)[35] and it would be nice to examine in some detail how it relates to the $W_\infty$ model of Ref.[29] as well as the $SU(2)$-symmetric matrix model of Ref.[32].

## References


[1] V. A. Kazakov, A. A. Migdal and I. K. Kostov, "Critical Properties Of Randomly Triangulated Planar Random Surfaces", Phys. Lett. B **157** (1985) 295.

[2] E. Brezin and V. A. Kazakov, "Exactly Solvable Field Theories Of Closed Strings", Phys. Lett. B **236** (1990) 144.

[3] M. R. Douglas and S. H. Shenker, "Strings In Less Than One-Dimension", Nucl. Phys. B **335** (1990) 635.

[4] D. J. Gross and A. A. Migdal, "Nonperturbative Two-Dimensional Quantum Gravity", Phys. Rev. Lett. **64** (1990) 127.

[5] T. Takayanagi and N. Toumbas, "A matrix model dual of type 0B string theory in two dimensions", JHEP **0307** (2003) 064 [arXiv:hep-th/0307083].

[6] M. R. Douglas, I. R. Klebanov, D. Kutasov, J. Maldacena, E. Martinec and N. Seiberg, "A new hat for the c = 1 matrix model", arXiv:hep-th/0307195.

[7] M. Kontsevich, "Intersection theory on the moduli space of curves and the matrix Airy function", Commun. Math. Phys. **147** (1992) 1.

[8] R. Penner, "Perturbative series and the moduli space of Riemann surfaces", J. Diff. Geom. **27** (1988) 35.

[9] E. Witten, "On the structure of the topological phase of two-dimensional gravity", Nucl. Phys. B **340** (1990) 281.

[10] J. Distler and C. Vafa, "A critical matrix model at $c = 1$", Mod. Phys. Lett. A **6** (1991) 259.

[11] J. McGreevy and H. Verlinde, "Strings from tachyons: The $c = 1$ matrix reloaded", arXiv:hep-th/0304224.

[12] I. R. Klebanov, J. Maldacena and N. Seiberg, "D-brane decay in two-dimensional string theory", JHEP **0307** (2003) 045 [arXiv:hep-th/0305159].

[13] D. Gaiotto and L. Rastelli, "A paradigm of open/closed duality: Liouville D-branes and the Kontsevich model", arXiv:hep-th/0312196.

[14] V. Fateev, A. B. Zamolodchikov and A. B. Zamolodchikov, "Boundary Liouville field theory. I: Boundary state and boundary two-point function", arXiv:hep-th/0001012.

[15] J. Teschner, "Remarks on Liouville theory with boundary", arXiv:hep-th/0009138.

[16] A. B. Zamolodchikov and A. B. Zamolodchikov, "Liouville field theory on a pseudosphere", arXiv:hep-th/0101152.

[17] J. L. Harer and D. Zagier, "The Euler characteristic of the moduli space of curves", Inv. Math. **85** (1986) 457.

[18] J. L. Harer, "The cohomology of the moduli space of curves", in "Theory of Moduli", Lecture Notes in Mathematics, Springer-Verlag (1988), E. Sernesi (Ed.).

[19] K. Strebel, "Quadratic Differentials", Springer-Verlag (1984).

[20] S. Mukhi, "Topological matrix models, Liouville matrix model and c = 1 string theory", arXiv:hep-th/0310287.

[21] D. Mumford, "Towards An Enumerative Geometry Of The Moduli Space Of Curves," in *Arithmetic And Geometry*, eds. M. Artin and J. Tate (Birkhauser, 1985).

[22] S. Morita, "Characteristic Classes of Surface Bundles," Invent. Math. **90** (1987) 551.

[23] E. Miller, "The Homology Of The Mapping Class Group," J. Diff. Geom. **24** (1986) 1.

[24] E. Witten, "Two-Dimensional Gravity And Intersection Theory On Moduli Space", Surveys Diff. Geom. **1** (1991) 243.

[25] E. Witten, "On the Kontsevich model and other models of two-dimensional gravity", IASSNS-HEP-91-24

[26] M. Adler and P. van Moerbeke, "A Matrix integral solution to two-dimensional W(p) gravity" Commun. Math. Phys. **147** (1992) 25.

[27] S. Kharchev, A. Marshakov, A. Mironov, A. Morozov and A. Zabrodin, "Towards unified theory of 2-d gravity", Nucl. Phys. B **380** (1992) 181 [arXiv:hep-th/9201013].

[28] I. R. Klebanov, "String theory in two dimensions", arXiv:hep-th/9108019.

[29] C. Imbimbo and S. Mukhi, "The topological matrix model of $c = 1$ string", Nucl. Phys. B **449** (1995) 553 [arXiv:hep-th/9505127].

[30] R. Dijkgraaf, G. W. Moore and R. Plesser, "The partition function of 2-D string theory", Nucl. Phys. B **394** (1993) 356 [arXiv:hep-th/9208031].

[31] S. Y. Alexandrov, V. A. Kazakov and I. K. Kostov, "2-D string theory as normal matrix model", Nucl. Phys. B **667** (2003) 90 [arXiv:hep-th/0302106].

[32] D. Ghoshal, S. Mukhi and S. Murthy, "Liouville D-branes in two-dimensional strings and open string field theory", JHEP **0411** (2004) 027 [arXiv:hep-th/0406106].

[33] R. Dijkgraaf and C. Vafa, "$N = 1$ supersymmetry, deconstruction, and bosonic gauge theories", arXiv:hep-th/0302011.

[34] M. Aganagic, R. Dijkgraaf, A. Klemm, M. Marino and C. Vafa, "Topological strings and integrable hierarchies", arXiv:hep-th/0312085.

[35] M. Aganagic, H. Ooguri, N. Saulina and C. Vafa, "Black holes, q-deformed 2d Yang-Mills, and non-perturbative topological strings", arXiv:hep-th/0411280.

# MATRIX MODELS AND 2D STRING THEORY

Emil J. Martinec
*Emil J. Martinec*
*University of Chicago*
*5640 S. Ellis Ave., Chicago, IL 60637, USA*
e-martinec@uchicago.edu

**Abstract** String theory in two-dimensional spacetime illuminates two main threads of recent development in string theory: (1) Open/closed string duality and (2) Tachyon condensation. In two dimensions, many aspects of these phenomena can be explored in a setting where exact calculations can be performed. These lectures review the basic aspects of this system.

## 1. Introduction

One of the most remarkable developments in string theory in recent years is the idea that, in certain circumstances (superselection sectors), it has a presentation as a large N gauge dynamics – gravitation is a collective phenomenon of the gauge theory, and closed strings are represented by loop observables of the gauge theory. The gauge theory in these situations provides an ansatz for the nonperturbative *definition* of the theory in that superselection sector.

By superselection sector, one means a choice of asymptotic behavior for the low-energy fields. A canonical example is string theory in $AdS_5 \times \mathbb{S}^5$ (for a review, see [1]), where the states of the theory all have a metric that asymptotes to the anti-de Sitter metric times a round sphere, and the self-dual five-form field strength of type IIB supergravity carries $N$ units of flux through the $\mathbb{S}^5$. The gauge theory equivalent is maximally supersymmetric $U(N)$ Yang-Mills theory in $D = 4$ spacetime dimensions. The correspondence equates states of geometry and matter in this superselection sector with states of the gauge theory. Both $AdS_5 \times \mathbb{S}^5$ and maximally supersymmetric gauge theory possess the same global superconformal symmetry ($SU(2, 2|4)$ in the language of supergroups), which then organizes the state space into representations of the superconformal algebra. For instance, one can match the one-particle states, and the operators that create them from the vacuum, by their representation properties. The operators that create and destroy strings are represented in the



*E. Brezin et al. (eds.), Applications of Random Matrices in Physics,* 403–457.

gauge theory description by Wilson loops, $\mathrm{tr}[\exp i\oint A]$, and their supersymmetric generalizations.

Upon the injection of a little energy, in gravity the generic state is a gas of supergravitons (the graviton and particles related to it by supersymmetry); if we put in a lot of energy, we expect a black hole to form. On the gauge theory side, at low energies the excitations are built from collections of gauge singlet operators (multiple 'single-particle creation operators') acting on the vacuum; at high energies, the gauge theory undergoes a "deconfinement transition" where energy is equipartitioned into all $N^2$ fields of the matrix field theory. The correspondence equates the transition from supergraviton gas to black hole on the geometry side, and the deconfinement transition on the gauge theory side [2].

Indeed, this equivalence first arose via the study of black holes carrying D-brane charge in string theory (for a review, see [3]). On the one hand, the dynamics on the branes is described at low energies by the lightest strings attached to the branes. The spacetime effective field theory, in which these strings are the quanta, is a Yang-Mills gauge theory with various matter fields. On the other hand, the branes source a geometry in which there is an increasing redshift of physics near the branes, as seen by asymptotic observers. Thus low energy also means gravitational physics near the branes. The gauge/gravity equivalence is the statement that these two descriptions have an overlapping region of validity, namely that of objects near the branes at low energies. In particular, geometrical excitations of the brane typically lead to horizon formation ('black' branes), whose thermodynamic properties (*c.f.* [3, 1]) can be compared to those of the gauge theory in the cases where they can be computed.

The loop variables describing strings in the gauge theory representation are often cumbersome to work with, and it remains a problem to dig out quasi-local gravitational and other closed string physics from this exact formulation. For instance, the local physics of the horizon and singularity of black holes and black branes are not well-understood in the gauge theory language (although there is some recent progress [4]). It would be useful to have a well-developed dictionary translating between gauge theory quantities and the standard perturbative formulation of string theory as a sum over surfaces. Generally, we don't know how to read off local physics beyond qualitative statements which are dictated by symmetries (in particular, by scaling arguments) [5–7].

Part of the reason that this dictionary is poorly developed is that the correspondence is a strong/weak coupling duality. The radius of curvature $R$ of both $AdS_5$ and $\mathbb{S}^5$, relative to the Planck scale $\ell_{\rm pl}$ of quantum gravity, is $N = (R/\ell_{\rm pl})^4$; relative to the scale $\ell_{\rm s}$ set by the string tension, it is $g^2_{\rm YM}N = (R/\ell_{\rm s})^4$. Thus for the spacetime to have a conventional interpretation as a geometry well-approximated by classical Einstein gravity, we should work in the gauge theory at both large $N$ and large effective ('t Hooft) coupling strength

$g_{\rm YM}^2 N$. Thus when stringy and quantum gravity fluctuations are suppressed, the gauge theory description is strongly coupled; and when the gauge theory is perturbative, the geometry has unsuppressed quantum and stringy effects.

Often in physics, useful information can be gathered by consideration of low-dimensional model systems, which hopefully retain essential features of dynamics, while simplified kinematics renders precise analysis possible. If one or another side of the duality is exactly solvable, then we can bypass the difficulty of strong/weak duality.

String theory in two spacetime dimensions provides just such an example of the gauge/gravity (or rather open string/closed string) correspondence, in which the gauge theory is an exactly solvable random matrix model, and the worldsheet description of string theory involves a conformal field theory (CFT) which has been solved by conformal bootstrap techniques.

The random matrix formulation of 2D string theory was discovered well before the recent developments involving D-branes; in fact it provided some of the motivation for the discovery of D-branes. The initial work on the matrix model is reviewed extensively in [8, 9]. The exact solution of Liouville theory was not developed at that time, and so precise comparison with worldsheet computations was rather limited in scope. The development of the conformal bootstrap for Liouville [10–15], reviewed in [16, 17], took place in the following decade, while much of string theory research was focussed on gauge/gravity equivalence. It has only been in the last year or so that these various threads of research have been woven together [18–22].

Our goal in these lectures will be to provide a self-contained overview this system, giving an introduction to the matrix model of 2D string theory, as well as the CFT techniques used to calculate the corresponding perturbative string amplitudes. We will then illustrate the map between these two presentations of 2D string theory.

Along the way, we will encounter a second major theme in recent string research – the subject of *tachyon condensation* (for reviews, see [23, 24]. A tachyon is simply terminology for an instability, a perturbation which grows exponentially instead of undergoing small oscillations. Loosely speaking, in the 'effective potential' of string theory, one has chosen to start the world at a local maximum of some component of the 'string field'. By condensing this mode, one learns about the topography and topology of this effective potential, and thus about the vacuum structure of string theory.

Much effort has gone into understanding the tachyons associated to the decay of unstable collections of D-branes in string theory. Here the unstable mode or modes are (open) strings attached to the brane or branes. For example, when one has a brane and an anti-brane, the initial stages of their mutual annihilation is described by the condensation of the lightest (in this case, tachyonic) open string stretching between the brane and the anti-brane. Eventually the brane decays completely into (closed) string radiation. One might wonder whether there is a region of overlapping validity of the two descriptions, just as in the gauge/gravity (open string/closed string) correspondences described above. We will see evidence that this is the case in 2D string theory. The random matrix presentation of 2D string theory was first introduced as an alternative way to describe the worldsheets of closed 2D strings, yet the evidence suggests that it is in fact a description of the open string tachyon condensate on unstable D-particles.

The lectures are aimed at a broad audience; along the way, many ideas familiar to the practicing string theorist are summarized in order to make the presentation as self-contained as possible. We begin with a brief overview of perturbative string theory as a way of introducing our primary subject, which is string theory in two-dimensional backgrounds.

## 2. An overview of string theory

String theory is a generalization of particle dynamics.[1] The sum over random paths gives a representation of the particle propagator

$$\begin{aligned} G(x,x') &= \langle x'|\tfrac{i}{\partial^2-m^2}|x\rangle \\ &= \int_0^\infty dT\,\langle x'|e^{iT(\partial^2-m^2)}|x\rangle \\ &= \int_{\substack{X(0)=x\\ X(1)=x'}} \frac{\mathcal{D}g\,\mathcal{D}X}{\mathrm{Diff}}\,\exp\Big[i\int_0^1 dt\sqrt{g}[g^{tt}\partial_t X^2+m^2]\Big]\,. \end{aligned} \tag{2.1}$$

In the second line, the use of the proper time (*Schwinger*) parametrization turns the evaluation of the propagator into a quantum mechanics problem, which can be recast as a path integral given by the last line. The introduction of intrinsic worldline gravity via the worldline metric $g_{tt}$, while not essential, is useful for the generalization to string theory. The worldline metric $g_{tt}$ acts as a Lagrange multiplier that enforces the constraint

$$T_{tt} = (\partial_t X)^2 - g_{tt}m^2 = 0\,; \tag{2.2}$$

apart from this constraint, the dynamics of worldline gravity is trivial. Indeed, we can fix a gauge $g_{tt} = T$,[2] and after rescaling $\tau = Tt$, equation (2.1) boils down to the standard path integral representation

$$G(x,x') = \int_{\substack{X(0)=x\\ X(T)=x'}} \mathcal{D}X\,\exp\Big[i\int_0^T d\tau\,[(\partial_\tau X)^2+m^2]\Big]\,. \tag{2.3}$$

We can generalize this construction in several ways. For instance, we can put the particle in a curved spacetime with metric $G_{\mu\nu}(X)$, and in a background potential $V(X)$ that generalizes the constant $m^2$;[3] also, we can couple a charged particle to a background electromagnetic field specified by the vector potential $A_\mu(X)$. The effect is to replace the free particle action in (2.1) by a generalized 'worldline nonlinear sigma model'

$$\mathcal{S}_{\text{worldline}} = \int dt \Big[\sqrt{g} g^{tt} G_{\mu\nu}(X) \partial_t X^\mu \partial_t X^\nu + A_\mu(X) \partial_t X^\mu - \sqrt{g} V(X)\Big] . \tag{2.4}$$

String theory introduces a second generalization, replacing the notion of dynamics of pointlike objects to that of extended objects such as a one-dimensional string. Perturbative string dynamics is governed by an action which is the analogue of (2.4)

$$\mathcal{S}_{\text{WS}} = \frac{1}{4\pi\alpha'} \int d^2\sigma \Big[\Big(\sqrt{g} g^{ab} G_{\mu\nu}(X) + \epsilon^{ab} B_{\mu\nu}(X)\Big) \partial_a X^\mu \partial_b X^\nu + \alpha' \sqrt{g} R^{(2)} \Phi(X) + \sqrt{g}\, V(X)\Big] \tag{2.5}$$

where $a, b = 0, 1$ and $\mu, \nu = 0, ..., D-1$ are worldsheet and target space indices, respectively. The quantity $\alpha' = \ell_s^2$ sets a length scale for the target space parametrized by $X^\mu$; it plays the role of $\hbar$ for the generalized nonlinear sigma model (2.5). The antisymmetric tensor gauge field $B_{\mu\nu}$ is the direct generalization of the vector potential $A_\mu$; the former couples to the area element $dX^\mu \wedge dX^\nu$ of the two-dimensional string worldsheet in the same way that the latter couples to the line element $dX^\mu$ of the particle worldline. In addition, because intrinsic curvature $R^{(2)}$ can be non-trivial in two dimensions, one has an additional coupling of the curvature density to a field $\Phi$ known as the *string dilaton*.

The dynamical principle of the worldsheet theory is the requirement that

$$\langle \cdots\ T_{ab}\ \cdots \rangle = 0 \tag{2.6}$$

in all correlation functions. The two traceless components of these equations play the same role as the constraint (2.2) – they enforce reparametrization invariance on the worldsheet. The trace component is a requirement that the 2d QFT of the worldsheet dynamics is *locally scale invariant*, *i.e.* that the beta functions vanish. For example, setting $B_{\mu\nu} = V = 0$, the conditions through one loop are

$$\begin{aligned} \beta_{G_{\mu\nu}} &= \alpha'(\mathcal{R}_{\mu\nu}(G) + \nabla_\mu \nabla_\nu \Phi) + O(\alpha'^2) = 0 \\ \beta_\Phi &= \frac{D-26}{6} + \alpha'\big(\tfrac{1}{2}\nabla^2\Phi + (\nabla\Phi)^2\big) + O(\alpha'^2) = 0 \end{aligned} \tag{2.7}$$

where $\mathcal{R}_{\mu\nu}(G)$ is the Ricci curvature of the spacetime metric $G$, and $\nabla$ is the spacetime gradient. Thus, a reason to be interested in string theory is that, in contrast to the point particle, the string carries with it the information about what spacetimes it is allowed to propagate in – namely, those that satisfy the

Einstein equations coupled to a scalar dilaton (and other fields, if we had kept them nonzero).

Since the local invariances combine the reparametrization group *Diff* and the group of local scale transformations $Weyl$, the appropriate replacement for (2.1) is

$$\mathcal{Z} = \int \frac{\mathcal{D}g\,\mathcal{D}X}{\text{Diff} \times \text{Weyl}} \exp[i\mathcal{S}_{\text{WS}}] \,. \tag{2.8}$$

We can soak up the local gauge invariance by (locally on the worldsheet) choosing coordinates in which $g_{ab} = \delta_{ab}$. One cannot choose such flat coordinates globally, however, as one sees from the Gauss-Bonnet identity $\int \sqrt{g} R^{(2)} = 4\pi(2-2h)$.[4] Nevertheless, one can relate any metric via the symmetries to one of a $6h-6$ parameter family of reference metrics $\hat{g}_{ab}(m_r)$, $r = 1, ..., 6h-6$. The parameters $m_r$ are called the *moduli* of the 2d surface.[5] A simple picture of these parameters is shown in figure 1.

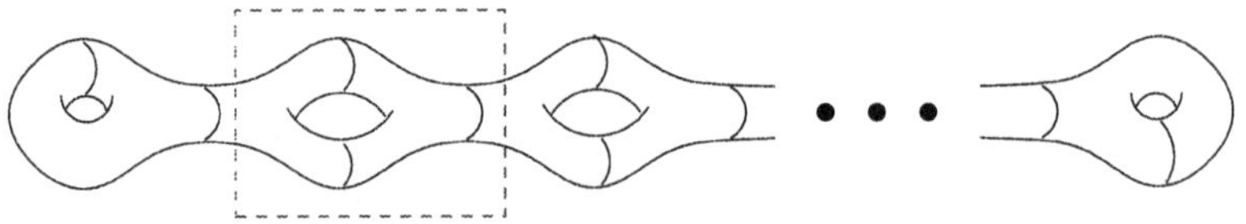

*Figure 1.* Each handle, except the end ones, contributes three closed string propagator tubes to the surface. Each tube has a length and a helical twist angle. The two end handles together contribute only three tubes, and so the number of moduli is $6h-6$.

Thus, after fixing all of the reparametrization and local scale invariance, the integration over metrics $\int \frac{\mathcal{D}g}{\text{Diff}\times\text{Weyl}}$ reduces to an integration over these moduli. The moduli are the string version of the Schwinger parametrization of the propagator (2.1) for a particle.

## 3. Strings in D-dimensional spacetime

A simple solution to the equations (2.7) uses ‘conformally improved’ free fields:[6]

$$G_{\mu\nu} = \eta_{\mu\nu} \quad , \qquad B_{\mu\nu} = V = 0 \quad , \qquad \Phi = n_\mu X^\mu \qquad \left(n^2 = \tfrac{26-D}{6\alpha'}\right) \,. \tag{3.9}$$

The geometry seen by propagating strings is flat spacetime, with a linear dilaton. The dilaton slope is timelike for $D > 26$ and spacelike for $D < 26$.

Just as the perturbative series for particles is a sum over Feynman graphs, organized in order of increasing number of loops in the graph, the perturbative expansion for strings is organized by the number of handles of the corresponding sum over worldsheets, weighted by the effective coupling $g_{\text{eff}}$ to the power $2h-2$ (where $h$ is the number of handles, often called the *genus* of the surface).

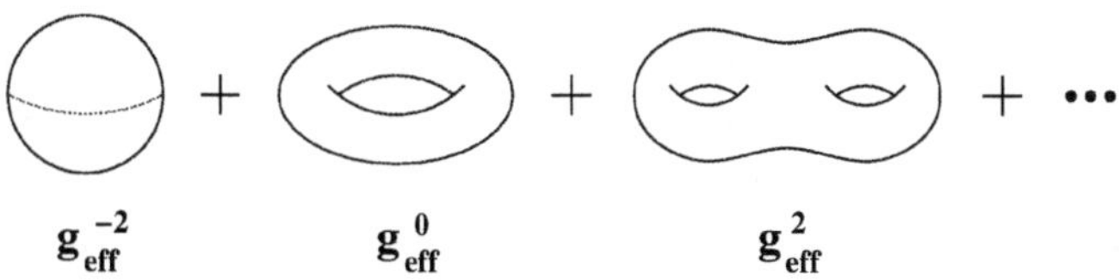


*Figure 2.* The string loop expansion.

go

Consider string worldsheets in the vicinity of the target space location $\hat{X}$. Using the Gauss-Bonnet identity, the term

$$\frac{1}{4\pi}\int \sqrt{g}R^{(2)}\,\Phi(X) \sim \Phi(\hat{X})(2-2h) \tag{3.10}$$

in the path integral over the (Euclidean) worldsheet action identifies the effective coupling as

$$g_{\text{eff}} = \exp[\Phi(\hat{X})]\ . \tag{3.11}$$

Thus we have strong coupling at large $\Phi = n \cdot X$, and we have to say what happens to strings that go there.

There is also a perturbative instability of the background. Perturbations of the spacetime background are scaling operators. Maintaining conformal invariance at the linearized level imposes marginality of the scaling operator. These marginal scaling operators are known as *vertex operators.* Consider for instance adding the potential term

$$V(X) = \int \frac{d^D k}{(2\pi)^D}\, v_k e^{ik\cdot X} \tag{3.12}$$

to the worldsheet action. The scale dimension of an individual Fourier component is determined by its operator product with the stress tensor[7]

$$T(z)\, e^{ik\cdot X(w)} \overset{z\to w}{\sim} \frac{\Delta}{(z-w)^2}\, e^{ik\cdot X(w)}\ . \tag{3.13}$$

Using the improved stress tensor[8]

$$T(z) = -\tfrac{1}{\alpha'}\partial_z X \cdot \partial_z X + n \cdot \partial_z^2 X \tag{3.14}$$

and evaluating the operator product expansion (3.13) via Wick contraction with the free propagator

$$X(z)X(w) \sim -\tfrac{\alpha'}{2}\log|z-w|^2\ , \tag{3.15}$$

one finds the scale dimension

$$\Delta = \tfrac{\alpha'}{4}k^2 + \tfrac{i\alpha'}{2}n \cdot k \quad . \tag{3.16}$$

Thus the condition of linearized scale invariance $\Delta = \bar{\Delta} = 1$ is a mass-shell condition for $V(X)$. This result should be no surprise – local scale invariance gives the equations of motion (2.7) of the background, so the linearized scale invariance condition should give the wave equation satisfied by small perturbations. The mass shell condition $\Delta = 1$ amounts to

$$(k + in)^2 = -n^2 - \tfrac{4}{\alpha'} \tag{3.17}$$

(recall $n^2 = \frac{26-D}{6\alpha'}$). Thus for $D < 2$, perturbations are "massive", and the string background is stable. For $D = 2$, the perturbations are "massless", leading to marginal stability. Finally, for $D > 2$ the perturbations are "tachyonic", and the background is unstable. The field $V(X)$ is conventionally called the *string tachyon* even though strictly speaking that characterization only applies to $D > 2$.

In the stable regime $D \leq 2$, a static background condensate $V(X)$ "cures" the strong coupling problem.[9] Let $n \cdot X = QX_1$ (recall $n^2 > 0$); then for $D < 2$

$$\begin{aligned} V_{\rm backgd} &= \mu\, e^{2bX_1} + \tilde{\mu}\, e^{2\tilde{b}X_1} \\ \left.\begin{matrix} b \\ \tilde{b} \end{matrix}\right\} &= \tfrac{Q}{2} \mp \sqrt{(\tfrac{Q}{2})^2 - \tfrac{1}{\alpha'}} = \tfrac{\sqrt{26-D} \mp \sqrt{2-D}}{\sqrt{24\alpha'}} \; . \end{aligned} \tag{3.18}$$

(note that $\tilde{b} = (b\alpha')^{-1}$). For $D = 2$ one has $b = \tilde{b} = 1/\sqrt{\alpha'}$, and so the two exponentials are not independent; rather

$$V^{(D=2)}_{\rm backgd} = \mu\, X_1 e^{2bX_1} + \tilde{\mu}\, e^{2bX_1} \; . \tag{3.19}$$

The exponential barrier self-consistently keeps perturbative string physics away from strong coupling for sufficiently large $\mu$.

For example, consider the scattering of a string tachyon of energy $E$ in $D = 2$. The string is a perturbation $\delta V(X) = \exp[-iEX^0 + ikX^1]$, with $ik = \pm iE + Q$ the solution to the on-shell condition $\Delta = 1$. The scattering is depicted in figure 4.

The worldsheet energy $\mathcal{E}_{\rm WS} = \alpha' E^2/2$ of the zero mode motion in $X^1$ of the string is determined by the stress tensor $T(z)$, equation (3.14); it is essentially the $X^1$ contribution to $\Delta$. The turning point of the motion is determined by this energy to be

$$V_{\rm backgd}(X_1^{\rm max}) = \mathcal{E}_{\rm WS} = \alpha' E^2/2 \; . \tag{3.20}$$

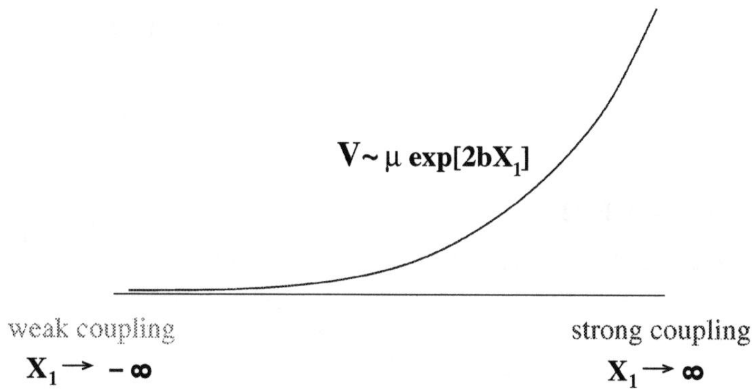


*Figure 3.* The tachyon background.

The effective coupling is largest at this point,

$$g_{\text{eff}} \sim e^{\Phi(X_1^{\text{max}})} \sim E^2/\mu \; ; \tag{3.21}$$

thus low energy scattering is self-consistently weakly coupled. The effective coupling is determined by the value of the dilaton at the turning point; we expect the scattering amplitude to be have a perturbative series in powers of $E^2/\mu$. Note that the high energy behavior is nonperturbative, however.

At this point we choose to relabel for $D = 2$ the spacetime coordinates

$$\phi \equiv X^1 \quad , \qquad X \equiv X^0 \tag{3.22}$$

in order to conform to standard notation in the subject, as well as to reduce the clutter of indices. Also, we will henceforth set $\alpha' = 1$ as a choice of units (*i.e.* we measure all spacetime lengths in "string units").

## 3.1 A reinterpretation of the background

The 2d QFT[10] of the "tachyon" background

$$\mathcal{S}_{\text{WS}} = \frac{1}{4\pi}\int \sqrt{g}\Big[g^{ab}\partial_a\phi\partial_b\phi + bQ\, R^{(2)}\phi + \mu\, e^{2b\phi}\Big] \tag{3.23}$$

has an alternative interpretation in terms of worldsheet intrinsic geometry [28], where $e^{2b\phi}g_{ab}$ is interpreted as a *dynamical metric*, and the remaining $D-1$ fields $X$ are thought of as "matter" coupled to this dynamical gravity. Let $\varphi = b\phi$; then the action becomes

$$\mathcal{S}_{\text{WS}} = \frac{1}{4\pi b^2}\int \sqrt{g}\Big[(\nabla\varphi)^2 + bQR^{(2)}\varphi + \mu b^2\, e^{2\varphi}\Big]\,. \tag{3.24}$$

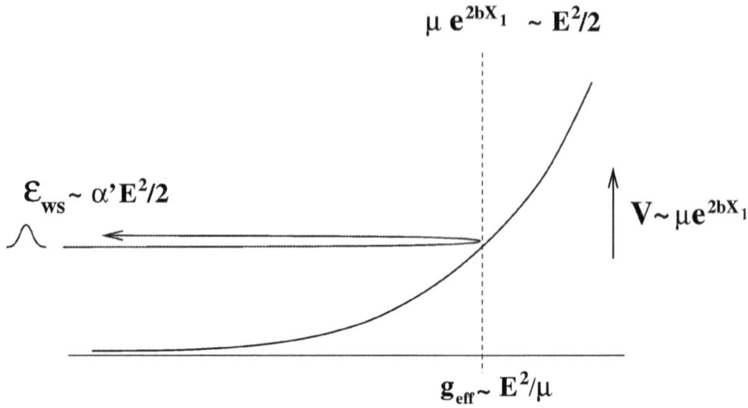


*Figure 4.* Scattering of a tachyon excitation off the tachyon background.

Note that $b$ plays the role of the coupling constant; the semi-classical limit is $b \to 0$ (and thus $Q = b^{-1} + b \to b^{-1}$). The equation of motion for $\varphi$ reads

$$\nabla_g^2\, 2\varphi - bQ\, R^{(2)}[g] = 2\mu b^2\, e^{2\varphi}\ . \tag{3.25}$$

Here $\nabla_g$ is the covariant derivative with respect to the intrinsic metric $g_{ab}$. Due to the properties of the curvature under local rescaling,

$$\nabla_g^2\, 2\varphi - R^{(2)}[g] = -e^{2\varphi} R^{(2)}[e^{2\varphi} g]\ , \tag{3.26}$$

the combination on the left-hand side of (3.25) is, in the semi-classical limit $b \to 0$, just the curvature of the dynamical metric $e^{2\varphi} g_{ab}$. The equation of motion can be written as the condition for constant curvature of this dynamical metric

$$R^{(2)}[e^{2\varphi} g] = -2\mu b^2\ , \tag{3.27}$$

known as the *Liouville equation*; the theory governed by the action (3.24) is the Liouville field theory. The equation (3.25) is the appropriate quantum generalization of the Liouville equation. The constant on the right-hand side of (3.27) is a cosmological constant for the 2d intrinsic fluctuating geometry.[11] Note that $\sqrt{g}e^{2\varphi}$ is the "dynamical area element", so that the potential term in the action (3.24) is a chemical potential for the dynamical intrinsic area of the worldsheet.

This interpretation of the static tachyon background in terms of fluctuating intrinsic geometry is only available for $D \leq 2$. For $D > 2$, the on-shell condition $\Delta = 1$ (equation (3.17)) is not solved by $V_{\rm backgd} = e^{2bX_1}$ for real $b$ (rather $b = \frac{1}{2}Q \pm i\lambda$), and so $\sqrt{g}V_{\rm backgd}$ is not the area of a dynamical surface.

## 3.2 KPZ scaling

The fact that the dynamical metric is integrated over yields useful information about the scaling of the partition and correlation functions with respect to the cosmological constant $\mu$, known as *KPZ scaling* [29–31]. Consider the shift $\varphi \to \varphi + \frac{\epsilon}{2}$ in the Liouville action (3.24) in genus $h$; this leads to

$$\mathcal{S}_h(\mu) \longrightarrow \mathcal{S}_h(e^{\epsilon}\mu) + (2-2h)\tfrac{Q}{2b}\,\epsilon\,. \tag{3.28}$$

However, this constant mode of $\varphi$ is integrated over in the Liouville partition function, and therefore $\mathcal{Z}_h(\mu)$ must be independent of $\epsilon$. We conclude

$$\mathcal{Z}_h(\mu) = \mathcal{Z}_h(e^{\epsilon}\mu)\exp[-(2-2h)\tfrac{Q}{2b}\,\epsilon] \quad\Longrightarrow\quad \mathcal{Z}_h(\mu) = c_h\,\mu^{(2-2h)Q/2b}\,. \tag{3.29}$$

For instance, for $D = 1$ (pure Liouville gravity, with no matter) one finds $\frac{Q}{2b} = \frac{5}{4}$, and so the genus expansion of the partition function is a series in $\mu^{-5/2}$. For $D = 2$, we have $\frac{Q}{2b} = 1$, and so the partition function is a series in $\mu^{-2}$.[12,13]

We could now pass to a discussion of correlation functions of this 2d Liouville QFT, and their relation to the scattering of strings. Instead, we will suspend this thread of development in favor of a random matrix formulation of the same physics. We will return to the quantization of Liouville theory later, when it is time to forge the link between these two approaches.

# 4. Discretized surfaces and 2D string theory

For spacetime dimension $D \leq 2$, we have arrived at an interpretation of the path integral describing string propagation in the presence of a background tachyon condensate as a sum over dynamical worldsheet geometries, in the presence of $D-1$ "matter fields".[14]

A discrete or lattice formulation of fluctuating worldsheet geometry can be given in terms of matrix Feynman graphs. Any tesselation of a surface built of regular polygons (see figure 5 for a patch of tesselated surface) has a dual[15] double-line "fatgraph", also depicted in figure 5. The double lines indicate the flow of matrix index contractions around the graph.

The partition function

$$\begin{aligned}\mathcal{Z}(g_i) &= \int \mathcal{D}^{N^2}M \exp[-\mathrm{tr}(\tfrac{1}{2}g_2M^2 + \mathcal{U}(M)]\\ \mathcal{U}(M) &= \tfrac{1}{3}g_3M^3 + \tfrac{1}{4}g_4M^4 + \dots \quad .\end{aligned} \tag{4.30}$$

serves as a generating function for fatgraphs, and thereby defines an ensemble of random surfaces. For example, consider a surface with triangles only, $g_{i>4} = 0$. Each face of the fatgraph gives a factor $N$ from the trace over the index loop bordering the face. Each vertex gives a factor $g_3$, and each propagator

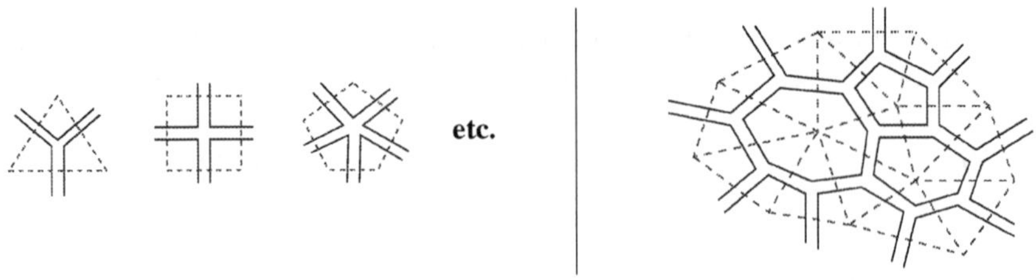


**(a)** **(b)**

*Figure 5.* (a) Regular polygons for tiling a surface, with dashed red edges; and the dual fatgraph vertices, with solid blue dual edges. (b) A patch of discrete surface tesselated with triangles, and the dual fatgraph.

$1/g_2$. The partition function

$$\mathcal{Z}(g) = \sum_{V,E,F} (g_3)^V (1/g_2)^E N^F \, d(V,E,F) \tag{4.31}$$

sums over the number $d(V,E,F)$ graphs with $V$ vertices, $E$ edges (propagators), and $F$ faces. Using the fact that each propagator shares two vertices, and each vertex ends three propagators, one has $2E = 3V$. The discrete version of the Gauss-Bonnet theorem (the Euler identity) is $V - E + F = 2 - 2h$. The partition function is thus

$$\mathcal{Z}(g) = \sum_{h=0}^{\infty} \sum_{A} N^{2-2h} \Big( \frac{g_3 N^{1/2}}{g_2^{3/2}} \Big)^A d(h,A) \tag{4.32}$$

where here and hereafter we write $V = A$, since the number of vertices $A$ is the discrete area of the surface. Large $N$ thus controls the topological expansion: $g_s^{\rm discrete} = 1/N$ is the string coupling of the discrete theory. The cosmological constant of the discrete theory is the free energy cost of adding area (triangles): $\mu_{\rm discrete} = -\log(g_3 N^{1/2}/g_2^{3/2})$.

Being a lattice theory, in order to compare with the continuum formulation of previous sections we need to take the continuum limit of the matrix integral. That is, we want to send the discrete area $A$ to infinity in units of the lattice spacing (or equivalently, send the lattice spacing to zero for a "typical" surface in the ensemble).

Taking this limit amounts to balancing the suppression of surface area by the 2d cosmological constant $\mu_{\rm discrete}$ against the entropy $d(h,A)$ of large Feynman graphs (roughly, if we want to add an extra vertex to a planar graph, there are of order $A$ places to put it). In other words, one searches for a phase transition or singularity in $\mathcal{Z}(g)$ where for some $g_{\rm crit}$ the partition sum is dominated by graphs with an asymptotically large number of vertices. Universality of this kind of critical phenomenon is the statement that the critical point is largely independent of the detailed form of the matrix potential $\mathcal{U}(M)$, for instance whether the dual tesselation uses triangles or squares in the microscopic theory (*i.e.* $M^3$ vs. $M^4$ interaction vertices in the graphical expansion).

Before discussing this phase transition, let us add in the matter. We wish to put discretized scalar field theory on the random surfaces generated by the path integral over $M$. The following modification does the job:

$$\mathcal{Z}= \int \mathcal{D}M \exp\Big[\mathrm{tr}\Big(\int dx \int dx' \, \tfrac{1}{2} M(x) G^{-1}(x-x') M(x') + \int dx \mathcal{U}(M(x))\Big)\Big] . \tag{4.33}$$

In the large $N$ expansion, we now have a propagator $G(x-x')$ in the Feynman rules (rather than $g_2^{-1} = const.$). Thus, on a given graph we have a product of propagators along the edges

$$\prod_{\rm edges} (\text{propagators}) = \prod_{\substack{i,j \\ \rm neighbors}} G(x_i - x_j) \; ; \tag{4.34}$$

the choice $G(x-x') = \exp[-(x-x')^2/\beta]$ leads to the discretized kinetic energy of a scalar field $X$

$$\prod_{\substack{i,j \\ \rm neighbors}} G(x_i - x_j) = \exp\Big[-\frac{1}{\beta} \sum_{\substack{i,j \\ \rm nghbrs}} (x_i - x_j)^2\Big] \tag{4.35}$$

which is the appropriate path integral weight for a scalar field on the lattice. The evaluation of the graph involves an integral $\prod_i \int dx_i$ over the location in $x$-space of all the vertices. In other words, we path integrate over the discretized scalar field with the probability measure (4.35).

Unfortunately, the gaussian kinetic energy that leads to this form of the propagator is not standard. Fortunately, for $D = 2$ (*i.e.* one scalar matter field) the choice $G(x-x') = \exp[-|x-x'|/\beta^{1/2}]$ turns out to be in the same universality class, and arises from a canonical kinetic energy for the matrix path integral

$$\begin{aligned} G^{-1}(p) = e^{\beta p^2} \quad &\longleftrightarrow \quad G(x) = \left(\tfrac{\pi}{\beta}\right)^{\frac{1}{2}} e^{-x^2/4\beta} \\ \tilde{G}^{-1}(p) = 1 + \beta p^2 \quad &\longleftrightarrow \quad \tilde{G}(x) = \tfrac{\pi}{\sqrt{\beta}} \, e^{-|x|/\beta^{1/2}} \end{aligned} \tag{4.36}$$

The continuum limit involves scalar field configurations which are slowly varying on the scale of the lattice spacing, which is enforced by taking $\beta p^2 \to 0$. But in this limit $G^{-1} \sim \tilde{G}^{-1}$ and so we expect the two choices to lead to the

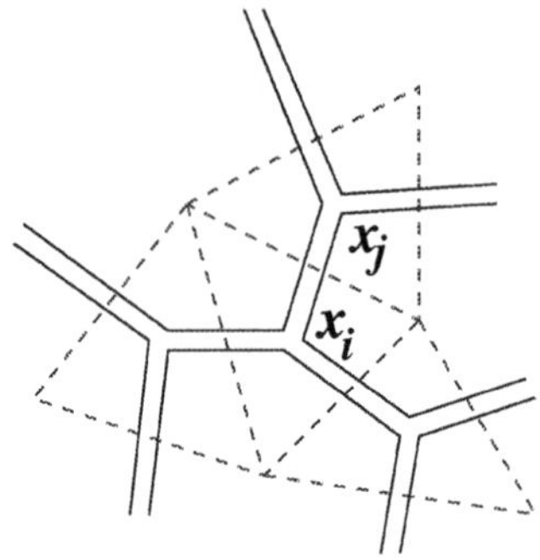


*Figure 6.* Fatgraph vertices live at points in $x$ space. The product over propagators weights the sum over configurations $\{x_i\}$ by a nearest-neighbor interaction determined by the propagator $G(x_i - x_j)$.

same continuum physics. But in $D = 2$ (*i.e.* one-dimensional $x$-space), $\tilde{G}(p)$ is the conventional Feynman propagator for $M$, and so we may write

$$\mathcal{Z} = \int \mathcal{D}M \exp\Big\{ - \int dx \; \mathrm{tr}\Big[ \frac{\beta}{2} \Big( \frac{dM}{dx} \Big)^2 - \mathcal{U}(M) \Big] \Big\} , \tag{4.37}$$

now with $\mathcal{U}(M) = -\frac{1}{2}M^2 - \frac{1}{3}gM^3$.

To analyze this path integral, it is most convenient to use the matrix analogue of polar coordinates. That is, let

$$M(x) = \Omega(x)\Lambda(x)\Omega^{-1}(x) \tag{4.38}$$

where $\Omega \in U(N)$ and $\Lambda = \mathrm{diag}(\lambda_1, \lambda_2, ..., \lambda_N)$. The integration measure $\mathcal{D}M$ becomes in these variables

$$\mathcal{D}M = \mathcal{D}\Omega \, \mathcal{D}\Lambda \, \Delta^2(\Lambda) \quad , \qquad \Delta(\Lambda) = \prod_{i<j} (\lambda_i - \lambda_j) \tag{4.39}$$

where $\mathcal{D}\Omega$ is the $U(N)$ group (Haar) measure.

A useful intuition to keep in mind is the analogous transformation from Cartesian to spherical coordinates for integration over the vector space $\mathbb{R}^n$. One uses the rotational invariance of the measure to write $d^n x = d\Omega_{n-1} dr \, r^{n-1}$, with $\Omega_{n-1}$ the space of angles which parametrize an orbit under the rotational group $O(n)$; $r$ parametrizes which orbit we have, and $r^{n-1}$ is the size of the orbit. The orbits degenerate at the origin $r = 0$, due to its invariance under $O(n)$, and this degeneration is responsible for the vanishing of the Jacobian factor $r^{n-1}$ on this degenerate orbit. Similarly, in the integration over matrices $\mathcal{D}M$ is the Cartesian measure on the matrix elements of $M$. The invariance of this measure under under unitary conjugation of $M$ allows us to pass to an integration over $U(N)$ orbits, parametrized by the diagonal matrix of eigenvalues $\Lambda$. The (*Vandermonde*) Jacobian factor $\Delta^2(\Lambda)$ characterizes the size of an orbit; the orbits degenerate whenever a pair of eigenvalues coincide, since the action of $SU(2) \subset U(N)$ (that rotates these eigenvalues into one another) degenerates at such points. The overall power of the Vandermonde determinant is determined by scaling (just as the power $r^{n-1}$ is fixed for the vector measure).

In these variables, the Hamiltonian for the matrix quantum mechanics (4.37) is

$$H = \sum_i \Big[-\frac{\beta}{2}\frac{1}{\Delta^2}\frac{\partial}{\partial\lambda_i}\Delta^2\frac{\partial}{\partial\lambda_i} + \mathcal{U}(M)\Big] + \frac{1}{2\beta}\sum_{i<j}\frac{\hat{\Pi}_{ij}\hat{\Pi}_{ji}}{(\lambda_i-\lambda_j)^2} \tag{4.40}$$

where $\hat{\Pi}_{ij}$ is the left-invariant momentum on $U(N)$, and the ordering has been chosen so that the operator is Hermitian with respect to the measure (4.39). The last term is the analogue of the angular momentum barrier in the Laplacian on $\mathbb{R}^n$ in spherical coordinates. Note that the kinetic operator for the eigenvalues can be rewritten

$$\sum_i \frac{1}{\Delta^2}\frac{\partial}{\partial\lambda_i}\Delta^2\frac{\partial}{\partial\lambda_i} = \sum_i \frac{1}{\Delta}\frac{\partial^2}{\partial\lambda_i^2}\Delta\,. \tag{4.41}$$

Wavefunctions for the $U(N)$ angular degrees of freedom will transform in representations of $U(N)$. The simplest possiblity is to choose the trivial representation, $\Psi_{U(N)}(\Omega) = 1$. In this $U(N)$ singlet sector, we can write the wavefunction as

$$\Psi(\Omega,\Lambda) = \Psi_{\text{eval}}(\Lambda) = \Delta^{-1}(\Lambda)\tilde{\Psi}(\Lambda) \tag{4.42}$$

and the Schrødinger equation becomes [33]

$$H\Psi_{\text{eval}}(\Lambda) = \Delta^{-1}(\Lambda)\sum_i\Big[-\frac{\beta}{2}\frac{\partial^2}{\partial\lambda_i^2} + \mathcal{U}(\lambda_i)\Big]\tilde{\Psi}(\Lambda)\,, \tag{4.43}$$

*i.e.* the eigenvalues are *decoupled* particles moving in the potential $\mathcal{U}(\lambda)$. The wavefunction $\Psi_{\text{eval}}$ is symmetric under permutation of the eigenvalues in the $U(N)$ singlet sector (these permutations are just the Weyl group action of $U(N)$); consequently $\tilde{\Psi}$ is totally antisymmetric under eigenvalue permutations – *the eigenvalues behave effectively as free fermions.*

## 4.1 An aside on non-singlets

What about non-singlet excitations? Gross and Klebanov [34, 8, 35] estimated the energy cost of non-singlet excitations and found it to be of order

$O(-\log\epsilon)$, where $\epsilon \to 0$ characterizes the continuum limit. Hence, angular excitations decouple energetically in the continuum limit. Alternatively, one can gauge the $U(N)$, replacing $\partial_x M$ by the covariant derivative $D_x M = \partial_x M + [A, M]$; the Gauss law of the gauge theory then projects onto $U(N)$ singlets.

The physical significance of non-singlet excitations is exhibited if we consider the theory in periodic Euclidean time $x \in \mathbb{S}^1$, $x \sim x + 2\pi R$, appropriate to the computation of the thermal partition function. In the matrix path integral, we must allow twisted boundary conditions for $M$ [34, 8, 35]:

$$M(x+2\pi R) = \Omega M(x)\Omega^{-1} \quad , \qquad \Omega \in U(N) \quad . \tag{4.44}$$

The matrix propagator is modified to

$$\langle M_i{}^k(x) M_j{}^l(x')\rangle = \sum_{m=-\infty}^{\infty} e^{-|x-x'+2\pi Rm|} \left(\Omega^m\right)_i{}^l \left(\Omega^{-m}\right)_j{}^k \, . \tag{4.45}$$

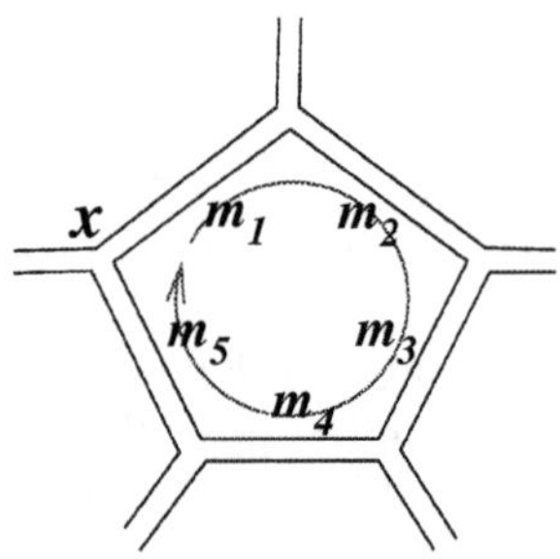


*Figure 7.* The product over twisted propagators around the face of a fatgraph allows monodromy for $x$, corresponding to a vortex insertion.

Consider a fixed set of $\{m_i\}$ and a fixed fatgraph. Following the propagators along the index line that bounds the face of a planar graph, figure 7, we see that the coordinate of a fatgraph vertex along the boundary shifts by

$$x \longrightarrow x + 2\pi\Big(\sum_i m_i\Big)R\,; \tag{4.46}$$

thus the sum over $\{m_i\}$ is a sum over vortex insertions on the faces of the graph (the vertices of the dual tesselation). The sum over twisted boundary conditions introduces vortices into the partition sum for the scalar matter field $X$. We can now understand the suppression of non-singlet wavefunctions as a reflection of the suppression of vortices in the 2d QFT of a periodic scalar below the Kosterlitz-Thouless transition.

## 4.2 The continuum limit

We are finally ready to discuss the continuum limit of the sum over surfaces. Recall that we wish to take $N \to \infty$, with the potential tuned to the vicinity of a phase transition – a nonanalytic point in the free energy as a function of the couplings in the potential $\mathcal{U}(M)$. We now know that the dynamics is effectively that of free fermionic matrix eigenvalues, moving in the potential $\mathcal{U}(\lambda)$. Consider $\mathcal{U}(\lambda) = -\frac{1}{2}\lambda^2 - g\lambda^3$, figure 8a.

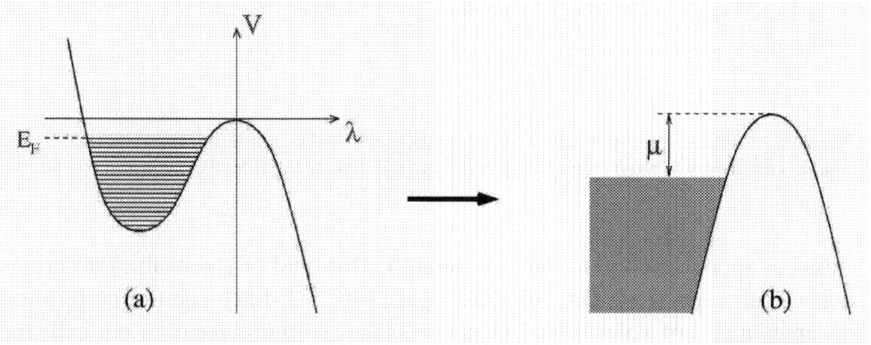


*Figure 8.* (a) Cubic eigenvalue potential. For small $g$, there are many metastable levels. (b) The scaling limit focusses on the vicinity of the local maximum of the potential.

There are many metastable levels in the well on the left of the local maximum of the potential. The coupling $g$ can be tuned so that there are more than $N$ such metastable single-particle states. As $N$ is sent to infinity, one can adjust $g \to 0$ so that there are always $N$ levels in the well. The metastable Fermi energy $E_F$ will be a function of $g$ and $N$. Consider an initial state where these states are populated up to some Fermi energy $E_F$ below the top of the barrier, and send $g \to 0$, $N \to \infty$, such that $E_F \to 0^-$. In other words, the phase transition we seek is the point where eigenvalues are about to spill over the top of the potential barrier out of the well on the left. The resulting situation is depicted in figure 8b, where we have focussed in on the quadratic maximum of the potential via the rescaling $\hat{\lambda} = \lambda/\sqrt{N}$, so that $\mathcal{U}(\hat{\lambda}) \sim -\frac{1}{2}\hat{\lambda}^2$. We hold $\mu = -NE_F$ fixed in the limit. The result is quantum mechanics of free fermions in an inverted harmonic oscillator potential, with Fermi level $-\mu < 0$. To avoid notational clutter, we will drop the hat on the rescaled eigenvalue, continuing to use $\lambda$ as the eigenvalue coordinate even though it has been rescaled by a factor of $\sqrt{N}$ from its original definition.

A useful perspective on the phase transition comes from consideration of the classical limit of the ensemble of eigenvalue fermions. The leading semi-

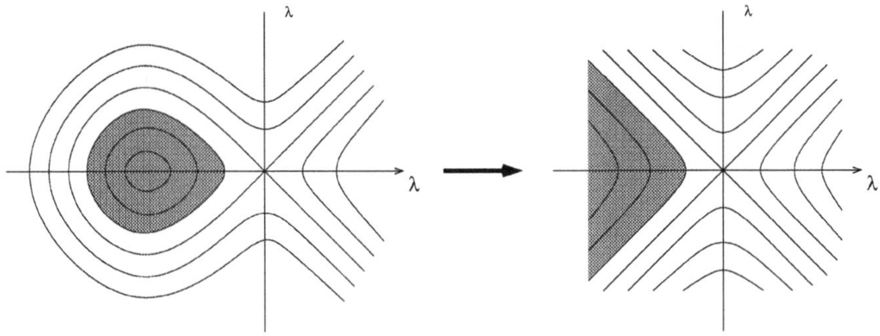


*Figure 9.* Phase space portrait of the classical limit of the free fermion ground state. The contours are orbits of fixed energy; the shaded region depicts the filled Fermi sea.

classical approximation to the degenerate Fermi fluid of eigenvalues describes it as an incompressible fluid in phase space [36, 37]. Each eigenvalue fermion occupies a cell of volume $2\pi\hbar$ in phase space, with one fermion per cell; the classical limit is a continuous fluid, which is incompressible due to Pauli exclusion. The metastable ground state, which becomes stable in this limit, has the fluid filling the interior of the energy surface in phase space of energy $E_F$; see figure 9.

The universal part of the free energy comes from the endpoint of the eigenvalue distribution near $\lambda \sim 0$. The limit $E_F \to 0^-$ leads to a change in this universal component, due to the singular endpoint behavior $\rho(\lambda) \sim \sqrt{\lambda^2 - E_F}$ of the eigenvalue density in this limit.

One should worry that the theory we have described is not well-defined, due to the fact that there is a finite rate of tunnelling of eigenvalues out of the metastable well. Single-particle wavefunctions in the inverted harmonic potential are parabolic cylinder functions

$$\psi_\omega(\lambda) = c_\omega D_{-\frac{1}{2}+i\omega}\big((1+i)\lambda\big) \overset{\lambda\to\infty}{\sim} \frac{1}{\sqrt{\pi\lambda}}\, e^{-i\lambda^2/2+i\omega\log|\lambda|} \,. \tag{4.47}$$

If we consider an incoming wave from the left with these asymptotics, with energy $E = -\omega < 0$, a WKB estimate of the tunnelling amplitude gives $T(\omega) \sim e^{-\pi\omega}$. Perturbation theory is an asymptotic expansion in $1/N \propto 1/\mu$ (from KPZ scaling), and since all filled levels have $\omega > \mu$, tunnelling effects behave as $e^{-cN}$ for some constant $c$ and can be ignored if one is only interested in the genus expansion. The genus expansion is the asymptotic expansion around $\mu \to \infty$, where tunnelling is strictly forbidden.[16] The worldsheet formalism is defined through the genus expansion; effects such as tunnelling are invisible at fixed genus.[17] Nonperturbatively (at finite $\mu$), the theory does not exist; yet we can make an asymptotic expansion around the metastable configuration of the matrix quantum mechanics, and compare the terms to the results of the worldsheet path integral. We will return to this point in section 8, where the analogous (and nonperturbatively stable) matrix model for the fermionic string is briefly discussed.

The claim is that the continuum limit of the matrix path integral just defined (valid at least in the asymptotic expansion in $1/\mu$) is in the *same universality class* as the $D = 2$ string theory defined via the worldsheet path integral for Liouville theory coupled to $c_{\rm matter} = 1$ (and Faddeev-Popov ghosts).

## 5. An overview of observables

Now that we have defined the model of interest, in both the continuum worldsheet and matrix formulations, the next issue concerns the observables of the theory – what physical questions can we ask? In this section we discuss three examples of observables: (i) macroscopic loop operators, which put holes in the string worldsheet; (ii) asymptotic scattering states, the components of the S-matrix; and (iii) conserved charges, which are present in abundance in any free theory (*e.g.* the energies of the particles are separately conserved).

### 5.1 Loops

Consider the matrix operator

$$\begin{aligned} W(\mathbf{z},x) &= -\frac{1}{N}\,{\rm tr}[\log(\mathbf{z}-M(x))] \\ &= +\frac{1}{N}\sum_{l=1}^{\infty}\frac{1}{l}\,{\rm tr}\big[(M(x)/\mathbf{z})^l\big]-\log\mathbf{z}\,. \end{aligned} \tag{5.48}$$

From the matrix point of view, $\exp[W(\mathbf{z},x)] = \det[\mathbf{z}-M(x)]$ is the characteristic polynomial of $M(x)$, and thus a natural collective observable of the eigenvalues. Note that $\mathbf{z}$ parametrizes the eigenvalue coordinate. As a collective observable of the matrix, this operator is rather natural – its exponential is the characteristic polynomial of the matrix $M(x)$, and hence encodes the information contained in the distribution of matrix eigenvalues.[18] On a discretized surface, $\frac{1}{l}{\rm tr}[M^l(x)]$ is the operator that punches a hole in a surface of lattice length $l$; see figure 10.[19] All edges bordering the hole are pierced by a propagator which leads to the point in time $x$ in target space, and the other end of each propagator also goes to the point $x$ in the continuum limit $\beta\to 0$. Thus the continuum theory has a Dirichlet condition for $x$ along the boundary.

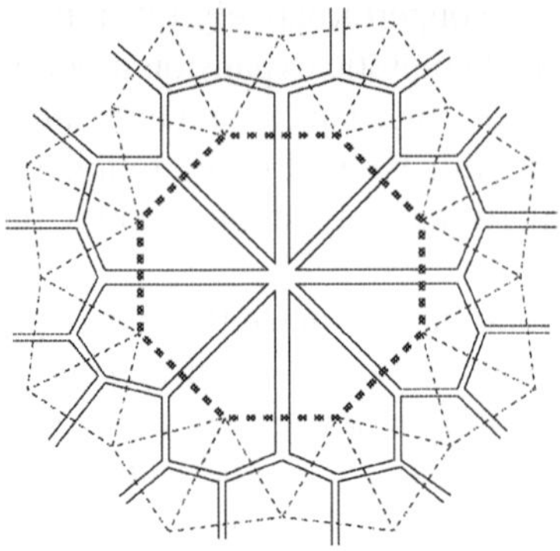

*Figure 10.* The operator $\mathrm{tr}[M^l(x)]/l$ inserts a boundary of lattice length $l$ into the fatgraph ($l = 8$ is depicted).

It is useful to rewrite the loop operator $W(\mathsf{z}, x)$ as follows:

$$\begin{aligned} W(\mathsf{z},x) &= -\lim_{\epsilon\to 0}\int_\epsilon^\infty \frac{d\ell}{\ell}\frac{1}{N}\,\mathrm{tr}\exp\bigl[-\ell\bigl(\mathsf{z}-M(x)\bigr)\bigr]+\log\epsilon \\ &= -\int\frac{d\ell}{\ell}\int d\lambda\, e^{-\ell(\mathsf{z}-\lambda)}\hat\rho(x,\lambda)+\log\epsilon \\ &= -\int\frac{d\ell}{\ell}\,e^{-\ell\mathsf{z}}\,\widetilde{W}(\ell,x)+\log\epsilon\quad , \end{aligned} \tag{5.49}$$

where in the first line we have simply introduced an integral representation for the logarithm, while in the second we have rewritten the trace over a function $f(M)$ of the matrix as an integral over the eigenvalue coordinate $\lambda$ of $f(\lambda)$ times the eigenvalue density operator $\hat\rho(x,\lambda)$. This defines the operator in the third line as

$$\widetilde{W}(\ell,x) = \int d\lambda\, e^{\ell\lambda}\hat\rho(x,\lambda)\ , \tag{5.50}$$

the Laplace transform of the eigenvalue density operator (recall that classically, the support of $\rho$ is along $(\lambda\in(-\infty,-\sqrt{2\mu}))$. The density operator is a bilinear of the fermion field operator

$$\begin{aligned} \hat\rho(x,\lambda) &= \hat\psi^\dagger\hat\psi(x,\lambda) \\ \hat\psi(x,\lambda) &= \int d\nu\, b_\nu\psi_\nu(\lambda)e^{-i\nu x} \end{aligned} \tag{5.51}$$

and its conjugate $\hat\psi^\dagger$ containing $b_\nu^\dagger$, with the anticommutation relation of mode operators

$$\{b_\nu^\dagger, b_{\nu'}\} = \delta(\nu-\nu')\ . \tag{5.52}$$

The mode wavefunctions are given in (4.47). The operator $\widetilde{W}$ is often called the *macroscopic loop operator.*

In the continuum formalism, we should consider the path integral on surfaces with boundary. The boundary condition on $X$ will be Dirichlet, as discussed above. For the Liouville field $\phi$, we use free (Neumann) boundary conditions, but with a boundary interaction

$$\mathcal{S}_L = \frac{1}{4\pi}\int \sqrt{g}[(\nabla_g\phi)^2 + QR^{(2)} + \mu\, e^{2b\phi}] + \sum_i \oint_{B_i} \mu_B^{(i)}\, e^{b\phi}\,. \tag{5.53}$$

Here, $\oint_{B_i} e^{b\phi} = \ell_{\rm bdy}^{(i)}$ is the proper length of the $i^{\rm th}$ boundary as measured in the dynamical metric; hence, $\mu_B^{(i)}$ is the *boundary cosmological constant* on that boundary component. The path integral over the dynamical metric sums over boundary lengths with the weight $e^{-\mathcal{S}_L}$, and therefore produces an integral transform with respect to the lengths of all boundaries. This transform has the same structure as the last line of (5.49). Let us truncate to zero modes along each boundary component, $\ell_{\rm bdy}^{(i)} = e^{b\phi_0^{(i)}}$. The path integral measure includes $\int d\phi_0^{(i)} = \int d\ell_{\rm bdy}^{(i)}/\ell_{\rm bdy}^{(i)}$, and the weight $e^{-\mathcal{S}_L}$ includes $e^{-\mu_B^{(i)}\ell_{\rm bdy}^{(i)}}\mathcal{P}(\ell_{\rm bdy}^{(i)})$, where $\mathcal{P}(\ell_{\rm bdy}^{(i)})$ is the probability measure for fixed boundary lengths. Comparison with (5.49) suggests we identify $\ell$ in $\widetilde{W}(\ell,x)$ as $\ell_{\rm bdy}$; $\mathsf{z} = \mu_B$; and $\mathcal{P}(\ell)$ is the correlator of a product of loop operators $\widetilde{W}(\ell,x)$.

Note in particular that the eigenvalue space of $\lambda$, which by (5.48) is the same as $\mathsf{z}$-space, is related to $\ell$-space (the Liouville coordinate $\phi$) by an *integral transform*. They are not the same! However, it is true that asymptotic plane waves in $\phi$ are the same as asymptotic plane waves in $\log\lambda$.

## 5.2 The S-matrix

Another observable is the S-matrix. The standard worldsheet prescription for string scattering amplitudes is to evaluate the integrated correlation functions of on-shell vertex operators. Asymptotic tachyon perturbations are produced by the operators

$$V_{i\omega}^{\rm in,out} = \alpha_\pm(\omega)\, e^{i\omega(x\mp\phi)}\, e^{Q\phi} \tag{5.54}$$

(whose dimension $\Delta = \bar{\Delta} = 1$ follows from (3.16)). The factor $e^{Q\phi}$ is just the local effective string coupling (3.11). The vibrational modes of the string are physical only in directions transverse to the string's worldsheet. Since the worldsheet occupies the only two dimensions of spacetime which are available, there are no transversely polarized string excitations and the only physical string states are the tachyon modes, which have only center-of-mass motion of the string. Actually, this statement is only true at generic momenta. For special momenta, there are additional states (in fact these momenta located at the poles in the relative normalization of $V_{i\omega}^{\rm matrix}$ and $V_{i\omega}^{\rm continuum}$). The effects of these extra states are rather subtle; for details, the reader is referred to [9]. The perturbative series for the tachyon S-matrix is

$$\mathbf{S}(\omega_i|\omega_j') = \sum_{h=0}^{\infty} \int \prod_r dm_r \left\langle \prod_i \int d^2 z_i \, V_{i\omega_i}^{(\rm in)} \prod_j \int d^2 w_j \, V_{i\omega_j'}^{(\rm out)} \right\rangle . \quad (5.55)$$

Actually, the statement that the tachyon is the only physical excitation is only true at generic momenta. For special momenta, there are additional states (in fact these momenta located at the poles in the relative normalization of $V_{i\omega}^{\rm matrix}$ and $V_{i\omega}^{\rm continuum}$, see section 6.3). The effects of these extra states are rather subtle; for details and further references, the reader is referred to [9].

In the matrix approach, the *in* and *out* modes are ripples (density perturbations) on the surface of the Fermi sea of the asymptotic form

$$\delta\hat{\rho}(\omega,\lambda) = \hat{\psi}^\dagger \hat{\psi}(\omega,\lambda) \overset{\lambda\to-\infty}{\sim} \frac{1}{2\lambda}\left(\alpha_+(\omega)e^{+i\omega\log|\lambda|} + \alpha_-(\omega)e^{-i\omega\log|\lambda|}\right) \quad (5.56)$$

as we will verify in the next section. The $\alpha_\pm(\omega)$ are right- and left-moving modes of a free field in $x \pm \log|\lambda|$, normalized as

$$\left[\alpha_\omega^\pm, \alpha_{\omega'}^\pm\right] = -\omega\delta(\omega+\omega') . \quad (5.57)$$

Thus, to calculate the S-matrix we should perform a kind of LSZ reduction of the eigenvalue density correlators [41]. Once again, as in the case of the macroscopic loop, the primary object is the density correlator.

The phase space fluid picture of the classical theory leads to an efficient method to compute the classical S-matrix [37, 42], and provides an appealing picture of the classical dynamics of the tachyon field.

## 5.3 Conserved charges

Since the dynamics of the matrix model is that of free fermions, there will be an infinite number of conserved quantities of the motion. For instance, the energies of each of the fermions is separately conserved. In fact, all of the phase space functions

$$q_{mn}(\lambda,p) = (\lambda+p)^{r-1}(\lambda-p)^{s-1}\, e^{-(r-s)x} \quad (5.58)$$

($p$ is the conjugate momentum to $\lambda$) are time independent for motion of a particle in the inverted oscillator potential, generated by $H = \frac{1}{2}(p^2-\lambda^2)$, ignoring operator ordering issues. These charges generate canonical transformations, and can be regarded as generators of the algebra of area-preserving polynomial vector fields on phase space (see [9] and references therein). Note that the time-independent operators with $m = n$ are simply powers of the energy, $q_{mm} = (-H)^{m-1}$. Formally, the operator

$$\hat{q}_{mn} = \int d\lambda\, \hat{\psi}^\dagger(\lambda) q_{mn}(\lambda, -i\partial_\lambda)\hat{\psi}(\lambda) \tag{5.59}$$

implements the corresponding transformation on the fermion field theory, ignoring questions of convergence. For $m = n$ we can be more precise: Energy should be measured relative to the Fermi energy,

$$\hat{q}_{mm} = \int_{-\mu}^{\infty} d\nu\,(-\mu-\nu)^{m-1} b_\nu^\dagger b_\nu - \int_{-\infty}^{-\mu} d\nu\,(-\mu-\nu)^{m-1} b_\nu b_\nu^\dagger\ ; \tag{5.60}$$

this expression is finite for finite energy excitations away from the vacuum state with Fermi energy $-\mu$.

The operators realizing these conserved charges in the worldsheet formalism were exhibited in [43] (for recent work, see [44, 45]). The charges $q_{12}$ and $q_{21}$ generate the full algebra of conserved charges, so it is sufficient to write expressions for them. They are realized on the worldsheet as operators $\mathcal{O}_{12}$ and $\mathcal{O}_{21}$

$$\begin{aligned} \mathcal{O}_{12} &= (\mathsf{cb} + \partial\phi - \partial x)(\bar{\mathsf{c}}\bar{\mathsf{b}} + \bar{\partial}\phi - \bar{\partial}x)e^{-x-\phi} \\ \mathcal{O}_{21} &= (\mathsf{cb} + \partial\phi + \partial x)(\bar{\mathsf{c}}\bar{\mathsf{b}} + \bar{\partial}\phi + \bar{\partial}x)e^{+x-\phi}\ . \end{aligned} \tag{5.61}$$

Here $\mathsf{b}(z)$, $\mathsf{c}(z)$ are the Faddeev-Popov ghosts for the local gauge choice $g_{ab} = \delta_{ab}$, *c.f.* [25, 26]. These operators have scale dimension $\Delta = \bar{\Delta} = 0$, and can be placed anywhere (unintegrated) on the two-dimensional worldsheet – moving them around changes correlators by gauge artifacts which decouple from physical quantities. The relation between matrix and continuum expressions for the conserved charges was worked out recently in [44, 45].

## 6. Sample calculation: the disk one-point function

An illustrative example which will allow us to compare these two rather different formulations of 2D string theory (and thereby check whether they are in fact equivalent) is the mixed correlator of one in/out state and one macroscopic loop. This correlator computes the process whereby an incoming tachyon is absorbed by the loop operator (or an outgoing one is created by the loop).

### 6.1 Matrix calculation

On the matrix side, we must evaluate the density-density correlator

$$\langle \mathrm{vac}|\hat{\rho}(\lambda_1, x_1)\,\hat{\rho}(\lambda_2, x_2)|\mathrm{vac}\rangle \tag{6.62}$$

and Laplace transform with respect to $\lambda_1$ to get the macroscopic loop, while performing LSZ reduction in $\lambda_2$. The evaluation of (6.62) proceeds via substitution of (5.51) and use of (5.52) as well as the vacuum property

$$
\begin{aligned}
b_\nu|\mathrm{vac}\rangle &= 0 \quad , \qquad \nu > \mu \\
b_\nu^\dagger|\mathrm{vac}\rangle &= 0 \quad , \qquad \nu < \mu
\end{aligned}
\tag{6.63}
$$

(note that we have not performed the usual redefinition of creation/annihilation operators below the Fermi surface). The result is [46]

$$
\langle\hat{\rho}(1)\hat{\rho}(2)\rangle = \int_\mu^\infty d\nu\, e^{-i\nu(x_2-x_1)}\psi_\nu^\dagger(\lambda_1)\psi_\nu(\lambda_2)\int_{-\infty}^\mu d\nu'\, e^{i\nu'(x_2-x_1)}\psi_{\nu'}(\lambda_1)\psi_{\nu'}^\dagger(\lambda_2).
\tag{6.64}
$$

The parabolic cylinder wavefunctions have the asymptotics (for $Y{=}\sqrt{\lambda^2-2\nu}\gg 1$, $\nu \gg 1$)

$$
\psi_\nu(\lambda) \sim \left[\frac{1}{\pi Y}\right]^{1/2} \sin\!\left(\tfrac{1}{2}\lambda Y + \nu\tau(\nu,\lambda) - \tfrac{\pi}{4}\right)
\tag{6.65}
$$

where

$$
\tau(\nu,\lambda) = -\int_{-2\sqrt{\nu}}^{-\lambda} \frac{d\lambda'}{\sqrt{\lambda'^2 - 2\nu}} = \log\!\left(\frac{-\lambda + \sqrt{\lambda^2 - 2\nu}}{\sqrt{2\nu}}\right)
\tag{6.66}
$$

is the WKB time-of-flight of the semiclassical fermion trajectory, as measured from the turning point of its motion.

At this point, we will make some approximations. We wish to compare the matrix and worldsheet field theory computations. However, the latter is only well-behaved in a low-energy regime, as we saw in section 3. Therefore we will approximate the energies in (6.64) as $\nu \sim \mu + \delta$, $\nu' \sim \mu - \delta'$, with $\delta, \delta' \ll \mu$, so that the density perturbation is very near the Fermi surface. In addition, substituting the parabolic cylinder wavefunction asymptotics (6.65) in (6.64), we drop all rapidly oscillating terms going like $\exp[\pm\frac{i}{2}\lambda^2]$; these terms should wash out of the calculation when we take $\lambda_2 \to \infty$ to perform the LSZ reduction.

With these approximations, one finds

$$
\psi_\nu(\lambda_2)\,\psi_{\nu'}^\dagger(\lambda_2) \overset{\lambda_2\to-\infty}{\sim} \frac{1}{4\pi\lambda_2}\left[\left(\sqrt{\tfrac{2}{\mu}}\,|\lambda_2|\right)^{i(\nu-\nu')} + \left(\sqrt{\tfrac{2}{\mu}}\,|\lambda_2|\right)^{-i(\nu-\nu')}\right] + O\!\left(\frac{\omega^2}{\mu}\right)
\tag{6.67}
$$

(recall $\omega = \nu - \nu'$). We wish to identify this with the in/out wave (5.56). Recall that initially the wavefunctions were multiplied by mode operators $b_{\nu'}^\dagger$, $b_\nu$; there is also a sum over energies. Comparing, we see that

$$
\alpha_\omega = \int_0^\omega d\varepsilon\, b_{\omega-\varepsilon}^\dagger b_\varepsilon \times \frac{1}{2\pi}\left(\frac{\mu}{2}\right)^{-i\omega/2}
\tag{6.68}
$$

which is (up to an overall phase, which we can absorb in the definition of the operators) just the standard *bosonization formula* for 2D fermions.[20]

As for the other part of the expression, the wavefunctions at $\lambda_1$, we make the same set of approximations, except that we use the full expression (6.65) rather than its $\lambda \to \infty$ limit. One finds

$$\psi_\nu^\dagger(\lambda_1)\psi_{\nu'}(\lambda_1) \sim \frac{1}{4\pi\sqrt{\lambda_1^2-2\mu}}\left[\left(\frac{-\lambda_1+\sqrt{\lambda_1^2-2\mu}}{\sqrt{2\mu}}\right)^{\overline{i(\nu-\nu')}}+\left(\frac{-\lambda_1+\sqrt{\lambda_1^2-2\mu}}{\sqrt{2\mu}}\right)^{\overline{-i(\nu-\nu')}}\right]+O\left(\frac{\omega^2}{\mu}\right). \tag{6.69}$$

Note that the terms of order $\omega^2/\mu$ that have been dropped are exactly of the form to be contributions of higher topologies of worldsheet. As we saw in the scattering of waves bouncing off the exponential Liouville wall in section 3, the effective string coupling (3.21) is $g_{\text{eff}} \sim \omega^2/\mu$.

Fixing the sum of the energies $\nu - \nu' = \omega$ (*e.g.* by Fourier transformation in $x$), the remaining energy integral is trivial and gives a factor of $\omega$. The macroscopic loop is finally obtained by Laplace transform with respect to $\lambda_1$; the answer is a Bessel function:[21]

$$\int_1^\infty \left[\left(\sqrt{t^2-1}+t\right)^{i\omega}+\left(\sqrt{t^2-1}+t\right)^{-i\omega}\right] e^{-ut}\,\frac{dt}{\sqrt{t^2-1}} = 2\,K_{i\omega}(u) \tag{6.70}$$

so that

$$\widetilde{W}_{i\omega}(\ell) \equiv {}_{\text{out}}\langle\text{vac}|\,\widetilde{W}(\ell,x)\,|\omega\rangle_{\text{in}} = 2\,\frac{\omega}{2\pi}\,K_{i\omega}(\sqrt{2\mu}\,\ell)\,. \tag{6.71}$$

The transformation (5.49) to $z$-space yields

$$\begin{aligned} W_{i\omega}(\mathbf{z}) &\equiv {}_{\text{out}}\langle\text{vac}|\,W(\mathbf{z},x)\,|\omega\rangle_{\text{in}} = \int_0^\infty \frac{d\ell}{\ell}\,e^{-\ell\sqrt{2\mu}\,\text{ch}\pi s}\,\widetilde{W}_{i\omega}(\ell) \\ &= 2\frac{\omega}{2\pi}\Gamma(i\omega)\Gamma(-i\omega)\,\cos(\pi s\omega) \end{aligned} \tag{6.72}$$

where we have parametrized $\mathbf{z} = \sqrt{2\mu}\,\text{ch}(\pi s)$.

The amplitude just calculated actually reveals quite a bit about the theory. We have learned that the corrections to the leading-order expressions (6.71), (6.72) are of order $\omega^2/\mu$, in agreement with the estimated higher order corrections in Liouville theory. It is a straightforward (if tedious) exercise to retain higher orders in the expansion, and thereby compute the corrections to the amplitude coming from surfaces with handles.

Another feature of Liouville theory we see appearing is its quantum wavefunction [47, 48]. In quantum theory, an operator $\mathcal{O}$ creates a state $\mathcal{O}|0\rangle$, whose overlap with the position eigenstate $|x\rangle$ is the wavefunction $\psi_{\mathcal{O}}(x) = \langle x|\mathcal{O}|0\rangle$. Similarly, we wish to interpret the state created by the macroscopic loop $\widetilde{W}(\ell,x)|\text{vac}\rangle$ as the position eigenstate in the space of $(\ell,x)$, whose overlap with the state $V_{i\omega}|\text{vac}\rangle$ is the wavefunction corresponding to the operator $V_{i\omega}$. This wavefunction is sometimes called the *Wheeler-de Witt* wavefunction.

In the continuum formulation the correlation function (6.72) involves one macroscopic loop of boundary cosmological constant $\mu_B$, and one tachyon perturbation $V_{i\omega}$, as depicted in figure 11.

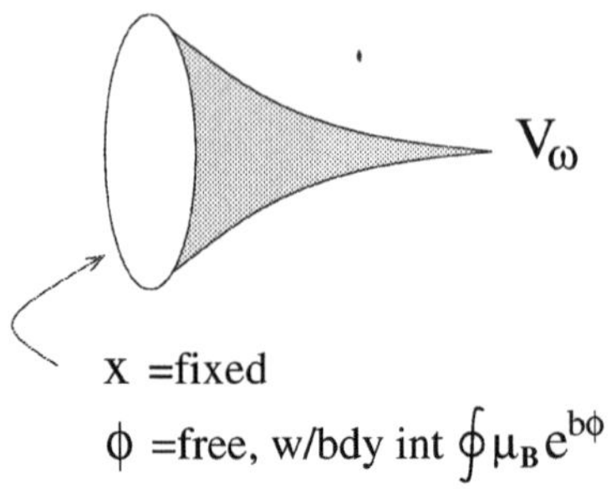


*Figure 11.* The disk one-point function of a tachyon perturbation is the leading-order contribution to the process whereby an incoming tachyon is absorbed by a macroscopic loop operator.

Indeed, if we butcher the theory by truncating to the spatial zero modes $\phi_0(\sigma_0) = \frac{1}{2\pi}\int d\sigma_1 \phi(\sigma_0, \sigma_1)$ on a worldsheet of cylindrical topology,[22] we arrive at Liouville quantum mechanics, whose Schrødinger equation reads

$$\left[-\frac{\partial^2}{\partial\phi_0^2} + 2\pi\mu\, e^{2b\phi_0} - \omega^2\right] \psi_\omega(\phi_0) = 0 \,. \tag{6.73}$$

The resulting wavefunctions

$$\psi_\omega(\phi_0) = \frac{2\,(\mu/2)^{-i\omega/2b}}{\Gamma(-i\omega/b)}\, K_{i\omega/b}\Big(\sqrt{2\mu}\, e^{b\phi_0}\Big) \tag{6.74}$$

are, up to normalization, identical to $\widetilde{W}(\ell = e^{b\phi_0})$.

## 6.2 Continuum calculation

There is actually more to be learned from the exact evaluation of this disk one-point correlator in the full Liouville plus matter CFT, as opposed to its quantum-mechanical zero mode truncation. In particular, one finds the precise relation between equivalent observables of the two formalisms. The non-trivial part is the calculation of the Liouville component, which rests on a conformal bootstrap for Liouville correlators on surfaces with boundary developed in [13–15], building on earlier work (reviewed in [16, 17]) on closed surfaces. We will only sketch the construction; the reader interested in more details should consult these references (and the references in these references).

The basic observation is the identity

$$\partial_z^2 \mathbf{V}_{-b/2}(z) = b^2\, T_{zz}\, \mathbf{V}_{-b/2}(z) \tag{6.75}$$

(and similarly for $\mathbf{V}_{-1/2b}$, *i.e.* $b \leftrightarrow 1/b$), where $\mathbf{V}_\alpha = e^{2\alpha\phi}$ are the exponential operators of Liouville field theory. This identity is consistent with the semiclassical limit $b \to 0$, $Q = b^{-1} + b \to b^{-1}$, since

$$\partial_z^2\, e^{-b\phi} = [b^2(\partial_z\phi)^2 - b\partial_z^2\phi]\, e^{-b\phi} = b^2 T_{zz}\, e^{-b\phi}\,. \tag{6.76}$$

Correlation functions with extra insertions of $T_{zz}$ are given in terms of those without such insertions, by the Ward identities of conformal symmetry. Thus, plugging (6.75) into a correlation function leads to second order differential equations on correlators involving $\mathbf{V}_{-b/2}$ (and similarly $\mathbf{V}_{-1/2b}$). Conformal invariance also dictates the structure of the correlator we wish to calculate,

$$\langle \mathbf{V}_\alpha(z)\rangle_{\mu_B} = \frac{U(\alpha)}{|z-\bar{z}|^{2\Delta_\alpha}} \tag{6.77}$$

where $z$ is a coordinate on the upper half-plane, see figure 12. This is equivalent to the correlator on the disk via the conformal transformation $z = -i\frac{w+i}{w-i}$; taking into account that the operator $\mathbf{V}_\alpha$ transforms like a tensor of weight $\Delta_\alpha$ in both $z$ and $\bar{z}$, one finds

$$\langle \mathbf{V}_\alpha(z)\rangle_{\mu_B,\,\text{disk}} = \frac{U(\alpha)}{(1-|w|^2)^{2\Delta_\alpha}}\,. \tag{6.78}$$

The nontrivial information lies in the overall coefficient $U(\alpha)$.

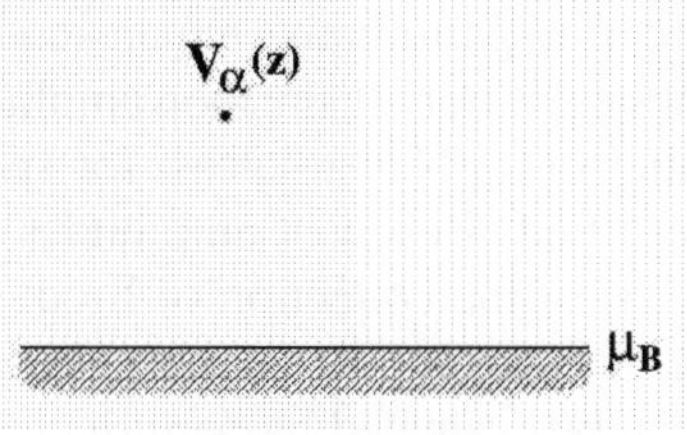


*Figure 12.*

In order to employ the Ward identity (6.75), we consider instead the two-point correlator

$$\langle \mathbf{V}_\alpha(z)\, \mathbf{V}_{-b/2}(w)\rangle\,. \tag{6.79}$$

The fact that $\mathbf{V}_{-b/2}$ satisfies a second order differential equation implies that only two scaling dimensions (up to integers) appear in its operator product expansion (OPE) with $\mathbf{V}_\alpha$, schematically

$$\mathbf{V}_\alpha \mathbf{V}_{-b/2} \sim C_+(\alpha)\Big[\mathbf{V}_{\alpha-b/2}\Big] + C_-(\alpha)\Big[\mathbf{V}_{\alpha+b/2}\Big]\,, \tag{6.80}$$

where the square brackets denote the operator together with all that can be obtained from it by the action of the conformal algebra, the so-called *conformal block*. The differential equation coming from (6.75), together with (6.80), yields

$$\langle \mathbf{V}_\alpha(z)\,\mathbf{V}_{-b/2}(w)\rangle = C_+(\alpha)U(\alpha - b/2)\mathcal{G}_+(\xi) + C_-(\alpha)U(\alpha + b/2)\mathcal{G}_-(\xi) \tag{6.81}$$

where $\mathcal{G}_\pm$ are hypergeometric functions of the cross-ratio $\xi = \frac{(z-w)(\bar{z}-\bar{w})}{(z-\bar{w})(\bar{z}-w)}$. What are the coefficients $C_\pm(\alpha)$? For $C_+(\alpha)$ the result of the OPE satisfies conservation of the "charge" of the exponential (the Liouville zero-mode momentum $p_\phi$). Even though this momentum is not conserved due to the presence of the tachyon wall, which violates translation invariance in $\phi$, if we nevertheless use free field theory to evaluate it we find trivially $C_+(\alpha) = 1$. We similarly use naive perturbation theory in powers of $\mu$ to evaluate $C_-(\alpha)$, bringing down the tachyon potential $\int \mu e^{2b\phi}$ in a power series expansion and evaluating the resulting integrated correlation functions using free field theory. Only the first term in the $\mu$ expansion contributes, and we find

$$\begin{aligned} C_-(\alpha) &= \left\langle \mathbf{V}_\alpha(0)\mathbf{V}_{-b/2}(1)\mathbf{V}_{Q-\alpha-b/2}(\infty)\Big(-\mu\int d^2z\,\mathbf{V}_b(z,\bar{z})\Big)\right\rangle_{\mathrm{FFT}} \\ &= -\pi\mu\,\frac{\gamma(2b\alpha - 1 - b^2)}{\gamma(-b^2)\gamma(2b\alpha)} \end{aligned} \tag{6.82}$$

where $\gamma(x) \equiv \frac{\Gamma(x)}{\Gamma(1-x)}$.[23]

Why are we allowed to use a perturbative expansion in $\mu$ and free field theory for evaluating these quantities? After all, the loop amplitude $\mu^{-i\omega/2}K_{i\omega}(\sqrt{2\mu}\,\ell)$ is certainly *not* polynomial in $\mu$. Nevertheless, for special "resonant" amplitudes this procedure is justified. Resonant amplitudes are those for which the sum of the exponents $\sum \alpha_i$ of the collection of Liouville operators $\mathbf{V}_{\alpha_i}$ adds up to a negative multiple of the exponent $2b$ of the Liouville potential. In such cases, the path integral can be evaluated by perturbation theory in $\mu$. This feature is related to the property that the integral over the constant mode of $\phi$ in the path integral is dominated by the region $\phi \to -\infty$, where the Liouville potential is effectively vanishing. The use of free field theory methods is then justified. The correlators that define $C_\pm$ satisfy this resonance condition. Note that we are not using free field theory to evaluate the full amplitude, but rather only to evaluate the operator product coefficients with the special degenerate operator $\mathbf{V}_{-b/2}$ (and similarly $\mathbf{V}_{-1/2b}$).

We now have partial information on the correlation function. To get a closed system of equations, we need a second relation on (6.79). For this purpose, we consider the OPE of $\mathbf{V}_{-b/2}(w)$ with its image across the boundary to make the identity operator, by taking $w \to \bar{w}$ (in the process, we need to transform to another basis for the hypergeometric functions $\mathcal{G}_\pm$ adapted to this particular degeneration). In this limit, the correlator (6.79) factorizes,

$$\langle \mathbf{V}_\alpha(z)\mathbf{V}_{-b/2}(w)\rangle_{\mu_B} \overset{w\to\bar{w}}{\sim} \langle \mathbf{V}_\alpha(z)\rangle_{\mu_B} \langle \mathbf{V}_{-b/2}(w)\rangle_{\mu_B} . \tag{6.83}$$

The first factor on the right-hand side is given simply in terms of $U(\alpha)$, and the second factor is yet another resonant amplitude, which we can evaluate in free field theory by bringing down the boundary cosmological constant interaction from the action:

$$(\mathrm{Im}\, w)^{2\Delta_\alpha} \Big\langle \mathbf{V}_{-b/2}(w) B_Q(\infty)\Big(-\mu_B \oint d\xi\, B_b(\xi)\Big)\Big\rangle_{\mathrm{FFT}} = -2\pi\mu_B \frac{\Gamma(1-2b^2)}{\Gamma^2(-b^2)} . \tag{6.84}$$

Here the integral over $\xi$ is along the boundary, which is the real axis; $B_\alpha$ is the operator $e^{2\alpha\phi}$ inserted on the boundary; and $B_Q$ represents the extrinsic curvature of the boundary at infinity.

Equating the two expressions (6.81) and (6.83), and using (6.82), (6.84), one arrives at a shift relation on $U(\alpha)$ [13–15],

$$-\frac{2\pi\mu_B}{\Gamma(-b^2)} U(\alpha) = \frac{\Gamma(-b^2+2b\alpha)}{\Gamma(-1-2b^2+2b\alpha)} U(\alpha - b/2) - \frac{\pi\mu\Gamma(-1-b^2+2b\alpha)}{\gamma(-b^2)\Gamma(2b\alpha)} U(\alpha + b/2) . \tag{6.85}$$

There is a similar shift relation obtained by use of $\mathbf{V}_{-1/2b}$. It is convenient to write $\mu_B$ in terms of a parameter $s$ via

$$\cosh^2(\pi b s) = \frac{\mu_B^2}{\mu} \sin(\pi b^2) ; \tag{6.86}$$

then the two discrete shift relations (obtained by use of both $\mathbf{V}_{-b/2}$ and $\mathbf{V}_{-1/2b}$) are solved by

$$U(\alpha) = \frac{2}{b}\Big(\pi\mu\gamma(b^2)\Big)^{\frac{Q-2\alpha}{2b}} \Gamma(2b\alpha - b^2)\Gamma(\frac{2\alpha}{b} - \frac{1}{b^2} - 1) \cosh[(2\alpha - Q)\pi s] . \tag{6.87}$$

For the vertex operators with $\alpha = \frac{1}{2}Q + \frac{i}{2}\omega$ appearing in the scattering amplitudes, this translates into

$$U(\alpha = \tfrac{1}{2}Q + \tfrac{i}{2}\omega) = 2i\omega\Big(\pi\mu\gamma(b^2)\Big)^{-i\omega/2b} \Gamma(ib\omega)\Gamma(i\omega/b) \cos(\pi s\omega) . \tag{6.88}$$

The shift operator relations don't fix the overall normalization of $U(\alpha)$. This normalization is obtained by demanding that the residues of the poles at $2\alpha = Q - nb$ (*i.e.*, $i\omega = -nb$) for $n = 1, 2, 3, \ldots$, agree with the "resonant amplitude" integrals for these special momenta.

Note that the full set of resonant amplitude integrals involve bringing down powers of both $\mu e^{2b\phi}$ and also $\tilde{\mu} e^{(2/b)\phi}$ from the action. One needs to use the complete set in order to provide sufficient constraints to fully determine the Liouville correlators. Hence *both* are present in the theory; moreover, one finds for consistency that their coefficients must be related:

$$\pi\tilde{\mu}\,\gamma(1/b^2) = [\pi\mu\,\gamma(b^2)]^{1/b^2}\ . \tag{6.89}$$

It turns out that this is more or less the relation implied by the analytic continuation of the amplitude for reflection off the Liouville potential

$$\tilde{\mu}/b = \mu\, b\,\mathcal{R}(\omega = i(Q-2b))\ . \tag{6.90}$$

The reflection amplitude $\mathcal{R}(\omega)$ for $V_{i\omega} \to V_{-i\omega}$ may be read off the two-point correlation function for tachyon vertex operators. A similar relation holds for the boundary cosmological constant; the boundary interaction is actually

$$\delta S_{\rm bdy} = \oint \left(\mu_B e^{b\phi} + \tilde{\mu_B} e^{(1/b)\phi}\right) \tag{6.91}$$

with

$$\cosh^2(\pi s/b) = \frac{\tilde{\mu}_B^2}{\tilde{\mu}}\,\sin(\pi/b^2)\ . \tag{6.92}$$

Thus there is a kind of strong/weak coupling duality in Liouville QFT, characterized by

$$b \leftrightarrow 1/b\quad ,\qquad \mu\leftrightarrow\tilde{\mu}\quad ,\qquad \mu_B\leftrightarrow\tilde{\mu_B} \tag{6.93}$$

(recall that $b\to 0$ was the weak coupling limit of Liouville theory). The parameter $s$ is invariant under this transformation.

## 6.3 Comparing the results

Finally, we are ready to compare the two approaches. First we must assemble the Liouville disk amplitude with the contributions of the free matter field $X$ and the Faddeev-Popov ghosts. There is a factor of $1/2\pi$ from gauge fixing the conformal isometries of the punctured disk (rotations around the puncture). The disk expectation value of the matter is

$$\left\langle e^{i\omega X}(z)\right\rangle_{\rm Dirichlet} = \frac{1}{|z-\bar{z}|^{2\Delta_\omega}} \tag{6.94}$$

(equivalently, $(1-|w|^2)^{-2\Delta_\omega}$ if we are working on the disk rather than the upper half-plane), which simply reflects the fact that the Dirichlet boundary condition on $X$ is a delta function (and hence its Fourier transform is one). The factors of $|z-\bar{z}|$ cancel among Liouville, matter, and ghosts (we must take $b\to 1$ in the Liouville part since $D=2$; this involves a multiplicative renormalization of $\mu$ and $\mu_B$ in order to obtain finite results). This cancellation of coordinate dependence merely reflects that we have correctly calculated a conformally invariant and therefore physical amplitude. Thus

$$\langle V_{i\omega}(z,\bar{z})\rangle_{\text{disk}} = \frac{1}{2\pi}\, 2i\omega\, \hat{\mu}^{-i\omega/2}(\Gamma(i\omega))^2 \cos(\pi s\omega) \tag{6.95}$$

where we have defined

$$\hat{\mu} = \pi\mu\gamma(b^2) \overset{b\to 1}{\sim} 2\pi\mu(1-b) \tag{6.96}$$

as the quantity to be held fixed in the $b \to 1$ limit.

Comparing to the matrix model result (6.72), we find the same result provided that we identify

$$\begin{aligned} V_{i\omega}^{\text{matrix}} &= (\hat{\mu})^{i\omega/2}\frac{\Gamma(-i\omega)}{\Gamma(i\omega)} V_{i\omega}^{\text{continuum}} \\ \tfrac{1}{2}\mu^{\text{matrix}} &= \hat{\mu}^{\text{continuum}} \\ \tfrac{1}{2}\mu_B^{\text{matrix}} &= \tfrac{1}{2}\mathbf{z}_{\text{matrix}} = \hat{\mu}_B^{\text{continuum}} \equiv 2\pi\mu_B^{\text{cont}}(1-b)\ . \end{aligned} \tag{6.97}$$

(the last relation amounts to $\hat{\mu_B} = \sqrt{\hat{\mu}}\,\mathrm{ch}(\pi s)$). Thus the exact evaluation of the worldsheet amplitude allows a precise mapping between the continuum and matrix approaches.

The energy-dependent phase in the relative normalizations of $V_{i\omega}$ results in a varying time delay of reflection for particles of different energy. It was shown in [49] that this time delay reproduces what one would expect based on the gravitational redshift seen by one particle after another has been sent in. Thus the so-called "leg-pole factor" $\frac{\Gamma(-i\omega)}{\Gamma(i\omega)}$ in equation (6.97) is an important physical effect, which is added by hand to the matrix model. It is not yet understood if there is a derivation of this factor from first principles in the matrix model.

Other amplitudes that have been computed on both sides of the correspondence and shown to agree include

- The tree level S-matrix [50, 51, 42],[24]
- The torus partition function [52, 53],
- The disk one-point function calculated above [46–48, 13, 14],
- The annulus correlation function for two macroscopic loops [47, 19].

One can also show that the properties of the ground ring of conserved charges defined in section 5.3 agree between the matrix and continuum formulations, at leading order in $1/\mu$ [22]. For instance, on the sphere one calculates using the Liouville OPE coefficients $C_\pm$ that

$$\langle \mathcal{O}_{12}\mathcal{O}_{21}\rangle = \langle \mathcal{O}_{22}\rangle = \langle -H\rangle = \mu\ . \tag{6.98}$$

This result is consistent with the fact that perturbative excitations live at the Fermi surface, where the energy is $H = \frac{1}{2}(p_\lambda^2 - \lambda^2) = -\mu$.

Thus the matrix approach is reproducing the quantum dynamics of Liouville CFT coupled to a free field. Note that the matrix approach is much more economical computationally, and we immediately see how to compute the higher order corrections (just go to higher order in $1/\mu$ in our approximations); for Liouville, we need to work *much* harder – we need to go back to the conformal bootstrap and compute correlation functions on the disk with handles, then integrate over the moduli space.

## 7. Worldsheet description of matrix eigenvalues

Finally, what about the eigenvalues themselves? They are gauge invariant observables which are manifest in the matrix formulation; what is their description in the continuum formalism? Note that this question bears on the continuum description of nonperturbative phenomena such as the eigenvalue tunnelling which leads to the nonperturbative instability of the model. Experience from string theory in higher dimensions (*e.g.* black hole microphysics) has taught us that D-brane dynamics provides a description of strong coupling physics. Therefore we should examine the D-branes of 2D string theory. The fact that the tension of D-branes is naively $O(1/g_s)$ means that they are the natural light degrees of freedom in the strong coupling region.

In the worldsheet description of dynamics, a D-brane is an object which puts boundaries on the worldsheet. The boundary conditions on the worldsheet fields $X^\mu$ tell us about the position of the brane and the boundary interactions in the worldsheet action specify the background fields localized on the brane. Perturbations of the boundary background fields are (marginal) scaling operators on the boundary. The theory thus has two sectors of strings – *open strings* that couple to worldsheet boundaries (D-branes), and *closed strings* that couple to the bulk of the worldsheet.

In a sense, the macroscopic loop is a spacelike D-brane – one with Dirichlet boundary conditions in the timelike direction $X$ and Neumann boundary conditions in the spacelike direction $\phi$. The boundary interaction $\mu_B \oint e^{b\phi}$ is a "boundary tachyon" that keeps $\phi_{\rm bdy}$ away from the strong coupling region $\phi \to \infty$ (at least for the appropriate sign of $\mu_B$). This D-brane is however a collective observable at fixed time $x$ of the matrix model, and not a dynamical object. A depiction of the D-brane interpretation of the calculation of section 6 is shown in figure 13a.

Instead, the matrix eigenvalue is localized in the spatial coordinate $\lambda$ and hence quasi-localized in $\phi$. Here it is important to recall that $\phi$ and $\lambda$ are related by the integral transform (5.49), and are thus not directly identified. Nevertheless, localized disturbances in $\phi$ bouncing off the exponential Liouville

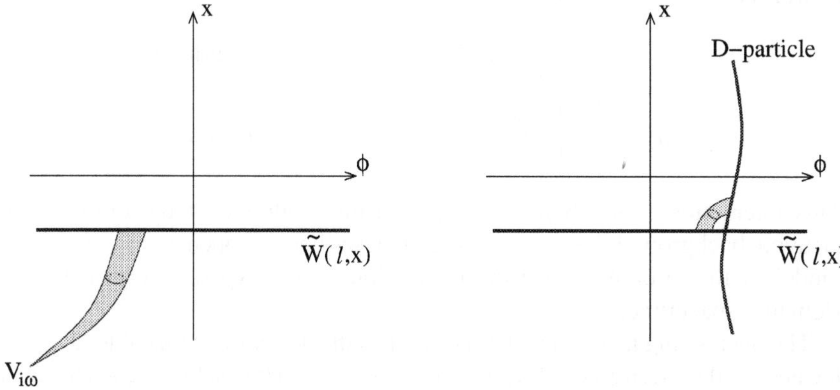


*Figure 13.* (a) A macroscopic loop is a spatial D-brane that absorbs and emits closed strings. (b) The loop is also a probe of the motion of D-particles.

wall are related to localized disturbances of the Fermi surface in $\lambda$ bouncing off the inverted oscillator barrier, so there is a rough equivalence.

Therefore, we consider Dirichlet boundary conditions for $\phi$. Since $\phi$ shifts under local scale transformations ($e^{2b\phi}g_{ab}$ is the dynamical metric), the Dirichlet boundary condition

$$\phi\Big|_{\text{bdy}} = \phi_0 \tag{7.99}$$

is not conformally invariant unless $\phi = \pm\infty$. Now $\phi = -\infty$ is the weak coupling asymptotic boundary of $\phi$ space, and corresponds to boundaries of zero size, which we usually think of as punctures in the worldsheet where local vertex operators are inserted. On the other hand, $\phi = +\infty$ is what we want, a boundary deep inside the Liouville wall at strong coupling.

In fact, we know a classical (constant negative curvature) geometry with this property:

$$ds^2 = e^{2b\phi}dzd\bar{z} = \frac{Q}{\pi\mu b}\frac{dzd\bar{z}}{(1-z\bar{z})^2}\,, \tag{7.100}$$

the Poincare disk (or Lobachevsky plane). Proper distances blow up toward the boundary: $\phi \to \infty$, as advertised.

For this D-brane to move in time, the boundary condition in $X$ should be Neumann. What sort of conformally invariant boundary interaction can we have? Since $\phi$ is fixed on the boundary, the interaction can only involve $X$;

conformal invariance then dictates

$$\delta S_{\text{bdy}} = \beta \oint \cos(X) \quad , \quad (X \text{ Euclidean}) \tag{7.101}$$

$$\delta S_{\text{bdy}} = \begin{cases} \beta \oint \cosh(X) \\ \beta \oint \sinh(X) \end{cases} \quad , \quad (X \text{ Lorentzian})$$

This interaction is the boundary, open string analogue of the closed string tachyon background $V(X)$ in (2.5); it describes an open string 'tachyonic mode' of the D-brane, since the interaction grows exponentially in Lorentz signature spacetime.[25]

The open string tachyon (7.101) describes the decay of an unstable D-particle located in the strong coupling region $\phi \to \infty$. The tachyon condensate in Lorentz signature looks promising to be the description of an eigenvalue in the matrix model, whose classical motion is

$$\begin{aligned} \lambda(x) &= \lambda_0 \cosh(x) \quad , \quad E = -\tfrac{1}{2}\lambda_0^2 < 0 \\ \lambda(x) &= \lambda_0 \sinh(x) \quad , \quad E = +\tfrac{1}{2}\lambda_0^2 > 0 \end{aligned} \tag{7.102}$$

depending on whether the eigenvalue passes over, or is reflected by, the harmonic barrier. Similarly, the Euclidean trajectory $\lambda(x) = \lambda_0 \cos(x)$ is oscillatory, appropriate to the computation of the WKB tunnelling of eigenvalues under the barrier.

How do we see that this is so? In [22] (building on earlier work [18, 20]) this result was demonstrated by computing the ground ring charges $\mathcal{O}_{12}$ and $\mathcal{O}_{21}$ on the disk, and showing that they give the classical motions above. Here we will employ a complementary method: We will probe the D-brane motion with the macroscopic loop. This will exhibit the classical motion quite nicely.

## 7.1 Lassoing the D-particle

The matrix model calculation of a macroscopic loop probing a matrix eigenvalue is trivial. Recall that the macroscopic loop is

$$W(\mathbf{z}, x_0) = -\frac{1}{N} \operatorname{tr} \log(\mathbf{z} - M(x_0)) = -\int d\lambda\, \hat{\rho}(\lambda, x_0) \log(\mathbf{z} - \lambda) \,. \tag{7.103}$$

An individual eigenvalue undergoing classical motion along the trajectory $\lambda(x_0)$ gives a delta-function contribution to the eigenvalue density

$$\delta\rho(\lambda, x_0) = \delta(\lambda - \lambda(x_0)) \tag{7.104}$$

where $\lambda(x_0) = -\lambda_0 \cos(x_0)$ for Euclidean signature, and $\lambda(x_0) = -\lambda_0 \cosh(x_0)$ for Lorentzian signature. Plugging into (7.103), we find

$$W_{\text{eval}}(\mathbf{z}, x_0) = -\log[\mathbf{z} - \lambda(x_0)] \,. \tag{7.105}$$

In the worldsheet formalism, the presence of a macroscopic loop introduces a second boundary, besides the one describing the D-particle. The leading order connected correlator of the loop and the D-particle is thus an annulus amplitude; the worldsheet and boundary conditions are depicted in figure 14, while the spacetime interpretation is shown in figure 13b. The parameter $\tau$ is an example of a modulus of the surface, the Schwinger parameter for the propagation of a closed string, which cannot be gauged away by either reparametrizations or local scale transformations; in the end, we will have to integrate over it.

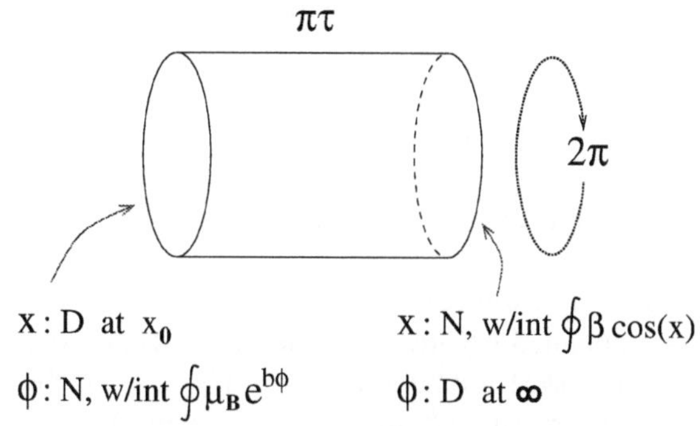


*Figure 14.* The annulus worldsheet describing a macroscopic loop probing a D-particle

There are two ways to think about this worldsheet as the propagation of a string. If we view worldsheet time as running around the circumference of the annulus, we think of the diagram as the one-loop vacuum amplitude of an open string, a string having endpoints. At one endpoint of the string, we classically have the boundary condition

$$\partial_n \phi = 2\pi \mu_B \, e^{b\phi} \quad , \qquad X = x_0 \tag{7.106}$$

describing the macroscopic loop; at the other end, we have

$$\phi = \infty \quad , \qquad \partial_n X = 2\pi\beta \, \sin(X) \tag{7.107}$$

describing the moving D-particle. On the other hand, we can think of the diagram as the propagation of a closed string for a worldsheet time $\pi\tau$, folded into "boundary states" $|B\rangle$ which implement the boundary conditions on the fields. These boundary states are completely determined by these conditions, *e.g.*

$$\begin{aligned} &(\partial_n \phi - 2\pi \mu_B \, e^{b\phi}) |B_N(\mu_B)\rangle_\phi = 0 \\ &(X - x_0)|B_D(x_0)\rangle_X = 0 \end{aligned} \tag{7.108}$$

and so on. Because the Liouville and matter fields do not interact, the boundary state factorizes into the tensor product of the boundary states for $X$ and for $\phi$. The Liouville partition function can then be written

$$\begin{aligned} \mathcal{Z}_L(q) &= \langle B_N, \mu_B | e^{-\pi\tau H} | B_D \rangle \\ &= \int d\nu \, \Psi^*_{\rm FZZT}(\nu, \mu_B) \Psi_{\rm ZZ}(\nu) \frac{q^{\nu^2}}{\eta(q)} \end{aligned} \tag{7.109}$$

where $q = \exp[-2\pi\tau]$, and the Dedekind eta function $\eta(q) = q^{1/24}\prod_{n=1}^{\infty}(1-q^n)$ represents the contribution to the partition function of all the Liouville oscillator modes. The quantities $\Psi_{\rm FZZT}$ and $\Psi_{\rm ZZ}$ are the zero mode parts of the Neumann and Dirichlet boundary state wavefunctions, respectively; see [13, 14] and [15], respectively. Explicitly,

$$\Psi_{\rm FZZT}(\nu, \mu_B) = \cos(2\pi\nu s) \left[\frac{\Gamma(1+2i\nu b)\Gamma(1+2i\nu/b)}{2^{1/4}(-2\pi i\nu)} \hat{\mu}^{-i\nu/b}\right] \tag{7.110}$$

$$\Psi_{\rm ZZ}(\nu) = 2\sinh(2\pi\nu/b)\sinh(2\pi\nu b) \left[\frac{\Gamma(1+2i\nu b)\Gamma(1+2i\nu/b)}{2^{1/4}(-2\pi i\nu)} \hat{\mu}^{-i\nu/b}\right]$$

Here, $s$ parametrizes $\mu_B$ as in equation (6.86). The "Neumann" wavefunction $\Psi_{\rm FZZT}$ is the one obtained before, from the macroscopic loop calculation; $\nu$ is the Liouville zero-mode momentum $\alpha = \frac{1}{2}Q + i\nu$ in the "closed string channel". This is not surprising; before we used the macroscopic loop to probe the wavefunction of a scattering state, now we are using it to probe a D-brane state to see if it has the properties of a matrix eigenvalue.

The authors of [15] showed that $\Psi_{\rm ZZ}(\nu)$ has the property that all operators behave like the identity operator as they approach the corresponding boundary (so that one approaches the constant negative curvature "vacuum" near the boundary of the Poincare disk). Ordinarily in Liouville theory, when an operator such as $\mathbf{V}_\alpha$ approaches the boundary $z = \bar{z}$ (*e.g.* with boundary condition (7.106)), it expands as a sum of boundary operators $B_\beta$. For the boundary state with wavefunction $\Psi_{\rm ZZ}(\nu)$, only the identity boundary operator $B_0 = \mathbb{1}$ appears in the limit $z \to \bar{z}$.

Now for the matter partition function. The annulus partition function with the requisite boundary conditions was computed in [54] for Euclidean $X$, with the result

$$\mathcal{Z}_X = \frac{1}{\sqrt{2}\,\eta(q)} \sum_{n=-\infty}^{\infty} q^{n^2/4} \cos[n\pi(\tfrac{1}{2}+\gamma)] \quad , \quad \sin(\pi\gamma) \equiv \cos(x_0)\sin(\pi\beta) \, . \tag{7.111}$$

Again the Dedekind eta function represents the contribution of the $X$ oscillator modes, and the sum results from the zero modes. The Faddeev-Popov ghost partition function is

$$\mathcal{Z}_{\rm gh} = \eta^2(q) \, , \tag{7.112}$$

cancelling the oscillator $\eta$ functions of $\phi$ and $X$. This is related to the fact that there are no transverse directions in which the string can oscillate – at generic momenta just the tachyon, with only center of mass motion of the string, is physical.

Combining all the contributions, we have

$$\begin{aligned} \mathcal{Z} &= \int_0^\infty d\tau\, \mathcal{Z}_L \cdot \mathcal{Z}_X \cdot \mathcal{Z}_{\rm gh} \\ &= \int_0^\infty d\tau \int d\nu \cos(2\pi\nu s) \sum_{n=-\infty}^{\infty} \cos[n\pi(\tfrac{1}{2}+\gamma)]\, q^{\nu^2+n^2/4} . \end{aligned} \tag{7.113}$$

Doing the $\tau$ integral, and the $\nu$ integral by residues,[26] one finds

$$\mathcal{Z} = 2\sum_{n=1}^{\infty} \frac{1}{n} \exp[-n\pi s] \cos[n\pi(\tfrac{1}{2}+\gamma)] . \tag{7.114}$$

The sum is readily performed, and after a little algebra, one obtains

$$\mathcal{Z} = -\log[2(\cosh(\pi s) + \sin(\pi\gamma))] \quad . \tag{7.115}$$

Define now

$$\lambda(x_0) = -\sqrt{2\mu}\, \sin(\pi\gamma) = -\sqrt{2\mu}\, \sin(\pi\beta)\, \cos(x_0) \equiv -\lambda_0 \cos(x_0) \tag{7.116}$$

and recall that $\mu_B = \sqrt{2\mu}\, {\rm ch}(\pi s) = {\rm z}$; then we have

$$\mathcal{Z} = -\log[{\rm z} - \lambda(x_0)] + \tfrac{1}{2}\log(\mu/2) . \tag{7.117}$$

The additive constant is ambiguous, and depends on how we regularize the divergent term in (7.113); nevertheless, it is independent of the boundary data for $X$ and $\phi$, and so does not affect the measurement of the D-particle motion.[27] Dropping this last term, we finally reproduce the result (so easily found) for the probe of eigenvalue motion in the matrix model, equation (7.105)!

Note that, even though the Dirichlet boundary condition on $\phi$ is in the strong coupling region $\phi \to \infty$, the wavefunctions are such that we obtain sensible results for the amplitude. Note also that the boundary interaction for the D-particle depends only on $X$, and thus the D-particle naively is not moving in $\phi$. This is a cautionary tale, whose moral is to compute physical observables! Nevertheless, when the D-particle reaches the asymptotic region of weak coupling, it should be moving in both $\phi$ and $\lambda$. Somehow the field space coordinate of the open string tachyon and the $\phi$ coordinate of spacetime become related in the course of the tachyon's condensation, and it remains to be understood how this occurs.

The consideration of multiple D-particles elicits the matrix nature of their open string dynamics. Now we must add to the description of the boundary state a finite dimensional (*Chan-Paton*) Hilbert space $\mathcal{H}_{\rm CP}$ describing which D-particle a given worldsheet boundary is attached to. Open string operators act as operators on this finite-dimensional Hilbert space, *i.e.* they are matrix-valued (equivalently, an open string is an element of $\mathcal{H}_{\rm CP} \otimes \mathcal{H}^*_{\rm CP}$ specifying the Chan-Paton boundary conditions at each end).

The open string tachyon is now a matrix field; the parameter $\beta$ in equation (7.101) is a matrix of couplings $\beta_{ij}$, $i, j = 1, ..., n$ for $n$ D-particles. There is an additional possible boundary interaction

$$\delta S_{\rm bdy} = \oint A_{ij}\, \partial_t X \tag{7.118}$$

which is a matrix gauge field on the collection of D-particles. We ignored it in our previous discussion because its role is to implement Gauss' law on the collection of D-particles, which is trivial in the case of a single D-particle. When several D-particles are present, however, this Gauss law amounts to a projection onto $U(n)$ singlet states. Thus the continuum description suggests that the $U(N)$ symmetry of the matrix mechanics is gauged, which as mentioned in section 4.1 projects the theory onto $U(N)$ singlet wavefunctions. Singlet sector matrix mechanics looks very much like the quantum mechanics of $N$ D-particles in 2D string theory.

Several ingredients of the relation between the continuum and matrix formulations remain to be understood. The probe calculation tells us that the open string tachyon condensate on the D-particle describes its leading order, classical trajectory. One should understand how higher order corrections lead to the quantum corrections for the wavefunction of a quantum D-particle, and show that this series matches the WKB series for the wavefunctions of the eigenvalue fermions of the matrix model. Also, the description in the continuum formulation of an eigenvalue as a D-particle is quite different from the ensemble of eigenvalues in the Fermi sea, whose collective dynamics is expressed via continuum worldsheets. Under what circumstances is eigenvalue dynamics that of D-particles, as opposed to that of closed string worldsheets? For instance, the $U(n)$ symmetry of a collection of $n$ D-particles should extend to the full $U(N)$ symmetry of the whole matrix. How do we see this larger symmetry in the continuum formulation? We cannot simply turn all the fermions into D-particles; there would then be nothing left to make the continuum worldsheets that attach to these branes. The continuum formalism is really adapted to describing a small number of matrix eigenvalues that have been separated from rest of the ensemble, thus leading to distinct treatment of the few separated ones as D-particles, and the vast ensemble of remaining ones as the threads from which continuum worldsheets are woven.

## 7.2 Summary

To summarize, we have an expanding translation table between matrix and continuum formalisms:

| **Continuum** | **Matrix** |
|---|---|
| Closed string vacuum | Fermi sea of matrix eigenvalues |
| Liouville potential $\mu\, e^{2b\phi}$ | Inverted oscillator barrier |
| Worldsheet cosmological constant $\mu$ | Fermi energy $-2\mu$ |
| Strings | Fermi surface density wave quanta |
| String S-matrix | Density wave S-matrix |
| D-branes w/$X$: D, $\phi$: N, $\mathcal{S}_{\rm bdy}=\mu_B \oint e^{b\phi}$ | Macroscopic loops $\mathrm{tr}[\log(\mathbf{z}-M(x))]$ |
| Boundary cosm. const. $\mu_B$ | Loop eigenvalue parameter $2\mathbf{z}$ |
| D-branes w/$\phi$: D, $X$: N, $\mathcal{S}_{\rm bdy}=\beta \oint \cos X$ | Eigenvalues outside the Fermi sea |
| Open string tachyon coupling $\beta$ | Eigenvalue energy $E=-\mu\, \sin^2 \pi\beta$ |
| Open string tachyon on $n$ D-particles | A block of the matrix $M$ |

A similar dictionary is known for the fermionic string, which will be described briefly in the next section. Here one has the added advantage that the model is nonperturbatively well-defined. In these models both sides of the duality are again calculable. One may hope that open/closed string duality can be worked out in complete detail in this example, and that it will lead to valuable insights into the general class of open/closed string dualities to which it belongs.

# 8. Further results

## 8.1 Fermionic strings

The remarkable agreement between the continuum and matrix formulations of 2D string theory leads us to believe that they are equivalent. However, in the case of the bosonic string, both are asymptotic expansions. Worldsheet perturbation theory is an asymptotic expansion, and it was our hope that, as in higher dimensional gauge/gravity correspondences, the matrix (gauge) theory formulation would provide a nonperturbative definition of the theory. But the nonperturbative instability of the vacuum to eigenvalue tunnelling across to the right-hand side of the oscillator barrier means that the theory does not really exist after all.

An obvious fix for this difficulty would be to fill up the other side of the barrier with fermions as well (see *e.g.* [56] for an example of this proposal). But this leads to an equally obvious question: We found agreement with continuum bosonic strings using just the fluctuations on one side of the oscillator barrier. What do the fluctuations on the other side describe? Perturbatively, they are a second, decoupled copy of the same dynamics. Nonperturbatively, the two sides of the barrier communicate, by tunnelling and by high energy processes that pass over the barrier. It is now understood [21, 22] that this stable version of the matrix model describes 2D fermionic string theory (the type 0B string, in the arcane terminology of the subject).

The fermionic string extends the construction of section 2 by supersymmetrizing the worldsheet theory: The spacetime coordinates $X^\mu(\sigma)$ of the worldsheet path integral gain superpartners $\psi^\mu(\sigma)$ which transform worldsheet spinors (and spacetime vectors). The local reparametrization and scale invariance condition generalizes to local supersymmetry and super-scale invariance; in other words, the stress tensor $T$ has a superpartner $G$, and the dynamical condition is that both must vanish in correlation functions.

If we perform the same exercise in the path integral formulation (2.1) of the particle propagator in flat spacetime, the quantization of the superpartner $\psi$ leads to equal time anticommutation relations

$$\{\psi^\mu, \psi^\nu\} = \delta^{\mu\nu} . \tag{8.119}$$

One realization of these anticommutation relations is to represent the $\psi$'s as Dirac matrices. The quantum mechanical Hilbert space contains not only the position wavefunction, but also a finite dimensional spin space in which the $\psi^\mu$ act – the particle being propagated is a spinor. Thus worldsheet supersymmetry is a way to introduce spacetime fermions into a worldline or worldsheet formalism.[28] A second realization of the anticommutation relations (8.119), using complex fermions, treats $\psi^*_\mu$ as a creation operator, and its conjugate $\psi^\mu$ as an annihilation operator. Starting from the fermion 'vacuum' $|0\rangle$, $\psi^\mu|0\rangle = 0$, the set of polarization states propagated along the particle worldline, $\{\psi^*_{\mu_1} \cdots \psi^*_{\mu_r}|0\rangle\}$, transform as a collection of antisymmetric tensors $C_{\mu_1 \ldots \mu_r}$ in spacetime.

The same story arises in the string generalization; the worldsheet fermions $\psi^\mu$ can realize a collection of antisymmetric tensors in spacetime, or under suitable conditions the propagating strings are spinors in spacetime. The so-called type 0 fermionic strings do not realize spacetime fermions, but do contain the antisymmetric tensor fields. We can divide the set of antisymmetric tensor fields into those with even rank and those with odd rank. The type 0A theory involves a projection onto odd rank tensors (with even rank field strength), while the type 0B theory contains even rank tensors (with odd rank field strength). In particular, the type 0B theory contains a 0-form or scalar potential $C$ in addition to the tachyon $V$. This scalar provides the needed extra degrees of freedom to represent, in the worldsheet formalism, the density oscillations on either side of the harmonic barrier in the matrix model with both sides filled.

To describe the vertex operator for this scalar requires a bit of technology [57]. One can think of the left- and right-moving worldsheet fermions $\psi(z)$, $\bar{\psi}(\bar{z})$ in terms of the 2d Ising model. In addition to the fermion operators, the

Ising model has order and disorder operators $\sigma(z,\bar{z})$ and $\mu(z,\bar{z})$, often called *spin fields*. In 2D string theory, we thus have the spacetime coordinate fields $X$, $\phi$ and their superpartners $\psi_X$, $\psi_\phi$, as well as the spin fields $\sigma_X$, $\sigma_\phi$, $\mu_X$, $\mu_\phi$, as the ingredients out of which we can build vertex operators (there may also be contributions from Faddeev-Popov ghosts if this is required by gauge invariance). The on-shell tachyon vertex is now

$$V_{i\omega} = (\psi_X \pm \psi_\phi)(\bar{\psi}_X \pm \bar{\psi}_\phi)\, e^{i\omega(X\pm\phi)}\, e^{Q\phi}\,, \tag{8.120}$$

and there is also the second (so-called *RR*) scalar, with vertex operator

$$C_{i\omega} = \Sigma_{\rm gh}(\sigma_X\sigma_\phi \pm \mu_X\mu_\phi)\, e^{i\omega(X\pm\phi)}\, e^{Q\phi}\,. \tag{8.121}$$

Here, $\Sigma_{\rm gh}$ is a spin field for the Faddeev-Popov ghosts arising from fixing local supersymmetry [57]. It was shown in [51] that the tree-level S-matrix amplitudes for the linear combinations

$$T_{L,R}(\omega) = \frac{\Gamma(-i\omega\sqrt{\alpha'/2})}{\Gamma(i\omega\sqrt{\alpha'/2})}\, V_{i\omega} \pm \frac{\Gamma(\frac{1}{2}-i\omega\sqrt{\alpha'/2})}{\Gamma(\frac{1}{2}+i\omega\sqrt{\alpha'/2})}\, C_{i\omega} \tag{8.122}$$

decouple from one another, *i.e.* the connected amplitudes involving both sets of operators $T_{L,R}$ vanish; and the amplitudes involving just one set are the same as for the bosonic string, up to a rescaling $\alpha' \to 2\alpha'$. This strongly suggests we identify $T_{L,R}$ as the asymptotic modes of density fluctuations on the left and right sides of the harmonic barrier in the symmetrically filled matrix model. Note that there are again energy-dependent phases involved in the relation between matrix model asymptotic states and continuum asymptotic states. One should think of the fields $V$ and $C$ as corresponding to the symmetric and antisymmetric perturbations of the Fermi sea of the matrix model, after these phases are stripped off. This identification is consistent with the fact that S-matrix amplitudes vanish for an odd number of parity-odd density perturbations; the $Z_2$ Ising symmetry causes the correlator of an odd number of spin fields to vanish as well. The two-to-one map of $\lambda$-space to $\phi$-space in the type 0B model highlights their nonlocal relation, a feature we have already seen several times.

This proposal passes checks analogous to the bosonic string – the tree level S-matrix, the torus partition function, and expectations of the ground ring operators on the sphere and on the disk, all agree between matrix and continuum approaches [22].

The $\mathbb{Z}_2$ symmetry that changes the sign of spin operators like $C_{i\omega}$, called *NSR parity*, also characterizes the boundary states, splitting them into $\mathbb{Z}_2$ even (*NS*) and odd (*R*) components. For instance, there are separate *NS* and *R* macroscopic loops. We may determine their functional form in the matrix model by repeating the calculation of section 7.1. The main differences will be that the calculation splits into these two boundary state sectors. The boundary state wavefunctions $\Psi_{NS}$ and $\Psi_R$ for both Dirichlet (ZZ) and Neumann (FZZT) branes appearing in (7.109) are given in [58, 59] (see also [22], sections 6 and 7). The matter partition function on the annulus [60] is essentially the same as equation (7.111), with the sum over $n$ restricted to even integers in the NS sector and odd integers in the R sector. The analysis then proceeds along the lines of section 7.1; one finds

$$\begin{aligned} \mathcal{Z}_{\rm NS} &= -\tfrac{1}{2}\log[\mu_B^2 - \lambda^2(x_0)] + \tfrac{1}{2}\log(\mu/2) \\ \mathcal{Z}_{\rm R} &= \tfrac{1}{2}\log\left[\frac{\mu_B - \lambda(x_0)}{\mu_B + \lambda(x_0)}\right] . \end{aligned} \tag{8.123}$$

These results prove a conjecture [21, 61] for the form of the macroscopic loop operators in the matrix model for the type 0B fermionic string.

The super-Liouville boundary state wavefunctions [58, 59] are also the major ingredients of the disk one-point functions that yield the wavefunctions corresponding to the operators $V_{i\omega}$ and $C_{i\omega}$. For the tachyon, one finds essentially the same result (6.95), while for the RR scalar $C$, one finds

$$\langle C_{i\omega}(z,\bar{z})\rangle_{\rm disk} = \frac{1}{2\pi}\,\hat{\mu}^{-i\omega/2}(\Gamma(\tfrac{1}{2}+i\omega))^2\,\cos(\pi s\omega)\,. \tag{8.124}$$

The corresponding integral transforms to loop length wavefunctions again yield Bessel functions [21, 22, 61].

There are also a few discrete symmetries that match on both sides of the correspondence. One example is the $\lambda \to -\lambda$ parity symmetry of the matrix model, which appears as the $\mathbb{Z}_2$ NSR parity symmetry which sends $C \to -C$ in the continuum theory. The continuum theory also has a symmetry under $\mu \to -\mu$;[29] in the matrix model, this is the symmetry of the Hamiltonian $H = \frac{1}{2}(p^2 - \lambda^2)$ under $p \leftrightarrow \lambda$, combined with an interchange of particles and holes. A few other checks, as well as a second 2D fermionic string model – the type 0A string, whose matrix model formulation involves the dynamics of open strings in a system of D-particles and their antiparticles – can be found in [22].

## 8.2 Remarks on tachyon condensation

The structure of the bosonic and fermionic matrix models of 2D string theory is a remarkable illustration of the effective picture of tachyon condensation on systems of unstable D-branes [23]. In perturbative string theory, a D-brane is a heavy, semiclassical object much like a soliton. The analogy to solitons is in fact quite precise [62–65]. Unstable D-branes are like solitons that do not carry a topological charge, and thus can decay to the vacuum (plus radiation). But being heavy, the initial stages of the decay are a collective process of instability of the 'soliton' field configuration. The quanta of this unstable mode are open string tachyons, and the initial stages of the decay are best described as the condensation of this tachyonic mode. An effective potential picture of this process is shown in figure 15a for an unstable brane in the bosonic string.

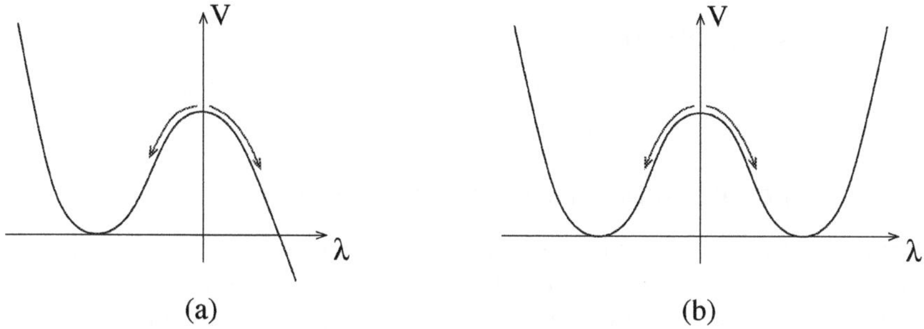


*Figure 15.* The effective potential for the open string tachyon on an unstable D-brane in the (a) bosonic, and (b) fermionic strings.

The heuristic picture of the effective potential identifies the local minimum to the left of the unstable point with the "closed string vacuum", and the difference in energy between local maximum and local minimum is the energy of the initial unstable brane. An initial state of the tachyon field $T$ localized at the unstable maximum of the effective potential is meant to describe the presence of the unstable brane, and condensation of $T$ describes its decay. Condensation to $\langle T \rangle < 0$ represents decay toward the closed string vacuum. The abyss to the right of the local maximum is meant to represent the fact that condensing the open string tachyon to $\langle T \rangle > 0$ leads to singularities at finite time in perturbative calculations [66] with no known string interpretation; it is not understood whether there is any stable, nonsingular state to which the system evolves when the open string tachyon condenses in that direction.

Qualitatively, this picture is identical to that of the matrix potential of figure 8a. The only difference is that the closed string vacuum is itself described via the open-closed string equivalence as a degenerate gas of D-particles – in a sense a collection of unstable D-branes that have "already decayed". The absence of eigenvalues to the right of the barrier means that there is no worldsheet interpretation for eigenvalues in this region, just like the region to the right of 15a.

A similar story applies to the fermionic string. The open string tachyon effective potential has two symmetric wells, as in figure 15b. On one hand, condensation in either direction of the open string tachyon on an unstable D-brane leads to its decay to the closed string vacuum; on the other hand, the matrix model for the type 0B string has just such a potential, with both wells filled by eigenvalue fermions, and a string interpretation of the physics on either side. The analogy also holds for the matrix model equivalent of the type 0A string.

## 9. Open problems

What remains to be understood? In this concluding section, let us list a few unresolved issues and directions for future work.

### 9.1 The open-closed string duality map

While there is a qualitative map (5.49) between $\lambda$-space and $\phi$-space at the level of zero modes, a precise map between the matrix model and the full Liouville field theory remains to be worked out. This would require a complete translation between quantities in the matrix model and the Liouville (plus free scalar) field theory. One indication of a missing ingredient is that the asymptotic states of the matrix model have the leg poles of the continuum formalism stripped off, see equation (6.97). The poles incorporate the effects of discrete physical states in the continuum formulation [51]. While, as argued above, only center-of-mass string motion is physical at generic, continuous momenta, there is an additional discrete spectrum of physical states at special momenta [67–69, 43]; a simple example is the zero-momentum graviton vertex operator $V_{\rm grav} = \partial X \bar{\partial} X$, which is manifestly physical since it is the action density for $X$. The continuum formalism knows how to incorporate gravitational effects, while these are currently put into the matrix model by hand; the matrix prescription for the S-matrix is to compute the LSZ-reduced density wave scattering amplitudes, and then multiply the result by a leg-pole factor $\frac{\Gamma(i\omega)}{\Gamma(-i\omega)}$ for each asymptotic state. It is this leg-pole factor which is responsible for perturbative gravitational effects [49, 50].

### 9.2 Gravitational effects

Perhaps a part of the explanation for this absence of gravitational and other discrete state effects in the matrix model is that, since the linear dilaton lifts the string "tachyon" mode to zero mass, it also raises the graviton to positive mass; its effects are subleading to the tachyon, and might be masked by or effectively absorbed into tachyon dynamics [70, 71].

Initially there was hope that the matrix model would teach us about nonperturbative gravity, and in particular lead to a solvable model of black hole dynamics. A second background solution to the string equations of motion

(2.6)-(2.7) appears to be a black hole [72, 73]

$$\begin{aligned} ds^2 &= d\phi^2 \pm \tanh^2(\tfrac{1}{\sqrt{2(k-2)}}\,\phi)\,dx^2 \\ \Phi &= \Phi_0 + \log[\cosh(\tfrac{1}{\sqrt{2(k-2)}}\,\phi)]\,, \end{aligned} \tag{9.125}$$

depending on whether we are interested in Euclidean or Lorentzian signature. The metric is written in Schwarzschild-like coordinates, where the horizon at $\phi = 0$ is infinitely redshifted relative to the asymptotic region $|\phi| \to \infty$. This sigma model (2.5) on this background describes an exact conformal field theory, the $SL(2,\mathbb{R})/U(1)$ gauged WZW model (the signature is determined by the conjugacy class of the $U(1) \subset SL(2,\mathbb{R})$ being gauged). The level $k$ of the $SL(2,\mathbb{R})$ current algebra symmetry of the WZW model is $k = 9/4$ for the bosonic string, and $k = 5/2$ for the fermionic string, in order that the slope $Q = \sqrt{\frac{2}{k-2}}$ of the asymptotically linear dilaton in (9.125) have the right value for 2D string theory. Note that the radius of curvature of the geometry is of order $1/\sqrt{k}$ in the vicinity of the horizon $\phi \sim 0$; therefore it is important to have an exact conformal field theory, since the corrections to the leading order equations of motion (2.7) are significant. Note also that the leading asymptotic perturbation $e^{2Q\phi}\partial X\bar\partial X$ of the metric away from flat spacetime, is the reflected version (the other on-shell value of Liouville momentum) of the special physical graviton operator $\partial X\bar\partial X$ discussed above. Thus the background can be thought of as the nonlinear completion of this linearized deformation. A shift in $\phi$ makes $e^{-2\Phi_0}$ the coupling in front of the asymptotic graviton in (9.125); as in higher dimensions, the coefficient of the leading asymptotic deformation of the metric away from flat spacetime is the mass of the black hole [74, 75], $\mu_{\rm bh} = e^{-2\Phi_0}$.

A great deal is known about this CFT. There is a conformal bootstrap, analogous to that of Liouville theory [76, 77]. The analogue of the two degenerate operators $V_{-b/2}$, $V_{-1/2b}$ of Liouville theory are the degenerate operators $\Phi_j$ of $SL(2,\mathbb{R})$ current algebra, having spin $j = -\frac{3}{2}$ and $j = -\frac{k}{2}$.

A rather remarkable conjecture [78] claims that the Euclidean $SL(2,\mathbb{R})/U(1)$ gauged WZW model is equivalent as a quantum field theory to another model, the so-called Sine-Liouville theory, whose action is

$$\mathcal{S}_{\rm SL} = \frac{1}{4\pi}\int\sqrt{g}\,\Big[g^{ab}\partial_a\phi\partial_b\phi + g^{ab}\partial_a X\partial_b X + QR^{(2)}\phi + \mu_{\rm sl}\cos R[X_l - X_r]\,e^{\frac{1}{Q}\phi}\,. \tag{9.126}$$

Here $Q^2 = \frac{2}{k-2}$; $X$ is compactified on a circle of radius $R = 2/Q$; and $X_l - X_r$ is the axial component of $X$, so that the potential in (9.126) acts as a generating function for vortices in the worldsheet partition function. In [77], this equivalence is argued to hold at the level of the conformal bootstrap for correlation functions. The "resonant amplitudes", which are those correlators dominated in the path integral by the asymptotic region $\phi \to -\infty$, involve only the operators $\Phi_{-\frac{r}{2}-\frac{s}{2}(k-2)-1}$, $r, s = 1, 2, \ldots$. These correlators must in general be perturbatively dressed by *both* the asymptotic graviton $\mu_{\rm bh}\, e^{2Q\phi}\partial X\bar{\partial}X$, which dresses $r$, and by the Sine-Liouville interaction $\mu_{\rm sl}\, e^{\frac{1}{Q}\phi}\cos R(X_l - X_r)$, which dresses $s$. Self-consistency requires the coefficients of these two interactions to be related [77]; one finds [77]

$$\pi\mu_{\rm bh}\,\frac{\Gamma(-Q^2/2)}{\Gamma(1+Q^2/2)} = \left(\pi\mu_{\rm sl}\,Q^2/2\right)^{Q^2}. \tag{9.127}$$

Again, as in Liouville theory there is a sense in which both dressing operators are present in the theory.

There is again a kind of strong/weak coupling duality, since the metric deformation is dominant at weak coupling ($\phi \to -\infty$) for $Q \ll 1$, while the Sine-Liouville coupling is dominant for $Q \gg 1$. Since $Q = 2$ for the 2D string, one has the sense that the Sine-Liouville description is somewhat more appropriate. In higher dimensions, when the curvature of a black hole reaches string scale, it undergoes a phase transition to a gas of strings [79] (the transition point is known as the *correspondence point*). The apparent dominance of the Sine-Liouville coupling may be an indication that the "black hole" of 2D string theory is actually on this other side of the correspondence point, where it is better thought of as a gas or condensate of strings.

The equivalence with Sine-Liouville leads to a natural candidate [80] for a matrix model equivalent to the Euclidean "black hole" – simply turn off the Liouville potential and turn on a condensate of vortices in the compactified Euclidean theory, *c.f.* section 4.1. The matrix description of the background thus has a closer affinity to the tachyon condensate of (9.126) than it has to the Euclidean black hole of (9.125).[30]

Yet another reason to suspect the absence of objects that could truly be characterized as black holes in 2D string theory, is the absence of nonsinglet states in the matrix model. As mentioned in the introduction, the appearance of black holes in the density of states in higher dimensional versions of the gauge/gravity equivalence is associated to a deconfinement transition. The thermodynamics one is led to [74, 75] on the basis of the classical gravity solution (9.125) yields a density of states $\rho = \exp[\sqrt{2}\,\pi E]$. Such a density of states will not come from the quantum mechanics of the degenerate Fermi gas of the singlet secctor of the matrix model, but might concievably come from the liberation of nonsinglet degrees of freedom of the matrix. However, this is absent from the matrix model – the $U(N)$ degrees of freedom are gauged away.

Indeed, a calculation [83] of nonperturbative high energy scattering in the matrix model – a process that in higher dimensions would certainly lead to the formation of black holes as long-lived intermediate states – reveals none of the features that would be predicted on the basis of the appearance of black holes being formed during the scattering process.

In short, low energy gravitational effects are put into the matrix model by hand, via the leg-pole factors. High energy gravitational effects such as black hole formation seem to be absent altogether. Does the matrix model incorporate any form of 2d gravity? If so, how? If not, why not?

## 9.3 Short-distance physics

Even though it would appear that black hole physics is absent from the matrix model, intriguing remnants of Planck scale (or more precisely, ultra-short distance) physics seem to be present. Namely, the spacing of eigenvalues in the matrix model is of order the D-particle Compton wavelength $L_c \sim e^{\Phi}\ell_s$.

The fact that loop length scales as $\ell \sim e^{b\phi}$, together with the integral transform (5.49), suggests that the eigenvalue coordinate scales as $\lambda \sim -e^{-b\phi}$ (in the sense of KPZ scaling). From equations (5.51) and (6.65) one determines $\langle\hat{\rho}\rangle \sim |\lambda|$ as $\lambda \to -\infty$. The eigenvalue spacing is $\delta\lambda \sim 1/\langle\hat{\rho}\rangle$, and thus $\delta\lambda/\lambda \sim \lambda^{-2}$. In terms of the Liouville coordinate, this spacing is $\delta\phi \sim e^{2\phi} = e^{\Phi}$, which is $L_c$! This result generalizes to the discrete series of $c < 1$ conformal field theories coupled to Liouville gravity, which are thought of as string theory in $D < 2$. Here we have $b = \sqrt{q/p}$, with $p, q \in \mathbb{Z}$ and $q < p$. The pair $(p, q)$ characterize the matter conformal field theory, with $c_{\rm matter} = 1 - 6\frac{(p-q)^2}{pq}$. These models have a realization as an integral over two random matrices [84–86] with the eigenvalue density scaling as $\rho(\lambda) \sim \lambda^{p/q}$. Tracing through the KPZ scaling, one finds $\delta\lambda/\lambda \sim \lambda^{-(1+1/b^2)}$, and once again the eigenvalue spacing is $\delta\phi \sim e^{Q\phi} = e^{\Phi}$. An appealing interpretation of this result is that spacetime has a graininess or discrete structure at the short distance scale $L_c$. It would be interesting to find some 'experimental' manifestation of this spacetime graininess.

## 9.4 Open string tachyons

In higher-dimensional spacetime, the canonical picture of the decay of unstable D-branes has the initial stages of the decay well-described by open string tachyon condensation; at late times the brane has decayed, open strings are absent, and the energy is carried off by a pulse of closed string radiation.

The qualitative picture is rather different in 2D string theory. Here the branes don't really decay; the open string tachyon merely describes their motion in spacetime, and there is an *equivalence* between two characterizations of the dynamics in terms of open or of closed strings.

The worldsheet formulation has elements of both open and closed string descriptions of D-brane decay. Closed string worldsheets represent the collective dynamics of the Fermi sea of "decayed" eigenvalues; eigenvalues extracted from the sea are represented as D-branes with explicit open string degrees of freedom. Thus the continuum description is naively overcomplete. For instance, one can compute the "radiation" of closed strings from the "decaying" D-brane representing an eigenvalue rolling off the potential barrier [18, 20]. One finds a closed string state

$$|\psi\rangle \sim \exp\Big[i\int d\omega\, v_p\, \alpha_p^\dagger\Big]|\mathrm{vac}\rangle \tag{9.128}$$

with the coefficient $v_p$ given to leading order by the disk expectation value of the tachyon vertex operator $V_{i\omega}$ with the boundary conditions (7.107). Roughly, the closed string tachyon bosonizes the eigenvalue fermion.

Of course, the eigenvalue doesn't decay, but stays in its wavepacket as it propagates to infinity. An eigenvalue fermion maintains its identity as it rolls to infinity; we are not forced to bosonize it. There appears to be some redundancy in the worldsheet description, unless different descriptions are valid in non-overlapping regimes (as is the case in other open/closed string equivalences); but then it remains to be seen what effects force us to describe the dynamics as that of D-branes or that of closed strings, and in what regimes those effects are important. A possible clue is the form of the D-brane boundary state, which fixes the boundary at $\phi = \infty$ throughout the motion, and instead describes the dynamics as occuring in the field space of the open string tachyon. On the other hand, we know that $\lambda \to -\infty$ corresponds to $\phi \to -\infty$, and therefore at late times an appropriate boundary state should have significant support in this weak coupling region. This suggests that the perturbative boundary state description of the rolling eigenvalue breaks down at finite time.[31] It was pointed out in [87] that the boundary state represents a source for closed strings that grows exponentially in time, so that one would expect the perturbative formalism to break down at a time of order $x \sim \log\mu$ (note that this is roughly the WKB time of flight from the top of the potential to the edge of the Fermi sea). Once again we run into the issues surrounding the quantization of the D-brane motion mentioned at the end of section 7.1.

A similar issue is the absence so far of a completely convincing worldsheet description of holes in the Fermi sea of eigenvalues (for a proposal based on analytic continuation of the boundary states, see [22, 88]). Holes lie within the Fermi sea instead of being separated from it, and so all the questions as to when and whether there is an open string description apply here as well. The worldsheet description of holes is an important missing entry in our translation table.

It is interesting that the open/closed string equivalence in this system is built out of objects that don't carry conserved charge, as opposed to standard examples like D3 branes providing the gauge theory dual to $AdS_5 \times \mathbb{S}^5$, which are charged sources for antisymmetric tensors $C_{(r)}$. It raises the question of whether there are other situations in string theory where there is an open-closed string equivalence in terms of uncharged objects. A good part of the program to understand open string tachyon condensation is driven by this question. Is the late-time dynamics of the open string tachyon condensate on unstable branes (sometimes called *tachyon matter*) an alternate description of (at least a self-contained subsector of) closed string dynamics? We have one system where the answer is yes, and it would be interesting to know if there are others, and if so whether such an equivalence holds generically (*c.f.* [18] for a discussion in the present context).

## 9.5 Closed string tachyons

Although the linear dilaton lifts the mass shell of the "tachyon" to zero in 2D spacetime, the spacelike tachyon condensate of 2D string theory may still contain clues to the properties of closed string tachyons in string theory. While much of the physics of open string tachyon condensation is relatively well understood by now, closed string tachyon condensation is still rather mysterious. The only controlled examples which have been studied involve closed string tachyons on localized defects [89–92] (for reviews, see [24, 93]). In these cases, the localized defect decays to flat spacetime with a pulse of radiation, much like the decay of D-branes via open string tachyon condensation. The condensation of delocalized tachyons is less well understood. The resulting backgrounds will have a cosmological character since the spacetime geometry will react to the stress-energy density of the evolving tachyon field.

Examples of this sort are just beginning to be studied in 2D string theory. In a sense, the closed string tachyon condensate is really only a stationary rather than a static background of the continuum theory. From the open string point of view, the custodian of this 2D cosmos must sit with a bucket of eigenvalues and keep throwing them in at a constant rate in order to preserve the Fermi sea. If this entity tires of its task, the Fermi sea drains away; the corresponding closed string background is then a time-dependent tachyon field. Properties of such backgrounds have been investigated in [81, 82, 94–97], and might serve as a paradigm for the general problem of closed string tachyon condensates.

## Acknowledgements

Thanks to the organizers of the school for their kind invitation to present these lectures, and to the Aspen Center for Physics for its hospitality during the assembly of these notes. I also wish to acknowledge support from DOE grant DE-FG02-90ER-40560.

# References


[1] O. Aharony, S. S. Gubser, J. M. Maldacena, H. Ooguri, and Y. Oz, "Large n field theories, string theory and gravity," *Phys. Rept.* **323** (2000) 183–386, hep-th/9905111.

[2] E. Witten, "Anti-de sitter space, thermal phase transition, and confinement in gauge theories," *Adv. Theor. Math. Phys.* **2** (1998) 505–532, hep-th/9803131.

[3] A. W. Peet, "Tasi lectures on black holes in string theory," hep-th/0008241.

[4] L. Fidkowski, V. Hubeny, M. Kleban, and S. Shenker, "The black hole singularity in ads/cft," *JHEP* **02** (2004) 014, hep-th/0306170.

[5] A. W. Peet and J. Polchinski, "Uv/ir relations in ads dynamics," *Phys. Rev.* **D59** (1999) 065011, hep-th/9809022.

[6] T. Banks, M. R. Douglas, G. T. Horowitz, and E. J. Martinec, "Ads dynamics from conformal field theory," hep-th/9808016.

[7] V. Balasubramanian, P. Kraus, A. E. Lawrence, and S. P. Trivedi, "Holographic probes of anti-de sitter space-times," *Phys. Rev.* **D59** (1999) 104021, hep-th/9808017.

[8] I. R. Klebanov, "String theory in two-dimensions," hep-th/9108019.

[9] P. H. Ginsparg and G. W. Moore, "Lectures on 2-d gravity and 2-d string theory," hep-th/9304011.

[10] H. Dorn and H. J. Otto, "Two and three point functions in liouville theory," *Nucl. Phys.* **B429** (1994) 375–388, hep-th/9403141.

[11] A. B. Zamolodchikov and A. B. Zamolodchikov, "Structure constants and conformal bootstrap in liouville field theory," *Nucl. Phys.* **B477** (1996) 577–605, hep-th/9506136.

[12] J. Teschner, "On the liouville three point function," *Phys. Lett.* **B363** (1995) 65–70, hep-th/9507109.

[13] V. Fateev, A. B. Zamolodchikov, and A. B. Zamolodchikov, "Boundary liouville field theory. i: Boundary state and boundary two-point function," hep-th/0001012.

[14] J. Teschner, "Remarks on liouville theory with boundary," hep-th/0009138.

[15] A. B. Zamolodchikov and A. B. Zamolodchikov, "Liouville field theory on a pseudosphere," hep-th/0101152.

[16] J. Teschner, "Liouville theory revisited," *Class. Quant. Grav.* **18** (2001) R153–R222, hep-th/0104158.

[17] Y. Nakayama, "Liouville field theory: A decade after the revolution," *Int. J. Mod. Phys.* **A19** (2004) 2771–2930, hep-th/0402009.

[18] J. McGreevy and H. Verlinde, "Strings from tachyons: The c = 1 matrix reloaded," *JHEP* **12** (2003) 054, hep-th/0304224.

[19] E. J. Martinec, "The annular report on non-critical string theory," hep-th/0305148.

[20] I. R. Klebanov, J. Maldacena, and N. Seiberg, "D-brane decay in two-dimensional string theory," *JHEP* **07** (2003) 045, hep-th/0305159.

[21] T. Takayanagi and N. Toumbas, "A matrix model dual of type 0b string theory in two dimensions," *JHEP* **07** (2003) 064, hep-th/0307083.

[22] M. R. Douglas, I. R. Klebanov, D. Kutasov, J. Maldacena, E. Martinec, and N. Seiberg, "A new hat for the c = 1 matrix model," hep-th/0307195.

[23] A. Sen, "Non-bps states and branes in string theory," hep-th/9904207.

[24] E. J. Martinec, "Defects, decay, and dissipated states," hep-th/0210231.

[25] M. B. Green, J. H. Schwarz, and E. Witten, "Supestring theory, vols. 1 and 2,". Cambridge, Uk: Univ. Pr. (1987) ( Cambridge Monographs On Mathematical Physics).

[26] J. Polchinski, "String theory. vols. 1 and 2,". Cambridge, UK: Univ. Pr. (1998).

[27] E. Braaten, T. Curtright, and C. B. Thorn, "An exact operator solution of the quantum liouville field theory," *Ann. Phys.* **147** (1983) 365.

[28] A. M. Polyakov, "Quantum geometry of bosonic strings," *Phys. Lett.* **B103** (1981) 207–210.

[29] V. G. Knizhnik, A. M. Polyakov, and A. B. Zamolodchikov, "Fractal structure of 2d-quantum gravity," *Mod. Phys. Lett.* **A3** (1988) 819.

[30] F. David, "Conformal field theories coupled to 2-d gravity in the conformal gauge," *Mod. Phys. Lett.* **A3** (1988) 1651.

[31] J. Distler and H. Kawai, "Conformal field theory and 2-d quantum gravity or who's afraid of joseph liouville?," *Nucl. Phys.* **B321** (1989) 509.

[32] J. Polchinski, "Critical behavior of random surfaces in one-dimension," *Nucl. Phys.* **B346** (1990) 253–263.

[33] E. Brezin, C. Itzykson, G. Parisi, and J. B. Zuber, "Planar diagrams," *Commun. Math. Phys.* **59** (1978) 35.

[34] D. J. Gross and I. R. Klebanov, "Vortices and the nonsinglet sector of the c = 1 matrix model," *Nucl. Phys.* **B354** (1991) 459–474.

[35] D. Boulatov and V. Kazakov, "One-dimensional string theory with vortices as the upside down matrix oscillator," *Int. J. Mod. Phys.* **A8** (1993) 809–852, hep-th/0012228.

[36] S. R. Das and A. Jevicki, "String field theory and physical interpretation of d = 1 strings," *Mod. Phys. Lett.* **A5** (1990) 1639–1650.

[37] J. Polchinski, "Classical limit of (1+1)-dimensional string theory," *Nucl. Phys.* **B362** (1991) 125–140.

[38] J. Zinn-Justin, "Perturbation series at large orders in quantum mechanics and field theories: Application to the problem of resummation," *Phys. Rept.* **70** (1981) 109.

[39] S. H. Shenker, "The strength of nonperturbative effects in string theory,". Presented at the Cargese Workshop on Random Surfaces, Quantum Gravity and Strings, Cargese, France, May 28 - Jun 1, 1990.

[40] V. Balasubramanian, M. Berkooz, A. Naqvi, and M. J. Strassler, "Giant gravitons in conformal field theory," *JHEP* **04** (2002) 034, `hep-th/0107119`.

[41] G. W. Moore, M. R. Plesser, and S. Ramgoolam, "Exact s matrix for 2-d string theory," *Nucl. Phys.* **B377** (1992) 143–190, `hep-th/9111035`.

[42] G. W. Moore and R. Plesser, "Classical scattering in (1+1)-dimensional string theory," *Phys. Rev.* **D46** (1992) 1730–1736, `hep-th/9203060`.

[43] E. Witten, "Ground ring of two-dimensional string theory," *Nucl. Phys.* **B373** (1992) 187–213, `hep-th/9108004`.

[44] A. Sen, "Rolling tachyon boundary state, conserved charges and two dimensional string theory," *JHEP* **05** (2004) 076, `hep-th/0402157`.

[45] A. Sen, "Symmetries, conserved charges and (black) holes in two dimensional string theory," `hep-th/0408064`.

[46] G. W. Moore, "Double scaled field theory at c = 1," *Nucl. Phys.* **B368** (1992) 557–590.

[47] G. W. Moore, N. Seiberg, and M. Staudacher, "From loops to states in 2-d quantum gravity," *Nucl. Phys.* **B362** (1991) 665–709.

[48] G. W. Moore and N. Seiberg, "From loops to fields in 2-d quantum gravity," *Int. J. Mod. Phys.* **A7** (1992) 2601–2634.

[49] M. Natsuume and J. Polchinski, "Gravitational scattering in the c = 1 matrix model," *Nucl. Phys.* **B424** (1994) 137–154, `hep-th/9402156`.

[50] P. Di Francesco and D. Kutasov, "Correlation functions in 2-d string theory," *Phys. Lett.* **B261** (1991) 385–390.

[51] P. Di Francesco and D. Kutasov, "World sheet and space-time physics in two-dimensional (super)string theory," *Nucl. Phys.* **B375** (1992) 119–172, `hep-th/9109005`.

[52] M. Bershadsky and I. R. Klebanov, "Genus one path integral in two-dimensional quantum gravity," *Phys. Rev. Lett.* **65** (1990) 3088–3091.

[53] N. Sakai and Y. Tanii, "Compact boson coupled to two-dimensional gravity," *Int. J. Mod. Phys.* **A6** (1991) 2743–2754.

[54] J. Callan, Curtis G., I. R. Klebanov, A. W. W. Ludwig, and J. M. Maldacena, "Exact solution of a boundary conformal field theory," *Nucl. Phys.* **B422** (1994) 417–448, `hep-th/9402113`.

[55] D. Kutasov, K. Okuyama, J. Park, N. Seiberg, and D. Shih, "Annulus amplitudes and zz branes in minimal string theory," *JHEP* **08** (2004) 026, `hep-th/0406030`.

[56] A. Dhar, G. Mandal, and S. R. Wadia, "Discrete state moduli of string theory from the c=1 matrix model," *Nucl. Phys.* **B454** (1995) 541–560, `hep-th/9507041`.

[57] D. Friedan, E. J. Martinec, and S. H. Shenker, "Conformal invariance, supersymmetry and string theory," *Nucl. Phys.* **B271** (1986) 93.

[58] T. Fukuda and K. Hosomichi, "Super liouville theory with boundary," *Nucl. Phys.* **B635** (2002) 215–254, `hep-th/0202032`.

[59] C. Ahn, C. Rim, and M. Stanishkov, "Exact one-point function of n = 1 super-liouville theory with boundary," *Nucl. Phys.* **B636** (2002) 497–513, `hep-th/0202043`.

[60] M. R. Gaberdiel and A. Recknagel, "Conformal boundary states for free bosons and fermions," *JHEP* **11** (2001) 016, `hep-th/0108238`.

[61] T. Takayanagi, "Notes on d-branes in 2d type 0 string theory," *JHEP* **05** (2004) 063, `hep-th/0402196`.

[62] E. Witten, "D-branes and k-theory," *JHEP* **12** (1998) 019, `hep-th/9810188`.

[63] A. Sen, "Descent relations among bosonic d-branes," *Int. J. Mod. Phys.* **A14** (1999) 4061–4078, `hep-th/9902105`.

[64] J. A. Harvey and P. Kraus, "D-branes as unstable lumps in bosonic open string field theory," *JHEP* **04** (2000) 012, `hep-th/0002117`.

[65] J. A. Harvey, P. Kraus, F. Larsen, and E. J. Martinec, "D-branes and strings as non-commutative solitons," *JHEP* **07** (2000) 042, `hep-th/0005031`.

[66] A. Sen, "Rolling tachyon," *JHEP* **04** (2002) 048, `hep-th/0203211`.

[67] B. H. Lian and G. J. Zuckerman, "2-d gravity with c = 1 matter," *Phys. Lett.* **B266** (1991) 21–28.

[68] P. Bouwknegt, J. G. McCarthy, and K. Pilch, "Brst analysis of physical states for 2-d gravity coupled to c <= 1 matter," *Commun. Math. Phys.* **145** (1992) 541–560.

[69] E. Witten and B. Zwiebach, "Algebraic structures and differential geometry in $2-d$ string theory," *Nucl. Phys.* **B377** (1992) 55–112, `hep-th/9201056`.

[70] T. Banks, "The tachyon potential in string theory," *Nucl. Phys.* **B361** (1991) 166–172.

[71] A. A. Tseytlin, "On the tachyonic terms in the string effective action," *Phys. Lett.* **B264** (1991) 311–318.

[72] E. Witten, "On string theory and black holes," *Phys. Rev.* **D44** (1991) 314–324.

[73] G. Mandal, A. M. Sengupta, and S. R. Wadia, "Classical solutions of two-dimensional string theory," *Mod. Phys. Lett.* **A6** (1991) 1685–1692.

[74] G. W. Gibbons and M. J. Perry, "The physics of 2-d stringy space-times," *Int. J. Mod. Phys.* **D1** (1992) 335–354, `hep-th/9204090`.

[75] C. R. Nappi and A. Pasquinucci, "Thermodynamics of two-dimensional black holes," *Mod. Phys. Lett.* **A7** (1992) 3337–3346, `gr-qc/9208002`.

[76] J. Teschner, "On structure constants and fusion rules in the sl(2,c)/su(2) wznw model," *Nucl. Phys.* **B546** (1999) 390–422, `hep-th/9712256`.

[77] A. Giveon and D. Kutasov, "Notes on ads(3)," *Nucl. Phys.* **B621** (2002) 303–336, `hep-th/0106004`.

[78] V. Fateev, A. B. Zamolodchikov, and A. B. Zamolodchikov, "unpublished,".

[79] G. T. Horowitz and J. Polchinski, "A correspondence principle for black holes and strings," *Phys. Rev.* **D55** (1997) 6189–6197, `hep-th/9612146`.

[80] V. Kazakov, I. K. Kostov, and D. Kutasov, "A matrix model for the two-dimensional black hole," *Nucl. Phys.* **B622** (2002) 141–188, `hep-th/0101011`.

[81] S. Y. Alexandrov, V. A. Kazakov, and I. K. Kostov, "Time-dependent backgrounds of 2d string theory," *Nucl. Phys.* **B640** (2002) 119–144, `hep-th/0205079`.

[82] J. L. Karczmarek and A. Strominger, "Matrix cosmology," *JHEP* **04** (2004) 055, `hep-th/0309138`.

[83] E. Martinec and K. Okuyama, "Scattered results in 2d string theory," `hep-th/0407136`.

[84] T. Tada, "(q,p) critical point from two matrix models," *Phys. Lett.* **B259** (1991) 442–447.

[85] J. M. Daul, V. A. Kazakov, and I. K. Kostov, "Rational theories of 2-d gravity from the two matrix model," *Nucl. Phys.* **B409** (1993) 311–338, `hep-th/9303093`.

[86] I. K. Kostov, "Strings with discrete target space," *Nucl. Phys.* **B376** (1992) 539–598, `hep-th/9112059`.

[87] T. Okuda and S. Sugimoto, "Coupling of rolling tachyon to closed strings," *Nucl. Phys.* **B647** (2002) 101–116, `hep-th/0208196`.

[88] D. Gaiotto, N. Itzhaki, and L. Rastelli, "On the bcft description of holes in the c = 1 matrix model," *Phys. Lett.* **B575** (2003) 111–114, `hep-th/0307221`.

[89] A. Adams, J. Polchinski, and E. Silverstein, "Don't panic! closed string tachyons in ale space-times," *JHEP* **10** (2001) 029, `hep-th/0108075`.

[90] C. Vafa, "Mirror symmetry and closed string tachyon condensation," `hep-th/0111051`.

[91] J. A. Harvey, D. Kutasov, E. J. Martinec, and G. Moore, "Localized tachyons and rg flows," `hep-th/0111154`.

[92] J. R. David, M. Gutperle, M. Headrick, and S. Minwalla, "Closed string tachyon condensation on twisted circles," *JHEP* **02** (2002) 041, `hep-th/0111212`.

[93] M. Headrick, S. Minwalla, and T. Takayanagi, "Closed string tachyon condensation: An overview," *Class. Quant. Grav.* **21** (2004) S1539–S1565, `hep-th/0405064`.

[94] J. L. Karczmarek and A. Strominger, "Closed string tachyon condensation at c = 1," *JHEP* **05** (2004) 062, `hep-th/0403169`.

[95] J. L. Karczmarek, A. Maloney, and A. Strominger, "Hartle-hawking vacuum for c = 1 tachyon condensation," `hep-th/0405092`.

[96] S. R. Das, J. L. Davis, F. Larsen, and P. Mukhopadhyay, "Particle production in matrix cosmology," *Phys. Rev.* **D70** (2004) 044017, `hep-th/0403275`.

[97] P. Mukhopadhyay, "On the problem of particle production in c = 1 matrix model," *JHEP* **08** (2004) 032, `hep-th/0406029`.

# MATRIX MODELS AS CONFORMAL FIELD THEORIES

Ivan K. Kostov
*Member of CNRS*
*C.E.A. - Saclay, Service de Physique Théorique*
*F-91191 Gif-sur-Yvette, France*
kostov@spht.saclay.cea.fr

**Abstract** In these notes we explain how the asymptotic properties of correlation functions of $U(N)$ invariant matrix integrals can be derived by means of conformal field theory. In the large $N$ limit such CFT describe gaussian field on a Riemann surface. Our basic example is the hermitian matrix model. We give an explicit operator construction of the corresponding collective field theory in terms of a bosonic field on a hyperelliptic Riemann surface, with special operators associated with the branch points. The quasiclassical expressions for the spectral kernel and the joint eigenvalue probabilities are then easily obtained as correlation functions of current, fermionic and twist operators.

## 1. Introduction and historical notes

The statistical ensembles of random matrices of large size (matrix models) have been introduced in 1951 by Wigner in order to analyze the spectral properties of complicated systems with chaotic behavior [1]. In this approach the Hamiltonian of a chaotic system is considered as a large matrix with random entries. Consequently, the analytical studies of random matrix ensembles carried out in the next 25 years (see the Mehta's book [2]) were oriented to the calculation of the spectral correlation functions or joint eigenvalue probabilities of the random matrix $\mathbf{M}$, which can be expressed as determinants of the spectral kernel

$$K(x, y) = \langle \det(x - \mathbf{M}) \det(y - \mathbf{M}) \rangle . \tag{1.1}$$

The spectral kernel can be evaluated by the method of orthogonal polynomials [2] . Its large $N$ asymptotics is characterized by the interposition of a smooth behavior and fast oscillations with wavelength $\sim 1/N$ (in a scale where the total range of the spectrum is kept finite). The smooth large distance behavior depends on the concrete form of the matrix potential. On the contrary, the microscopic behavior characterized by oscillations depends only on the symmetry group (the unitary group in the case) and fall into several universality



*E. Brezin et al. (eds.), Applications of Random Matrices in Physics,* 459–487.

classes. It is the microscopic behavior of the spectral correlations, which is interesting from the point of view of applications to chaotic systems. For a review see, for example, [4] .

The discovery by ‘t Hooft of the $1/N$ expansion [3] gave a meaning of the smooth part of the spectral correlators and opened the possibility of using random matrix models to solve various combinatorial problems, the simplest of which is the enumeration of planar graphs [6, 7]. In this kind of problems the solution is encoded in the $1/N$ expansion of the loop correlation functions, which are the correlation functions of the collective field variable

$$W(z) = \mathrm{tr}\left(\frac{1}{z - \mathbf{M}}\right). \tag{1.2}$$

In the large $N$ limit the correlation functions of the resolvent (1.2) are meromorphic functions with cuts along the intervals where the spectral density is nonzero. The discontinuity along the cuts gives the smooth part of the joint eigenvalue probabilities.

The first exact results in the large $N$ limit were obtained by direct application of the saddle point method [6] , but later it was recognized that a more powerful method is provided by the so called loop equations [5] , whose iterative solution allows one to reconstruct order by order the $1/N$ expansion. The most efficient iterative procedure proved to be the "moment's description" [8] .

In the early 90’s, after the publication of the seminal papers [9] , the two resolution techniques in random matrix models (orthogonal polynomials and loop equations) developed rapidly and were recognized as particular cases of well developed mathematical methods. The method of orthogonal polynomials was reformulated in terms of the theory of $\tau$ functions of integrable hierarchies (see, for example, the review article [10] ). On the other hand it was observed that the loop equations generate a representation of (half of) the Virasoro algebra and are therefore equivalent to the requirement that the theory is *conformal invariant* [11, 12]. This allowed to describe a class of large $N$ matrix models near criticality in terms of the Hilbert space of twisted bosonic fields, and apply the well developed formalism of the conformal field theory. Then it was observed that the two approaches are closely related since the $\tau$ functions associated with the matrix models can be formulated as fermionic theories with conformal invariance [13] .

In the string theory applications, the description in terms of traces has been used almost exclusively. In the last years it has been realized that the observables of the type (1.1) are interesting by themselves, because they describe the non-perturbative effects in string theories [14–20]. In the CFT description, the two types of variables are related by the two-dimensional bosonization, which has its analog in the higher-dimensional string theories.

# 2. Hermitian matrix integral: saddle points and hyperelliptic curves

## 2.1 Definition of the hermitian matrix integral

The partition function of the unitary ensembleis defined as the integral

$$\mathcal{Z}_N = \int d\mathbf{M} \exp\left(\sum_{n\geq 0} t_n \operatorname{Tr} \mathbf{M}^n\right) \tag{2.3}$$

where $d\mathbf{M}$ denotes the translational invariant measure in the space of $N \times N$ hermitian matrices $\mathbf{M} = \{M_{ij}\}_{i,j=1}^N$. The measure is normalized as

$$d\mathbf{M} = \frac{1}{\mathrm{Vol}[U(N)]} \prod_{k=1}^{N} \frac{dM_{kk}}{2\pi} \prod_{k<j} 2\, d\mathrm{Re}\, M_{kj}\, d\mathrm{Im}\, M_{kj} \tag{2.4}$$

where $\mathrm{Vol}[U(N)] = \prod_{k=1}^{N} \frac{(2\pi)^k}{k!}$ is the volume of the unitary group. The integrand depends on the matrix variable $\mathbf{M}$ only through its eigenvalues $x_1, ..., x_N$ and the integral (2.3) is actually reduced to

$$\mathcal{Z}_N = \int \prod_{i=1}^{N} dx_i\, e^{-V(x_i)} \prod_{i<j} (x_i - x_j)^2, \tag{2.5}$$

$$V(x) = -\sum_{n\geq 0} t_n x_i^n. \tag{2.6}$$

The "Dyson gas" (2.5) can be interpreted as the partition function of $N$ two-dimensional electric charges with repulsive logarithmic interaction, confined by the potential $V(x)$. The charges are restrained on the real axis due to the hermiticity consition for the random matrix.

## 2.2 Collective field and saddle point equations in the large $N$ limit

When the number of charges is large, it makes sense to introduce the collective field $\rho(x)$ representing the (non-normalized) charge density

$$\rho(x) = \sum_i \delta(x - x_i). \tag{2.7}$$

In the thermodinamical limit $N \to \infty$ the state of the Dyson gaz is described by the classical density $\rho_c(x)$, which is determined by the saddle point of the integral (2.5).

In order to write the saddle-point equations one should first get rid of the constraint $\int dx\rho(x) = N$, which can be done by introducing an additional field, the lagrange multiplier $\alpha$. Then the Coulomb gas integral (2.5) turns to the functional integral[1]

$$\mathcal{Z}_N = \int D\rho\, d\alpha\, e^{-\mathcal{S}(\rho,\alpha)} \tag{2.8}$$

where the effective action $\mathcal{S}$ is given by

$$\mathcal{S}[\rho,\alpha] = \int dx\rho(x)V(x) - \int dxdy\rho(x)\rho(y)\log|x-y| + \alpha[\int dx\rho(x) - N] \tag{2.9}$$

The classical free energy is proportional to $N^2$ and is equal to minus the saddle point value of the effective action:

$$\mathcal{F} \equiv \log \mathcal{Z} = -\mathcal{S}_c. \tag{2.10}$$

The saddle-point $\delta S/\delta\rho(x) = 0$ give on the support of the eigenvalue density

$$2\varphi_c(x) = \alpha \qquad \text{if} \quad \rho(x) > 0 \tag{2.11}$$

where

$$\varphi_c(x) = \tfrac{1}{2}\sum_{n\geq 0} t_n x^n + \int_{-\infty}^{\infty} dx'\rho_c(x')\log|x-x'|. \tag{2.12}$$

The function $-2\varphi_c(z)$ considered in the complex plane is the effective action for one eigenvalue in the mean field of the other eigenvalues, or the effective electric potential at the point $z$. The latter is a sum of the external potential $V(z)$ and the one produced by the charged eigenvalue liquid. The Laurent expansion of the effective potential at infinity is

$$\varphi_c(z) = -\tfrac{1}{2}t_n z^n + N\log z - \sum_{n\geq 1}\tfrac{1}{n}W_n z^{-n} \tag{2.13}$$

where $W_n$ are the moments of the random matrix

$$W_n = \int dx\, x^n \rho(x) = \partial_{t_n}\mathcal{F}. \tag{2.14}$$

The electric charge $N$, the dipole charge $W_1$ and the multipole charges $W_2, W_3, \ldots$ contain all the information about the charge distribution $\rho_c(x)$. The electric force

$$H_c(z) = \partial_z\varphi_c(z) = -\tfrac{1}{2}V(z) + W(z) \tag{2.15}$$

is a meromorphic function with a cut along the support of $\rho_c$.

Since the effective action (2.9) depends on $N$ only through the term $-\alpha N$, the the saddle point value of the lagrange multiplier gives the derivative of the free energy with respect to the total charge:

$$\partial_N \mathcal{F} = \alpha. \tag{2.16}$$

Thus the derivative of the free energy in $N$ is given by the value of the effective potential on the support of the eigenvalue density.

If the external potential $V(x)$ has several local minima, the support of $\rho_c$ may consist of several intervals $[a_{2k-1}, a_{2k}]$, $k = 1, ..., p$, associated with the different potential wells and the total charge is distributed as

$$N = \sum_{k=1}^{p} N_k. \tag{2.17}$$

In this case one should introduce a lagrange multiplier $\alpha_k$ associated with the number $N_k$ of eigenvalues trapped in the $k$-th well and the last term in the effective action (2.9) should be replaced by

$$\sum_{k=1}^{p} \alpha_k \left[ \int_{a_{2k-1}}^{a_{2k}} dx \rho(x) - N_k \right]. \tag{2.18}$$

The saddle-point equations then state that the effective potential (2.12) is constant along each of the intervals $[a_{2k-1}, a_{2k}]$ and its value there is equal to the derivative of the the conditions

$$2\varphi_c(x) = \alpha_k, \qquad x \in [a_{2k-1}, a_{2k}] \tag{2.19}$$

where

$$\alpha_k = \partial_{N_k} \mathcal{F}. \tag{2.20}$$

The free energy $\mathcal{F}_N = \log \mathcal{Z}_N$ is equal to (minus) the saddle point value of the action. Once one knows the derivatives $\alpha_k = \partial_{N_k}\mathcal{F}$ and $W_n = \partial_n \mathcal{F}$ (we use the shorthand notation $\partial_n \equiv \partial/\partial t_n$), the free energy is evaluated as

$$\begin{aligned} \mathcal{F}_N = -\mathcal{S}_c &= -\tfrac{1}{2} \int dx \rho_c(x) V(x) + \tfrac{1}{2} \sum_k \alpha_k N_k \\ &= \tfrac{1}{2} \sum_{n=0}^{\infty} t_n W_n + \tfrac{1}{2} \sum_{k=1}^{p} \alpha_k N_k, \end{aligned} \tag{2.21}$$

The expression (2.21) for the free energy is equivalent to the differential constraint

$$\left( 2 + N\partial_N - \sum_n t_n \partial_n \right) \mathcal{F} = 0,$$

which is a consequence of the scaling

$$\mathcal{F}_N(t_0, t_1, ...) = N^2 \mathcal{F}_0(\tfrac{1}{N}t_0, \tfrac{1}{N}t_1, ...) \tag{2.22}$$

satisfied by the leading term in the large $N$ expansion of the free energy.

## 2.3 Loop equations and Virasoro constraints

A more refined approach to solve the matrix model (2.3) is based is infinite set of identities for the correlation functions of the collective field variable

$$W(z) = \sum_{i=1}^{N} \frac{1}{z - x_i} = \mathrm{tr}\left(\frac{1}{z - \mathbf{M}}\right), \tag{2.23}$$

known as loop equations. The loop equations hold for any value of $N$ and allow to calculate iteratively the coefficients of the $1/N$ expansion of any observable.

The loop equations are consequence of the translational invariance of the integration measure in (2.5). By inserting the derivative $\partial/\partial x_i$ into the integral for the expectation value of the resolvent (2.23), one finds

$$\Big\langle \sum_{i=1}^{N} \Big( \frac{\partial}{\partial x_i} + 2 \sum_{j(\neq i)} \frac{1}{x_i - x_j} + \sum_{n\geq 0} n t_n x_i^{n-1} \Big) \frac{1}{z - x_i} \Big\rangle = 0 \tag{2.24}$$

where $\langle \quad \rangle$ means the average with respect to this partition function (2.5) . Using the identity

$$\sum_i \frac{1}{(z - x_i)^2} + 2 \sum_{i\neq j} \frac{1}{z - x_i} \frac{1}{x_i - x_j} = \sum_{i,j} \frac{1}{z - x_i} \frac{1}{z - x_j} \tag{2.25}$$

(2.24) can be written as

$$\Big\langle W^2(z) + \sum_{i=1}^{N} \frac{1}{z - x_i} \sum_{n\geq 0} n t_n x_i^{n-1} \Big\rangle = 0. \tag{2.26}$$

Expressed in terms of the collective field

$$\varphi(z) = -\tfrac{1}{2} V(z) + \sum_{i=1}^{N} \ln(z - x_i), \tag{2.27}$$

the loop equations (2.26) take the elegant form

$$\oint_{\infty} \frac{dz'}{2\pi i} \Big\langle \frac{T(z) - T(z')}{z - z'} \Big\rangle = 0, \qquad T(z) = [\partial\varphi(z)]^2, \tag{2.28}$$

which is equivalent to the requirement that the expectation value of the observable $T(z)$ is an entire function of $z$.

The collective field $\varphi(z)$ can be represented by a *free chiral boson*

$$\langle\, \hat{\varphi}(z) \,\rangle = \mathcal{Z}_N^{-1}\, \hat{\varphi}(z) \cdot \mathcal{Z}_N \tag{2.29}$$

using the fact that the insertion of the operator $\operatorname{tr} \mathbf{M}^n = \sum_i x_i^n$ is equivalent to taking a partial derivative with respect to the coupling $t_n$. The chiral boron is a sum of a positive and negative frequency parts

$$\hat{\varphi}(z) = \hat{\varphi}_+(z) + \hat{\varphi}_-(z), \tag{2.30}$$

$$\hat{\varphi}_+(z) = \ln z\, \partial_0 - \sum_{n\geq 1} \frac{z^{-n}}{n} \partial_n, \qquad \hat{\varphi}_-(z) = \tfrac{1}{2} \sum_{n\geq 0} t_n z^n \tag{2.31}$$

which satisfy the the canonical commutation relation

$$[\hat{\varphi}_+(z), \hat{\varphi}_-(z')] = \tfrac{1}{2} \log\left(z - z'\right) \tag{2.32}$$

and will play a key role in the following. We have replaced $N$ by $\partial_0 = \partial/\partial t_0$ in the definition (2.31), using the fact that the dependence on the coupling $t_0$ is trivial, $\partial_0 \mathcal{Z}_N = N \mathcal{Z}_N$. Then the resolvent of the random matrix is represented by the positive frequency part of the current $\partial\tilde{\varphi}$:

$$\langle\, W(z) \mathcal{O}_1 \mathcal{O}_2 ... \,\rangle = \mathcal{Z}_N^{-1}\, \partial_z \hat{\varphi}_+(z) \cdot \langle\, \mathcal{O}_1 \mathcal{O}_2 ... \,\rangle\, \mathcal{Z}_N. \tag{2.33}$$

From the point of view of the statistical mechanics of planar graphs, the field $\hat{\varphi}_+$ is the *loop insertion operator* creating on the graph a boundary, or loop, with boundary parameter $z$.

Expressed in terms of the chiral boson, the loop equation (2.26) states that the singular at $z = 0$ part of the mode expansion of the energy-momentum tensor

$$\hat{T}(z) = [\partial\hat{\varphi}(z)]^2 = \tfrac{1}{2} \sum_{n\in\mathbb{Z}} \hat{L}_n\, z^{-n-2} \tag{2.34}$$

vanishes, which is equivalent to a set of linear differential constraints on the partition function (2.3):

$$\hat{L}_n \cdot \mathcal{Z}_N = 0 \qquad (n \geq -1). \tag{2.35}$$

Here $\hat{L}_n$ are the Virasoro operators

$$\hat{L}_n = \sum_{k=0}^{n} \frac{\partial}{\partial t_k} \frac{\partial}{\partial t_{n-k}} + \sum_{k=0}^{\infty} k t_k \frac{\partial}{\partial t_{n+k}}. \tag{2.36}$$

satisfying the algebra

$$[\hat{L}_m, \hat{L}_n] = (m-n)\hat{L}_{m+n}. \tag{2.37}$$

The differential constraints (2.35) mean that the integral (2.3) as a function of $N$ and the coupling constants $\{t_0, t_1, t_2, ...\}$ realizes a highest weight representation of the Virasoro algebra. Therefore the matrix model (2.3) is defines a chiral CFT.

## 2.4 The classical solution as a hyperelliptic curve

The matrix model (2.3) can be considered as a zero-dimensional QFT with Planck constant $\hbar = 1/N$. In the classical, or dispersionless, limit $\hbar \to 0$ the correlations of the collective field vanish and the collective field is a $c$-function

$$\varphi_c(z) = N \lim_{N\to\infty} \tfrac{1}{N} \langle \varphi(z) \rangle . \tag{2.38}$$

We prefer to use the non-normalized quantities, which makes the formulas look simpler. The Planck constant is thus encoded in the expectation value of the collective field, which is of order $N$. The classical current $H_c(z) = \partial\varphi_c$ satisfies the classical loop equation

$$T_c \equiv \partial\varphi_c^2(z) = \{\text{entire function of } z\} \tag{2.39}$$

with general solution

$$H_c(z) \equiv \partial\varphi_c(z) = -M(z)\, y(z), \tag{2.40}$$

where $M(z)$ is an entire function of $z$ and the function $y(z)$ satisfies the quadratic equation

$$y^2 = \prod_{k=1}^{2p} (z - a_k). \tag{2.41}$$

Assuming that $a_1 < a_2 < ... < a_{2p}$, the classical spectral density

$$\rho_c(x) = \tfrac{1}{2\pi i}[H_c(x - i0) - H_c(x + io)] \tag{2.42}$$

is supported by the intervals $[a_{2k-1}, a_{2k}]$ where the Riemann surface of the meromorphic function $H_c(z)$ has cuts.

In this way each classical solution defines a *hyperellyptic curve* with equation (2.41), which is also known as spectral curve of the matrix model. The hyperellyptic curve is characterized by a set of canonical $A$ and $B$ cycles (Fig.1). The cycle $A_k$ encircles the cut $[a_{2k-1}, a_{2k}]$ and the cycle $B_k$ encircles the points $a_{2k}, ..., a_{2p-1}$, passing through the $k$-th and the $p$-th cuts. It will be convenient to add also a $p$-th $B$-cycle passing through the $p$-th cut and

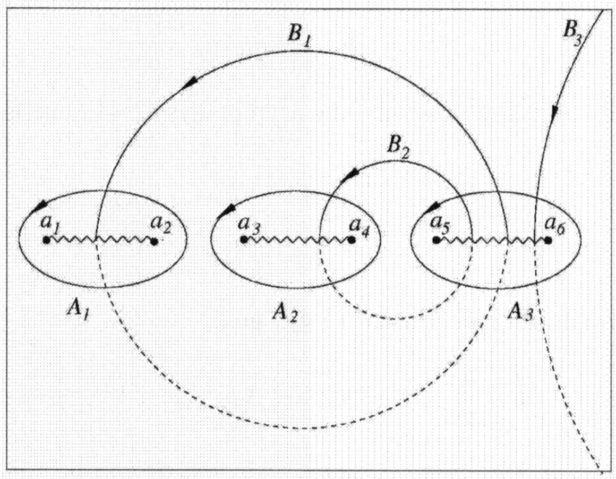


*Figure 1.* The $A$ and $B$ cycles for a genus 2 spectral curve ($p = 3$).

relating the points $z = \infty$ on the first and the second sheet of the Riemann surface of $H_c$, so that for any $k, j = 1, ..., p$

$$A_k \circ B_j = \delta_{k,j}. \tag{2.43}$$

Another useful choice of the $B$-cycles is as

$$\hat{B}_k = B_k + B_p.$$

The cycle $\hat{B}_k$ connects the infinite points on the first and the second sheets and passes through the $k$-th cut.

The function $M(z)$, which is polynomial for polynomial potential, and the endpoints $a_{2k-1}$ and $a_{2k}$ of the cuts are determined by the asymptotics

$$H_c(z) = -\tfrac{1}{2} \sum_{n \geq 1} n t_n z^{n-1} + N z^{-1} + ... \tag{2.44}$$

and the normalization conditions

$$\oint_{A_k} \frac{dz}{2\pi i} H_c(z) = N_k, \quad k = 1, ..., p, \tag{2.45}$$

which mean that the filling numbers $N_k$ are the electric charges associated with the cycles $A_k$. The explicit expression for $H_c(z)$ and $M(z)$ are given by the contour integrals

$$H_c(z) = -\tfrac{1}{2} \oint_{\infty} \frac{dz'}{2\pi i} \frac{y(z)}{y(z')} \frac{\partial V(z) - \partial V(z')}{z - z'}, \tag{2.46}$$

$$M(z) = -\tfrac{1}{2} \oint_{\infty} \frac{dz'}{2\pi i} \frac{1}{z - z'} \frac{\partial V(z')}{y(z')}. \tag{2.47}$$

Expanding (2.46) in $1/z$ and using the asymptotics (2.44), one finds

$$\oint_{\infty} \frac{dz}{2\pi i} \frac{z^k \, \partial V(z)}{y(z)} = 0 \qquad (k = 0, ..., p-1) \tag{2.48}$$

while the normalization conditions (2.4)C give

$$- \oint_{A_k} \frac{dz}{2\pi i} M(z) \, y(z) = N_k, \quad k = 1, ..., p. \tag{2.49}$$

We have seen that the filling numbers $N_k$, which give the ellectric charges associated with the connected components of the eigenvalue distribution, are associated with the $A$-cycles of the hyperellyptic curve (2.41). The variables dual to the electric charges are given by the derivatives $\partial\mathcal{F}/\partial N_k$ and are equal, according to (2.19)-(2.20) , to the values of the effective potential $\varphi_c$ at the cuts

$$\varphi_c(x) = \tfrac{1}{2} \frac{\partial \mathcal{F}}{\partial N_k}, \qquad x \in [2k-1, 2k]. \tag{2.50}$$

Integrating (2.50) to a distant cutoff point $z = \Lambda$ on the upper and the lower sheets, these conditions can be written in terms of contour integrals of the classical current along the $\hat{B}$-cycles:

$$\int_{\hat{B}_k} H_c(z) dz = -\frac{\partial \mathcal{F}}{\partial N_k} - N \log \Lambda + \tfrac{1}{2} \partial V(\Lambda). \tag{2.51}$$

The first term on the r.h.s. gives the quantum correction to the bare potential at the point $z = \Lambda$. The identities (2.51) were extensively used in the application of matrix models to $N = 2$ SYM developed by Dijkgraaf and Vafa.

Let us notice that the geometrical meaning of the derivatives of the observ-abes in $N_j$, which describe the reaction of the system when one of the eigen-values is sent to infinity, have more direct geometrical meaning in terms of the hyperelliptic curve. For example, the derivatives of the current

$$\partial_{N_j} H_c(z) = \partial \omega_j / \partial z \tag{2.52}$$

form a basis of holomorphic abelian differentials $d\omega_j = dz \partial_z \omega_j$ associated with the cycles $A_k$. Indeed, (2.4)C imply

$$\tfrac{1}{2\pi i} \oint_{A_k} d\omega_j = \delta_{kj}, \qquad k, j = 1, ..., p. \tag{2.53}$$

The $p$ functions $\partial\omega_k/\partial z$ behave as $1/z$ at infinity and are completely determined by the positions $a_{2k-1}, a_{2k}$ of the cuts. Their explicit form is

$$d\omega_j(z) = \sum_{k=1}^{p} [A^{-1}]_{jk}\, d\omega^k, \qquad d\omega^k = \frac{z^{k-1}}{y(z)}, \tag{2.54}$$

where the matrix $A_{jk}$ is given by

$$A^{jk} = \oint_{A_k} \frac{dz}{2\pi i} \frac{z^{j-1}}{y(z)}. \tag{2.55}$$

The integrals of the holomorphic differentials $d\omega_j$ along the $B$-cycles give the period matrix of the hyperelliptic curve

$$\tau_{kj} = \frac{1}{2\pi i} \oint_{B_k} d\omega_j \qquad (k, j = 1, ..., p-1). \tag{2.56}$$

The integrals along the infinite $\hat{B}$ cycles can be considered as the period matrix of the genus $g = p - 1$ hyperelliptic curve with two punctures at $z = \infty$ on the upper and lower sheets

$$\hat{\tau}_{kj} = \frac{1}{2\pi i} \oint_{\hat{B}_k} d\omega_j \qquad (k, j = 1, ..., p). \tag{2.57}$$

Then (2.51) means that the period matrix is given by the matrix of the second derivatives of the free energy in the charges:

$$\hat{\tau}_{kj} = -\frac{\partial^2 \mathcal{F}}{\partial N_k \partial N_j}. \tag{2.58}$$

*The one-cut solution ($p = 1$)*
In this case the only abelian differential is ($a_1 = a, a_2 = b$)

$$\partial_N H_c(z) = \partial_z \omega = \frac{1}{\sqrt{(z-a)(z-b)}}, \tag{2.59}$$

which is integrated to

$$\omega(z) = 2 \ln \tfrac{\sqrt{z-a}+\sqrt{z-b}}{\sqrt{b-a}}. \tag{2.60}$$

The effective potential is

$$\partial_N \varphi(z) = \tfrac{1}{2}\partial_N^2 \mathcal{F} + \omega(z) \tag{2.61}$$

where we used that $\omega(b) = 0$ and $\partial_N \varphi(b) = \frac{1}{2}\partial_N^2 \mathcal{F}$. Expanding at $z \to \infty$ and comparing with (2.13) we find for the susceptibility (the second derivative of the free energy in $N$)

$$\partial_N^2 \mathcal{F} = -2 \log \tfrac{b-a}{4}. \tag{2.62}$$

## 3. The hermitian matrix model as a chiral CFT

Instead of solving directly the Virasoro constraints (2.35), one can use apparatus of the conformal field theory. The basic observation is that the matrix model is a system of $N$ free fermions associated with the $N$ eigenvalues of the random matrix. There are different fermionic representations of the matrix integral depending on the choice of the fermionic wave functions. The traditional representation is that of orthogonal polynomials [55, 22]. A more natural one from the point of view of the conformal symmetry is based on a two-component compex Neveu-Schwarz (*i.e.* periodic around the origin) chiral fermion. In this way the formal solution of the loop equations will be given in terms of a chiral fermion or, through bosonization, of a chiral bosonic field with Liouville-like interaction, which is by its definition conformal invariant.

We will use the bosonic representation to explore the large $N$ limit of the operator solution of the loop equations through the conformal field theory and reproduce the $1/N$ expansion for the free energy and the correlation functions. The bosonic field can be considered as the collective field describing the Dyson gas of eigenvalues. This field is defined on the complex plane cut along the support of the spectral density. It can be considered as a theory with boundary, which in turn can be described as a *chiral* CFT defined on the corresponding hyperelliptic Riemann surface determined by the classical background (the spectral density in the thermodynamical limit). In this way the cuts are replaced by local singularities, the branch points of the Riemann surface, to which one associates *twist* operators. To satisfy the conformal invariance the twist operators should be dressed to *star* operators. The spectral correlations can be determined directly from the correlation functions of this bosonic field.

### 3.1 Neveu-Schwarz chiral fermions

Here we give the definition and the and basic properties of the free chiral fermion with Neveu-Schwarz boundary condition at infinity. We will consider a two-component fermion $\psi(z) = \{\psi^{(a)}(z)\}_{a=1,2}$ with energy-momentum tensor

$$T(z) = -\tfrac{1}{2} \sum_{a=1,2} \psi^{(a)} \partial \psi^{(a)}. \tag{3.63}$$

The chiral fermion is mode expanded in the local coordinate $1/z$ near $z = \infty$ as

$$\psi^{(a)}(z) = \sum_{r \in \mathbb{Z}+\frac{1}{2}} \psi_r^{(a)} \, z^{-r-\frac{1}{2}}, \quad \psi^{*(a)}(z) = \sum_{r \in \mathbb{Z}+\frac{1}{2}} \psi_r^{*(a)} \, z^{-r-\frac{1}{2}} \tag{3.64}$$

where the fermion modes satisfy the canonical anticommutation relations

$$[\psi_r^{(a)}, \psi_{r'}^{*(b)}]_+ = \delta_{a,b}\delta_{r+r',0} \tag{3.65}$$

with all other anticommutators equal to zero. The left and right NS vacuum states are defined by the requirement that the fermion operators with $r \geq \frac{1}{2}$ annihilate the right vacuum $|0\rangle$ and the fermion operators with $r \leq -\frac{1}{2}$ annihilate the left vacuum $\langle 0|$. The corresponding normal ordering

$$\psi_r^{(a)}\psi_s^{*(b)} =: \psi_r^{(a)}\psi_s^{*(b)} : + \langle 0|\psi_r^{(a)}\psi_s^{*(b)}|0\rangle$$

consists in moving the fermions with positive index to the right. As a consequence we have the following OPE:

$$\psi^{(a)}(z)\psi^{*(b)}(z') =: \psi^{(a)}(z)\psi^{*(b)}(z') : + \frac{\delta_{a,b}}{z - z'}. \tag{3.66}$$

The diagonal fermion currents have mode expansions

$$J^{(a)}(z) =: \psi^{(a)}(z)\psi^{*(a)}(z) := \sum_{n\in\mathbb{Z}} z^{-n-1}J_n^{(a)}, \tag{3.67}$$

where the amplitudes $J_n^{(a)} = \sum_r : \psi_r^{(a)}\psi_{-r+n}^{*(a)}$ : satisfy the commutation relations

$$[J_m^{(a)}, J_n^{(b)}] = m\delta_{a,b}\delta_{m+n,0}$$

of the $u(1) \otimes u(1)$ current algebra. We will need the commutation relations between current and fermion operators:

$$[J_n^{(a)}, \psi^{(b)}(z)] = z^n\delta_{a,b}\psi^{(b)}(z),$$
$$[J_n^{(a)}, \psi^{*(b)}(z)] = -z^n\delta_{a,b}\psi^{*(b)}(z). \tag{3.68}$$

Besides the Fock space vacuums of zeri charge $|0\rangle$ and $\langle 0|$ one can construct vacuum states with $u(1) \times u_(1)$ charge $\vec{N} = \{N^{(1)}, N^{(2)}\}$. These states are eigenvectors of the operators $H_0^{(a)}$:

$$\langle \vec{N}|J_0^{(a)} = N^{(a)}\langle \vec{N}|, \qquad J_0^{(a)}|\vec{N}\rangle = N^{(a)}|\vec{N}\rangle.$$

The four fermion bilinears $\psi^{(a)}\psi^{*(b)}$ generate a $u(2) = su(2)\times u(1)$ current algebra. The $su(2)$ subalgebra is generated by

$$H = \tfrac{1}{2}(J^{(1)} - J^{(2)}), \quad E_+ = \psi^{(1)}\psi^{*(2)}, \quad E_- = \psi^{(2)}\psi^{*(1)} \tag{3.69}$$

and the $u(1)$ piece by

$$\tilde{H} = \tfrac{1}{2}(J^{(1)} + J^{(2)}). \tag{3.70}$$

We will be interested only in the case of zero $u(1)$ charge, $\tilde{N} = \frac{1}{2}(N^{(1)} + N^{(2)}) = 0$. Then the vacuum states are classified by the weight $N = \frac{1}{2}(N^{(1)} - N^{(2)}) = N^{(1)} = -N^{(2)}$ of the $su(2)$ cartan element $H_0$:

$$\langle N|H_n = \delta_{n,0}N, \quad H_n|N\rangle = \delta_{n,0}N \qquad (n \geq 0)$$

while the operators $E_+$ and $E_-$ raise and lower $N$ by one unit. The left vacuum state $\langle N|$ of weight $N$ is related to the vacuum with zero charge by

$$\langle N| = \langle 0| \prod_{r=\frac{1}{2}}^{N-\frac{1}{2}} (\psi_r^{(2)} \psi_r^{*(1)}) \tag{3.71}$$

## 3.2 Fock space representation of the matrix model partition function

It follows from the OPE (3.66) that the matrix element of a product of $N$ operators $E_+$ between the left vacuum of charge $N$ and the right vacuum of charge 0 is equal to the square of the Vandermonde determinant:

$$\Big\langle N \Big| E_+(z_1)...E_+(z_n) \Big| 0 \Big\rangle = \delta_{n,N} \prod_{i<j} (z_i - z_j)^2. \tag{3.72}$$

The r.h.s. is the probability measure in (2.5) with $t_n = 0$.

The probability measure for arbitrary couplings $t_n$ will be constructed as a deformation of the scalar product (3.72). First we write an intermediate formula

$$\mathcal{Z}_N = \langle N| e^{Q_+^{\{t\}}} |0\rangle \tag{3.73}$$

with

$$Q_+^{\{t\}} = \int_{-\infty}^{\infty} dx\, e^{\sum_n t_n x^n}\, E_+(x), \tag{3.74}$$

which is an immediate consequence of (3.72). Then we represent the operator (3.74) as a deformation of the “bare" operator

$$Q_+ = \int_{-\infty}^{\infty} dx\, E_+(x) \tag{3.75}$$

generated by the ‘hamiltonians’ $H_n$ $(n \geq 1)$ representing the modes of the $su(2)$ cartan current

$$H(z) = \sum_n H_n\, z^{-n-2}. \tag{3.76}$$

The evolution along the 'times' $t_n (n \geq 0)$ is governed by the operator

$$U_{\{t\}} = \exp\left(\sum_{n\geq 0} t_n H_n\right). \tag{3.77}$$

Using the formulas for the 'time' evolution of the fermion operators

$$\begin{aligned} U_{\{t\}}\, \psi^{(2)}(z)\, U^{-1}_{\{t\}} &= e^{\frac{1}{2}\sum_n t_n z^n} \psi^{(2)}(z), \\ U_{\{t\}}\, \psi^{*(1)}(z)\, U^{-1}_{\{t\}} &= e^{\frac{1}{2}\sum_n t_n z^n} \psi^{*(1)}(z) \end{aligned} \tag{3.78}$$

it is easy to show that (3.74) and (3.75) are related by the

$$Q_+^{\{t\}} = U_{\{t\}}\, Q_+\, U^{-1}_{\{t\}}. \tag{3.79}$$

Plugging this in (3.73) and using the fact that $H_n^{(a)}|N >= 0$ for any $n > 0$, we get the fermionic Fock space representation of the partition function (2.5) :

$$\mathcal{Z}_N = \langle N | U_{\{t\}}\, e^{Q_+} |0\rangle. \tag{3.80}$$

## 3.3 CFT derivation of the Virasoro constraints

The Fock space representation of the partition function leads to a one-line proof of the Virasoro constraints (2.35). The energy-momentum tensor (3.63) is a sum of two commuting pieces associated with the $\hat{su}(2)$ and $\hat{u}$ subalgebras, $T =: H^2 : + : \tilde{H}^2 :$. It is sufficient to examine only the first term, since the second one commutes automatically with $E_\pm$. It is easy to see that the Virasoro generators $L_n$ defined by

$$: H^2(z) := \tfrac{1}{2} \sum_{n\in\mathbb{Z}} L_n z^{-n-2}$$

commute with the operator (3.75) when $n \geq -1$. Indeed, since the potential is assumed to diverge as $x \to \pm\infty$, the boundary terms of the integral

$$[L_n, Q_+] = \int_{-\infty}^{\infty} dx [L_n, E_+(\lambda)] = \int_{-\infty}^{\infty} dx \frac{d}{dx}\left(x^{n+1} E_+(x)\right) \tag{3.81}$$

can be neglected, and the result is zero. Therefore

$$\langle N | U_{\{t\}}\ : H^2(z) :\ e^{Q_+} |0\rangle = \{\text{regular function at } z = 0\}. \tag{3.82}$$

The Virasoro constraints (2.35) are obtained by commuting $L_n$, $n \geq -1$, with $U_{\{t\}}$ and using the identities

$$[\partial_n, U_{\{t\}}] = U_{\{t\}} H_n, \quad t_n U_{\{t\}} = [U_{\{t\}}, H_{-n}].$$

## 3.4 Fock space representation in terms of a free boson

The 2D bosonization formulas

$$\psi^{(a)} =: e^{\varphi^{(a)}} :$$
$$\psi^{*(a)} =: e^{-\varphi^{(a)}} : \tag{3.83}$$

give a representation of the partition function (2.5) in terms of a two-component chiral boson $\varphi(z) = \{\varphi^{(a)}(z)\}_{a=1,2}$ with mode expansion at $z = \infty$

$$\varphi^{(a)}(z) = \hat{q}^{(a)} + J_0^{(a)} \ln z - \sum_{n\neq 0} \frac{J_n^{(a)}}{n} z^{-n}, \tag{3.84}$$

$$[H_m^{(a)}, H_n^{(b)}] = n\delta_{a,b}\delta_{n+m,0}, \quad [H_0^{(a)}, \hat{q}^{(b)}] = \delta_{a,b}. \tag{3.85}$$

The four $u(2)$ currents are represented in terms of the boson field as

$$J^{(a)} = \partial\varphi^{(a)}, \quad E_+ =: e^{\varphi^{(1)}-\varphi^{(2)}} :, \quad E_- =: e^{\varphi^{(2)}-\varphi^{(1)}} : \tag{3.86}$$

where the bosonic normal ordering is defined in the usual way:

$$: J_n^{(a)} J_{-n}^{(a)} :=: J_{-n}^{(a)} J_n^{(a)} := J_{-n}^{(a)} J_n^{(a)} \quad (n > 0)$$

and

$$: \hat{q}^{(a)} J_0^{(a)} : = : J_0^{(a)} \hat{q}^{(a)} := \hat{q}^{(a)} J_0^{(a)}.$$

The boson Fock space is generated by the oscillators with negative frequencies applied to the vacuum vector $|0\rangle$ such that

$$J_n^{(a)} |0\rangle = 0 \quad (n \geq 0). \tag{3.87}$$

The left vacuum $\langle 0|$ is similarly defined, with the normalization $\langle 0|0\rangle = 1$, and the charged state $\langle N|$ is constructed as

$$\langle N| = \langle 0|e^{N(\hat{q}^{(1)}-\hat{q}^{(2)})}. \tag{3.88}$$

Then the partition function is given by the same formula (3.80).

As the $u(1)$ current $\tilde{H}$ decouples, the problem can be reformulated in terms of a single bosonic field

$$\phi = \tfrac{1}{\sqrt{2}}(\varphi^{(1)} - \varphi^{(2)}). \tag{3.89}$$

This field is the CFT realization of the free scalar boson (2.31) that appeared through the loop equations. The components of the $\hat{su}(2)$ current are expressed in terms of the field $\phi$ as

$$H(z) = \frac{1}{\sqrt{2}}\, \partial\phi(z), \quad E_{\pm} =: e^{\pm\sqrt{2}\phi} : \tag{3.90}$$

which yields the bosonic realizations for the energy-momentum tensor

$$T(z) = \sum_{n} L_n z^{-n-2} = \tfrac{1}{2} : \partial\phi(z)\partial\phi(z) : \tag{3.91}$$

the evolution operator (3.77)

$$U_{\{t\}} = \exp\left( \frac{1}{\sqrt{2}} \sum_{n\geq 0} t_n \oint_{\infty} \frac{dz}{2\pi i} z^n \partial_z \phi(z) \right) \tag{3.92}$$

and the "screening operator" $Q_+$

$$Q_+ = \int_{-\infty}^{\infty} dx \ : e^{\sqrt{2}\phi(x)} : \tag{3.93}$$

The current algebra is invariant under the discrete translations of of the field $\varphi^{(a)}$:

$$\phi^{(a)} \to \phi^{(a)} + i\pi \quad \text{or} \quad \phi \to \phi + i\pi\sqrt{2} \tag{3.94}$$

as well as the transformations

$$\phi^{(1)} \leftrightarrow \phi^{(2)} \quad \text{or} \quad \phi \to -\phi. \tag{3.95}$$

In this way the target space of the field $\phi$ is a $\mathbb{Z}_2$ orbifold compactified at the self-dual radius $R_{\text{s.d.}} = \frac{1}{\sqrt{2}}$. The geometrical meaning of the orbifold symmetry (3.95) will become clear when we consider the quasiclassical limit of the bosonic field.

## 3.5 The observables of the matrix model in terms of CFT fields

A complete set of observables in the matrix model is given by the joint distribution probabilities for $n$ eigenvalues ($1 \leq n \leq N$)

$$P(x_1, ..., x_n) = \frac{(N-n)!}{N!} \Big\langle \prod_{k=1}^{n} \delta(x_k - M \Big\rangle \tag{3.96}$$

normalized as $\int dx_1 ... dx_n \; P(x_1, ..., x_n) = 1$. Using the fermionic representation (3.80) one can express the probabilities (3.96) as expectation values of fermionic bilinears $E_+ = \psi^{*(1)}\psi^{(2)}$:

$$P(x_1, ..., x_n) = \frac{(N-n)!}{N!} \left\langle \prod_{k=1}^{n} E_+(x_k) \right\rangle \tag{3.97}$$

where by definition

$$\langle \mathcal{O} \rangle = \frac{\langle N | U_{\{t\}} \mathcal{O} e^{Q_+} | 0 \rangle}{\langle N | U_{\{t\}} e^{Q_+} | 0 \rangle}. \tag{3.98}$$

In particular, the non-normalized spectral density is the expectation value of the fermionic current

$$\rho(x) \equiv NP(x) = \langle \psi^{*(1)}(x) \psi^{(2)}(x) \rangle = \langle \, E_+(x) \, \rangle \, . \tag{3.99}$$

As in any free fermionic system, the expectation value in (3.97) is expressed in terms of the two-point fermion correlator

$$K(x, x') = \left\langle \psi^{*(1)}(x) \psi^{(2)}(x') \right\rangle \tag{3.100}$$

known as spectral kernel. This leads to the determinantal representation of the joint eigenvalue probabilities:

$$P(x_1, ..., x_n) = \frac{(N-n)!}{N!} \det{}_{n \times n} K(x_i, x_j). \tag{3.101}$$

All the observables in the matrix models can be expressed through the fermionic two-point function (3.100).

The second representation of $P(\lambda_1, ..., \lambda_n)$, which is the more convenient one from the point of view of the perturbative $1/N$ expansion, is through the correlators of the Cartan current $H(z)$. The $U(N)$-invariant correlation functions in the matrix model are obtained through the identification

$$2 \, \mathrm{Tr} \, [(z - M)^{-1}] = \sqrt{2} \, \partial \phi_+(z), \tag{3.102}$$

where $\phi_+(z)$ is the regular at $z = \infty$ part of the Laurent series of the operator $\phi(z)$:

$$\phi_+(z) = -\sqrt{2} \sum \frac{H_n}{n} z^{-n}, \qquad \phi_-(z) = \phi(z) - \phi_+(z).$$

The consistency of the two representations is guaranteed by an infinite number of Ward identities. For example, using the operator product expansion for the currents $H$ and $E_+$, one finds

$$H(z) \, e^{Q_+} |0\rangle = \int_{-\infty}^{\infty} \frac{dx \; E_+(x)}{z - x} \, e^{Q_+} \Big| 0 \Big\rangle. \tag{3.103}$$

This equation is the operator counterpart of the representation of the expectation value of the resolvent as a spectral integral

$$W(z) = \int dx \, \frac{\rho(x)}{z - x}.$$

# 4. Quasiclassical expansion: CFT on a hyperelliptic Riemann surface

## 4.1 Quasiclassical expansion

In the limit $N \to \infty$ the bosonic field develops a large (of order $N$) vacuum expectation value

$$\varphi_c^{(a)}(z) = \langle \varphi^{(a)}(z) \rangle$$

The classical currents

$$H_c^{(a)} = \partial \varphi_c^{(a)}(z)$$

solve the classical Virasoro constraints (2.39) and are given by the two branches of the meromorphic function (2.40) . Circling around a branch point exchanges $H^{(1)}(z)$ and $H^{(2)}(z)$. Therefore one can speak of a *single* field $\varphi(z)$ defined on the two-fold branched covering of the complex plane given by the hyperelliptic Riemann surface $y^2 = \prod_{k=1}^{2p}(z - a_k)$, the two sheets of which are sewed along the $p$ cuts $[a_{2k-1}, a_{2k}]$. The double-valued field $\varphi(z)$ is holomorphic on the Riemann surface, $\bar{\partial}\varphi = 0$.

The real part of the holomorphic function $\phi(z)$ is related to the effective potential $\Gamma(z, \bar{z})$ for an eigenvalue placed at the point $z = x+iy$ in the complex plane:

$$\Gamma(x, y) = -2 \,\mathrm{Re}\, \varphi_c^{(1)}(x + iy). \tag{4.104}$$

The effective potential $\Gamma$ is a sum of the external potential $V$ and the one created by the classical distribution of the electric charges. It satisfies the Laplace equation

$$\partial\bar{\partial}\Gamma(x, y) = -2\pi \rho_c(x)\, \delta(y) \tag{4.105}$$

with the boundary condition at infinity

$$\Gamma(x, y) = \mathrm{Re}\, V(x + iy) - N \ln(x^2 + y^2) + ...$$

As we have seen, the effective potential is constant along each connected component $[a_{2k-1}, a_{2k}]$ of the support of the eigenvalue density

$$\Gamma(x, 0) \equiv \Gamma_k = -\tfrac{\partial}{\partial N_k}\mathcal{F}, \quad x \in [a_{2k-1}, a_{2k}].$$

The fluctuations around the classical solution are small and the large $N$ expansion arises as the quasiclassical expansion in the collective QFT. The perturbative piece of the free energy $\mathcal{F}_N = \ln \mathcal{Z}_N$ is given by

$$\mathcal{F}_N[V] = \sum_{g=0}^{\infty} N^{2-2g} \mathcal{F}^{(g)}[V/N]. \tag{4.106}$$

As in any QFT there the $1/N$ expansion is asymptotic and should be corrected by non-perturbative $e^{-N}$ terms. The $1/N$ expansion is also called genus expansion because the term $\mathcal{F}^{(g)}$ represents the sum of all connected Feynman diagrams of genus $g$.

The leading term is simply the action for for the classical bosonic field is, according to (2.21),

$$\begin{aligned} N^2\mathcal{F}^{(0)} &= \int dx dx' \rho(x)\rho(x') \ln(\lambda - \lambda') - \int d\lambda \rho(\lambda) V(\lambda) \\ &= -\tfrac{1}{2} \sum_{k=1}^{p} N_k \Gamma_k + \tfrac{1}{2} \sum_{n \geq 0} t_n W_n. \end{aligned} \tag{4.107}$$

The higher terms in the $1/N$ expansion can be determined from the conformal invariance. The standard way is to write the loop equations (the Virasoro constraints) and solve them order by order in the genus expansion. Alternatively one can construct perturbatively the operator solution of the Virasoro constraints on the background given by the classical solution.

## 4.2 The two-point correlator of the resolvent for the one-cut solution

We would like to calculate the connected correlation function of the resolvent

$$W(z, z') = \left\langle \operatorname{Tr} \frac{1}{z - \mathbf{M}} \operatorname{Tr} \frac{1}{z' - \mathbf{M}} \right\rangle. \tag{4.108}$$

where $\langle \quad \rangle$ means expectation value in the matrix model. This correlator is given, up to a sign, by the correlation function of the two branches of the current $H(z) = \partial\varphi(z)$ associated with the upper and the lower sheet:

$$\begin{aligned} W(z, z') &= \tfrac{1}{2} \langle \partial\phi_+(z) \partial\phi_+(z') \rangle \\ &= \tfrac{1}{2} \left( \langle \partial\phi(z) \partial\phi(z') \rangle - \frac{1}{(z - z')^2} \right). \\ &= - \left\langle \partial\varphi^{(1)}(z) \partial\varphi^{(2)}(z') \right\rangle_c. \end{aligned} \tag{4.109}$$

The correlator for the resolvent of the matrix model In order to evaluate (4.109) , we need the expression for the two-point correlator of the gaussian field $\phi(z)$

defined on the Riemann surface with a cut along the interval $[a,b]$. We will perform a conformal transformation to the uniformization variable $\omega$

$$\omega(z) = 2\ln\frac{\sqrt{z-a}+\sqrt{z-b}}{\sqrt{b-a}}, \qquad z = \tfrac{1}{2}[a+b+(b-a)\cosh\omega], \tag{4.110}$$

calculate the correlator (4.109) in the $\omega$-space ans then transform the expression back to the $z$ space.

The Riemann surface of $y(z)$ is mapped to the cylinder $\omega \equiv \omega + 2\pi i$. The first sheet is mapped to $\mathrm{Re}\,\omega > 0$ and the second one to $\mathrm{Re}\,\omega < 0$. The contour going along the two edges of the cut are mapped to the circle $[-i\pi, i\pi]$. The branched points become the fixed points $\omega = 0$ and $\omega = \pm i\pi$ of the $\mathbb{Z}_2$ symmetry $\omega \to -\omega$. Thus $\varphi^{(2)}(\omega) = \varphi^{(1)}(-\omega)$.

The two-point function for the field $\varphi(\omega)$ on the Riemann surface of $y(z)$ is defined uniquely by its short-distance singularity $\varphi(\omega)\varphi(\omega') \approx \log(\omega - \omega')$ and the periodicity:

$$\langle\, \varphi(\omega)\varphi(\omega')\,\rangle_c = \ln\left(2\sinh\frac{\omega-\omega'}{2}\right). \tag{4.111}$$

The correlator of the field $\varphi(z)$ is obtained from (4.111) by substituting $\omega = \omega(z)$ and $\omega' = \omega(z')$. Therefore

$$W(z,z') = -\frac{\partial^2}{\partial z \partial z'} \ln\left(2\sinh\frac{\omega+\omega'}{2}\right). \tag{4.112}$$

Integrating in $z$ and $z'$ we get

$$\langle\, \mathrm{Tr}\,\log(z-M)\,\mathrm{Tr}\,\log(z'-M)\,\rangle_c = -\ln\left(2\sinh\frac{\omega+\omega'}{2}\right) + \frac{\omega+\omega'}{2} = \log\frac{1}{1-e^{-\omega-\omega'}} \tag{4.113}$$

where the integration constants are fixed so that the correlation functions vanishes when one of its arguments goes to infinity. In terms of the spectral parameters $z$ and $z'$

$$\langle\, \mathrm{tr}\,\log(z-M)\,\mathrm{tr}\,\log(z'-M)\,\rangle_c = -\log\left(2\frac{\sqrt{(z-a)(z'-b)}+\sqrt{(z-b)(z'-a)}}{(\sqrt{z-a}+\sqrt{z-b})(\sqrt{z'-a}+\sqrt{z'-b})}\right) \tag{4.114}$$

Differentiating twice we find the correlation function of the resolvent

$$\left\langle \mathrm{tr}\,\frac{1}{z-M}\;\mathrm{tr}\,\frac{1}{z'-M}\right\rangle_c = \frac{\frac{\sqrt{(z-a)(z'-b)}}{\sqrt{(z-b)(z'-a)}} + \frac{\sqrt{(z'-a)(z-b)}}{\sqrt{(z'-b)(z-a)}} - 2}{4(z-z')^2}. \tag{4.115}$$

## 4.3 The quasiclassical expression for the spectral kernel

The quasiclassical evaluation of the spectral kernel can be done as well, but in this case it is not so easy to control the approximation. Again we split the bosonic field $\varphi$ into a classical and quantum parts and consider the quantum part as a free field on the Riemann surface. The kernel (3.100) is expressed in the free field approximation as a correlation function of vertex operators

$$K(\lambda, \mu) = \langle : e^{\varphi^{(1)}(x)} :: e^{-\varphi^{(2)}(y)} : \rangle. \tag{4.116}$$

First we will calculate the correlation function

$$\mathcal{G}(z, z') = \langle : e^{\varphi(z)} :: e^{-\varphi(z')} : \rangle \tag{4.117}$$

for arbitrary complex arguments. Again we perform a conformal transformation to the $\omega$ variable

$$\mathcal{G}(z, z') \left( \frac{dz}{d\omega} \frac{dz'}{d\omega'} \right)^{1/2} = \frac{e^{\varphi_c(z) - \varphi_c(z')}}{2\sinh\frac{\omega - \omega'}{2}} \tag{4.118}$$

where $\frac{dz}{d\omega} = \frac{b-a}{2}\sinh\omega$. This correlation function is related to this of the bosonic field (up to $1/N$ terms in the exponent)

$$\mathcal{G}(z, z') = \frac{1}{b-a} \frac{e^{\varphi_c(z) - \varphi_c(z')}}{\sinh\frac{\omega - \omega'}{2} \sqrt{\sinh\omega \sinh\omega'}} = e^{\langle \varphi(z) \rangle - \langle \varphi(z') \rangle + \frac{1}{2} \langle [\varphi(z) + \varphi(z')]^2 \rangle}$$

In terms of the original variables $z$ and $z'$

$$\mathcal{G}(z, z') = \frac{b-a}{2} \frac{e^{\varphi_c(z) - \varphi_c(z')}}{\left( \sqrt{(z-a)(z'-b)} - \sqrt{(z-b)(z'-a)} \right) [(z-b)(z'-a)(z-a)(z'-b)]^{1/4}} \tag{4.119}$$

The kernel $K(x, x')$ is defined on the cut $[a, b]$ as the average over the four values of this function on both sides of the cut $[a, b]$

$$K(x, x') = \frac{1}{4\pi i} \sum_{\epsilon, \epsilon' = \pm} \mathcal{G}(x + i\epsilon 0, x' + i\epsilon' 0). \tag{4.120}$$

(The average should be taken because of the ambiguity due to presence of the cut; this ambiguity of $\mathcal{G}$ appears only in the large $N$ limit.) One finds explicitly, for $x, x' \in [a, b]$

$$K(x,x') = \frac{1}{2\pi} \frac{\frac{\sin \text{Im}\left(\frac{\varphi_C(x)-\varphi_C(x')}{i}\right)}{\sqrt{(x-a)(b-x')}-\sqrt{(b-x)(x'-a)}} + \frac{\cos \text{Im}\left(\frac{\varphi_C(z)+\varphi_C(z')}{i}\right)}{\sqrt{(x-a)(b-x')}+\sqrt{(b-x)(x'-a)}}}{[(x-a)(b-x)(x'-a)(b-x')]^{1/4}}. \tag{4.121}$$

This expression coinsides with the one obtained in [23] and in [24, 25] by solving the appropriate loop equations.

When $x, x' \in [a, b]$ and $x - x' \sim 1/N$, then $\varphi(x) - \varphi(x') = -i\pi \int_{x'}^{x} \rho(\xi) d\xi$ and one obtains, neglecting the second term which strongly oscillates with $x + x'$, the well known short distance behavior

$$K(x,x') = \frac{\sin[\pi N(x-x')\rho(\frac{x+x'}{2}])}{\pi(x-x')}. \tag{4.122}$$

Here it is assumed that $x$ is not too close to the branch point. In the case when one of the arguments is outside the eigenvalue interval, the exponent is real and negative and the spectral kernel decays rapidly.

## 4.4 Twist and star operators

The operator construction is technically equivalent to solving the loop equations in the vicinity of a branch point, which has been exploited in the early 90's to find an operator solution for the pure 2d quantum gravity [11] . It is analogous to the solution of the loop equations using the 'method of moments' [8].

Consider the bosonic current $H(z) = \frac{1}{\sqrt{2}} \partial\phi(z)$ near a branch point at $z = a$ around which the current changes sign. The mode expansion near the branch point goes in the half-integer powers of $z - a$:

$$\partial\phi(z) = \partial\phi_c(z) + \sum_{r \in \mathbf{Z}+\frac{1}{2}} \alpha_r (z - a_i)^{-r-1}. \tag{4.123}$$

where

$$[\alpha_r, \alpha_s] = r\delta_{r+s,0} \tag{4.124}$$

and he right Fock vacuum $|0_a\rangle$ and its conjugated $\langle 0_a|$ are defined by

$$\alpha_r|0_a\rangle = \langle 0_a|\alpha_{-r} = 0 \quad (r > 0). \tag{4.125}$$

The expansion of the classical current is

$$\partial\phi_c(z) = \sum_{r \geq \frac{1}{2}} \mu_r(a_k) \cdot (z - a_k)^{r-1}. \tag{4.126}$$

This defines the coefficients $\mu_r(a_k)$. For example, $\mu_{3/2}(a_k)$ is related to the polynomial $M(z)$ in (2.40) by

$$\mu_{3/2}(a_k) = -\sqrt{2}\, M(a_k) \prod_{j(\neq k)} (a_k - a_j)^{1/2}.$$

The vacuum state $|0_a\rangle$ can be thought of as the result of the action of a *twist operator* of dimension $\frac{1}{16}$ on the $SL(2)$-invariant vacuum[2] :

$$|0_a\rangle = \sigma(a)|0_a\rangle.$$

Thus instead of thinking of $\varphi(x)$ as a field living on the hyperelliptic Riemann surface, we can also think of it on the complex $z$-plane in the presence of $2p$ twist operators $\sigma(a_k)$ associated with the branch points $a_k$.

We have to look for an operator which creates a conformally invariant state near the branch point. The twist operator itself does not satisfy all the Virasoro constraints, in particular it does not satisfy $L_{-1}$ and $L_0$. Therefore we will look for a new operator which satisfies all constraints. Such operators are called star operators [31] , and they are constructed from the modes of the twisted bosonic field near the branch point.

## 4.5 Evaluation of the term $\mathcal{F}^{(1)}$ for a multicut background

Up to $1/N^2$ correction one finds simply an extra multiplicative factor

$$S(a_i) = \left[\mu_{3/2}(a_i)\right]^{-1/24} \sigma(a_i). \tag{4.127}$$

This factor compensates the anomaly of the constraint $L_0$ associated with the dimension $\frac{1}{16}$ of the twist operator. Therefore with this accuracy

$$\mathcal{Z}_N = e^{N^2\mathcal{F}^{(0)}} \langle S(a_1)\dots S(a_{2p})\rangle = e^{N^2\mathcal{F}^{(0)}} \left(\prod_{k=1}^{2p} \mu_{\frac{3}{2}}(a_k)\right)^{-\frac{1}{24}} \mathcal{Z}_{\text{twist}} \tag{4.128}$$

where $\mathcal{Z}_{\text{twist}}$ is the correlation function of $2p$ twist operators

$$\mathcal{Z}_{\text{twist}} = \langle \sigma(a_1)\dots\sigma(a_{2p})\rangle = (\det A)^{-1/2} \prod_{k<j} (a_k - a_j)^{-1/8}$$

Thus we obtain for the genus one free energy

$$\mathcal{F}^{(1)} = -\frac{1}{24}\sum_{k=1}^{2p} \ln \mu_{\frac{3}{2}}(a_k) - \tfrac{1}{2}\det A - \frac{1}{8}\sum_{k<j}(a_k - a_j) \tag{4.129}$$

The same expression is found by solving directly the loop equations [8, 32]. Thus we get up to $1/N^2$ terms

$$\mathcal{Z}_N = e^{\frac{1}{2}\sum t_n W_n + i\pi N\cdot\tau\cdot N} \left(\det A\right)^{-\frac{1}{2}} \prod_{j<k}^{2p} (a_j - a_k)^{-\frac{1}{8}}. \tag{4.130}$$

Using the identity

$$\det A_{ij} = \theta^2[0](0|\tau) \left[ \prod_{j,l=1}^{p} (a_{2j-1} - a_{2l-1})(a_{2j} - a_{2l}) \right]^{-1/2}$$

where $\theta[0](u|\tau) = \sum_{m\in\mathbb{Z}^p} e^{i\pi m\cdot\tau\cdot m + 2\pi i m\cdot u}$ is the theta function on the Riemann surface with zero characteristics, we write this in the form (see eq. (3.32) of [21])

$$\mathcal{F}^{(1)} = -\frac{1}{24}\sum_{k=1}^{2p} \ln \mu_{\frac{3}{2}}(a_k) + \frac{1}{8}\log\left(\det{}_{kj}\left[\frac{1}{a_{2k-1} - a_{2j}}\right]\right) - \log\theta[0](0|\tau).$$

Note that $\mathcal{Z}_{\text{twist}}$ is is equal to the chiral determinant of a $c = 1$ free boson CFT on the Riemann surface. The full partition function of the free field is given by

$$\det{}^{-1/2}\Delta = \det{}^{-1/2}(\tau - \bar{\tau})\, |\mathcal{Z}_{\text{twist}}|^2$$

where $\Delta$ id the Laplace operator on the Riemann surface. Here the period matrix is of rang $p$ and corresponds to a genus $p$ curve with an extra handle associates with a pinched cycle at $z = \infty$.

# 5. Generalization to chains of random matrices

The most natural generalization of the CFT construction is given by the $ADE$ matrix models [33] , which were introduced as a nonperturbative microscopic realization of the rational string theories with $C < 1$. Each one of these models is associated with a rank $r$ classical simply laced Lie algebra (that is, of type $A_r, D_r, E_{6,7,8}$) or its affine extension, and represents a system of $r$ coupled random matrices.

Here we will discuss only the models of the $A$-series, for which there exists a simple fermionic representation. The model associated with $A_r = su(r+1)$ represents a chain of $r$ Hermitian matrices $M_a$ of size $N_a \times N_a$ ($a = 1, ..., r$), interacting by means of $r - 1$ auxiliary gauge-field-like rectangular complex matrix variables $A_{\tilde{a}}$ ($\tilde{a} = 1, ..., r - 1$) of size $N_{\tilde{a}} \times N_{\tilde{a}+1}$. In this way the $M_a$ and $A_{\tilde{a}}$ are associated respectively with the nodes and the links of the

Dynkin diagram of $A_r$. The partition function of the matrix chain is given by the following integral

$$\mathcal{Z}_{\vec{N}}[\vec{V}] = \int \prod \prod_{a=1}^{r} dM_a \, e^{-\operatorname{tr} V^a(M_a)} \int \prod_{\tilde{a}=1}^{r-1} dA_{\tilde{a}} dA_{\tilde{a}}^\dagger e^{-\operatorname{tr} A_{\tilde{a}} A_{\tilde{a}}^\dagger M_{\tilde{a}} - \operatorname{tr} A_{\tilde{a}}^\dagger A_{\tilde{a}} M_{\tilde{a}+1}}. \tag{5.131}$$

After integrating with respect to the $A$-matrices and the angular variables of the $M$-matrices, the partition function reduces to an integral with respect of the eigenvalues $x_{ai}$, $i = 1, ..., N_a$, of the hermitian matrices $M_a$

$$\mathcal{Z}_{\vec{N}}[\vec{V}] = \int \prod_{a=1}^{r} \prod_{i} dx_{ai} \, e^{-\sum_i V^a(x_{ai})} \prod_{i<j} (x_{ai} - x_{aj})^2 \prod_{\tilde{a}=1}^{r-1} \prod_{i,j} \frac{1}{(x_{\tilde{a}i} + x_{\tilde{a}+1,j})}. \tag{5.132}$$

*Remark:* the integral with respect to the $A$-matrices exists only if all eigenvalues of the matrices $M_a$ are positive. This can be achieved by an appropriate choice of the potential. We will thus assume that the integration is restricted to the positive real axis $x_a > 0$[3].

Let $\mathcal{C}_+$ be a contour representing the boundary of the half-plane $\operatorname{Re} z > 0$. Using the fact that the eigenvalue integration is restricted to the positive real axis, we can write the following loop equations for each $a$

$$\left\langle W_a(z)^2 + \oint_{\mathcal{C}_+} \frac{dz'}{2\pi i} \frac{1}{z - z'} W(z') \Big[ \sum_b G^{ab} W_b(-z') - \partial_z V^a(z) \Big] \right\rangle_{N,t} = 0. \tag{5.133}$$

The representation of this integral as a Fock space expectation value is a generalization of the $su(2)$ construction of Sect.3. Now we consider $r + 1$ fermion fields $\psi^{(a)}(z)$, $a = 1, ..., r + 1$, whose modes in the expansion satisfy the anticommutation relations $[\psi_r^{(a)}, \psi_s^{*(b)}]_+ = \delta_{rs}\delta_{ab}$, so that the bilinears $J^{(a)}(z) =: \psi^{*(a)}(z)\psi^{(a)}(z) :$, $J_a^+(z) = \psi^{*(a)}(z)\psi^{(a+1)}(-z)$ and $J_a^-(z) = \psi^{*(a+1)}(z)\psi^{(a)}(-z)$ generate an algebra related to the $u(r+1)$ current algebra[4]. As before, the $u(1)$ current $\tilde{J} = \frac{1}{r+1}\sum_a J^{(a)}$ completely decouples so that the Cartan currents of $su(r)$ are given by the differences $J^a(z) = J^{(a)}(z) - J^{(a+1)}(-z)$.

Let us express the potentials $V^a(z)$ as differences $V^a(z) = V^{(a)}(z) - V^{(a+1)}(-z)$, and define the Hamiltonians $H^{(a)}[V^{(a)}] = \oint \frac{dz}{2\pi i} V^{(a)}(z) J^{(a)}(z)$. It is easy to see that the partition function (5.132) can be written similarly to (3.80) as

$$\mathcal{Z}_N[\vec{V}] = \langle \vec{l} | \prod_{a=1}^{r+1} e^{H^{(a)}[V^{(a)}]} \prod_{a=1}^{r} e^{Q_a^+} |0\rangle, \tag{5.134}$$

where the vacuum state of charge $\vec{l} = (l^{(1)}, ..., l^{(r+1)})$ is defined by mnfio with $l^{(a+1)} - l^{(a)} = 2N_a$, and the "screening operators" are given by $Q_a^+ = \int_0^\infty dx\, [J^{(a)}(z) - J^{(a+1)}(-z)]$.

The bosonic representation is obtained according to (3.83) in terms of the $r+1$ bosonic fields $\varphi^{(a)}$, $a = 1, ..., r+1$. They are split into $r$ fields associated with the with the $su(r)$ and the $u(1)$ parts

$$\phi^a(z) = \varphi^{(a)}(z) - \varphi^{(r+1)}(-z)\ (a = 1, ..., r),\ \ \tilde{\phi}(z) = \frac{1}{r+1}\sum_a \varphi^{(a)}((-)^{a-1}z). \tag{5.135}$$

It is convenient to define another set of fields $\phi_a$ by

$$\phi^a(z) = 2\phi_a(z) - \sum_b G^{ab}\phi_b(-z) \tag{5.136}$$

where $G^{ab} = (\vec{\alpha}^a \cdot \vec{\alpha}^b)$ is the adjacency matrix of the $A_r$ Dynkin diagram. The fields $\phi^a$ and $\phi_a$ are related as the contravariant and covariant components of a vector field in the base of the simple roots $\vec{\alpha}^a$ of the $su(r)$. Note however the reflection $z \to -z$ in the second term of (5.135)n.

If we define in a similar way the covariant components $V_a$ of the potential by

$$V^a(z) = 2V_a(z) - \sum_b G^{ab}V_b(-z) \tag{5.137}$$

then

$$\phi_a(z) = -V_a(z) + W_a(z),\ \ W_a(z) = \sum_{i=1}^{N_a} \frac{1}{z - x_{ai}}. \tag{5.138}$$

The loop equations (5.133) read, in terms of the contravariant components $\phi^a$

$$\left\langle \int_{C_+} \frac{dz'}{z - z'}[\partial\phi^a(z')]^2 \right\rangle = 0\ \ (a = 1, ..., r). \tag{5.139}$$

The Riemann surface defined by the classical solution represents an $(r+1)$-fold covering of the complex plane which has in the simplest case two cuts $[a, b]$ and $[-b, -a]$ on the real axis $(0 < a < b)$. The fields $\varphi^{(a)}$ represent the values on the different sheets of a single meromorphic field $\varphi(z)$. The quasiclassical expressions for the correlation functions and the free energy are obtained by introducing $2r$ twist operators

$$\sigma_k^+[(-)^k a] =: e^{\frac{1}{2}\phi^k[(-)^k a]} :,\ \ \sigma_k^-[(-)^k b] =: e^{-\frac{1}{2}\phi^k[(-)^k b]} : \tag{5.140}$$

associated with the points $\pm a, \pm b$. In the limit $a \to 0, b \to \infty$ the $r$ twist operators located at the points $b$ and $-b$ merge into a $\mathbb{Z}_r$ twist operator at the origin while the rest form a $\mathbb{Z}_r$ twisted boundary condition at infinity. In this case the bosonic description is given by the $\mathbb{Z}_n$ twisted field considered in refs. [11] .

## Notes

1. This representation of the functional integral is actually true only in the large $N$ limit.

2. The notion of a twist operator and the description of the gaussian field on a Riemann surface in terms of twist operators has been first introduced by Al. Zamolodchikov [21] and Dixon *at al* [26] (see also [27–30] ).

3. Another way to achieve this is to take $M_a = B_a B_a^\dagger$ where $B_a$ are complex $N_a \times N_a$ matrices.

4. The difference originates in the minus sign in the definition of the currents $J_a^\pm$ corresponding to the simple roots of $u(r+1)$.

## References

[1] E. Wigner, *Proc. Cambridge Philos. Soc.* 47 (1951) 790.

[2] M.L. Mehta, *Random Matrices*, 2nd ed. , New York, Academic Press, 1991.

[3] G. 't Hooft, *Nucl. Phys.* **BB** (7) 2 (1974) 461.

[4] J.J.M. Verbaarschot, *Universal Behavior of Dirac Spectra*, hep-th/9710114

[5] A. A. Migdal, *Phys. Rep.* 102 (1983) 199, F. David, *Mod. Phys. Lett.* **A** (5) (1990) 1019.

[6] E. Brézin C. Itzykson, G. Parisi and J.-B. Zuber, *Comm. Math. Phys.* **5** (9) (1978) 35

[7] G. 't Hooft, *Counting planar diagrams with various restrictions*, hep-th/9808113

[8] J. Ambjorn, L. Chekhov, C. Kristjansen and Yu. Makeenko, *Nucl. Phys.* **BB** (4) 04 (1993) 127 and erratum *Nucl. Phys.* **B4** (4) 9 (1995) 681.

[9] E. Brézin and V. Kazakov, *Phys. Lett.* **BB** (2) 36 (1990) 144; M. Douglas and S. Shenker, *Nucl. Phys.* **BB** (3) 35 (1990) 635; D. Gross and A.A. Migdal, *Phys. Rev. Lett.* **6** (4) (1990) 127; *Nucl. Phys.* **BB** (3) 40 (1990) 333.

[10] P. Di Francesco, P. Ginsparg and J. Zinn-Justin, *Phys.Rev.*254 (1995) 1-133.

[11] M. Fukuma, H. Kawai and R. Nakayama, *Int. J. Mod. Phys.* A 6 (1991) 1385; R. Dijkgraaf, H. Verlinde and E. Verlinde, *Nucl. Phys.* **BB** (3) 48 (1991) 435.

[12] A. Marshakov, A. Mironov et al., *Phys. Lett.* **B2** (6) 5 B (1991) 99; A. Mironov and S. Pakuliak, *Int. Journ. Mod. Phys.* A8 (1993) 3107; S. Kharchev, A. Marshakov, A. Mironov, A. Morozov and S. Pakuliak, *Nucl. Phys.* **BB** (4) 04 (1993) 717.

[13] A. Morozov, Matrix Models as Integrable Systems, hep-th/9502091

[14] J. Polchinski, "Combinatorics Of Boundaries In String Theory," *Phys. Rev.* **D50** (1994) 6041, hep-th/9407031.

[15] M. Fukuma, S. Yahikozawa, "Comments on D-Instantons in $c < 1$ Strings, *Phys. Lett.* **B460** (1999) 71-78, hep-th/9902169.

[16] J. McGreevy and H. Verlinde, "Strings from tachyons: The c = 1 matrix reloated," hep-th/0304224.

[17] I. R. Klebanov, J. Maldacena and N. Seiberg, "D-brane decay in two-dimensional string theory," hep-th/0305159.

[18] E. J. Martinec, "The annular report on non-critical string theory," arXiv:hep-th/0305148.

[19] V. A. Kazakov and I. K. Kostov, "Instantons in non-critical strings from the two-matrix model," arXiv:hep-th/0403152.

[20] M. Hanada, M. Hayakawa, N. Ishibashi, H. Kawai, T. Kuroki, Y. Matsuo and T. Tada, "Loops versus matrices: The nonperturbative aspects of noncritical string," arXiv:hep-th/0405076.

[21] Al. B. Zamolodchikov, *Nucl. Phys.* **BB** (2) 85 [FS19] (1987) 481

[22] C. Itzykson and J.-B. Zuber, J. Math. Phys. 21 (1980) 411.

[23] E. Brézin and S. Hikami, *Nucl. Phys.* **BB** (4) 22 (515) 1994.

[24] E. Brézin, S. Hikami and A. Zee, Nucl. Phys. **B464** (1996) 411.

[25] P. Zinn-Justin, *Comm. Math. Phys.* **1** (9) 4, 631 (1988).

[26] L. Dixon, D. Friedan, E. Martinec and S. Shenker, "The conformal field theory of orbifolds", *Nucl. Phys.* **BB282** (1977) 13.

[27] V. G. Knizhnik, *Comm. Math. Phys.* **1** (1) 2 (1987) 567; M. Bershadsky and A. Radul, *Int. Journ. Mod. Phys.* A. Vol. 2 (1987) 165.

[28] M. Bershadsky and A. Radul, Int. J. Mod. Phys. **A2** (1987) 165.; K. Miki, Phys. Lett. **B191** (1987) 127.

[29] E. Verlinde and H. Verlinde, Nucl. Phys. **B288** (1987) 357.

[30] R. Dijkgraaf, E. Verlinde and H. Verlinde, Commun. Math. Phys. **115** (1988) 649.

[31] G. W. Moore, "Matrix Models Of 2-D Gravity And Isomonodromic Deformation," lectures at 1990 Cargese Workshop on *Random Surfaces, Quantum Gravity and Strings,* Prog. Theor. Phys. Suppl. **102**, 255 (1990).

[32] L. Chekhov, "Genus one correlation to multi-cut matrix model solutions," arXiv:hep-th/0401089.

[33] I. Kostov, *Phys. Lett.* **BB** (2) 97 (1992)74.

# LARGE $N$ ASYMPTOTICS OF ORTHOGONAL POLYNOMIALS FROM INTEGRABILITY TO ALGEBRAIC GEOMETRY

B. Eynard
*Service de Physique Theorique de Saclay,*
*F-91191 Gif-sur-Yvette Cedex, France.*
eynard@wasa.saclay.cea.fr

## 1. Introduction

Random matrices play an important role in physics and mathematics [28, 19, 6, 14, 25, 34, 13]. It has been observed more and more in the recent years how deeply random matrices are related to integrability ($\tau$-functions), and algebraic geometry.

Here, we consider the computation of large n asymptotics for orhogonal polynomials as an example of a problem where the concepts of integrability, isomonodromy and algebraic geometry appear and combine.

The method presented here below, is not, to that date, rigorous mathematicaly. It is based on the asumption that an integral with a large number of variables can be approximated by a saddle-point method. This asumption was never proven rigorously, it is mostly based on "physical intuition". However, the results given by that method have been rigorously proven by another method, namely the Riemann–Hilbert method [7, 8, 11, 12]. The method presented below was presented in many works [17, 16, 2, 20, 18].

## 2. Definitions

Here we consider the 1-Hermitean matrix model with polynomial potential:

$$\begin{aligned} Z_N &:= \int_{H_N} dM\, \mathrm{e}^{-N\mathrm{tr}\, V(M)} \\ &= \int_{\mathbb{R}^N} dx_1 \dots dx_N \ (\Delta(x_1, \dots, x_N))^2 \prod_{i=1}^{N} \mathrm{e}^{-NV(x_i)} \end{aligned} \tag{2-1}$$



*E. Brezin et al. (eds.), Applications of Random Matrices in Physics,* 489–513.

where $\Delta(x_1,\ldots,x_N) := \prod_{i>j}(x_i - x_j)$, and the $x_i$'s are the eigenvalues of the matrix $M$, and $V(x)$ is a polynomial called the potential:

$$V(x) = \sum_{k=0}^{\deg V} g_k x^k \tag{2-2}$$

REMARK 2.1 *All the calculations which are presented below, can be extended to a more general setting, with no big fundamental changes:*

*- one can consider $V'(x)$ any rational fraction [3] instead of polynomial, in particular one can add logarithmic terms to the potential $V(x)$.*

*- one can consider arbitrary paths (or homology class of paths) of integrations $\Gamma^N$ insteaf of $\mathbb{R}^N$, in particular finite segments [1] ...*

*- one can study non hermitean matrix models [20], where the Vandermonde $\Delta^2$ is replaced by $\Delta^\beta$ where $\beta = 1, 2, 4$.*

*- one can consider multi-matrix models, in particular 2-matrix model [2, 17, 16].*

## 3. Orthogonal polynomials

Consider the family of monic polynomials $p_n(x) = x^n + O(x^{n-1})$, defined by the orthogonality relation:

$$\int_{\mathbb{R}} p_n(x) p_m(x) \mathrm{e}^{-NV(x)} dx = h_n \delta_{nm} \tag{3-3}$$

It is well known that the partition function is given by [30]:

$$Z_N = N! \prod_{n=0}^{N-1} h_n \tag{3-4}$$

Such an orthogonal family always exists if the integration path is $\mathbb{R}$ or a subset of $\mathbb{R}$, and if the potential is a real polynomial. In the more general setting, the orthogonal polynomials "nearly always" exist (for arbitrary potentials, the set of paths for which they don't exist is enumerable).

We define the kernel:

$$K(x,y) := \sum_{n=0}^{N-1} \frac{p_n(x) p_n(y)}{h_n} \tag{3-5}$$

One has the following usefull theorems:

THEOREM 1 *Dyson's theorem [15]: any correlation function of eigenvalues, can be written in terms of the kernel $K$:*

$$\rho(\lambda_1,\ldots,\lambda_k) = \det(K(\lambda_i,\lambda_j)) \tag{3-6}$$

Thus, if one knows the orthogonal polynomials, then one knows all the correlation functions.

THEOREM 2 *Christoffel-Darboux theorem [30, 32]: The kernel $K(x,y)$ can be written:*

$$K(x,y) = \gamma_N \frac{p_N(x)p_{N-1}(y) - p_N(y)p_{N-1}(x)}{x-y} \tag{3-7}$$

Thus, if one knows the polynomials $p_N$ and $p_{N-1}$, then one knows all the correlation functions.

Our goal now, is to find large $N$ "strong" asymptotics for $p_N$ and $p_{N-1}$, in order to have the large $N$ behaviours of any correlation functions.

**Notation:** we define the wave functions:

$$\psi_n(x) := \frac{1}{\sqrt{h_n}} p_n(x)\, \mathrm{e}^{-\frac{N}{2}V(x)} \tag{3-8}$$

they are orthonormal:

$$\int \psi_n(x)\psi_m(x) = \delta_{nm} \tag{3-9}$$

## 4. Differential equations and integrability

It can be proven that $(\psi_n, \psi_{n-1})$ obey a differential equation of the form [9, 7, 30, 33, 5]:

$$-\frac{1}{N}\frac{\partial}{\partial x}\begin{pmatrix}\psi_n(x)\\ \psi_{n-1}(x)\end{pmatrix} = \mathcal{D}_n(x)\begin{pmatrix}\psi_n(x)\\ \psi_{n-1}(x)\end{pmatrix} \tag{4-10}$$

where $\mathcal{D}_n(x)$ is a $2\times 2$ matrix, whose coefficients are polynomial in $x$, of degree at most $\deg V'$. (In case $V'$ is a rational function, then $\mathcal{D}$ is a rational function with the same poles).

$(\psi_n, \psi_{n-1})$ also obeys differential equations with respect to the parameters of the model [7, 5], i.e. the coupling constants, i.e. the $g_k$'s defined in 2-2:

$$\frac{1}{N}\frac{\partial}{\partial g_k}\begin{pmatrix}\psi_n(x)\\ \psi_{n-1}(x)\end{pmatrix} = \mathcal{U}_{n,k}(x)\begin{pmatrix}\psi_n(x)\\ \psi_{n-1}(x)\end{pmatrix} \tag{4-11}$$

where $\mathcal{U}_{n,k}(x)$ is a $2\times 2$ matrix, whose coefficients are polynomial in $x$, of degree at most $k$.

It is also possible to find some discrete recursion relation in $n$ (see [5]).

The compatibility of these differential systems, i.e. $\frac{\partial}{\partial x}\frac{\partial}{\partial g_k} = \frac{\partial}{\partial g_k}\frac{\partial}{\partial x}$, $\frac{\partial}{\partial g_j}\frac{\partial}{\partial g_k} = \frac{\partial}{\partial g_k}\frac{\partial}{\partial g_j}$, as well as compatibility with the discrete recursion, imply **integrability**, and allows to define a $\tau$-function [27, 5].

We define the spectral curve as the locus of eigenvalues of $\mathcal{D}_n(x)$:

$$E_n(x,y) := \det(y\mathbf{1} - \mathcal{D}_n(x)) \tag{4-12}$$

REMARK 4.1 In the 1-hermitean-matrix model, $\mathcal{D}_n$ is a $2\times 2$ matrix, and thus $\deg_y E_n(x,y) = 2$, i.e. the curve $E_n(x,y) = 0$ is an **hyperelliptical curve**. In other matrix models, one gets algebraic curves which are not hyperelliptical.

REMARK 4.2 What we will se below, is that the curve $E_N(x,y)$ has a large $N$ limit $E(x,y)$, which is also an hyperelliptical curve. In general, the matrix $\mathcal{D}_N(x)$ has no large $N$ limit.

## 5. Riemann-Hilbert problems and isomonodromies

The $2\times 2$ system $\mathcal{D}_N$ has 2 independent solutions:

$$-\frac{1}{N}\frac{\partial}{\partial x}\begin{pmatrix}\psi_n(x)\\ \psi_{n-1}(x)\end{pmatrix} = \mathcal{D}_n(x)\begin{pmatrix}\psi_n(x)\\ \psi_{n-1}(x)\end{pmatrix},$$

$$-\frac{1}{N}\frac{\partial}{\partial x}\begin{pmatrix}\phi_n(x)\\ \phi_{n-1}(x)\end{pmatrix} = \mathcal{D}_n(x)\begin{pmatrix}\phi_n(x)\\ \phi_{n-1}(x)\end{pmatrix} \tag{5-13}$$

where the wronskian is non-vanishing: $\det\begin{pmatrix}\psi_n(x) & \phi_n(x)\\ \psi_{n-1}(x) & \phi_{n-1}(x)\end{pmatrix} \neq 0.$

We define the matrix of fundamental solutions:

$$\Psi_n(x) := \begin{pmatrix}\psi_n(x) & \phi_n(x)\\ \psi_{n-1}(x) & \phi_{n-1}(x)\end{pmatrix} \tag{5-14}$$

it obeys the same differential equation:

$$-\frac{1}{N}\frac{\partial}{\partial x}\Psi_n(x) = \mathcal{D}_n(x)\,\Psi_n(x) \tag{5-15}$$

Here, the second solution can be constructed explicitely:

$$\phi_n(x) = \mathrm{e}^{+\frac{N}{2}V(x)} \int \frac{dx'}{x-x'}\,\psi_n(x')\mathrm{e}^{-\frac{N}{2}V(x')} \tag{5-16}$$

Notice that $\phi_n(x)$ is discontinuous along the integration path of $x'$ (i.e. the real axis in the most simple case), the discontinuity is simply $2i\pi\psi_n(x)$. In terms of fundamental solutions, one has the jump relation:

$$\Psi_n(x+i0) = \Psi_n(x-i0)\begin{pmatrix}1 & 2i\pi\\ 0 & 1\end{pmatrix} \tag{5-17}$$

Finding an invertible piecewise analytical matrix, with given large $x$ behaviours, with given jumps on the borders between analytical domains, is called a **Riemann–Hilbert problem** [7, 8, 4].

It is known that the Riemann–Hilbert problem has a unique solution, and that if two R-H problems differ by $\epsilon$ (i.e. the difference between jumps and behaviours at $\infty$ is bounded by $\epsilon$), then the two solutions differ by at most $\epsilon$ (roughly speeking, harmonic functions have their extremum on the boundaries). Thus, this approach can be used [7, 11, 12] in order to find large $N$ asymptotics of orthogonal polynomials: The authors of [7] considered a guess for the asymptotics, which satisfies another R-H problem, which differs from this one by $O(1/N)$.

Notice that the jump matrix in 5-17 is independent of $x$, of $n$ and of the potential, it is a constant. The jump matrix is also called a monodromy, and the fact that the monodromy is a constant, is called **isomonodromy** property [27].

Consider an invertible, piecewise analytical matrix $\Psi_n(x)$, with appropriate behaviours[1] at $\infty$, which satisfies 5-17, then, it is clear that the matrix $-\frac{1}{N}\Psi_n'(x)(\Psi_n(x))^{-1}$, has no discontinuity, and given its behaviour at $\infty$, it must be a polynomial. Thus, we can prove that $\Psi_n(x)$ must satisfy a differential system $\mathcal{D}_n(x)$ with polynomial coefficients. Similarly, the fact that the monodromy is independent of $g_k$ and $n$ implies the deformation equations, as well as the discrete recursion relations.

Thus, the isomonodromy property, implies the existence of compatible differential systems, and integrability [6, 24, 26, 27, 33, 5].

## 6. WKB–like asymptotics and spectral curve

Let us look for a formal solution of the form:

$$\Psi_N(x) = A_N(x)\,\mathrm{e}^{-NT(x)}B_N \tag{6-18}$$

where $T(x) = \mathrm{diag}(T_1(x), T_2(x))$ is a diagonal matrix, and $B_N$ is independent of $x$. The differential system $\mathcal{D}_N(x)$ is such that:

$$\begin{aligned}\mathcal{D}_N(x) &= -\frac{1}{N}\Psi_N'\Psi_N^{-1} = A_N(x)T'(x)A_N^{-1}(x) - \frac{1}{N}A_N'(x)A_N^{-1}(x)\\ &= A_N(x)T'(x)A_N^{-1}(x) + O(\frac{1}{N})\end{aligned} \tag{6-19}$$

this means, that, under the asumption that $A_N(x)$ has a large $N$ limit $A(x)$, $T_1'(x)$ and $T_2'(x)$ are the large $N$ limits of the eigenvalues of $\mathcal{D}_N(x)$.

With such an hypothesis, one gets for the orthogonal polynomials:

$$\psi_N(x) \sim A_{11}\mathrm{e}^{-NT_1(x)}B_{1,1} + A_{12}\mathrm{e}^{-NT_2(x)}B_{2,1} \tag{6-20}$$

We are now going to show how to derive such a formula.

# 7. Orthogonal polynomials as matrix integrals

## 7.1 Heine's formula

THEOREM 3 *Heine's theorem [32]. The orthogonal polynomials* $p - n(x)$ *are given by:*

$$\begin{aligned} p_n(\xi) &= \frac{\int dx_1 \dots dx_N \, \prod_{i=1}^N (\xi - x_i) \, (\Delta(x_1, \dots, x_N))^2 \, \prod_{i=1}^N \mathrm{e}^{-NV(x_i)}}{\int dx_1 \dots dx_N \, (\Delta(x_1, \dots, x_N))^2 \, \prod_{i=1}^N \mathrm{e}^{-NV(x_i)}} \\ &= \langle \det(\xi \mathbf{1} - M) \rangle \end{aligned} \qquad (7\text{-}21)$$

i.e. the orthogonal polynomial is the average of the characteristic polynomial of the random matrix.

Thus, we can define the orthogonal polynomials as matrix integrals, similar to the partition function $Z$ define in 2-1.

## 7.2 Another matrix model

Define the potential:

$$V_h(x) := V(x) - h \ln(\xi - x) \qquad (7\text{-}22)$$

and the partition function:

$$Z_n(h,T) := \mathrm{e}^{-\frac{n^2}{T^2} F_n(h,T)} := \int dx_1 \dots dx_n \, (\Delta(x_1, \dots, x_n))^2 \prod_{i=1}^n \mathrm{e}^{-\frac{n}{T} V_h(x_i)} \qquad (7\text{-}23)$$

i.e. $Z_N(0,1) = Z$ is our initial partition function.

Heine's formula reads:

$$p_n(\xi) = \frac{Z_n(\frac{1}{N}, \frac{n}{N})}{Z_n(0, \frac{n}{N})} = \mathrm{e}^{-N^2 (F_n(\frac{1}{N}, \frac{n}{N}) - F_n(0, \frac{n}{N}))} \qquad (7\text{-}24)$$

The idea, is to perform a Taylor expansion in $h$ close to 0 and $T$ close to 1.

**Taylor expansion.** We are interested in $n = N$ and $n = N - 1$, thus $T = \frac{n}{N} = 1 + \frac{n-N}{N} = 1 + O(1/N)$ and $h = 0$ or $h = 1/N$, i.e. $h = O(1/N)$:

$$T = 1 + O(1/N) \quad , \quad h = O(1/N) \qquad (7\text{-}25)$$

Roughly speaking:

$$\begin{aligned} p_n(\xi) &\sim \mathrm{e}^{-N^2 \, h \frac{\partial F}{\partial h} + (T-1) h \frac{\partial^2 F}{\partial h \partial T} + \frac{h^2}{2} \frac{\partial^2 F}{\partial h^2} + O(1/N^3)} \\ &\sim \mathrm{e}^{-N \frac{\partial F}{\partial h}} \, \mathrm{e}^{-(n-N) \frac{\partial^2 F}{\partial h \partial T}} \, \mathrm{e}^{-\frac{1}{2} \frac{\partial^2 F}{\partial h^2}} \, (1 + O(1/N)) \end{aligned} \qquad (7\text{-}26)$$

where all the derivatives are computed at $T = 1$ and $h = 0$.

**Topological expansion.** Imagine that $F_n$ has a $1/n^2$ expansion of the form:

$$F = F^{(0)} + \frac{1}{n^2}F^{(1)} + O(\frac{1}{n^3}) \tag{7-27}$$

where all $F^{(0)}$ and $F^{(1)}$ are analytical functions of $T$ and $h$, than one needs only $F^{(0)}$ in order to compute the asymptotics 7-26.

Actualy, that hypothesis is not always true. It is wrong in the so called "mutlicut" case. But it can be adapted in that case, we will come back to it in section 11.2. For the moment, let us conduct the calculation only with $F^{(0)}$.

# 8. Computation of derivatives of $F^{(0)}$

We have defined:

$$Z_n(h,T) = \mathrm{e}^{-\frac{n^2}{T^2}F_n(h,T)} = \int dM_{n\times n}\mathrm{e}^{-\frac{n}{T}\mathrm{tr}\,V(M)}\,\mathrm{e}^{h\frac{n}{T}\ln(\xi-M)} \tag{8-28}$$

this implies that:

$$-\frac{n^2}{T^2}\frac{\partial F_n}{\partial h} = \left\langle \frac{n}{T}\mathrm{tr}\,\ln(\xi - M)\right\rangle_{V_h} \tag{8-29}$$

i.e.

$$\frac{\partial F_n}{\partial h} = -\frac{T}{n}\langle \mathrm{tr}\,\ln(\xi - M)\rangle_{V_h} \tag{8-30}$$

It is a primitive of $-\frac{T}{n}\langle \mathrm{tr}\,\ln(x - M)\rangle_{V_h}$, which behaves as $-\frac{T}{n}\ln x + O(1/x)$ at large $x$. Therefore, we define the resolvent $W(x)$:

$$W(x) := \frac{T}{n}\left\langle \mathrm{tr}\,\frac{1}{x-M}\right\rangle_{V_h} \tag{8-31}$$

Notice that it depends on $\xi$ through the potential $V_h$, i.e. through the average $< . >$. And we define the effective potential:

$$V_{\text{eff}}(x) = V_h(x) - 2T\ln x - 2\int_\infty^x (W(x') - \frac{T}{x'})dx' \tag{8-32}$$

which is a primitive of $V_h'(x) - 2W(x)$. Thus , we have:

$$\frac{\partial F_n}{\partial h} = \frac{1}{2}\left(V_{\text{eff}}(\xi) - V_h(\xi)\right) \tag{8-33}$$

We also introduce:

$$\Omega(x) := \frac{\partial W(x)}{\partial T} \quad , \quad \ln \Lambda(x) := \ln x + \int_\infty^x (\Omega(x') - \frac{1}{x'}) dx' = -\frac{1}{2}\frac{\partial}{\partial T} V_{\text{eff}}(x) \tag{8-34}$$

$$H(x,\xi) := \frac{\partial W(x)}{\partial h} \quad , \quad \ln H(\xi) := \int_\infty^\xi H(x',\xi) dx' \tag{8-35}$$

i.e.

$$\frac{\partial^2 F_n}{\partial h^2} = -\ln H(\xi) \quad , \quad \frac{\partial^2 F_n}{\partial h \partial T} = -\ln \Lambda(\xi) \tag{8-36}$$

With these notations, the asymptotics are:

$$\psi_n(\xi) \sim \sqrt{H(\xi)}\ (\Lambda(\xi))^{n-N}\ \mathrm{e}^{-\frac{N}{2} V_{\text{eff}}(\xi)}\ (1 + O(1/N)) \tag{8-37}$$

Now, we are going to compute $W$, $\Lambda$, $H$, etc, in terms of geometric properties of an hyperelliptical curve.

REMARK 8.1 This is so far only a sketch of the derivation, valid only in the 1-cut case. In general, $F_n$ has no $1/n^2$ expansion, and that case will be addressed in section 11.2.

REMARK 8.2 These asymptoics are of the form of 6-18 in section.6, and thus, $\frac{1}{2}V'(x) - W(x)$ is the limit of the eigenvalues of $\mathcal{D}_N(x)$.

## 9. Saddle point method

There exists many ways of computing the resolvent and its derivatives with respect to $h$, $T$, or other parameters. The loop equation method is a very good method, but there is not enough time to present it here. There are several saddle-point methods, which all coincide to leading order. We are going to present one of them, very intuitive, but not very rigorous on a mathematical ground, and not very appropriate for next to leading computations. However, it gives the correct answer to leading order.

Write:

$$Z_n(h,T) = \mathrm{e}^{-\frac{n^2}{T^2} F_n(h,T)} = \int dx_1 \dots dx_n \mathrm{e}^{-\frac{n^2}{T^2} \mathcal{S}(x_1,\dots,x_n)} \tag{9-38}$$

where the action is:

$$\mathcal{S}(x_1,\dots,x_n) := \frac{T}{n} \sum_{i=1}^n V_h(x_i) - 2\frac{T^2}{n^2} \sum_{i>j} \ln (x_i - x_j) \tag{9-39}$$

The saddle point method consists in finding configurations $x_i = \overline{x}_i$ where $\mathcal{S}$ is extremal, i.e.

$$\forall i = 1, \dots n, \qquad \left.\frac{\partial \mathcal{S}}{\partial x_i}\right|_{x_j = \overline{x}_j} = 0 \tag{9-40}$$

i.e., we have the **saddle point equation**:

$$\forall i = 1, \dots n, \qquad V_h'(\overline{x}_i) = 2\frac{T}{n}\sum_{j\neq i}\frac{1}{\overline{x}_i - \overline{x}_j} \tag{9-41}$$

The saddle point approximation[2] consists in writting:

$$Z_n(h,T) \sim \frac{1}{\sqrt{\det\left(\frac{\partial S}{\partial x_i \partial x_j}\right)}}\, \mathrm{e}^{-\frac{n^2}{T^2}S(\overline{x}_1,\dots,\overline{x}_n)}\,(1+O(1/n)) \tag{9-42}$$

where $(\overline{x}_1,\dots,\overline{x}_n)$ is the solution of the saddlepoint equation which minimizes $\Re S$.

REMARK 9.1 The saddle point equation may have more than one minimal solution $(\overline{x})$.
- in particular if $\xi \in \mathbb{R}$, there are two solutions, complex conjugate of each other.
- in the multicut case, there are many saddlepoints with near-minimal action.

In all cases, one needs to sum over all the saddle points. Let us call $\{\overline{x}\}_I$, the collection of saddle points. We have:

$$Z_n \sim \sum_I \frac{C_I}{\sqrt{S''(\{\overline{x}\}_I)}}\, \mathrm{e}^{-\frac{n^2}{T^2}S(\{\overline{x}\}_I)}\,(1+O(1/n)) \tag{9-43}$$

Each saddle point $\{\overline{x}\}_I$ corresponds to a particular minimal $n$-dimensional integration path in $\mathbb{C}^n$,noted $\Gamma_I$, and the coefficients $C_I \in \mathbb{Z}$ are such that:

$$\mathbb{R}^n = \sum_I C_I \Gamma_I \tag{9-44}$$

## 10. Solution of the saddlepoint equation

We recall the saddle point equation:

$$\forall i = 1, \dots n, \qquad V_h'(\overline{x}_i) = 2\frac{T}{n}\sum_{j\neq i}\frac{1}{\overline{x}_i - \overline{x}_j} \tag{10-45}$$

We introduce the function:

$$\omega(x) := \frac{T}{n}\sum_{j=1}^{n}\frac{1}{x - \overline{x}_j} \tag{10-46}$$

in the large $N$ limit, $\omega(x)$ is expected to tend toward the resolvent, at least in the case there is only one minimal saddle point. Indeed, the $\overline{x}_i$'s are the position of the eigenvalues minimizing the action, i.e. the most probable positions of eigenvalues of $M$, and thus 10-46 should be close to $\frac{T}{n}\mathrm{tr}\,\frac{1}{x-M}$.

## 10.1 Algebraic method

Compute $\omega^2(x) + \frac{T}{n}\omega'(x)$, you find:

$$\begin{aligned}
\omega^2(x) + \frac{T}{n}\omega'(x) &= \frac{T^2}{n^2}\sum_{i=1}^{n}\sum_{j=1}^{n}\frac{1}{(x-\overline{x}_i)(x-\overline{x}_j)} - \frac{T^2}{n^2}\sum_{i=1}^{n}\frac{1}{(x-\overline{x}_i)^2} \\
&= \frac{T^2}{n^2}\sum_{i\neq j}^{n}\frac{1}{(x-\overline{x}_i)(x-\overline{x}_j)} \\
&= \frac{T^2}{n^2}\sum_{i\neq j}^{n}\left(\frac{1}{x-\overline{x}_i} - \frac{1}{x-\overline{x}_j}\right)\frac{1}{\overline{x}_i-\overline{x}_j} \\
&= \frac{2T^2}{n^2}\sum_{i=1}^{n}\frac{1}{x-\overline{x}_i}\sum_{j\neq i}^{n}\frac{1}{\overline{x}_i-\overline{x}_j} \\
&= \frac{T}{n}\sum_{i=1}^{n}\frac{V_h'(\overline{x}_i)}{x-\overline{x}_i} \\
&= \frac{T}{n}\sum_{i=1}^{n}\frac{V_h'(x)-(V_h'(x)-V_h'(\overline{x}_i))}{x-\overline{x}_i} \\
&= V_h'(x)\omega(x) - \frac{T}{n}\sum_{i=1}^{n}\frac{V_h'(x)-V_h'(\overline{x}_i)}{x-\overline{x}_i} \\
&= (V'(x) - \frac{h}{x-\xi})\omega(x) - \frac{T}{n}\sum_{i=1}^{n}\frac{V'(x)-V'(\overline{x}_i)}{x-\overline{x}_i} + h\frac{\omega(\xi)}{x-\xi}
\end{aligned}$$

(10-47)

i.e. we get the equation:

$$\omega^2(x) + \frac{T}{n}\omega'(x) = V'(x)\omega(x) - P(x) - h\frac{\omega(x)-\omega(\xi)}{x-\xi} \qquad (10\text{-}48)$$

where $P(x) := \frac{T}{n}\sum_{i=1}^{n}\frac{V'(x)-V'(\overline{x}_i)}{x-\overline{x}_i}$ is a polynomial in $x$ of degree at most $\deg V - 2$.

In the large $N$ limit, if we assume[3] that we can drop the $1/NW'(x)$ term, we get an algebraic equation, which is in this case an hyperelliptical curve. In particular at $h = 0$ and $T = 1$:

$$\omega(x) = \frac{1}{2}\left(V'(x) - \sqrt{V'^2(x) - 4P(x)}\right) \qquad (10\text{-}49)$$

The properties of this algebraic equation have been studied by many authors, and the $T$ and $h$ derivatives, as well as other derivatives were computed in various works. Here, we briefly sketch the method. See [29, 28, 21] for more details.

## 10.2 Linear saddle point equation

In the large $N$ limit, both the average density of eigenvalues, and the density of $\overline{x}$ tend towards a continuous compact support density $\overline{\rho}(x)$. In that limit, the resolvent is given by:

$$\omega(x) = T \int_{\text{supp}\,\overline{\rho}} \frac{\overline{\rho}(x')\,dx'}{x - x'} \tag{10-50}$$

i.e.

$$\forall x \in \text{supp}\,\overline{\rho}, \qquad \overline{\rho}(x) = -\frac{1}{2i\pi T}(\omega(x + i0) - \omega(x - i0)) \tag{10-51}$$

and the saddle point equation 10-45, becomes a linear functional equation:

$$\forall x \in \text{supp}\,\overline{\rho}, \qquad V_h'(x) = \omega(x + i0) + \omega(x - i0) \tag{10-52}$$

The advantage of that equation, is that it is linear in $\omega$, and thus in $\overline{\rho}$. The nonlinearity is hidden in supp $\overline{\rho}$.

**Example: One cut.** If the support of $\overline{\rho}$ is a single interval:

$$\text{supp}\,\overline{\rho} = [a, b] \quad , \quad a < b \tag{10-53}$$

then, look for a solution of the form:

$$\omega(x) = \frac{1}{2}\left(V_h'(x) - M_h(x)\sqrt{(x - a)(x - b)}\right) \tag{10-54}$$

The saddle point equation 10-52 implies that $M_h(x + i0) = M_h(x - i0)$, i.e. $M_h$ has no discontinuities, and because of its large $x$ behaviour, as well as its behaviours near $\xi$, it must be a rational function of $x$, with a simple pole at $x = \xi$. $M_h$, $a$ and $b$ are entirely determined by their behaviours near poles, i.e.:

$$\omega(x) \underset{x\to\infty}{\sim} \frac{T}{x} \tag{10-55}$$

$$\omega(x) \underset{x\to\xi}{\sim} \text{regular} \quad \longrightarrow M_h(x) \underset{x\to\xi}{\sim} -\frac{h}{x - \xi} \tag{10-56}$$

Thus, one may write:

$$\omega(x) = \frac{1}{2}\left(V'(x) - M(x)\sqrt{(x - a)(x - b)} - \frac{h}{x - \xi}\left(1 - \frac{\sqrt{(x - a)(x - b)}}{\sqrt{(\xi - a)(\xi - b)}}\right)\right) \tag{10-57}$$

where $M(x)$ is now a polynomial (which still depends on $h$ and $T$ and the other parameters), it is such that:

$$M(x) = \operatorname*{Pol}_{x \to \infty} \frac{V'(x)}{\sqrt{(x-a)(x-b)}} \tag{10-58}$$

The density is thus:

$$\overline{\rho}(x) = \frac{1}{2\pi T} M_h(x) \sqrt{(x-a)(b-x)} \quad , \quad \text{supp}\, \overline{\rho} = [a, b] \tag{10-59}$$

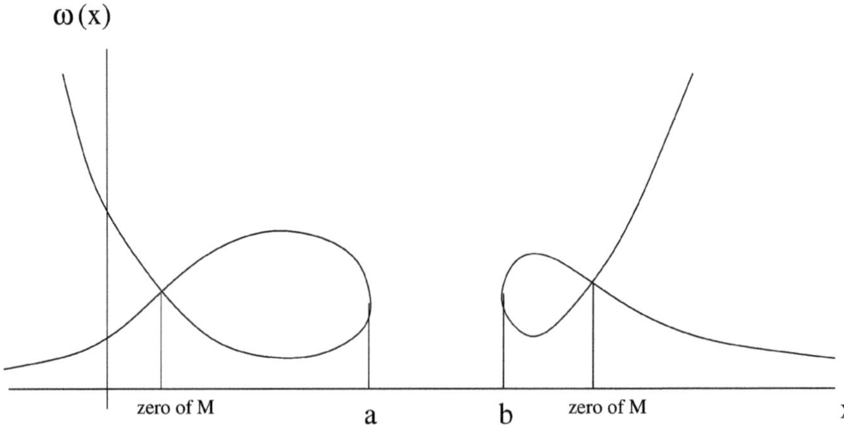


**Multi-cut solution.** Let us assume that the support of $\overline{\rho}$ is made of $s$ separated intervals:

$$\text{supp}\, \overline{\rho} = \cup_{i=1}^{s} [a_i, b_i] \tag{10-60}$$

then, for any sequence of integers $n_1, n_2, \ldots, n_s$ such that $\sum_{i_1}^{s} n_i = n$, it is possible to find a solution for the saddle point equation. That solution obeys 10-52, as well as the conditions:

$$\forall i = 1, \ldots, s \quad , \quad \int_{a_i}^{b_i} \rho(x) dx = T \frac{n_i}{N} \tag{10-61}$$

The solution of the saddle point equation can be described as follows: let the polynomial $\sigma(x)$ be defined as:

$$\sigma(x) := \prod_{i=1}^{s} (x - a_i)(x - b_i) \tag{10-62}$$

The solution of the saddle point equation 10-52, is of the form:

$$\omega(x) = \frac{1}{2}\left(V_h'(x) - M_h(x)\sqrt{\sigma(x)}\right) \tag{10-63}$$

where $M_h(x)$ is a rational function of $x$, with a simple pole at $x = \xi$. $M_h$, and $\sigma(x)$ are entirely determined by their behaviours near poles, i.e.:

$$\omega(x) \underset{x\to\infty}{\sim} \frac{T}{x} \tag{10-64}$$

$$\omega(x) \underset{x\to\xi}{\sim} \text{regular} \quad \longrightarrow M_h(x) \underset{x\to\xi}{\sim} -\frac{h}{x-\xi} \tag{10-65}$$

and by the conditions that:

$$\forall i = 1, \ldots, s \quad , \quad \int_{a_i}^{b_i} M_h(x)\sqrt{\sigma(x)}dx = 2i\pi T\frac{n_i}{n} \tag{10-66}$$

## 10.3 Algebraic geometry: hyperelliptical curves

Consider the curve given by:

$$\omega(x) = \frac{1}{2}\left(V_h'(x) - M_h(x)\sqrt{(x-a)(x-b)}\right) \tag{10-67}$$

It has two sheets, i.e. for each $x$, there are two values of $\omega(x)$, depending on the choice of sign of the square-root.

- In the physical sheet (choice $+\sqrt{}$), it behaves near $\infty$ like $\omega(x) \sim T/x$

- In the second sheet (choice $-\sqrt{}$), it behaves near $\infty$ like $\omega(x) \sim V_h'(x)$

Since $\omega(x)$ is a complex valued, analytical function of a cmplex variable $x$, the curve can be thought of as the embedding of a Riemann surface into $\mathbb{C}\times\mathbb{C}$.

I.e. we have a Riemann surface $\mathcal{E}$, with two (monovalued) functions defined on it: $p \in \mathcal{E}$, $\to x(p) \in \mathbb{C}$, and $p \in \mathcal{E}$, $\to \omega(p) \in \mathbb{C}$. For each $x$, there are two $p \in \mathcal{E}$ such that $x(p) = x$, and this is why there are two values of $\omega(x)$.

Each of the two sheets is homeomorphic to the complex plane, cut along the segments $[a_i, b_i]$, and the two sheets are glued together along the cuts. The complex plane, plus its point at infinity, is the Riemann sphere. Thus, our curve $\mathcal{E}$, is obtained by taking two Riemann spheres, glued together along $s$ circles. It is a genus $s-1$ surface.

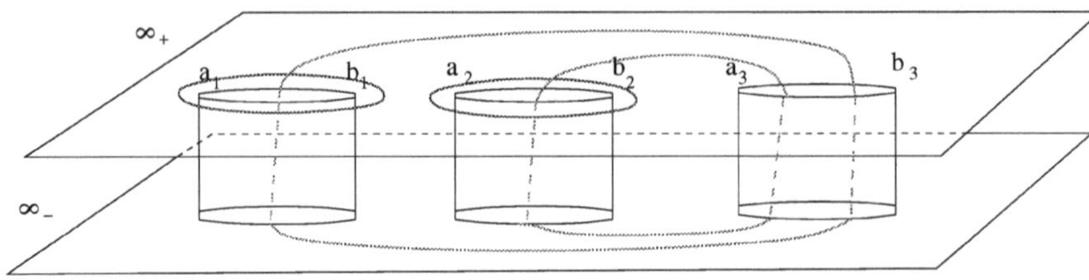


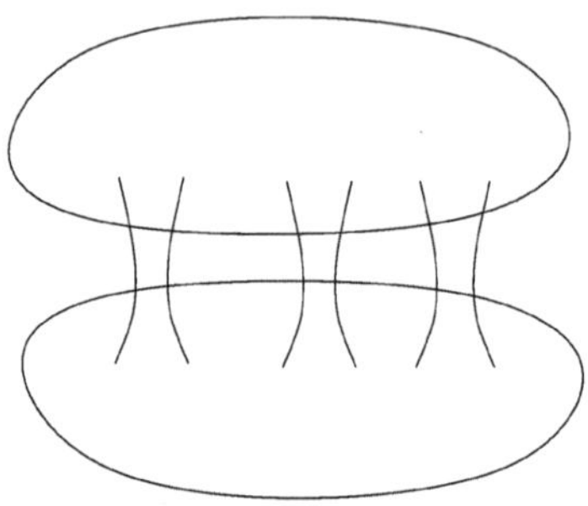

## 10.4 Genus zero case (one cut)

If the curve as genus zero, it is homeomorphic to the Riemann sphere $\mathcal{E} = \mathbb{C}$. One can always choose a rational parametrization:

$$x(p) = \frac{a+b}{2} + \gamma(p+1/p) \quad , \quad \gamma = \frac{b-a}{4} \tag{10-68}$$

$$\sqrt{(x-a)(x-b)} = \gamma(p-1/p) \tag{10-69}$$

so that $\omega$ is a rational function of $p$.

That representation maps the physical sheet onto the exterior of the unit circle, and the second sheet onto the interior of the unit circle. The unit circle is the image of the two sides of the cut $[a,b]$, and the branchpoints $[a,b]$ are maped to $-1$ and $+1$. Changing the sign of the square root is equivalent to changing $p \to 1/p$.

The branch points are of course the solutions of $dx/dp = 0$, i.e. $dx(p) = 0$:

$$dx(p) = \gamma\left(1 - \frac{1}{p^2}\right) dp \quad , \quad dx(p) = 0 \leftrightarrow p = \pm 1 \leftrightarrow x(p) = a, b \tag{10-70}$$

There are two points at $\infty$, $p = \infty$ in the physical sheet, and $p = 0$ in the second sheet.

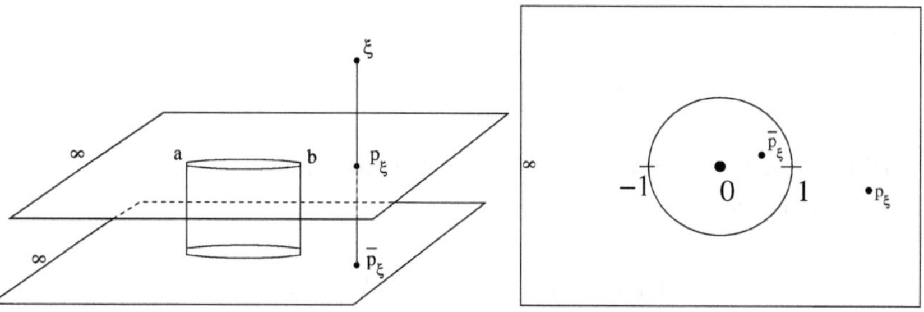


Since the resolvent $\omega(p)$ is a rational function of $p$, it is then entirely determined by its behaviour near its poles. the poles are at $p = \infty$, $p = 0$, $p = p_\xi$ and $p = \overline{p}_\xi$ (the two points of $\mathcal{E}$ such that $x(p) = \xi$, such that $p_\xi$ is in the physical sheet, and $\overline{p}_\xi$ is in the second sheet): The boundary conditions:

$$\left\{\begin{array}{l} \omega(p) \underset{p\to\infty}{\sim} \dfrac{T}{x(p)} \\ \omega(p) \underset{p\to 0}{\sim} V'(x(p)) - \dfrac{T}{x(p)} - \dfrac{h}{x(p)} \\ \omega(p) \underset{p\to\overline{p}_\xi}{\sim} -\dfrac{h}{x(p)-\xi} \\ \omega(p) \underset{p\to p_\xi}{\sim} \text{regular} \end{array}\right. \tag{10-71}$$

***T* derivative.** Now, let us compute $\partial\omega(p)/\partial T$ at $x(p)$ fixed. Eq. 10-71 becomes:

$$\left\{\begin{array}{l} \dfrac{\partial\omega(p)}{\partial T} \underset{p\to\infty}{\sim} \dfrac{1}{x(p)} \\ \dfrac{\partial\omega(p)}{\partial T} \underset{p\to 0}{\sim} -\dfrac{1}{x(p)} \\ \dfrac{\partial\omega(p)}{\partial T} \underset{p\to\overline{p}_\xi}{\sim} \text{regular} \\ \dfrac{\partial\omega(p)}{\partial T} \underset{p\to p_\xi}{\sim} \text{regular} \end{array}\right. \tag{10-72}$$

Moreover, we know that $\omega(x)$ has a square-root behaviour near $a$ and $b$, in $\sqrt{(x-a)(x-b)}$, and $a$ and $b$ depend on $T$, thus $\partial\omega/\partial T$ may behave in $((x-a)(x-b))^{-1/2}$ near $a$ and $b$, i.e. $\partial\omega/\partial T$ may have simple poles at $p = \pm 1$.

Finaly, $\partial\omega(p)/\partial T$, has simple poles at $p = 1$ and $p = -1$, and vanishes at $p = 0$ and $p = \infty$, the only possibility is:

$$\left.\frac{\partial\omega(p)}{\partial T}\right|_{x(p)} = \frac{p}{\gamma(p^2-1)} = \frac{1}{p}\frac{dp}{dx} \tag{10-73}$$

which is better written in terms of differential forms:

$$\left.\frac{\partial \omega(p)}{\partial T}\right|_{x(p)} dx(p) = \frac{dp}{p} = d\ln p \tag{10-74}$$

the RHS is independent of the potential, it is universal.

With the notation 8-34, we have:

$$\Omega(p)dx(p) = \frac{dp}{p} \quad , \quad \Lambda(p) = \gamma p \tag{10-75}$$

***h* derivative.** The $h$ derivative is computed in a very similar way.

$$\begin{cases} \frac{\partial \omega(p)}{\partial h} \underset{p\to\infty}{\sim} O(p^{-2}) \\ \frac{\partial \omega(p)}{\partial h} \underset{p\to 0}{\sim} -\frac{1}{x(p)} \\ \frac{\partial \omega(p)}{\partial h} \underset{p\to \overline{p}_\xi}{\sim} -\frac{1}{x(p)-\xi} \\ \frac{\partial \omega(p)}{\partial h} \underset{p\to p_\xi}{\sim} \text{regular} \end{cases} \tag{10-76}$$

implies that $\partial\omega/\partial h$ can have poles at $p=\pm 1$ and at $p=\overline{p}_\xi$, and vanishes at $p=0$. The only possibility is:

$$\left.\frac{\partial \omega(p)}{\partial h}\right|_{x(p)} = \frac{-p\overline{p}_\xi}{\gamma(p-\overline{p}_\xi)(p^2-1)} \tag{10-77}$$

i.e.

$$\left.\frac{\partial \omega(p)}{\partial h}\right|_{x(p)} dx(p) = \frac{dp}{p} - \frac{dp}{p-\overline{p}_\xi} = d\ln\frac{p}{p-\overline{p}_\xi} \tag{10-78}$$

which again is universal.

With the notation 8-35, we have:

$$H(p,p_\xi)dx(p) = \frac{dp}{p} - \frac{dp}{p-\frac{1}{p_\xi}} \quad , \quad H(p_\xi) = \ln\left(\frac{p_\xi}{p_\xi-\overline{p}_\xi}\right) = -\ln\left(\frac{1}{\gamma}\frac{dx}{dp}(\xi)\right) \tag{10-79}$$

## 10.5 Higher genus

For general genus, the curve can be parametrized by $\theta$-functions. Like rational functions for genus 0, $\theta$-functions are the building blocks of functions defined on a compact Riemann surface, and any such function is entirely determined by its behaviour near its poles, as well as by its integrals around irreducible cycles. All the previous paragraph can be extended to that case.

Let $\infty_+$ and $\infty_-$ be the points at infinity, i.e. the two poles of $x(p)$, with $\infty_+$ in the physical sheet and $\infty_-$ in the second sheet. Let $p = p_\xi$ and $p = \overline{p}_\xi$ be the two points of $\mathcal{E}$ such that $x(p) = \xi$, and with $p_\xi$ in the physical sheet, and $\overline{p}_\xi$ in the second sheet.

The differential form $\omega(p)dx(p)$ is entirely determined by:

$$\begin{cases} \omega(p)dx(p) \underset{p\to\infty_+}{\sim} T\dfrac{dx(p)}{x(p)} & , \operatorname*{Res}_{\infty_+} \omega(p)dx(p) = -T \\ \omega(p)dx(p) \underset{p\to\infty_-}{\sim} dV(x(p)) - T\dfrac{dx(p)}{x(p)} - h\dfrac{dx(p)}{x(p)} & , \operatorname*{Res}_{\infty_-} \omega(p)dx(p) = T + h \\ \omega(p)dx(p) \underset{p\to\overline{p}_\xi}{\sim} -h\dfrac{dx(p)}{x(p)-\xi} & , \operatorname*{Res}_{\overline{p}_\xi} \omega(p)dx(p) = -h \\ \omega(p)dx(p) \underset{p\to p_\xi}{\sim} \text{regular} & , \operatorname*{Res}_{p_\xi} \omega(p)dx(p) = 0 \\ \oint_{\mathcal{A}_i} \omega(p)dx(p) = T\dfrac{n_i}{n} = \dfrac{n_i}{N} & \end{cases} \tag{10-80}$$

Since $\partial\omega/\partial T, h$ can diverge at most like $(x-a_i)^{-1/2}$ near a branch point $a_i$, and $dx(p)$ has a zero at $a_i$, the differential form $\partial\omega dx/\partial T, h$ has no pole at the branch points.

## 10.6 Introduction to algebraic geometry

We introduce some basic concepts of algebraic geometry. We refer the reader to [22, 23] for instance.

THEOREM 4 *Given two points $q_1$ and $q_2$ on the Riemann surface $\mathcal{E}$, there exists a unique differential form $dS_{q_1,q_2}(p)$, with only two simple poles, one at $p = q_1$ with residue $+1$ and one at $p = q_2$ with residue $-1$, and which is normalized on the $\mathcal{A}_i$ cycles, i.e.*

$$\begin{cases} \operatorname*{Res}_{p\to q_1} dS_{q_1,q_2}(p) = +1 \\ \operatorname*{Res}_{p\to q_2} dS_{q_1,q_2}(p) = -1 \\ \oint_{\mathcal{A}_i} dS_{q_1,q_2}(p) = 0 \end{cases} \tag{10-81}$$

*dS is called an "abelian differential of the third kind".*

Starting from the behaviours near poles and irreducible cycles 10-80, we easily find:

$$\Omega(p)dx(p) = \left.\frac{\partial\omega(p)dx(p)}{\partial T}\right|_{x(p)} = -dS_{\infty_+,\infty_-}(p) \tag{10-82}$$

$$H(p,p_\xi)dx(p) = \left.\frac{\partial\omega(p)dx(p)}{\partial h}\right|_{x(p)}$$
$$= -dS_{\overline{p}_\xi,\infty_-}(p) = dS_{p_\xi,\infty_+}(p) - d\ln\left(x(p)-x(p_\xi)\right) \quad (10\text{-}83)$$

THEOREM 5 *On an algebraic curve of genus $g$, there exist exactly $g$ linearly independent "holomorphic differential forms" (i.e. with no poles), $du_i(p)$, $i=1,\ldots,g$. They can be chosen normalized as:*

$$\oint_{\mathcal{A}_i} du_j(p) = \delta_{ij} \quad (10\text{-}84)$$

For hyperelliptical surfaces, it is easy to see that if $L(x)$ is a polynomial of degree at most $g-1=s-2$, the differential form $\frac{L(x)}{\sqrt{\prod_{i=1}^{s}(x-a_i)(x-b_i)}}dx$ is regular at $\infty$, at the branch points, and thus has no poles. And there are $g$ linearly independent polynomials of degree at most $g-1$. The irreducible cycles $\mathcal{A}_i$ is a contour surrounding $[a_i,b_i]$ in the positive direction.

DEFINITION 6 *The matrix of periods is defined by:*

$$\tau_{ij} := \oint_{\mathcal{B}_i} du_j(p) \quad (10\text{-}85)$$

*where the irreducible cycles $\mathcal{B}_i$ are chosen canonicaly conjugated to the $\mathcal{A}_i$, i.e. $\mathcal{A}_i\cap\mathcal{B}_j=\delta_{ij}$. In our hyperelliptical case, we choose $\mathcal{B}_i$ as a contour crossing $[a_i,b_i]$ and $[a_s,b_s]$.*

*The matrix of periods is symmetric $\tau_{ij}=\tau_{ji}$, and its imaginary part is positive $\Im\tau_{ij}>0$. It encodes the complex structure of the curve.*

The holomrphic forms naturaly define an embedding of the curve into $\mathbb{C}^g$:

DEFINITION 7 *Given a base point $q_0\in\mathcal{E}$, we define the Abel map:*

$$\begin{array}{rcl}\mathcal{E} & \longrightarrow & \mathbb{C}^g \\ p & \longrightarrow & \vec{u}(p) = (u_1(p),\ldots,u_g(p)) \quad , \quad u_i(p) := \int_{q_0}^{p} du_i(p)\end{array} \quad (10\text{-}86)$$

*where the integration path is chosen so that it does not cross any $\mathcal{A}_i$ or $\mathcal{B}_i$.*

DEFINITION 8 *Given a symmetric matrix $\tau$ of dimension $g$, such that $\Im\tau_{ij}>0$, we define the $\theta$-function, from $\mathbb{C}^g\to\mathbb{C}$ by:*

$$\theta(\vec{u},\tau) = \sum_{\vec{m}\in\mathbb{Z}^g} \mathrm{e}^{i\pi\vec{m}^t\tau\vec{m}}\,\mathrm{e}^{2i\pi\vec{m}^t\vec{u}} \quad (10\text{-}87)$$

*It is an even entire function. For any $\vec{m}\in\mathbb{Z}^g$, it satisfies:*

$$\theta(\vec{u}+\vec{m}) = \theta(\vec{u}) \quad , \quad \theta(\vec{u}+\tau\vec{m}) = \mathrm{e}^{-i\pi(2\vec{m}^t\vec{u}+\vec{m}^t\tau\vec{m})}\,\theta(\vec{u}) \quad (10\text{-}88)$$

DEFINITION 9 *The theta function vanishes on a codimension 1 submanifold of $\mathbb{C}^g$, in particular, it vanishes at the odd half periods:*

$$\vec{z} = \frac{\vec{m}_1 + \tau\,\vec{m}_2}{2} \;,\; \vec{m}_1 \in \mathbb{Z}^g \;,\; \vec{m}_2 \in \mathbb{Z}^g \;,\; (\vec{m}_1^t \vec{m}_1) \in 2\mathbb{Z}+1 \;\longrightarrow\; \theta(\vec{z}) = 0 \tag{10-89}$$

*For a given such odd half-period, we define the characteristic $\vec{z}$ $\theta$-function:*

$$\theta_{\vec{z}}(\vec{u}) := \mathbf{e}^{i\pi m_2 \vec{u} +}\, \theta(\vec{u} + \vec{z}) \tag{10-90}$$

*so that:*

$$\theta_{\vec{z}}(\vec{u}+\vec{m}) = \mathbf{e}^{i\pi \vec{m}_2^t \vec{m}}\, \theta_{\vec{z}}(\vec{u}) \;,\quad \theta_{\vec{z}}(\vec{u}+\tau\vec{m}) = \mathbf{e}^{-i\pi \vec{m}_1^t \vec{m}}\, \mathbf{e}^{-i\pi(2\vec{m}^t\vec{u}+\vec{m}^t\tau\vec{m})}\, \theta_{\vec{z}}(\vec{u}) \tag{10-91}$$

*and*

$$\theta_{\vec{z}}(\vec{0}) = 0 \tag{10-92}$$

DEFINITION 10 *Given two points $p, q$ in $\mathcal{E}$, as well as a basepoint $p_0 \in \mathcal{E}$ and an odd half period $z$, we define the prime form $E(p,q)$:*

$$E(p,q) := \frac{\theta_{\vec{z}}(\vec{u}(p) - \vec{u}(q))}{\sqrt{dh_{\vec{z}}(p)dh_{\vec{z}}(q)}} \tag{10-93}$$

*where $dh_{\vec{z}}(p)$ is the holomorphic form:*

$$dh_{\vec{z}}(p) := \sum_{i=1}^{g} \left.\frac{\partial \theta_{\vec{z}}(\vec{u})}{\partial u_i}\right|_{\vec{u}=\vec{0}} du_i(p) \tag{10-94}$$

THEOREM 11 *The abelian differentials can be written:*

$$dS_{q_1,q_2}(p) = d \ln \frac{E(p,q_1)}{E(p,q_2)} \tag{10-95}$$

With these definitions, we have:

$$\Lambda(p) = \gamma \frac{\theta_{\vec{z}}(\vec{u}(p) - \vec{u}(\infty_-))}{\theta_{\vec{z}}(\vec{u}(p) - \vec{u}(\infty_+))} \;,\quad \gamma := \lim_{p\to\infty_+} \frac{x(p)\,\theta_{\vec{z}}(\vec{u}(p) - \vec{u}(\infty_+))}{\theta_{\vec{z}}(\vec{u}(\infty_+) - \vec{u}(\infty_-))} \tag{10-96}$$

$$H(p_\xi) = \frac{\theta_{\vec{z}}(\vec{u}(p_\xi) - \vec{u}(\infty_-))\theta_{\vec{z}}(\vec{u}(\infty_+) - \vec{u}(\overline{p}_\xi))}{\theta_{\vec{z}}(\vec{u}(p_\xi) - \vec{u}(\overline{p}_\xi))\theta_{\vec{z}}(\vec{u}(\infty_+) - \vec{u}(\infty_-))} = -\gamma \frac{\theta_{\vec{z}}(\vec{u}(\infty_+) - \vec{u}(\infty_-))}{\theta_{\vec{z}}(\vec{u}(p_\xi) - \vec{u}(\infty_+))^2} \frac{dh_{\vec{z}}(p_\xi)}{dx(p_\xi)} \tag{10-97}$$

# 11. Asymptotics of orthogonal polynomials

## 11.1 One-cut case

In the one-cut case, (i.e. genus zero algebraic curve), and if $V$ is a real potential, there is only one dominant saddle point if $\xi \notin [a,b]$, and two conjugated dominant saddle points if $x \in [a,b]$. More generaly, there is a saddle point corresponding to each determination of $p_\xi$ such that $x(p_\xi) = \xi$. i.e. $p_\xi$ and $\overline{p}_\xi = 1/p_\xi$. The dominant saddle point is the one such that $\Re(V_{\text{eff}}(p_\xi) - V(\xi))$ is minimal. The two cols have a contribution of the same order if:

$$\Re V_{\text{eff}}(p_\xi) = \Re V_{\text{eff}}(\overline{p}_\xi) \tag{11-98}$$

i.e. if $\xi$ is such that:

$$\Re \int_{\overline{p}_\xi}^{p_\xi} W(x)dx = 0 \tag{11-99}$$

If the potential is real, it is easy to see that the set of points which satisfy 11-99 is $[a, b]$, in general, it is a curve in the complex plane, going from $a$ to $b$, we call it the cut $[a, b]$ (similar curves were studied in [31]).

Then we have:

- For $x \notin [a, b]$, we write $\xi = \frac{a+b}{2} + \gamma(p_\xi + 1/p_\xi)$, $\gamma = \frac{b-a}{4}$:

$$p_n(\xi) \sim \sqrt{H(p_\xi)}\ (\Lambda(p_\xi))^{n-N}\ \mathrm{e}^{-\frac{N}{2}(V_{\text{eff}}(p_\xi)-V(\xi))}(1+O(1/N)) \tag{11-100}$$

  i.e.

$$p_n(\xi) \sim \sqrt{\frac{\gamma}{x'(p_\xi)}}\ (\gamma\, p_\xi)^{n-N}\ \mathrm{e}^{-\frac{N}{2}(V_{\text{eff}}(p_\xi)-V(\xi))}(1+O(1/N)) \tag{11-101}$$

- For $x \in [a, b]$, i.e. $p$ is on the unit circle $p = \mathrm{e}^{i\phi}$, $\xi = \frac{a+b}{2} + 2\gamma\cos\phi$:

$$\begin{aligned} p_n(\xi) \ &\sim\ \sqrt{H(p_\xi)}\ (\Lambda(p_\xi))^{n-N}\ \mathrm{e}^{-\frac{N}{2}(V_{\text{eff}}(p_\xi)-V(\xi))}(1+O(1/N)) \\ &+ \sqrt{H(\overline{p}_\xi)}\ \left(\Lambda(\overline{p}_\xi)\right)^{n-N}\ \mathrm{e}^{-\frac{N}{2}(V_{\text{eff}}(\overline{p}_\xi)-V(\xi))}(1+O(1/N)) \end{aligned} \tag{11-102}$$

  i.e.

$$\begin{aligned} p_n(\xi) \sim \frac{\gamma^{n-N}}{\sqrt{2\sin\phi(\xi)}}\, 2\cos\left(N\pi\int_a^\xi \rho(x)dx - (n-N+\frac{1}{2})\phi(\xi) + \alpha\right) \\ (1+O(1/N)) \end{aligned} \tag{11-103}$$

  i.e. we have an oscillatory behaviour

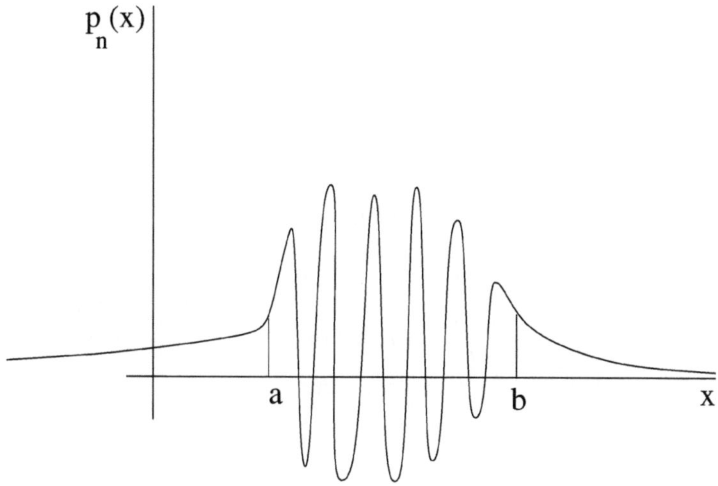


## 11.2 Multi-cut case

In the multicut case, in addition to having saddle-points corresponding to both determinantions of $p_\xi$, we have a saddle point for each filling fraction configuration $n_1, \ldots, n_s$ with $\sum_{i=1}^{s} n_i = n$. We write:

$$\epsilon_i = \frac{n_i}{N} \tag{11-104}$$

The saddle point corresponding to filling fractions which differ by a few units, contribute to the same order, and thus cannot be neglected. One has to consider the sommation over filling fractions [10].

Thus, one has to consider the action of a saddle point as a function of the filling fractions. We leave as an exercise for the reader to prove that the derivatives of $F$ are given by:

$$\frac{\partial F}{\partial \epsilon_i} = -\oint_{\mathcal{B}_i} W(x)dx \tag{11-105}$$

and:

$$\frac{\partial^2 F}{\partial \epsilon_i \partial T} = -2i\pi(u_i(\infty_+) - u_i(\infty_-)) \tag{11-106}$$

$$\frac{\partial^2 F}{\partial \epsilon_i \partial h} = -2i\pi(u_i(p_\xi) - u_i(\infty_+)) \tag{11-107}$$

$$\frac{\partial^2 F}{\partial \epsilon_i \partial \epsilon_j} = -2i\pi\tau_{ij} \tag{11-108}$$

The last relation implies that $\Re F$ is a convex function of $\epsilon$, thus it has a unique minimum:

$$\vec{\epsilon}^* \quad , \quad \Re \left.\frac{\partial F}{\partial \epsilon_i}\right|_{\vec{\epsilon}=\vec{\epsilon}^*} = 0 \tag{11-109}$$

We write:

$$\zeta_i := -\frac{1}{2i\pi} \left.\frac{\partial F}{\partial \epsilon_i}\right|_{\vec{\epsilon}=\vec{\epsilon}^*} \quad , \quad \zeta_i \in \mathbb{R} \tag{11-110}$$

We thus have the Taylor expansion:

$$\begin{aligned} F(T,h,\vec{\epsilon}) \sim\; & F(1,0,\vec{\epsilon}^*) - 2i\pi\vec{\zeta}^t(\vec{\epsilon}-\vec{\epsilon}^*) + (T-1)\frac{\partial F}{\partial T} + \frac{h}{2}(V_{\text{eff}}(p_\xi) - V(\xi)) \\ & + \frac{(T-1)^2}{2}\frac{\partial^2 F}{\partial T^2} - (T-1)h\ln\Lambda(p_\xi) - \frac{h^2}{2}\ln H(p_\xi) \\ & -2i\pi(\vec{\epsilon}-\vec{\epsilon}^*)^t\tau(\vec{\epsilon}-\vec{\epsilon}^*) - 2i\pi(T-1)(\vec{\epsilon}-\vec{\epsilon}^*)^t(\vec{u}(\infty_+)-\vec{u}(\infty_-)) \\ & -2i\pi h(\vec{\epsilon}-\vec{\epsilon}^*)^t(\vec{u}(p_\xi)-\vec{u}(\infty_+)) + \dots \end{aligned} \tag{11-111}$$

Thus:

$$\begin{aligned} Z \quad & \sim \sum_I C_I \mathrm{e}^{-N^2 F(\{x\}_I)} \\ & \sim \sum_{p=p_\xi,\overline{p}_\xi} \mathrm{e}^{-N^2 F(1,0,\vec{\epsilon}^*)} \\ & \quad \mathrm{e}^{N^2\left(-(T-1)\frac{\partial F}{\partial T} - \frac{h}{2}(V_{\text{eff}}(p)-V(\xi)) - \frac{(T-1)^2}{2}\frac{\partial^2 F}{\partial T^2} + (T-1)h\ln\Lambda(p) + \frac{h^2}{2}\ln H(p)\right)} \\ & \quad \sum_{\vec{n}} \mathrm{e}^{i\pi(\vec{n}-N\vec{\epsilon}^*)^t\tau(\vec{n}-N\vec{\epsilon}^*)} \mathrm{e}^{2i\pi N\vec{\zeta}^t(\vec{n}-N\vec{\epsilon}^*)} \\ & \quad \mathrm{e}^{2i\pi N(T-1)(\vec{n}-N\vec{\epsilon}^*)^t(\vec{u}(\infty_+)-\vec{u}(\infty_-))} \mathrm{e}^{2i\pi Nh(\vec{n}-N\vec{\epsilon}^*)^t(\vec{u}(p)-\vec{u}(\infty_+))} \end{aligned}$$

(11-112)

In that last sum, because of convexity, only values of $\vec{n}$ which don't differ from $N\vec{\epsilon}^*$ form more than a few units, contribute substantialy. Therefore, up to a non perturbative error (exponentialy small with $N$), one can extend the sum over the $n_i$'s to the whole $\mathbb{Z}^g$, and recognize a $\theta$-function (see 10-87):

$$\begin{aligned} Z \sim & \sum_{p=p_\xi,\overline{p}_\xi} \mathrm{e}^{-N^2 F(1,0,\vec{\epsilon}^*)} \\ & \mathrm{e}^{N^2\left((T-1)\frac{\partial F}{\partial T} + \frac{h}{2}(V_{\text{eff}}(p)-V(\xi)) + \frac{(T-1)^2}{2}\frac{\partial^2 F}{\partial T^2} + (T-1)h\ln\Lambda(p) + \frac{h^2}{2}\ln H(p)\right)} \\ & \mathrm{e}^{i\pi N^2\vec{\epsilon}^{*t}\tau\vec{\epsilon}^*_{\,}} \mathrm{e}^{-2i\pi N^2\vec{\zeta}^t\vec{\epsilon}^*} \mathrm{e}^{-2i\pi N^2(T-1)\vec{\epsilon}^{*t}(\vec{u}(\infty_+)-\vec{u}(\infty_-))} \mathrm{e}^{-2i\pi N^2 h\vec{\epsilon}^{*t}(\vec{u}(p)-\vec{u}(\infty_+))} \\ & \theta(N(\vec{\zeta}-\tau\vec{\epsilon}^*) + N(T-1)(\vec{u}(\infty_+)-\vec{u}(\infty_-)) + Nh(\vec{u}(p)-\vec{u}(\infty_+)), \tau) \end{aligned}$$

(11-113)

with $T-1 = \frac{n-N}{N}$ and $h=0$ or $h=1/N$, we get the asymptotics:

$$p_n(\xi) \quad \sim \sum_{x(p)=\xi} \sqrt{H(p)}\,(\Lambda(p))^{n-N}\,\mathrm{e}^{-\frac{N}{2}(V_{\text{eff}}(p)-V(\xi))}\,\mathrm{e}^{-2i\pi N\vec{\epsilon}^{*t}(\vec{u}(p)-\vec{u}(\infty_+))}$$

$$\frac{\theta(N(\vec{\zeta}-\tau\vec{\epsilon}^*)+(n-N)(\vec{u}(\infty_+)-\vec{u}(\infty_-))+(\vec{u}(p)-\vec{u}(\infty_+)),\tau)}{\theta(N(\vec{\zeta}-\tau\vec{\epsilon}^*)+(n-N)(\vec{u}(\infty_+)-\vec{u}(\infty_-)),\tau)} \tag{11-114}$$

Again, depending on $\xi$, we have to choose the determination of $p_\xi$ which has the minimum energy. If we are on a cut, i.e. if condition 11-99 holds, both determinations contribute. To summarize, outside the cuts, the sum 11-114 reduces to only one term, and along the cuts, the sum 11-114 contains two terms.

## 12. Conclusion

We have shown how the asymptotics of orthogonal polynomials (a notion related to integrability) is deeply related to algebraic geometry. This calculation can easily be extended to many generalizations, for multi-matrix models [17, 16, 2, 18], non-hermitean matrices ($\beta = 1, 4$) [20], rational potentials [3], ...

## Aknowledgements

The author wants to thank the organizer of the Les Houches summer school Applications of Random Matrices in Physics June 6-25 2004.

# References

[1] M. Bertola, "Bilinear semi–classical moment functionals and their integral representation", J. App. Theory (at press), math.CA/0205160

[2] M. Bertola, B. Eynard, J. Harnad, "Heuristic asymptotics of biorthogonal polynomials", Presentation by B.E. at AMS Northeastern regional meeting, Montreal May 2002.

[3] M. Bertola, B. Eynard, J. Harnad, "Semiclassical orthogonal polynomials, matrix models and isomonodromic tau functions" SPHT T04/019. CRM-3169 (2004), xxx, nlin.SI/0410043.

[4] Differential systems for bi-orthogonal polynomials appearing in two-matrix models and the associated Riemann-Hilbert problem. (M. Bertola, B.E., J. Harnad) , 60 pages, SPHT 02/097, CRM-2852. Comm. Math. Phys. 243 no.2 (2003) 193-240, xxx, nlin.SI/0208002.

[5] Partition functions for Matrix Models and Isomonodromic Tau Functions. (M. Bertola, B.E., J. Harnad) , 17 pages, SPHT 02/050, CRM 2841 , J. Phys. A Math. Gen. 36 No 12 (28 March 2003) 3067-3083. xxx, nlin.SI/0204054.

[6] P.M. Bleher and A.R. Its, eds., "Random Matrix Models and Their Applications", *MSRI Research Publications* **40**, Cambridge Univ. Press, (Cambridge, 2001).

[7] P. Bleher, A. Its, "Semiclassical asymptotics of orthogonal polynomials, Riemann-Hilbert problem, and universality in the matrix model" *Ann. of Math.* (2) **150**, no. 1, 185–266 (1999).

[8] P. Bleher, A. Its, "On asymptotic analysis of orthogonal polynomials via the Riemann-Hilbert method", Symmetries and integrability of difference equations (Canterbury, 1996), 165–177, *London Math. Soc. Lecture Note Ser.*, **255**, Cambridge Univ. Press, Cambridge, 1999.

[9] S.S. Bonan, D.S. Clark, "Estimates of the Hermite and the Freud polynomials", *J. Approx. Theory* **63**, 210-224 (1990).

[10] G. Bonnet, F. David, and B. Eynard, "Breakdown of Universality in multi-cut matrix models", J. Phys **A33**, 6739 (2000), xxx, cond-mat/0003324.

[11] P. Deift, T. Kriecherbauer, K. T. R. McLaughlin, S. Venakides, Z. Zhou, "Uniform asymptotics for polynomials orthogonal with respect to varying exponential weights and applications to universality questions in random matrix theory", *Commun. Pure Appl. Math.* **52**, 1335–1425 (1999).

[12] P. Deift, T. Kriecherbauer, K. T. R. McLaughlin, S. Venakides, Z. Zhou, "Strong asymptotics of orthogonal polynomials with respect to exponential weights", *Commun. Pure Appl. Math.* **52**, 1491–1552, (1999).

[13] P. Deift, *Orthogonal Polynomials and Random Matrices: a Riemann-Hilbert Approach, Courant* (New York University Press, ., 1999).

[14] P. Di Francesco, P. Ginsparg, J. Zinn-Justin, "2D Gravity and Random Matrices", *Phys. Rep.* **254**, 1 (1995).

[15] F.J. Dyson, "Correlations between the eigenvalues of a random matrix", Comm. Math. Phys. 19 (1970) 235-250.

[16] Correlation functions of eigenvalues of multi-matrix models, and the limit of a time dependent matrix. (B.E.) , 27 pages. SPHT 98/001 DTP 97-59 , Journal of Physics A 40 (1998) 8081, xxx, cond-mat/9801075.

[17] Eigenvalue distribution of large random matrices, from one matrix to several coupled matrices. (B.E.) , 31 pages. SPHT 97031. Nuc. Phys. B506,3 633-664 (1997). xxx, cond-mat/9707005.

[18] B. Eynard, "Polynômes biorthogonaux, problème de Riemann–Hilbert et géométrie algébrique", Habilitation à diriger les recherches, université Paris VII, (2005).

[19] B. Eynard "An introduction to random matrices", lectures given at Saclay, October 2000, notes available at http://www-spht.cea.fr/articles/t01/014/.

[20] B. Eynard, "Asymptotics of skew orthogonal polynomials", *J. Phys A.* **34** (2001) 7591, cond-mat/0012046.

[21] B. Eynard, "Master loop equations, free energy and correlations for the chain of matrices", *JHEP11(2003)018*, xxx, hep-th/0309036.

[22] H.M. Farkas, I. Kra, "Riemann surfaces" 2nd edition, Springer Verlag, 1992.

[23] J. Fay, "Theta Functions on Riemann Surfaces", Lectures Notes in Mathematics, Springer Verlag, 1973.

[24] A. Fokas, A. Its, A. Kitaev, "The isomonodromy approach to matrix models in 2D quantum gravity", *Commun. Math. Phys.* **147**, 395–430 (1992).

[25] T. Guhr, A. Mueller-Groeling, H.A. Weidenmuller, "Random matrix theories in quantum physics: Common concepts", *Phys. Rep.* **299**, 189 (1998).

[26] A.R. Its, A.V. Kitaev, and A.S. Fokas, "An isomonodromic Approach in the Theory of Two-Dimensional Quantum Gravity", *Usp. Matem. Nauk*, **45**, 6 (276), 135-136 (1990), (Russian), translation in *Russian Math. Surveys*, **45**, no. 6, 155-157 (1990).

[27] M. Jimbo, T. Miwa and K. Ueno, "Monodromy Preserving Deformation of Linear Ordinary Differential Equations with Rational Coefficients I, II, III.", *Physica* **2D**, 306-352 (1981), *Physica* **2D**, 407-448 (1981); *ibid.*, **4D**, 26–46 (1981).

[28] V.A. Kazakov, A. Marshakov, "Complex Curve of the Two Matrix Model and its Tau-function", *J.Phys.* **A36** (2003) 3107-3136, hep-th/0211236.

[29] I. Krichever, "The $\tau$-Function of the Universal Whitham Hierarchy, Matrix Models and Topological Field Theories", Comm. Pure Appl. Math. **47** (1994), no. 4, 437–475.

[30] M.L. Mehta, *Random Matrices*, 2nd edition, (Academic Press, New York, 1991).

[31] G. Moore, "Geometry of the string equations", *Comm. Math. Phys.* **133**, no. 2, 261–304 (1990).

[32] G. Szegø, "Orthogonal Polynomials", American Mathematical Society, Providence, 1967.

[33] C.A. Tracy and H. Widom, "Fredholm determinants, differential equations and matrix models", *Commun. Math. Phys.* **161**, 289-309 (1994).

[34] P. Van Moerbeke, Random Matrices and their applications, MSRI-publications **40**, 4986 (2000).

9 781402 045295

MIX
Papier aus verantwortungsvollen Quellen
Paper from responsible sources
FSC® C105338

If you have any concerns about our products,
you can contact us on
**ProductSafety@springernature.com**

In case Publisher is established outside the EU,
the EU authorized representative is:
**Springer Nature Customer Service Center GmbH**
**Europaplatz 3, 69115 Heidelberg, Germany**

Printed by Libri Plureos GmbH
in Hamburg, Germany